DATA ANALYSIS
IN BIOCHEMISTRY AND BIOPHYSICS

# Data Analysis in Biochemistry and Biophysics

*Magar E. Magar*

Department of Statistics
University of Wisconsin
Madison, Wisconsin

*ACADEMIC PRESS 1972 New York and London*

ACADEMIC PRESS, INC.
111 Fifth Avenue, New York, New York 10003

*United Kingdom Edition published by*
ACADEMIC PRESS, INC. (LONDON) LTD.
24/28 Oval Road, London NW1

LIBRARY OF CONGRESS CATALOG CARD NUMBER: 72-182652

PRINTED IN THE UNITED STATES OF AMERICA

# CONTENTS

## 2. *MATRICES*



## 3. *ANALYSIS*

PURE ANALYSIS



APPLIED ANALYSIS



## 4. *MINIMIZATION OF FUNCTIONS*

## 5. *STATISTICS I*



## 6. *STATISTICS II*



## 7. *ABSORPTION SPECTRA OF MIXTURES*



## 8. *ANALYSIS OF NUCLEIC ACID SPECTRA*

## 9. *MATRIX RANK ANALYSIS*



## 10. *OPTICAL ROTATORY DISPERSION AND CIRCULAR DICHROISM OF PROTEINS*



## 11. *OPTICAL ROTATORY DISPERSION AND CIRCULAR DICHROISM OF NUCLEIC ACIDS AND THEIR COMPONENTS*

## 12. AGGREGATING SYSTEMS



## 13. ALLOSTERIC EFFECTS AND OTHER COOPERATIVE PHENOMENA IN PROTEIN–LIGAND EQUILIBRIA



## 14. ENZYME KINETICS

## 15. *RAPID REACTIONS: TRANSIENT ENZYME KINETICS AND OTHER RAPID BIOLOGICAL REACTIONS*



## 16. *TRACER TECHNIQUES IN COMPARTMENTALIZED SYSTEMS*

# FOREWORD

It is difficult to overstate the importance of numerical and statistical analysis to the physical and biological sciences. This importance tends to be amplified the more the system under observation is subject to influences whose control by the investigator is incomplete or wholly absent. When one leaves the exact sciences and considers such areas as biology or psychology, the methods of numerical and statistical analysis are often absolutely essential for any form of quantitative interpretation of data, as well as for the development of rational conclusions.

It is to be expected therefore that this book will appeal to a broad spectrum of workers in different disciplines and will provide them with the fundamentals of numerical and statistical analysis appropriate for their needs. Quite apart from its utility, the subject is of considerable intellectual interest for its own sake and may appeal to many readers on this basis.

The advent of modern computational methods has greatly extended the applicability and feasibility of the procedures described herein. It is to be expected that the future will see their increased use in the traditional fields, as well as their extension to new areas.

While the evolution of the subject will of course continue in the future, it is my belief that further progress will be based upon what has already been achieved and that the material in this book is likely to retain its utility and timeliness for many years.

Robert F. Steiner

# PREFACE

This monograph covers topics of current interest to molecular biologists, biochemists, biophysicists, physiologists, and researchers in other related sciences. The main objective is to show how one may derive the maximum amount of quantitative and statistical information from data generated in studies of enzyme kinetics, protein–ligand equilibria, optical rotatory dispersion, circular dichroism, concentration-dependent aggregating systems, osmometry, ultracentrifugation, light scattering, low angle X-ray scattering, molecular weight distributions, gel filtration, absorption spectroscopy, stopped flow experiments, chemical relaxation methods and tracer techniques in multicompartment systems, and related areas. The methods described for the treatment of these topics can be extended to several other areas in molecular biology, biochemistry, and physiology.

The subject of matrix rank analysis is also treated. Attempts are made to determine the number of components contributing to various systems when they are subjected to various experimental conditions.

The main emphasis of this book is on the analysis and determination of parameters in various models occurring in biochemistry, biophysics, and molecular biology. Considerations of distinguishing between models are also examined.

The book will appeal to all researchers who are concerned with the biochemical and physiological topics mentioned above, and to scientists in other fields, who can, with a small amount of mathematical manipulation, bring their equations into the same form, or a similar form, as those used here. This monograph will also have a significant appeal to statisticians, biomathematicians, and computer programmers. It has been my experience that the underlying reason for tackling a biological problem is rarely probed by the biomathematician and statistician. This is a pity because the mathematically inclined reader can aid the biologist in numerous ways. Such a reader would achieve this by rederiving some equations; by saving time through suggesting another experimental design; or by explaining to the experimenter that, from the point of view of the statistical analysis he has performed on his results, the conclusion that has been drawn increases the probability of the validity of a certain hypothesis, or perhaps renders that hypothesis quite worthless. Consequently, if one, as a mathematically inclined reader, wishes to probe beyond the immediate reason of why any specific analysis is being done, then this monograph will be of far more than a limited interest and will, I believe, give a good introduction into the current and potential applications of a number of mathematical tools the reader is well acquainted with.

I have attempted to deal with all the necessary mathematical and statistical background information, however those scientists who do not wish to go through all these details can find the direct information they require. This is possible since the end result of each analysis is stated at the beginning of the chapter treating any given topic. In fact, I submit with a great deal of confidence that nonmathematical investigators working with any of the above topics or techniques, who find the equations so distasteful that they do not even want to look at them, can take this monograph and his problem to the nearest programmer–computer center, statistician, or biomathematician and obtain at once the numerical answers he requires. One of the prime objectives of this book is to increase communication between biologists and people in computer centers or departments of statistics.

For those biologists who wish to go through the entire book, only a minimal knowledge of linear algebra, statistics, and calculus is required. The rest is supplied in the book.

In the early chapters I cover the required mathematical background and in the later chapters I deal with the individual biochemical and physiological topics. In one or two instances I do not cover all the relevant mathematical background, but adequate references are given for obtaining the desired information.

I feel there is a urgent need for a book of this kind, since it is not inconceivable that a great many topics besides the ones treated here will require the analytical techniques used in this book. A great deal of empirical data in the literature have not been subjected to as extensive a numerical and statistical analysis as they require. In my treatment of the various subjects, I attempt to obtain the pertinent parameters of any model, in what I feel to be the best manner. I also attempt to determine the extent of reliability of the parameters and show how various hypotheses about the systems being analyzed can be tested.

To my knowledge some of the topics treated in this book have not been treated by these methods before.

## ACKNOWLEDGMENTS

The second, third, and fourth chapters were written in collaboration with Dr. John E. Fletcher of the National Institutes of Health. He also contributed immensely to Chapter 13. I am also grateful to Professor R. F. Steiner for his foreword, for numerous discussions, and for his contributions to Chapter 13 and other parts of the book. Thanks are due to Professor I. Tinoco, Jr. for reading Chapter 11 and to Dr. Gary Felsenfeld for reading Chapter 8. The author of course is solely responsible for all errors in the book. Finally, I would like to thank the staff of Academic Press for their cooperation and patience during the completion of this work.

# *CHAPTER 1*

# SCOPE OF PROBLEMS INVESTIGATED

## 1. Introduction

Often during the course of scientific investigation our results can be expressed as straight lines or curves or more general functions. This book deals with some of the functions arising in molecular biology, chemistry, and other related fields.

When an investigator, on the basis of consideration of the mechanism of a process, produces an equation which describes the process he naturally wishes to obtain the maximum amount of information from the equation. One of the tasks of this monograph is to extract this information from the equation which describes the process or the model. Using experimental data and the equation purporting to explain it, we attempt to obtain the maximum amount of information by determining the parameters of the model. We also investigate whether or not the model fits the data.

To illustrate what we have just said, consider the combination of the enzyme with a single substrate. The Michaelis–Menten equation is an attempt to explain the process. This equation may be written as

$$1/v = (1/V_{\max}) + (K_m/V_{\max})(1/S) \qquad (1\text{-}1)$$

where $v$ is the velocity of the reaction, $V_{\max}$ is a constant denoting the maximum velocity, $K_m$ is another constant which sometimes indicates the strength of binding of the enzyme to the substrate, and $S$ is the substrate concentration. The information we would like is the value of the parameters of Eq. (1-1) namely $V_{\max}$ and $K_m$, and their statistical reliability. In addition, having gathered a set of data of the velocity at various substrate concentrations, we would like to say whether or not Eq. (1-1) adequately describes the data or whether or not the data are consistent with the model which produced Eq. (1-1).

Those familiar with Eq. (1-1) know how to use it to obtain $V_{\max}$ and $K_m$. This they do by simply plotting $1/v$ against $1/S$. This plot should be a straight line if the data are to be consistent with Eq. (1-1) and the intercept of that line on the axis is equal to $1/V_{\max}$ and the slope is equal to $(K_m/V_{\max})$.

This simple example illustrates to a large extent what is involved in a large portion of this book. Much more complicated problems will be handled but basically the questions to be answered are the same. Here we want to determine the parameters of the model in the "best" possible way and we want to find out whether or not the data is consistent with the model. We want to carry out our task in as objective a manner as possible, which means that we do not, when using Eq. (1-1), draw the straight line to suit the fancy of our notions of what the parameter should be nor do we draw a straight line when it is inconsistent with most statistical criteria to draw one.

Passing from this simple example, we shall give the reader some idea of what is involved in other cases by giving some other illustrative examples to show our general attack on the problem.

## 2. A Problem from Absorption Spectroscopy

Our first example comes from absorption spectroscopy. Consider the well documented absorption spectra of the noninteracting compounds $\alpha$ and $\beta$. Both of these spectra are clearly curves as illustrated in Fig. 1-1 or, to enter rapidly in the spirit of this book, we may say that both spectra are *functions* of wavelength. Next, suppose we have a mixture of the compounds $\alpha$ and $\beta$ and we would like to determine the concentration of $\alpha$

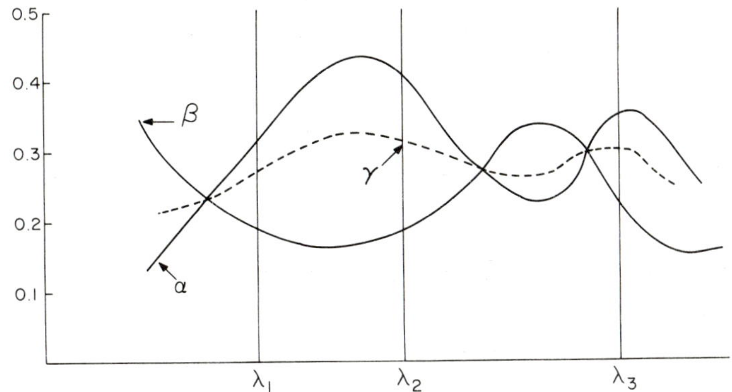


**Fig. 1-1.** The spectrum of $\alpha$, $\beta$, and $\gamma$, where $\gamma = 0.4\alpha + 0.6\beta$.

and $\beta$ in this mixture from the absorption spectrum of this mixture. From the Beer–Lambert relationship we have

$$A = \varepsilon c l \tag{1-2}$$

where $A$ is the absorption, $\varepsilon$ is the extinction coefficient, $c$ the concentration, and $l$ the length of the light path which we assume to be unity so that our relation becomes

$$A = \varepsilon c \tag{1-3}$$

Since this relationship should hold for all compounds, let us label the pertinent equations with a subscript, so for compound $\alpha$ we have

$$A_\alpha = \varepsilon_\alpha c_\alpha \tag{1-4}$$

and for compound $\beta$ we have

$$A_\beta = \varepsilon_\beta c_\beta \tag{1-5}$$

The wavelength where the absorption is taken must be specified because each extinction coefficient depends on the wavelength. Therefore, our relationship becomes for compound $\alpha$ at the wavelength $\lambda_i$

$$A_{\alpha\lambda_i} = \varepsilon_{\alpha\lambda_i} c_\alpha \tag{1-6}$$

and a similar relationship is obtained for $\beta$. The absorption at wavelength $\lambda_i$ of a mixture $\gamma$ of $\alpha$ and $\beta$, which we shall call $A_{\gamma\lambda_i}$, can be written as

$$A_{\alpha\lambda_i} + A_{\beta\lambda_i} = A_{\gamma\lambda_i} = \varepsilon_{\alpha\lambda_i} c_\alpha + \varepsilon_{\beta\lambda_i} c_\beta \tag{1-7}$$

To be more specific, let us say that we are measuring the absorption at $\lambda_1$ and $\lambda_2$ so our equations become

$$A_{\gamma\lambda_1} = \varepsilon_{\alpha\lambda_1} c_\alpha + \varepsilon_{\beta\lambda_1} c_\beta \tag{1-8a}$$

for the absorption at $\lambda_1$ and a similar equation for the absorption at $\lambda_2$

$$A_{\gamma\lambda_2} = \varepsilon_{\alpha\lambda_2} c_\alpha + \varepsilon_{\beta\lambda_2} c_\beta \tag{1-8b}$$

Looking at Eqs. (1-8a) and (1-8b), the quantities which we know are $A_{\gamma\lambda_1}$ and $A_{\gamma\lambda_2}$, the measured absorptions of the mixture $\gamma$, and $\varepsilon_{\alpha\lambda_1}$ and $\varepsilon_{\alpha\lambda_2}$, the extinction coefficients of the compound $\alpha$ at known wavelengths $\lambda_1$ and $\lambda_2$. Similarly the extinction coefficients $\varepsilon_{\beta\lambda_1}$ and $\varepsilon_{\beta\lambda_2}$ are also known. The quantities $\varepsilon_{\alpha\lambda_1}$, $\varepsilon_{\alpha\lambda_2}$, $\varepsilon_{\beta\lambda_1}$, and $\varepsilon_{\beta\lambda_2}$ are known because we already characterized the spectra of $\alpha$ and $\beta$. If those spectra had not been characterized, we could not proceed any further with our analysis to determine the concentration of $\alpha$ and $\beta$ in the mixture for we would not have known in terms of what components our mixture is supposed to be analyzed. Since $A_{\gamma\lambda_1}$, $A_{\gamma\lambda_2}$, $\varepsilon_{\alpha\lambda_1}$, $\varepsilon_{\alpha\lambda_2}$, $\varepsilon_{\beta\lambda_1}$ and $\varepsilon_{\beta\lambda_2}$ are all known, then Eqs. (1-8a) and (1-8b) represent two equations in the two unknowns $c_\alpha$ and $c_\beta$, and can be solved by the usual techniques. The reader can verify for himself by using Fig. 1-1 that the curve $\gamma$ is such that $c_\alpha = 0.4$ and $c_\beta = 0.6$, thus the mixture is four parts $\alpha$ and six parts $\beta$.

Exploring the matter further, if instead of using the measurement at $\lambda_1$, we used the measurement at $\lambda_3$, we would have obtained Eq. (1-8c) instead of (1-8a)

$$A_{\gamma\lambda_3} = \varepsilon_{\alpha\lambda_3} c_\alpha + \varepsilon_{\beta\lambda_3} c_\beta \tag{1-8c}$$

and it is easy to see that we could have used Eqs. (1-8b) and (1-8c) to solve for $c_\alpha$ and $c_\beta$ or we could have used (1-8a) and (1-8c) to solve for $c_\alpha$ and $c_\beta$. The question to ask now is, will the solution of all pairs of equations such as (1-8a), (1-8b), and (1-8c) yield the same result. The answer is yes, they will yield the same result provided there is *no experimental error* and therein lies the larger portion of our tale for there always is experimental error.

On account of the experimental error encountered, Eqs. (1-8a), (1-8b), and (1-8c) are not accurate representations of the situation, and they have to be modified to include the experimental error. This may be done by writing

$$\begin{aligned} A_{\gamma\lambda_1} &= \varepsilon_{\alpha\lambda_1} c_\alpha + \varepsilon_{\beta\lambda_1} c_\beta + \delta_1 \\ A_{\gamma\lambda_2} &= \varepsilon_{\alpha\lambda_2} c_\alpha + \varepsilon_{\beta\lambda_2} c_\beta + \delta_2 \\ A_{\gamma\lambda_3} &= \varepsilon_{\alpha\lambda_3} c_\alpha + \varepsilon_{\beta\lambda_3} c_\beta + \delta_3 \end{aligned} \tag{1-9}$$

where $\delta_1$, $\delta_2$, and $\delta_3$ are the experimental errors encountered at the given wavelengths. If we wish to be more general so as to use as much as possible of our available information, it would be wise to use as many wavelenghts as possible. Suppose we have made our measurements at $m$ wavelengths, our equations become

$$\begin{aligned} A_{\gamma\lambda_1} &= \varepsilon_{\alpha\lambda_1} c_\alpha + \varepsilon_{\beta\lambda_1} c_\beta + \delta_1 \\ A_{\gamma\lambda_2} &= \varepsilon_{\alpha\lambda_2} c_\alpha + \varepsilon_{\beta\lambda_2} c_\beta + \delta_2 \\ &\vdots \\ A_{\gamma\lambda_m} &= \varepsilon_{\alpha\lambda_m} c_\alpha + \varepsilon_{\beta\lambda_m} c_\beta + \delta_m \end{aligned} \tag{1-10}$$

Having the above $m$ equations, it becomes pertinent to ask how do we solve them in order to find $c_\alpha$ and $c_\beta$. A wise thing to do would be to minimize the sum of the error or to solve the equation with the condition that

$$| \delta_1 + \delta_2 + \cdots + \delta_m | \tag{1-11}$$

is a minimum. [That is what is actually done in linear programming, but generally with a number of other restrictions besides the condition in Eq. (1-11).] Another thing to do is to solve the equations with the condition that $(\delta_1^2 + \delta_2^2 + \cdots + \delta_m^2)$ is a minimum. This is a solution in the least square sense and will, in general, be highly recommended. It will be seen, however, that the least square solution will not always be capable of giving us an answer to the problem. Thus, it might happen (probably in a more complicated case than the one given above) that the least square solutions might yield a value of say $c_\alpha$ or $c_\beta$ which is negative. Since a negative concentration is an inadmissible solution, the least square solution has failed† to provide us with a satisfactory answer, and we must resort to other means.

Before indicating how we propose to tackle some of the problems posed by Eq. (1-10), let us generalize these equations to cover the case of a mixture that we seek to analyze in terms of say $n$ components and not just two components. Labeling the components $(1, 2, \ldots, n)$ our equations become

$$\begin{aligned} A_{\gamma\lambda_1} &= \varepsilon_{1\lambda_1} c_1 + \varepsilon_{2\lambda_1} c_2 + \varepsilon_{3\lambda_1} c_3 + \cdots + \varepsilon_{n\lambda_1} c_n + \delta_1 \\ A_{\gamma\lambda_2} &= \varepsilon_{1\lambda_2} c_1 + \varepsilon_{2\lambda_2} c_2 + \varepsilon_{3\lambda_2} c_3 + \cdots + \varepsilon_{n\lambda_2} c_n + \delta_2 \\ &\vdots \\ A_{\gamma\lambda_m} &= \varepsilon_{1\lambda_m} c_1 + \varepsilon_{2\lambda_m} c_2 + \varepsilon_{3\lambda_m} c_3 + \cdots + \varepsilon_{n\lambda_m} c_n + \delta_m \end{aligned} \tag{1-12}$$

† Failure here does not mean that the method is inadequate; it generally means that other considerations are involved. See the word of caution at the end of this chapter.

where $A_{\gamma\lambda_i}$ is the absorption of the mixture at wavelength $\lambda_i$; $c_1, c_2, \ldots, c_n$ are the concentration of the components 1 to $n$ in the mixture and $\varepsilon_{j\lambda_i}$ is the extinction coefficient of $j$th component at the $i$th wavelength.

Equation (1-12), although derived in connection with absorption spectra, can, by a suitable redefinition of the variables, apply to a large number of other topics. Or stating it in another manner, the equations to be analyzed will have the same form as Eq. (1-12) in absorption spectroscopy, optical rotatory dispersion, circular dichroism (of proteins and nucleic acids), concentration-dependent aggregating systems, and in several other problems which we tackle in this book.

## 3. The Use of Function Minimization and Matrices

At this stage of the proceeding it is pertinent to introduce two notions. One, concerning the nature of the book and another a helpful mathematical tool to solve equations such as (1-8a) and (1-8b). A great deal more will be said about these notions later on. We only give a rough introduction here to acquaint and motivate the reader as to what is involved.

### 3.1 Function Minimization

Looking at Eqs. (1-9) we note first that we can write any one of them (say the first one) in terms of the experimental error, as follows

$$\delta_1 = A_{\gamma\lambda_1} - \varepsilon_{\alpha\lambda_1}c_\alpha + \varepsilon_{\beta\lambda_1}c_\beta \tag{1-13}$$

If we do the same thing for all three equations, then sum up all the experimental error and equate to $\phi(c_\alpha, c_\beta)$ we obtain

$$\begin{aligned}\phi(c_\alpha, c_\beta) = \delta_1 + \delta_2 + \delta_3 = A_{\gamma\lambda_1} + A_{\gamma\lambda_2} + A_{\gamma\lambda_3} &- (\varepsilon_{\alpha\lambda_1}c_\alpha + \varepsilon_{\beta\lambda_1}c_\beta) \\ - (\varepsilon_{\alpha\lambda_2}c_\alpha + \varepsilon_{\beta\lambda_2}c_\beta) &- (\varepsilon_{\alpha\lambda_3}c_\alpha + \varepsilon_{\beta\lambda_3}c_\beta)\end{aligned} \tag{1-14}$$

Stating our problem again, we want to determine $c_\alpha$ and $c_\beta$ so as to make the experimental error as small as possible, or to minimize some function of the experimental error. Making $\phi(c_\alpha, c_\beta)$ or $\delta_1 + \delta_2 + \delta_3$ or the right-hand side of Eq. (1-14) as small as possible amounts to the same thing. Concentrating on the right-hand side of Eq. (1-14), we note that all the quantities there are known except $c_\alpha$ and $c_\beta$. We can also say that the right-hand side of Eq. (1-14) is a function of $c_\alpha$ and $c_\beta$ which we have called $\phi(c_\alpha, c_\beta)$.[†] The question, as far as we are concerned, is what values of $c_\alpha$ and $c_\beta$ minimize the right-hand side of Eq. (1-14) which, as we just stated, is a function of $c_\alpha$ and $c_\beta$. The question posed in this manner illustrates the need to study function minimization.

Suppose we knew nothing about function minimization and wanted to find $c_\alpha$ and $c_\beta$ such that $\phi(c_\alpha, c_\beta)$ is a minimum. What can we do? One thing we can do is to guess at the value of $c_\alpha$ and $c_\beta$ (say $c_\alpha = 0.1$ and $c_\beta = 0.9$), put them in the right-hand side of Eq. (1-14), and compute $\phi(c_\alpha, c_\beta)$. This procedure produces a value for $\phi(c_\alpha, c_\beta)$ since all the quantities on the right-hand side of (1-14) are known. Next, take another guess at $c_\alpha$ and $c_\beta$ (say $c_\alpha = 0.2$ and $c_\beta = 0.8$); this produces another value of $\phi(c_\alpha, c_\beta)$. One can keep doing this kind of thing until one finds the values of $c_\alpha$ and $c_\beta$ that give the smallest value of $\phi(c_\alpha, c_\beta)$. What we are doing here is *searching* for the minimum, but in an extremely haphazard way. This procedure is most inefficient. In this book, we show how we can arrive at the values of $c_\alpha$ and $c_\beta$ which minimize $\phi(c_\alpha, c_\beta)$ much more efficiently, using the various techniques of function minimization.

As a problem to the reader we recommended the computation of $\phi(c_\alpha, c_\beta)$ for various values of $c_\alpha$ and $c_\beta$ taking the following values for the absorption and extinction coefficients.

| | $\varepsilon_\alpha$ | $\varepsilon_\beta$ | $A_\gamma$ |
|---|---|---|---|
| $\lambda_1$ | 0.322 | 0.29 | 0.27 |
| $\lambda_2$ | 0.40 | 0.183 | 0.313 |
| $\lambda_3$ | 0.35 | 0.22 | 0.30 |

As an additional problem the reader is invited to look through the biochemical literature for a Lineweaver and Burke plot and take the values of $1/v$ and $1/S$ for that plot and attempt to obtain $V_{\text{max}}$ and $K_{\text{m}}$ to minimize $\phi(V_{\text{max}}, K_{\text{m}})$ written as follows

$$\phi(V_{\text{max}}, K_{\text{m}}) = \sum_{i=1}^{n} \left( \frac{1}{v_i} - \frac{1}{V_{\text{max}}} + \frac{K_{\text{m}}}{V_{\text{max}}} \frac{1}{S_i} \right) \tag{1-15}$$

[†] The definition of a function is given and explained in Chapter 3. We assume for the time being that the reader will grant us that the right hand side of Eq. (1-14) is a function of $c_\alpha$ and $c_\beta$.

where $n$ is the number of points labeled $i = 1, 2, \ldots, n$. We hope that the reader will find that in the case where the points for $1/v$ and $1/S$ all fall on a straight line, $\phi(V_{\mathrm{max}}, K_{\mathrm{m}})$ will equal zero for the proper values of $V_{\mathrm{max}}$ and $K_{\mathrm{m}}$. The readers who are acquainted with least squares straight line fits might realize that the function we minimize in order to obtain $V_{\mathrm{max}}$ and $K_{\mathrm{m}}$ by the least squares procedure is

$$\phi(V_{\mathrm{max}}, K_{\mathrm{m}}) = \sum_{i=1}^{n} \left( \frac{1}{v_i} - \frac{1}{V_{\mathrm{max}}} + \frac{K_{\mathrm{m}}}{V_{\mathrm{max}}} \frac{1}{S_i} \right)^2 \qquad (1\text{-}16)$$

## 3.2 Matrices

The entire subject of matrices is treated in Chapter 2. However, we will introduce them here briefly. Looking at Eqs. (1-8a) and (1-8b) we note we can write them as

$$A_{\gamma\lambda_1} = \varepsilon_{\alpha\lambda_1} c_\alpha + \varepsilon_{\beta\lambda_1} c_\beta \qquad (1\text{-}8a)$$

$$A_{\gamma\lambda_2} = \varepsilon_{\alpha\lambda_2} c_\alpha + \varepsilon_{\beta\lambda_2} c_\beta \qquad (1\text{-}8b)$$

or make an array of them like this

$$\begin{bmatrix} A_{\gamma\lambda_1} \\ A_{\gamma\lambda_2} \end{bmatrix} = \begin{bmatrix} \varepsilon_{\alpha\lambda_1} & \varepsilon_{\beta\lambda_1} \\ \varepsilon_{\alpha\lambda_2} & \varepsilon_{\beta\lambda_2} \end{bmatrix} \begin{bmatrix} c_\alpha \\ c_\beta \end{bmatrix} \qquad (1\text{-}17)$$

Putting the equations in the above array is to put Eqs. (1-8a) and (1-8b) in their equivalent matrix notation. People acquainted with the theory of matrices will have no trouble in noting the equivalence of Eqs. (1-8a) and (1-8b) and the array in (1-17). Accepting our word for the time being that (1-17) is equivalent to (1-8a) and (1-8b), the reader might ask why we are complicating things by introducing a different notation. The answer to this is that while it is easy to manipulate Eqs. (1-8a) and (1-8b) to obtain the values of the coefficients $c_\alpha$ and $c_\beta$, the task becomes more difficult if the reader had to manipulate, say, four equations in four unknowns. By displaying the complicated equations in arrays similar to (1-17) and learning the rule of manipulation of such arrays (matrix algebra) the reader will be in a position to handle complicated equations with much greater facility. Thus, taking pains to learn a few rules of matrix algebra gives the reader a powerful tool to handle complicated equations. This brief digression we hope serves to give the reader some feel for the necessity of studying the subjects in Chapter 2 and Chapter 3. Further practical examples of some of the systems we handle can now be given to give the reader some idea of the scope of the topics we treat.

## 4. An Example from Concentration-Dependent Aggregating Systems

For this example let us consider a concentration-dependent aggregating system whose equilibria and association constants may be written

$$\begin{aligned} M_1 + M_1 &\rightleftharpoons M_2, & K_2 &= M_2/(M_1)^2 \\ M_2 + M_1 &\rightleftharpoons M_3, & K_3 &= M_3/M_2M_1 \\ &\vdots & &\vdots \\ M_{n-1} + M_1 &\rightleftharpoons M_n, & K_n &= M_n/M_{n-1}M_1 \end{aligned} \qquad (1\text{-}18)$$

where $M_1, M_2, \ldots, M_n$ are the concentration of the monomer, dimer, $n$-mer respectively and $K_1, K_2, \ldots, K_n$ are the association constants of the various equilibria indicated in Eq. (1-18).

According to Kreuzer (1943) and Ts'o and Chan (1964) the stoichiometric molality $M$ by definition can be written as

$$M = M_1 + 2K_2(M_1)^2 + 3K_2K_3(M_1)^3 + \cdots + nK_2K_3 \cdots K_n(M_1)^n \qquad (1\text{-}19)$$

Those authors have also pointed out that the activity $\gamma$ can be written as

$$\gamma = M_1/M \qquad (1\text{-}20)$$

and the activity at the concentration $(c_i)$ is

$$\gamma(c_i) = M_1(c_i)/M(c_i) \qquad (1\text{-}20a)$$

By measuring the osmotic coefficient $\phi$ of this aggregating system at different concentrations of the aggregating species, we can obtain the activity coefficient at different concentrations of the aggregating species by using the Gibbs–Duhem equation

$$\ln \gamma = (\phi - 1) + \int_0^M (\phi - 1)\, d \ln M \qquad (1\text{-}21)$$

which for any given concentration $(c_i)$ can be represented as follows

$$\ln \gamma(c_i) = [\phi(c_i) - 1] + \int_0^{M(c_i)} [\phi(c_i) - 1]\, d \ln M(c_i) \qquad (1\text{-}21a)$$

Then by using the relation in Eq. (1-20), we can obtain the value of $M_1(c_i)$, the monomer concentration at any given concentration of the aggregating species. Or, for every $M$ at any given concentration we have a corresponding value for $M_1$, the monomer concentration at the same concentration. Consequently, we can write Eq. (1-19) as many times (say $m$ times) as we have measurements at differing concentrations of $M$. These are written as follows where we have included errors encountered in each measurement

$$\begin{aligned} M(c_1)/M_1(c_1) &= 1 + 2K_2[M_1(c_1)]^2 + 3K_2K_3[M_1(c_1)]^3 \\ &\quad + \cdots + nK_2K_3 \cdots K_n[M_1(c_1)]^n + \delta_1 \\ M(c_2)/M_1(c_2) &= 1 + 2K_2[M_1(c_2)]^2 + 3K_2K_3[M_1(c_2)]^3 \\ &\quad + \cdots + nK_2K_3 \cdots K_n[M_1(c_2)]^n + \delta_2 \\ &\vdots \\ M(c_m)/M_1(c_m) &= 1 + 2K_2[M_1(c_m)]^2 + 3K_2K_3[M_1(c_m)]^3 \\ &\quad + \cdots + nK_2K_3 \cdots K_n[M_1(c_m)]^n + \delta_m \end{aligned} \tag{1-22}$$

where $M(c_i)$ is the concentration of the aggregation species at the $i$th concentration and $M_1(c_i)$ is the concentration of the monomer at the $i$th concentration. Examination of Eq. (1-22) indicates that we are analyzing the function (of concentration) $M(c_i)/M_1(c_i)$ in terms of the following functions (also of concentration) $[M_1(c_i)]$, $[M_1(c_i)]^2$, $\ldots$, $[M_1(c_i)]^n$. If we think in terms of functions, only the similarity between Eqs. (1-12) and (1-22) becomes apparent. In the first instance we are analyzing the absorption of a mixture (which is a function of wavelength) in terms of the absorption of what one believes to be its components (which are also functions of wavelength), and in the second instance we are analyzing a function of concentration in terms of other functions of concentration.

In Eq. (1-22) the quantities we are interested are the association constants. Equation (1-22) may be solved for those quantities such that some function of the experimental error is minimized. We also note that we have products of the association constants in Eq. (1-22) but these do not present any problem, for if we write

$$\alpha_1 = K_2, \qquad \alpha_2 = K_2K_3, \qquad \alpha_{n-1} = K_2K_3 \cdots K_n$$

and make the appropriate substitutions in Eqs. (1-22), then solving our equation for $\alpha_1, \alpha_2, \ldots, \alpha_n$, we readily determine all the $K$ or the association constants.

## 5. Linear and Nonlinear Models

At this stage we must characterize the nature of parameters occurring in various models. We distinguish between two cases which cover all the problems which we treat. In the first case the parameters occur linearly and in the second the parameters occur nonlinearly. This distinction is best illustrated by examples for linear models.†

$$Y = \alpha_1 f_1(x) + \alpha_2 f_2(x) \tag{1-23}$$

$$Y = \alpha_1 f_1(x) + \alpha_2 \tag{1-24}$$

$$Y = \alpha_1[f_1(x)/f_2(x)] + \alpha_2 \tag{1-25}$$

In the above examples $\alpha_1$, $\alpha_2$, etc., are the parameters of the model and $f_1(x), f_2(x)$, etc., are either known functions or independent variables under the investigator's control. Our examples show us that for a model to be linear in the parameters, the parameters must occur once in any term where a term can be formed with one or more than one of our known functions or independent variables. Further, the parameter must appear as the first power of the coefficient. It does not matter how complicated the functions in a given term are as long as they are known.

As examples of nonlinear models we have†

$$Y = 1/[\alpha_1 + \alpha_2 f_1(x) + \alpha_3 f_2(x)] \tag{1-26}$$

$$Y = \alpha_1 f_1(x, \alpha_2) + \alpha_3 f_2(x, \alpha_4) \tag{1-27}$$

$$Y = \exp[\alpha_1 + \alpha_2 f_1(x) + \alpha_3 f_2(x)] \tag{1-28}$$

To explain the Eq. (1-27) further we can write $f_1(x, \alpha_2)$ and $f_2(x, \alpha_4)$ explicitly in two different ways as examples:

$$Y = \alpha_1 \exp(-\alpha_2 x_1) + \alpha_3 \exp(-\alpha_4 x_1) \tag{1-29a}$$

$$Y = \alpha_1 \sin \alpha_2 x_1 + \alpha_3 \sin \alpha_4 x_1 \tag{1-29b}$$

and so on. The thing to note in Eqs. (1-29a) and (1-29b) is that the $\alpha$'s occur more than once in a term.

† $Y$ and $f(x)$ are used interchangeably to denote the response of the system or the dependent variable, and $x_1, x_2, \ldots, x_i, \ldots$, are used interchangeably with $f_1(x)$, $f_2(x), \ldots, f_i(x), \ldots$, to denote independent variables or variables that are either known or are under the investigator's control. See explanation on the general linear problem and general nonlinear problem for examples.

## 6. The General Problem for Linear Models

The general formulation of our problem in a large number of applications may be written as follows

$$
\begin{aligned}
f(x_1) &= \alpha_1 f_1(x_1) + \alpha_2 f_2(x_1) + \cdots + \alpha_n f_n(x_1) + \delta_1 \\
f(x_2) &= \alpha_1 f_1(x_2) + \alpha_2 f_2(x_2) + \cdots + \alpha_n f_n(x_2) + \delta_2 \\
&\vdots \\
f(x_m) &= \alpha_1 f_1(x_m) + \alpha_2 f_2(x_m) + \cdots + \alpha_n f_n(x_m) + \delta_m
\end{aligned}
\tag{1-30}
$$

where $m > n$, which means that the number of data points are more than the number of parameters. The quantities $f(x_1), f(x_2), \ldots, f(x_m)$ are the responses of the measurements of the system under consideration at the points $x_1, x_2, \ldots, x_m$. In the case of absorption, optical rotatory dispersion, circular dichroism, and spectroscopy $x_1, x_2, \ldots, x_m$ will be wavelengths and $f(x_1), f(x_2), \ldots, f(x_m)$ are the absorption, optical rotation, or circular dichroism of the system under consideration. For reconcentration dependent aggregating systems, $x_1, x_2, \ldots, x_m$ will be various concentrations of the aggregating species, and $f(x_1), f(x_2), \ldots, f(x_m)$ will be the measured response (activity or molecular weight) of the system under consideration. The $f_1(x_1), f_1(x_2), \ldots, f_n(x_m)$ are the *known functions* or components in terms of which we seek to analyze the response of the system under consideration. The $\delta_1, \delta_2, \ldots, \delta_n$ are the errors involved in making the measurements. As in our illustrative examples, we generally want to solve Eq. (1-30) for the parameter $\alpha_1, \alpha_2, \ldots, \alpha_n$ with the provision of minimizing the sum of the errors of the sum or the square of the errors.

## 7. The General Problem for Nonlinear Models

The best way to illustrate that problem is to give some examples. One of the most nonlinear problems is the analysis of the response of a system in terms of the sum of exponentials. This type of equation occurs in tracer techniques, rapid enzyme reactions, fluorescent and phosphorescent decay, and other relaxation processes. The equation representing such processes can be written as

$$
f(t) = \sum_{i=1}^{n} \alpha_i \exp(\beta_i t) + \delta \tag{1-31}
$$

where $f(t)$ is a time dependent response and $\alpha_i$ and $\beta_i$ are the parameters.

Once again $\delta$ is the experimental error. To analyze Eq. (1-31) we take as many different measurements of the time dependent response $f(t)$ and we end up with Eq. (1-31) being written as many times (say $m$ times) as we have measurements. Our objective is to determine the $\alpha_i$ and $\beta_i$ such that some function of the experimental error (the $\delta$) is minimized. Another example of nonlinear models is the case of the contributing bands of circular dichroic spectra. For example suppose we have a circular dichroism spectrum which we want to resolve into its $n$ Gaussian components. Fundamentally we want to determine the parameters $[\theta^0]_{\lambda_i}$, $\lambda_i^0$, $(\Delta_i^0)^2$ from the following equations

$$
\begin{aligned}
[\theta]_{\lambda_1} &= \sum_{i=1}^{n} [\theta_i^0]_{\lambda_i} \exp[-(\lambda_1 - \lambda_1^0)^2/(\Delta_i^0)^2] + \delta_1 \\
[\theta]_{\lambda_2} &= \sum_{i=1}^{n} [\theta_i^0]_{\lambda_i} \exp[-(\lambda_2 - \lambda_i)^2/(\Delta_i^0)^2] + \delta_2 \\
&\vdots \\
[\theta]_{\lambda_m} &= \sum_{i=1}^{n} [\theta_i^0]_{\lambda_i} \exp[-(\lambda_m - \lambda_i)^2/(\Delta_i^0)^2] + \delta_m
\end{aligned} \tag{1-32}
$$

such that $\sum_{i=1}^{m} | \delta_i |$ or $\sum_{i=1}^{m} \delta_i^2$ is a minimum. Where $[\theta]_{\lambda_j}$ is the circular dichroism at wavelength $\lambda_j$, $[\theta_i^0]_{\lambda_i}$ is the extremum of the $i$th transition, $\lambda_i$ the wavelength of the $i$th transition, $\Delta_i^0$ is the half-width of the $i$th transition, $n$ the number of transitions involved, and $\delta_1, \delta_2, \ldots, \delta_m$ are the errors made in the measurements at $\lambda_1, \lambda_2, \ldots, \lambda_m$. It is readily perceived that the form of Eqs. (1-32) shows that the parameters $[\theta_i^0]$, $\lambda_i$ and $\Delta_i^0$ occur nonlinearly.

The above examples are only a limited number of the problems that can be handled. Inasmuch as any model is either linear or nonlinear in its parameters, the scopes of the methods is virtually limitless.

## 8. The Use of Statistics

The larger portion of this monograph deals with the methods of solution of equations such as (1-30) as well as some others which we mention later in the introduction. The error encountered in making our measurement will generally be considered to be random with certain distributions. When such is the case, statistical methods must be used to evaluate the conclusions we obtain from the solution of our equation. In general, we require statistical methods for the following main reasons.

In view of the fact that we will be dealing with functions, and since those functions will probably be more accurate in certain intervals of the independent variable, it is clear that greater weight must be given to data gleaned from the portions of our curves with the least experimental error. For example, it is clear that in the case of optical rotatory dispersion and circular dichroism, experimental data gathered in the 225–240 m$\mu$ region is more accurate than data gathered from the 195–225 m$\mu$ region. Consequently, more emphasis must be given to the data obtained from longer wavelengths. If we are dealing with aggregating systems, then the measured osmotic coefficient or weight average molecular weight will, in general, be more accurately given at higher concentration of the aggregating species. Still another example of this type would be experiments done with macromolecules and ligands, and here we find that the higher the ligand concentration the better we are able to estimate the extent of binding and so on. Judicious weighing of the data before analyzing it is very important. It is not very often that we see in the biochemical literature the inclusion of the standard error of the measurements on any given system. It might be thought that we are asking a little too much of investigators to report the standard error of all their measurements, since experiments will have to be repeated several times before the standard error is established with some precision. However, it is necessary to point out that in conducting most of the analyses found in this monograph, the investigator will not feel at ease when he or she gives equal weight to observations which they know to be quite unequal in their accuracy. This point cannot be overemphasized and is, in fact, quite fundamental if we wish to strengthen the validity of the conclusions drawn from our analyses.

The reason why statistical theory is required is that throughout this monograph we will be attempting to compute various numbers, such as the concentration of a component of a mixture, the percentage of $\alpha$-helix in a protein, the number of bases stacked in an RNA molecule, the association constant of a monomer–dimer equilibrium, the association constant of a protein and ligand, and so on. The techniques used here not only estimate the number but also its internal confidence. Thus we will say that a protein has $25 \pm 3\%$ $\alpha$-helix and not just that the protein has $25\%$ $\alpha$-helix.

Statistical theory is also required in order to test hypotheses about the parameters we estimate. More specifically, suppose we are analyzing data from aggregating systems to determine the various equilibrium constants by measuring the osmotic coefficients as shown in Eq. (1-22).

A number of possibilities arise with an equation of this type. The particles may not aggregate indefinitely, or if they do, there will be a highest aggregating species of significant concentration. Another possibility may be that the system may not aggregate in the manner discribed but as monomer–dimer–tetramer. In either of the cases stated above or in any scheme of aggregation, various hypotheses about the parameters will have to be tested. For example, if the system does not aggregate beyond a certain $n$-mer, then $K_{n+1}$ should be insignificant. If the system aggregates as monomer–dimer–tetramer, then $K_3$ should equal zero. Consequently, we will be required to test statistically the various association constants we compute in order to determine if they are significantly different from zero. Such procedures will tell us what the highest aggregating species of significant concentration is and will indicate the absence of certain aggregates.

Taking another example where we might have to test hypotheses about parameters, suppose we have a circular dichroism spectra which we wish to analyze into its Gaussian component. The question as to how many components must be used to analyze the spectrum comes up since it is obvious that the more Gaussian we use, the better the fit. To answer the question, we must apply statistical tests to determine the number of Gaussian curves that will best fit the data and infer from that the number of contributing bands.

As a general rule we are concerned with determining what model fits the data and this leads to the testing of hypotheses. A topic of much import which is barely touched on in this volume is experimental design to distinguish between models. Suppose an investigator has two or three models which he thinks might fit the data. In general, before one goes about gathering data, it is a good idea to consider how to design the experiment so that it will be possible to distinguish between one's models? Or, stating it differently, where should one take experimental points so that it will be possible to tell the difference between models? Experimental design helps with the accurate determination of the parameters and in this and other ways helps to distinguish between the models. It is always a good idea to use experimental design before one does the experiment, because not to do so often results in the identification of imprecise and highly correlated parameters where no amount of statistical help regardless how sophisticated can remedy the situation. Unfortunately this subject has only been touched upon here, but it is one of the more important uses of statistics.

## 9. The Problem of Constraining the Parameters

Turning back for a moment to the solution of our equations we find that in solving for the parameters $\alpha_1, \alpha_2, \ldots, \alpha_n$, in a large number of cases certain restrictions will have to be placed on them. For instance, in cases where we are analyzing the absorption of a mixture in terms of its components or in cases where we are trying to determine the association constants of some macromolecule with a ligand, the parameters $\alpha_1, \alpha_2, \ldots, \alpha_n$ must all be greater than or equal to zero since negative concentrations and association constants are meaningless. The solution of Eq. (1-30) must take these restrictions in consideration. In addition to the restriction or constraint that $\alpha_i \geq 0$ for all $i$, it might be necessary to constrain the parameters such that their total must equal some number. For example, if you know the total concentration of a certain mixture which has $m$ components then

$$\sum_{i=1}^{m} \alpha_i = k \tag{1-33}$$

where $k$ is the total concentration. In the case of analyzing the optical rotatory dispersion of proteins in terms of standard reference conformations, we might well require that

$$\sum_{i=1}^{m} \alpha_i = 100\% \tag{1-34}$$

or that the percentage of $\alpha$-helix, coil, $\beta$-form, etc., must all equal 100%. Still another example where we require constraints on our parameters is in case of enzyme kinetics or reaction rates where it might be required on the basis of independent information on the system that certain rate constants or equilibrium constants must lie in a certain range, i.e., not exceed or fall below certain numbers. We will find that while the least-square and regression analysis will, in general, provide us with the satisfying way of dealing with our problem, it will very often fail when constraints are put on the parameters. In such cases, linear programming and other iterative techniques will be found to be capable of handling almost all the problems which require constraints on the parameters.

## 10. Determining Unknown Components: Matrix Rank Analysis

So far, we have been dealing with the analysis of systems in terms of known components. Suppose, however, we did not know what our components are and we wanted to determine their spectra or shape. To handle this problem, we include a chapter on matrix rank analysis (Chapter 9) which provides us (in certain cases) with an answer to this question. In general, the matrix rank analysis will tell us the number of (linearly independent) components in a given system when the system is measured under *different conditions*. For example, suppose we measure the absorption spectrum of the amino acid tyrosine under different conditions of pH (say at pH 1-13). If we conduct a matrix rank analysis on this series of spectra, we might be able to detect the number of different spectra corresponding to the species **I–IV**. If the spectra of **I–IV** are all different

| I | II | III | IV |
|---|---|---|---|
| $NH_3^+$—CH—COOH | $NH_3^+$—CH—$COO^\ominus$ | $NH_3^+$—CH—$COO^\ominus$ | $NH_2^+$—CH—$COO^\ominus$ |
| $CH_2$ | $CH_2$ | $CH_2$ | $CH_2$ |
| OH | OH | OH | $O^\ominus$ |

(are not linear combinations of each other), then matrix rank analysis will tell us there are four components. With some ingenuity on our part it may also tell us the shapes of their spectra. If any two components are identical or linear combinations of any of the four spectra, the matrix rank analysis will show that there are only three components and tell us what the spectra are. In using matrix rank analysis, it becomes necessary, finally to identify the components elucidated by this method with suitable models. In the specific case of tyrosine, we might be able to identify one component with the spectrum taken at pH 1. This will probably be the spectrum **I**. Again the spectrum at pH 13 may well correspond to one of the components we have identified from our matrix rank analysis and this would probably be structure **IV**. It must be emphasized that the spectra which we determine from matrix rank analysis must, finally be identified with some entity because it does not do us much good if we analyze a system under different conditions and, on finding the number of components of such a system, not know to what they refer. Once we know what they refer to, we can then use them to analyze other systems. The final identification of a spectrum obtained from matrix rank analysis with a physical entity may be a simple procedure or may demand considerable experimental ingenuity. The matrix rank analysis method can be set schematically as follows for the general case and for specific example.

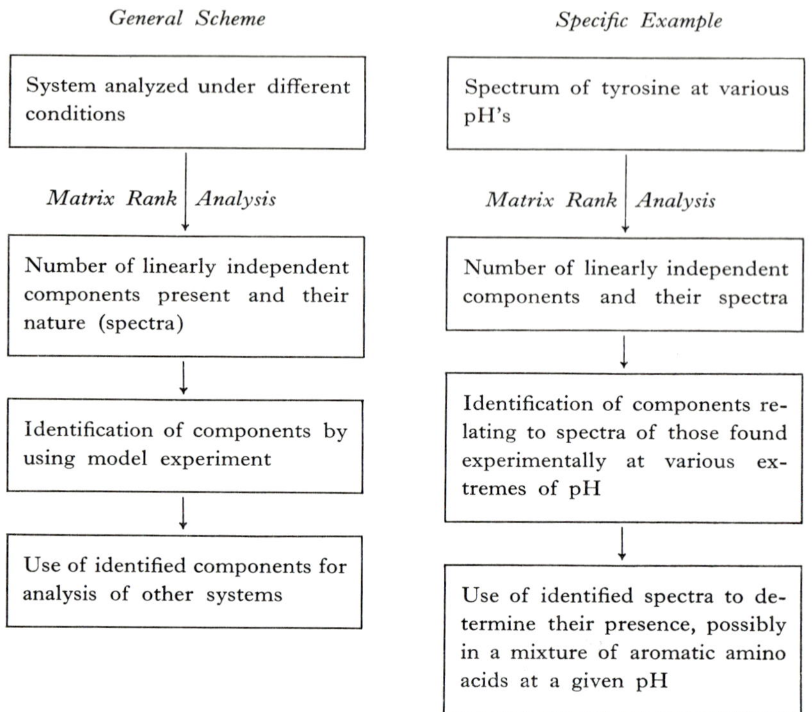


In Chapter 9, on matrix rank analysis, we give two pertinent applications, one from the work of McMullen *et al.* (1967), on the analysis of the spectra of RNA and one from the work of Ainsworth and Bingham (1968) on the analysis of the reaction rates of partially oxidized hemoglobin. In addition to the matrix rank analysis, we shall, in the same chapter, propose a different method for the determination of the nature of components in a system. This method will involve making certain assumptions about the system before the analysis can proceed and will be compared to the matrix rank analysis method in the appropriate chapter.

## 11. A Word of Caution

The above presents briefly the nature of the problems we are going to attack. At this stage a word of caution is in order. The methods we use to analyze data will ultimately yield us parameters that will fit the experimental data with great precision. In such event the investigator would be justified in saying that the model and its parameters are consistent with the data. However, it is necessary to point out two important circumstances that might render the investigator's results quite implausible. The first of those instances is when the experimental error is very large. In such cases it does not serve any purpose to go ahead with these analyses. As John Tukey has said, "If it is not worth doing, it is not worth doing well." The second instance is a little more difficult to detect, and that is when the investigator uses a model which does not reflect the realities of the situation. When an investigator using a certain model believes that the parameters obtained bear no resemblence to what is felt to be the physical realities of the situation, then consideration of another model may be in order. In concluding this chapter we must emphasize that the mathematical and statistical methods to be presented cannot in the last analysis be faulted. In obtaining the answer to any specific problem they are only as good as the precision of the accumulated data and the model used to analyze them.

## References

Ainsworth, S., and Bingham, W. S. (1968). *Biochem. Biophys. Acta* **160**, 10.
Kreuzer (1943). *Z. Phys. Chem.* (*Leipzig*) **B53**, 213.
McMullen, D. M., Jaskunas, S. R., and Tinoco, I., Jr. (1967). *Biopolymers* **5**, 589.
Ts'o, P. O. P., and Chan, S. I. (1964). *J. Amer. Chem. Soc.* **86**, 4176.

*CHAPTER 2*

# MATRICES

## 1. Introduction

This chapter is the first of five designed to build the required mathematical statistical background to handle the equations encountered in this type of work and to provide a framework within which we may systematically analyze data. From the introductory chapter we see that we are faced with experimental systems of linear equations, which could be written in the notation of matrices thus facilitating the manipulation of the equations and providing a compact notation.

## 2. Matrices

A matrix is a rectangular array of numbers enclosed by a pair of brackets and is depicted in its most general form as

$$\begin{bmatrix} a_{11} & a_{12} & a_{13} & \cdots & a_{1n} \\ a_{21} & a_{22} & a_{23} & \cdots & a_{2n} \\ \vdots & \vdots & \vdots & & \vdots \\ a_{m1} & a_{m2} & a_{m3} & \cdots & a_{mn} \end{bmatrix} \tag{2-1}$$

This array has $m$ rows and $n$ columns and $a_{ij}$ denotes the element in the $i$th row and the $j$th column. Thus we note that all elements in the second row have 2 as first subscript and all elements in the third column have 3 as second subscript. Similarly for other rows and columns. Sometimes matrices are written for simplicity as $(a_{ij})$. The matrix **A** denoted by (2-1) is called an $n \times m$ matrix.

For example

$$\begin{bmatrix} 5 & -2 & 1 \\ 3 & 6 & -4 \end{bmatrix} \tag{2-2}$$

$$\begin{bmatrix} 2 & 3 & -7 \\ 6 & 1 & 11 \\ 4 & -2 & 8 \end{bmatrix} \tag{2-3}$$

the matrix (2-2) is a $3 \times 2$ matrix and the elements $a_{13} = 1$ and $a_{22} = 6$. The matrix (2-3) is a $3 \times 3$ matrix and the elements $a_{32} = -2$, $a_{12} = 3$, and so on.

When $n = 1$ for all elements in a given matrix the matrix reduces to a single column and we refer to it as a column *vector*. This is illustrated by

$$\begin{bmatrix} a_{11} \\ a_{12} \\ \cdot \\ \cdot \\ \cdot \\ a_{m1} \end{bmatrix} \tag{2-4}$$

Similarly, if $m = 1$, the matrix reduces to a single row which we refer to as a row *vector*. This is illustrated by

$$[a_{11}, a_{12}, a_{13}, \ldots, a_{1n}] \tag{2-5}$$

## 3. Rules of Matrix Algebra

### 3.1 Equality of Matrices

Two matrices are equal if they have the same number of elements and if their corresponding elements are equal. Thus given a typical $a_{ij}$ element of the matrix **A** and the corresponding element $b_{ij}$ of a matrix **B** then the matrices are equal if and only if

$$a_{ij} = b_{ij} \qquad \text{for} \qquad i = 1, \ldots, n, \quad j = 1, \ldots, m \tag{2-6}$$

### 3.2. Addition and Subtraction of Matrices

Matrices can be added or substracted in the following manner. The sum and difference of a matrix **A** of order $m \times n$ with element $a_{ij}$ and a matrix **B** of order $m \times n$ with elements $b_{ij}$ is another matrix **C** of order $m \times n$ with elements $c_{ij}$ such that each element of **C** is the sum or difference of the corresponding element of **A** and **B**. Thus

$$\mathbf{C} = \mathbf{A} \pm \mathbf{B} = (a_{ij} \pm b_{ij})$$

For example if

$$\mathbf{A} = \begin{bmatrix} 2 & 4 & 6 \\ 0 & 2 & 8 \end{bmatrix}$$

and

$$\mathbf{B} = \begin{bmatrix} 1 & 2 & 0 \\ 1 & 2 & 5 \end{bmatrix}$$

then

$$\mathbf{A} + \mathbf{B} = \begin{bmatrix} 3 & 6 & 6 \\ 1 & 4 & 13 \end{bmatrix}$$

and

$$\mathbf{A} - \mathbf{B} = \begin{bmatrix} 1 & 2 & 6 \\ -1 & 0 & 3 \end{bmatrix}$$

Matrices not having the same order cannot be added or subtracted.

### 3.3 Multiplication of Matrices

We illustrate multiplication of matrices first specifically by multiplication of two vectors. The product of two vectors **ab** denoted by

$$\mathbf{a} = [a_{11}, a_{12}, a_{13}, \ldots, a_{1m}] \tag{2-7}$$

$$\mathbf{b} = \begin{bmatrix} b_{11} \\ b_{21} \\ \cdot \\ \cdot \\ \cdot \\ b_{m1} \end{bmatrix} \tag{2-8}$$

is a $1 \times 1$ matrix or a single element $c$ denoted by

$$c = [a_{11}b_{11} + a_{12}b_{12} + \cdots + a_{1m}b_{1m}] = \sum_{k=1}^{m} a_{1k}b_{1k} \tag{2-9}$$

*Example.*

$$[1, 2, 4]\begin{bmatrix} 3 \\ 0 \\ -1 \end{bmatrix} = (1)(3) + (2)(0) + 4(-1) = 3 + 0 + (-4) = -1$$

For matrices in general each element of any given row is multiplied by the corresponding element in the corresponding column. The product of **AB** of an $m \times n$ matrix $\mathbf{A} = (a_{ij})$ and an $n \times s$ matrix $\mathbf{B} = (b_{ij})$ is an $m \times s$ matrix $\mathbf{C} = (c_{ij})$ whose elements $c_{ij}$ are given below.

$$c_{ij} = a_{i1}b_{1j} + a_{i2}b_{2j} + \cdots + a_{in}b_{nj} = \sum_{k=1}^{n} a_{ik}b_{kj}, \qquad \begin{matrix} i = 1, 2, \ldots, m \\ j = 1, 2, \ldots, s \end{matrix} \tag{2-10}$$

For the matrices **A** and **B** to be *comformable* for multiplication, the number of columns of **A** must *equal* the number or rows of **B**. Examples of matrix multiplication are shown below. If

$$\mathbf{A} = \begin{bmatrix} a_{11} & a_{12} \\ a_{21} & a_{22} \end{bmatrix} \quad \text{and} \quad \mathbf{B} = \begin{bmatrix} b_{11} & b_{12} \\ b_{21} & b_{22} \end{bmatrix}$$

then

$$\mathbf{AB} = \begin{bmatrix} a_{11}b_{11} + a_{12}b_{21} & a_{11}b_{12} + a_{12}b_{22} \\ a_{21}b_{11} + a_{22}b_{21} & a_{21}b_{12} + a_{22}b_{22} \end{bmatrix}$$

and

$$\mathbf{BA} = \begin{bmatrix} b_{11}a_{11} + b_{12}a_{21} & b_{11}a_{12} + b_{12}a_{22} \\ b_{21}a_{11} + b_{22}a_{21} & b_{21}a_{12} + b_{22}a_{22} \end{bmatrix}$$

We notice that the elements of **AB** differ from the corresponding elements of **BA**. A number of our results follow from this example.

In general $\mathbf{AB} \neq \mathbf{BA}$, and it is said that matrix multiplication is *noncommutative*. Furthermore, $\mathbf{AB} = 0$ does not necessarily imply that $\mathbf{A} = 0$ or $\mathbf{B} = 0$ and $\mathbf{AB} = \mathbf{AC}$ does not necessarily mean $\mathbf{B} = \mathbf{C}$.

## 4. Special Matrices

When dealing with matrices we frequently encounter some special matrices. A *null matrix* or *zero matrix* has all its elements equal to zero, e.g.

$$\begin{bmatrix} 0 & 0 & 0 \\ 0 & 0 & 0 \end{bmatrix} \tag{2-11}$$

or

$$\begin{bmatrix} 0 & 0 & 0 \\ 0 & 0 & 0 \\ 0 & 0 & 0 \end{bmatrix} \tag{2-12}$$

A *square matrix* has the same number or rows and columns. Matrices (2-3) and (2-12) are square matrices. A *diagonal matrix* is a square matrix whose only nonzero elements are on the main diagonal. Matrix (2-13) depicted below is a diagonal matrix.

$$\begin{bmatrix} a_{11} & 0 & 0 & 0 \\ 0 & a_{22} & 0 & 0 \\ 0 & 0 & a_{33} & 0 \\ 0 & 0 & 0 & a_{44} \end{bmatrix} \tag{2-13}$$

If in a diagonal matrix $a_{11} = a_{22} = \cdots = a_{nn} = k$ where $k$ is some constant, then the diagonal matrix is called a *scalar matrix*. If in addition $k = 1$, the matrix is called the *identity matrix* and is denoted by $\mathbf{I}_n$ thus

$$\mathbf{I}_2 = \begin{bmatrix} 1 & 0 \\ 0 & 1 \end{bmatrix} \tag{2-14}$$

and

$$\mathbf{I}_3 = \begin{bmatrix} 1 & 0 & 0 \\ 0 & 1 & 0 \\ 0 & 0 & 1 \end{bmatrix} \tag{2-15}$$

### 4.1 Square Matrices

If $\mathbf{A}$ and $\mathbf{B}$ are square matrices such that

$$\mathbf{AB} = \mathbf{BA} \tag{2-16}$$

then $\mathbf{A}$ and $\mathbf{B}$ are said to commute. For example the identity matrix $\mathbf{I}_n$ commutes with the square matrix of order $n$, i.e.

$$\mathbf{AI}_n = \mathbf{I}_n\mathbf{A} \tag{2-17}$$

in addition each matrix commutes with itself $\mathbf{AA} = \mathbf{AA}$. Now if we define $\mathbf{A}^2 = \mathbf{AA}$, $\mathbf{A}^3 = \mathbf{A}^2\mathbf{A}$, $\mathbf{A}^4 = \mathbf{A}^3\mathbf{A}$, ..., $\mathbf{A}^n = \mathbf{A}^{n-1}\mathbf{A}$, it is easy to show, by induction, that $\mathbf{A}^n\mathbf{A}^m = \mathbf{A}^m\mathbf{A}^n = \mathbf{A}^{n+m}$. If for a matrix $\mathbf{A}$ we have $\mathbf{A}^{k+1} = \mathbf{A}$, where $k$ is a positive integer, then $\mathbf{A}$ is said to be *nilpotent* with index $k$. If in particular $k = 1$ so that $\mathbf{A}^2 = \mathbf{A}$, then $\mathbf{A}$ is called *indempotent*.

### 4.2 Transposition

The transpose of a matrix is defined as the matrix obtained by interchanging the rows and columns of the original matrix. For example if

$$\mathbf{A} = \begin{bmatrix} 2 & 0 & 5 \\ 1 & 6 & -2 \end{bmatrix}$$

then $\mathbf{A}'$ or $\mathbf{A}$ transpose may be written

$$\mathbf{A}' = \begin{bmatrix} 2 & 1 \\ 0 & 6 \\ 5 & -2 \end{bmatrix}$$

It can be shown that the transpose of a product of two matrices is the product in reverse order of their transposes, thus

$$(\mathbf{AB})' = \mathbf{B}'\mathbf{A}' \tag{2-18}$$

### 4.3 Symmetric Matrices

A square matrix $\mathbf{A}$ such that $\mathbf{A}' = \mathbf{A}$ is a symmetric matrix. Thus the matrix

$$\mathbf{A} = \begin{bmatrix} 1 & -2 & 4 \\ -2 & 1 & 3 \\ 4 & 3 & 1 \end{bmatrix} \tag{2-19}$$

is symmetric.

## 5. Determinants of Square Matrices

We shall denote the determinant of a square matrix $\mathbf{A}$ by $|\mathbf{A}|$

### 5.1 Permutations

The subject of permutations is a necessary preliminary to determinants. If we take the three integers 1, 2, 3, three at a time we can have the following $3! = 6$ permutations, namely

123, 132, 213, 321, 312, 231

If we take five integers at a time we have $5! = 120$ permutations some of which may be written as follows

53214, 41325, 12345, etc.

In any permutation when a larger integer precedes a smaller one we have an inversion. In any permutation when the number of inversions is odd (even) then the permutation is odd (even). Thus in the permutation 12345 there are no inversions so it is even. In the permutation 41253, we have 4 preceding 1, 4 preceding, 2 and 4 preceding 3 and 5 preceding 3 for a total of four inversions, so this permutation is also even. On the other hand the permutation 41235 is odd because 4 precedes 1, 4 precedes 2, and 4 precedes 3.

### 5.2 Determinants of 2 × 2 and 3 × 3 Matrices

With this discussion of permutations we consider a determinant of a $2 \times 2$ matrix and its value.

$$\begin{vmatrix} a_{11} & a_{12} \\ a_{21} & a_{22} \end{vmatrix} = a_{11}a_{22} - a_{12}a_{21} \tag{2-20}$$

and the determinant of a $3 \times 3$ matrix and its value

$$\begin{vmatrix} a_{11} & a_{12} & a_{13} \\ a_{21} & a_{22} & a_{23} \\ a_{31} & a_{32} & a_{33} \end{vmatrix} = a_{11}a_{22}a_{33} - a_{11}a_{23}a_{32} + a_{12}a_{23}a_{31} - a_{12}a_{21}a_{32} + a_{13}a_{21}a_{32} - a_{13}a_{22}a_{31} \tag{2-21}$$

From the expansion of these determinants we notice that half the terms occurring in the value of the determinant are negative and the other half are positive. Next, we note that each term is a product of the elements $a_{ij}$ and in each of these terms the first subscripts of the elements appear once and only once in the term and the second subscripts of the elements appear once and only once. Therefore if we order the elements in any given term so that *first* subscripts of the elements are in their natural order (as we have done in the above illustrative examples, then the second subscripts will be some permutation $(\alpha\beta\gamma \cdots \nu)$ of that natural order. For example in the second term in the expansion of the determinant of the $3 \times 3$ matrix, the first subscript of the first element is 1, the first subscript of the second element is 2, and the first subscript of the third element is 3, that is the first subscripts are ordered in their natural order. The second subscript of first, second, and third elements are 1, 3 and 2, respectively or some permutation of their natural order. With these considerations in mind we *define* the determinant of an $n \times n$ matrix $\mathbf{A}$ as the following function of the elements $a_{ij}$ of $\mathbf{A}$

$$|\mathbf{A}| = \Sigma(\pm\, a_{1\alpha}a_{2\beta} \cdots a_{n\nu})$$

where the summation is taken over all $n!$ permutations of the second subscripts. What sign we give to any term is prescribed by the following rule: If the permutation of the second subscripts is even, the sign of the term is positive, and if the permutation is odd the sign is negative.

As examples we evaluate a $2 \times 2$ and a $3 \times 3$ determinant. To evaluate the following $2 \times 2$ determinant

$$\begin{vmatrix} 2 & 3 \\ -2 & 1 \end{vmatrix}$$

we use Eq. (2-20) to give

$$2 - (-6) = 8$$

The readers can verify that the following is true for the $3 \times 3$ determinant by using Eq. (2-21).

$$\begin{vmatrix} -2 & 5 & -3 \\ -8 & -4 & -7 \\ -3 & -7 & 1 \end{vmatrix} = 119.$$

## 5.3 Properties of Determinants

We will illustrate the properties of determinants with $n = 3$ in order that the reader may verify them by direct expansion.

a. Transposition of a matrix does not change the value of its determinant and consequently all properties of rows can also be applied to columns, e.g.

$$\begin{vmatrix} a_{11} & a_{12} & a_{13} \\ a_{21} & a_{22} & a_{23} \\ a_{31} & a_{32} & a_{33} \end{vmatrix} = \begin{vmatrix} a_{11} & a_{21} & a_{31} \\ a_{12} & a_{22} & a_{32} \\ a_{13} & a_{23} & a_{33} \end{vmatrix} \tag{2-23}$$

b. If each element in a row (column) is zero, the corresponding value of the determinant is zero, e.g.

$$\begin{vmatrix} a_{11} & 0 & a_{13} \\ a_{21} & 0 & a_{23} \\ a_{31} & 0 & a_{33} \end{vmatrix} = 0 \tag{2-24}$$

c. Interchanging any two rows (columns) in a matrix reverses the sign of its determinant, e.g.

$$\begin{vmatrix} a_{11} & a_{12} & a_{13} \\ a_{21} & a_{22} & a_{23} \\ a_{31} & a_{32} & a_{33} \end{vmatrix} = - \begin{vmatrix} a_{11} & a_{13} & a_{12} \\ a_{21} & a_{23} & a_{22} \\ a_{31} & a_{33} & a_{32} \end{vmatrix} \tag{2-25}$$

d. If a row (column) of a matrix is multiplied by a number $p$ the value of the determinant of the matrix is multiplied by $p$, e.g.

$$\begin{vmatrix} a_{11} & pa_{12} & a_{13} \\ a_{21} & pa_{22} & a_{33} \\ a_{31} & pa_{32} & a_{33} \end{vmatrix} = p \begin{vmatrix} a_{11} & a_{12} & a_{13} \\ a_{21} & a_{22} & a_{23} \\ a_{31} & a_{32} & a_{33} \end{vmatrix} \tag{2-26}$$

e. If two rows (columns) of a matrix are equal the value of determinant of the matrix is zero, e.g.

$$\begin{vmatrix} a_{11} & a_{11} & a_{31} \\ a_{21} & a_{21} & a_{32} \\ a_{31} & a_{31} & a_{33} \end{vmatrix} = 0 \tag{2-27}$$

f. If each element of a row (column) of a matrix is expressed as the sum of two or more terms then the determinant can be expressed as two or more determinants, e.g.

$$\begin{vmatrix} a_{11} + a_{11} & a_{12} & a_{13} \\ a_{21} + a_{21} & a_{22} & a_{32} \\ a_{31} + a_{31} & a_{32} & a_{33} \end{vmatrix} = \begin{vmatrix} a_{11} & a_{12} & a_{13} \\ a_{21} & a_{22} & a_{23} \\ a_{31} & a_{32} & a_{33} \end{vmatrix} + \begin{vmatrix} a_{11} & a_{12} & a_{13} \\ a_{21} & a_{22} & a_{23} \\ a_{31} & a_{32} & a_{33} \end{vmatrix} \tag{2-28}$$

g. If to each element of a row or column of a matrix we add $n$ times the corresponding element of another row or column, the value of the determinant is not changed, e.g.

$$\begin{vmatrix} a_{11} + ra_{12} & a_{12} & a_{13} \\ a_{21} + ra_{22} & a_{22} & a_{23} \\ a_{31} + ra_{32} & a_{32} & a_{33} \end{vmatrix} = \begin{vmatrix} a_{11} & a_{12} & a_{13} \\ a_{21} & a_{22} & a_{23} \\ a_{31} & a_{32} & a_{33} \end{vmatrix} \tag{2-29}$$

The reader can easily demonstrate all these properties on the $2 \times 2$ matrix.

### 5.4 First Minors and Cofactors

*Definition.* The first minor of a matrix of order $n$ is the determinant of a matrix of order $n - 1$ obtained by the deletion of the row and column in which the element resides. Thus the first minor of the element $a_{23}$ in a matrix $\mathbf{A}$ of order 3 where

$$\mathbf{A} = \begin{bmatrix} a_{11} & a_{12} & a_{13} \\ a_{21} & a_{22} & a_{23} \\ a_{31} & a_{31} & a_{33} \end{bmatrix}$$

is the minor

$$| \mathbf{M}_{23} | = \begin{vmatrix} a_{11} & a_{12} \\ a_{31} & a_{32} \end{vmatrix}$$

More generally we label the minor of $a_{ij}$ obtained by the deletion of the $i$th row and $j$th column by $| \mathbf{M}_{ij} |$. The signed minor or *cofactor* of the element $a_{ij}$ is $(-1)^{i+j} | \mathbf{M}_{ij} |$. Minors and cofactors are used to simplify the expansion and evaluation of determinants.

### 5.5 Minors and Algebraic Complements

If we consider the matrix

$$\mathbf{A} = \begin{bmatrix} a_{11} & a_{12} & \cdots & a_{1n} \\ a_{21} & a_{22} & \cdots & a_{2n} \\ \vdots & \vdots & & \vdots \\ a_{1n} & a_{2n} & \cdots & a_{nn} \end{bmatrix} \tag{2-30}$$

and delete the first $m$ rows and $m$ columns, then we are left with a matrix or order $(n - m) \times (n - m)$ namely

$$\mathbf{A}_{m+1\ m+2\cdots n}^{m+1\ m+2\cdots n} = \begin{bmatrix} a_{m+1,m+1} & a_{m+2,m+2} & \cdots & a_{m+1,n} \\ a_{m+2,m+1} & a_{m+2,m+2} & \cdots & a_{m+2,n} \\ \vdots & & & \vdots \\ a_{n,m+1} & \cdots & \cdots & a_{n,n} \end{bmatrix} \tag{2-31}$$

The above matrix is the complementary minor of the matrix

$$\mathbf{A}_{12\cdots m}^{12\cdots m} = \begin{bmatrix} a_{11} & a_{12} & \cdots & a_{1m} \\ a_{21} & a_{22} & \cdots & a_{2m} \\ \vdots & \vdots & & \vdots \\ a_{m1} & a_{m2} & \cdots & a_{mm} \end{bmatrix} \tag{2-32}$$

and is also a complementary minor of $\mathbf{A}$. The two matrices (2-31) and (2-32) are complements of each other. We are not restricted to the deletion of the first $m$ rows and columns but we can pick any square matrix from our original matrix. For example in a $5 \times 5$ matrix we have the following pair of complementary minors.

$$\mathbf{A}_{52}^{13} = \begin{bmatrix} a_{21} & a_{23} \\ a_{51} & a_{53} \end{bmatrix} \tag{2-33a}$$

and

$$\mathbf{A}_{134}^{245} = \begin{bmatrix} a_{11} & a_{14} & a_{15} \\ a_{32} & a_{34} & a_{25} \\ a_{42} & a_{44} & a_{45} \end{bmatrix} \tag{2-33b}$$

The superscripts of **A** are the columns from which we obtain the elements and the subscripts give the rows. The above minors are obtained from the 5 × 5 matrix in the following manner.

$$\mathbf{A}_{12345}^{12345} = \begin{bmatrix} a_{11} & a_{12} & a_{13} & a_{14} & a_{15} \\ a_{21} & a_{22} & a_{23} & a_{24} & a_{25} \\ a_{31} & a_{32} & a_{33} & a_{34} & a_{35} \\ a_{41} & a_{42} & a_{43} & a_{44} & a_{45} \\ a_{51} & a_{52} & a_{53} & a_{54} & a_{55} \end{bmatrix} \tag{2-34}$$

The elements of the first minor are circled. The row and column in which each of these occur are crossed off leaving us with the elements of its complementary minor. To define the algebraic complement we let $p$ be the sum of subscripts and superscripts. The algebraic complement of

$$|\mathbf{A}_{134}^{245}|$$

is

$$(-1)^p \, |\mathbf{A}_{25}^{13}| = (-1) \, |\mathbf{A}_{25}^{13}|$$

since $p = 1 + 3 + 2 + 5$. For a 5 × 5 matrix there are 25 possible 1 square minors, 100 possible 2 square minors, 100 possible 3 square minors, 25 possible 4 square minors, and one 5 square minor, which is **A** itself.

While we shall not go into the details of evaluating determinants here, in the case where $n = 2$ or 3 this is easily done as we have shown in (2-20) and (2-21). As $n$ gets larger we employ various devices for evaluation by using the properties of minors, cofactors, and algebraic complements. For example when $n = 4$ the value of the following determinant is

$$\begin{vmatrix} a_{11} & a_{12} & a_{13} & a_{14} \\ a_{21} & a_{22} & a_{23} & a_{24} \\ a_{31} & a_{32} & a_{33} & a_{34} \\ a_{41} & a_{42} & a_{43} & a_{44} \end{vmatrix} = a_{11}\,|\mathbf{M}_{11}| - a_{12}\,|\mathbf{M}_{12}| + a_{13}\,|\mathbf{M}_{13}| - a_{14}\,|\mathbf{M}_{14}| \tag{2-35}$$

In the above instance we have multiplied the elements of any one row or column by their cofactors. In the Laplace expansion for the evaluation of determinants and other methods, we use the properties of minors and their algebraic complements to evaluate determinants (see any text on determinants for further detail, several are listed at the end of the chapter).

## 6. Matrix Rank and Elementary Transformations

### 6.1 Matrix Rank

A nonzero matrix **A** has *rank* $r$ if the determinant of at least one of its $r$-square minors is different from zero while every $(r + 1)$-square minor, if any, has determinant zero. For instance the matrix

$$\mathbf{A} = \begin{bmatrix} 1 & 2 & 3 \\ 2 & 3 & 4 \\ 3 & 5 & 7 \end{bmatrix}$$

has rank 2 because the determinant of the $3 \times 3$ matrix is zero, but the determinant of the minor

$$\begin{vmatrix} 1 & 2 \\ 2 & 3 \end{vmatrix} = -1 \neq 0$$

An $n$-square matrix **A** is *nonsingular* if rank $r = n$. If this condition is not fulfilled then the matrix is *singular*.

### 6.2 Elementary Transformations

Operations on matrices which do not alter the order or the rank of a matrix are called elementary transformations. These operations are reminiscent of the operations with determinants and are the following:

a. Interchanging any two rows.
b. Interchange any two columns.
c. Multiplication of all the elements in a single row or column by a non zero scalar.
d. Addition to the elements of any row (column) the elements of another row (column) multiplied by a scalar.

This last elementary transformation may be demonstrated as follows for a row operation. Here the matrix

$$\begin{bmatrix} 2 & 4 & -2 & 8 \\ 4 & 8 & 6 & 10 \\ -2 & -4 & 12 & -14 \end{bmatrix} \tag{2-36}$$

is transformed into

$$\begin{bmatrix} 2 & 4 & -2 & 8 \\ 0 & 0 & 10 & -6 \\ -2 & -4 & 12 & -14 \end{bmatrix} \tag{2-37}$$

by subtracting the elements of the first row multiplied by $-2$ from the second row.

### 6.3 Equivalent Matrices

*Definition.* Two matrices are equivalent if they can be obtained one from the other by a series of elementary transformations.

## 7. The Inverse of a Matrix

### 7.1 The Inverse of a Matrix

*Definition.* If $\mathbf{A}$ and $\mathbf{B}$ are *n-square* matrices of rank $n$ such that $\mathbf{AB} = \mathbf{BA} = \mathbf{I}_n$ then $\mathbf{B}$ is defined as the inverse of $\mathbf{A}$ and is written as $\mathbf{A}^{-1}$. Similarly $\mathbf{A}$ is the inverse of $\mathbf{B}$ and is written as $\mathbf{B}^{-1}$.

### 7.2 The Adjoint of a Matrix

*Definition.* The classical adjoint of a square matrix is defined as the square matrix formed by the matrix of cofactors of the elements of the original matrix. Example:

$$\mathbf{A} = \begin{bmatrix} 1 & 2 & 3 \\ 1 & 3 & 4 \\ 1 & 4 & 3 \end{bmatrix} \tag{2-38}$$

has the following adjoint

$$\text{adj } \mathbf{A} = \begin{bmatrix} \begin{vmatrix} 3 & 4 \\ 4 & 3 \end{vmatrix} & -\begin{vmatrix} 2 & 3 \\ 4 & 3 \end{vmatrix} & \begin{vmatrix} 2 & 3 \\ 3 & 4 \end{vmatrix} \\ -\begin{vmatrix} 1 & 4 \\ 1 & 3 \end{vmatrix} & \begin{vmatrix} 1 & 3 \\ 1 & 3 \end{vmatrix} & -\begin{vmatrix} 1 & 3 \\ 1 & 4 \end{vmatrix} \\ \begin{vmatrix} 1 & 3 \\ 1 & 4 \end{vmatrix} & -\begin{vmatrix} 1 & 2 \\ 1 & 4 \end{vmatrix} & \begin{vmatrix} 1 & 2 \\ 1 & 3 \end{vmatrix} \end{bmatrix} = \begin{bmatrix} -7 & 6 & -1 \\ 1 & 0 & -1 \\ 1 & -2 & 1 \end{bmatrix} \tag{2-39}$$

For adjoints it can be shown that

$$\mathbf{A}(\text{adj } \mathbf{A}) = |\,\mathbf{A}\,|\,\mathbf{I} \tag{2-40}$$

This is illustrated for the $3 \times 3$ case as follows

$$\begin{bmatrix} a_{11} & a_{12} & a_{13} \\ a_{21} & a_{22} & a_{23} \\ a_{31} & a_{32} & a_{33} \end{bmatrix} \begin{bmatrix} |M_{11}| & |M_{21}| & |M_{31}| \\ |M_{12}| & |M_{22}| & |M_{32}| \\ |M_{13}| & |M_{23}| & |M_{33}| \end{bmatrix} = \begin{bmatrix} |A| & 0 & 0 \\ 0 & |A| & 0 \\ 0 & 0 & |A| \end{bmatrix} = |A|\,\mathbf{I} \tag{2-41}$$

In general it is true that

$$\mathbf{A}(\text{adj }\mathbf{A}) = (\text{adj }\mathbf{A})\mathbf{A} = |\mathbf{A}|\,\mathbf{I} \tag{2-42}$$

Going back to singular $n$-square matrices we note that the product of two square matrices is singular if one of them is singular, and if $\mathbf{A}$ is singular then $\mathbf{A}(\text{adj }\mathbf{A}) = (\text{adj }\mathbf{A})\mathbf{A} = 0$. However, if $\mathbf{A}$ is nonsingular we can divide Eq. (2-42) throughout by $|\mathbf{A}|$ which gives us

$$\mathbf{AR} = \mathbf{RA} = \mathbf{I} \tag{2-43}$$

where the matrix $\mathbf{R} = |\mathbf{A}|^{-1} \text{adj }\mathbf{A} = (\text{adj }\mathbf{A})/|\mathbf{A}|$. From Eq. (2-43) we note that the matrix $\mathbf{R}$ is the inverse of $\mathbf{A}$. This follows from our previous definition of the inverse of a matrix. From the above representation of an inverse matrix it is clear that an $n$-square matrix $\mathbf{A}$ has an inverse if it is nonsingular. In addition, if the inverse exists it is unique. In the special case of a diagonal matrix

$$\begin{bmatrix} d_1 & & & 0 \\ & d_2 & & \\ & & \ddots & \\ 0 & & & d_n \end{bmatrix}$$

the inverse is given by

$$\begin{bmatrix} 1/d_1 & & & 0 \\ & 1/d_2 & & \\ & & \ddots & \\ 0 & & & 1/d_n \end{bmatrix}$$

### 7.3 Theorems on Rank and Singularity

A number of useful theorems concerning rank and singularity are the following:

*Theorem* 2-1. If $\mathbf{x}$ and $\mathbf{y}$ are vectors and if $\mathbf{A}$ is nonsingular and if the equation $\mathbf{y} = \mathbf{A}\mathbf{x}$ holds, then $\mathbf{x} = \mathbf{A}^{-1}\mathbf{y}$.

*Theorem* 2-2. The rank of the product of two matrices **A** and **B** is less than equal to the rank of **A** and is less than or equal to the rank of **B**.

*Theorem* 2-3. The rank of $\mathbf{A} + \mathbf{B}$ is less than or equal to the rank of **A** plus the rank of **B**.

## 8. Vector Algebra

### 8.1 Vectors

We illustrate this section by first considering the two dimensional vectors. The ordered pair of real numbers $(a_1, a_2)$ can be associated with a point $A$ in the plane (Fig. 2-1). We can write this pair as $[a_1, a_2]$ which denotes a two dimensional vector or 2-vector.

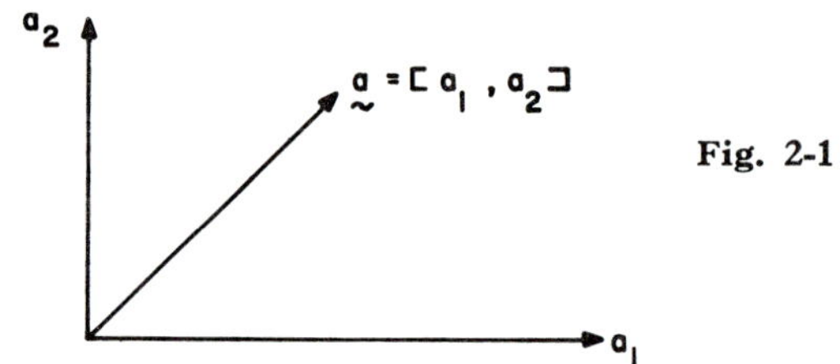


**Fig. 2-1**

### 8.2 Vector Operations

#### 8.2.1 *Addition.*

Suppose we represent two 2-vectors $\mathbf{a}_1$ and $\mathbf{a}_2$ as follows:

$$\mathbf{a}_1 = [a_{11}, a_{12}]$$
$$\mathbf{a}_2 = [a_{21}, a_{22}] \tag{2-44}$$

If we wish to add them we can do so graphically as in Fig. 2-2. The result of this addition can be written algebraically as follows

$$\mathbf{a}_3 = \mathbf{a}_1 + \mathbf{a}_2 = [a_{11} + a_{21}, a_{12} + a_{21}] \tag{2-45}$$

A result previously noted in the matrix addition.

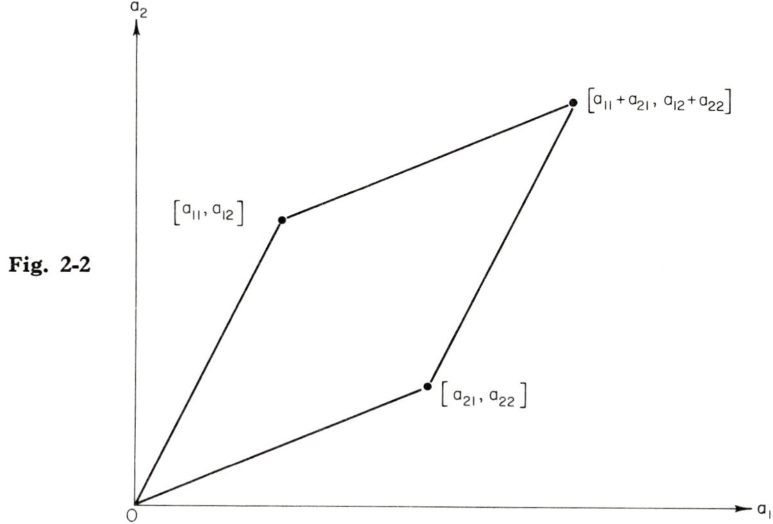


**Fig. 2-2**

### 8.2.2 *Scalar Multiplication*

The operation of scalar multiplication is represented by

$$k\mathbf{a}_1 = [ka_{11}, ka_{12}] \tag{2-46}$$

where $k$ is a scalar constant.

The general definition of an $n$-vector or an $n$-dimensional vector is the ordered set of $n$ elements written:

$$\mathbf{a} = [a_1, a_2, \ldots, a_n] \tag{2-47}$$

where $a_1, a_2, \ldots, a_n$ are the first, second, ..., and $n$th components respectively. A zero-vector has all its components equal to zero. Vectors as we have shown before can be written as row vectors or as column vectors.

## 8.3 Inner Products and Orthogonal Vectors

*Definition.* The inner product of the vectors $\mathbf{a} = [a_1, a_2, \ldots, a_n]$ and $\mathbf{b} = [b_1, b_2, \ldots, b_n]$ is defined as the scalar

$$\mathbf{a} \cdot \mathbf{b} = a_1b_1 + a_2b_2 + \cdots + a_nb_n$$

*Definition.* Two vectors are said to be orthogonal if their inner product is zero.

*Examples.* For the vectors $\mathbf{a}_1 = [2, 2, 2]$, $\mathbf{a}_2 = [4, 2, 4]$, and $\mathbf{a}_3 = [2, -4, 2]$

$$\mathbf{a}_1 \cdot \mathbf{a}_2 = 2(4) + 2(2) + 2(4) = 10$$
$$\mathbf{a}_1 \cdot \mathbf{a}_3 = 2(2) + (-4)2 + 2(2) = 0$$

The above definition tells us that $\mathbf{a}_1$ and $\mathbf{a}_3$ are orthogonal.

### 8.4 Linear Dependence of Vectors

Consider the set of $l$ $n$-vectors.

$$\begin{aligned} \mathbf{a}_1 &= [a_{11}, a_{12}, \ldots, a_{1n}] \\ \mathbf{a}_2 &= [a_{21}, a_{22}, \ldots, a_{2n}] \\ &\vdots \\ \mathbf{a}_l &= [a_{l1}, a_{l2}, \ldots, a_{ln}] \end{aligned} \tag{2-48}$$

*Definition.* The above set of vectors are said to be linearly dependent provided there exists $l$ elements $k_1, k_2, \ldots, k_l$ not all zero such that

$$k_1\mathbf{a}_1 + k_2\mathbf{a}_2 + k_3\mathbf{a}_3 + \cdots + k_l\mathbf{a}_l = 0 \tag{2-49}$$

If in order to satisfy the relationship in Eq. (2-49) all the $k$ must be zero, then the vectors are called linearly independent. A vector $\mathbf{a}_{l+1}$ is said to be expressible as a linear combination of $\mathbf{a}_1, \mathbf{a}_2, \ldots, \mathbf{a}_l$, if there exists elements $k_1, k_2, \ldots, k_l$ not equal to zero such that

$$\mathbf{a}_{l+1} = k_1\mathbf{a}_1 + k_2\mathbf{a}_2 + \cdots + k_l\mathbf{a}_l \tag{2-50}$$

### 8.5 Theorems on Linear Dependence

*Theorem* 2-4. If a set of vectors $\mathbf{a}_1, \mathbf{a}_2, \ldots, \mathbf{a}_l$ are linearly independent and if we add another vector $\mathbf{a}_{l+1}$ which makes the set linearly dependent then $\mathbf{a}_{l+1}$ can be expressed as a linear combination of $\mathbf{a}_1, \mathbf{a}_2, \ldots, \mathbf{a}_l$.

If in a set of $l$ vectors we have a subset of $n$ vectors which are linearly dependent, where $n < l$, then the entire set of $l$ vectors are linearly dependent. If the rank of the matrix

$$\mathbf{A} = \begin{bmatrix} a_{11} & a_{12} & \cdots & a_{1n} \\ a_{21} & a_{22} & \cdots & a_{2n} \\ \vdots & \vdots & & \vdots \\ a_{l1} & a_{l2} & \cdots & a_{ln} \end{bmatrix}, \qquad l \geq n \tag{2-51}$$

associated with the $l$ vectors is $r < l$ then there are only $r$ vectors in that set which are linearly independent. The rest of the $l - r$ vectors can be expressed as linear combinations of those $r$ vectors.

## 9. Linear Equations

### 9.1 Linear Equations

Suppose we have a set of linear equations in $n$ unknowns written thus

$$\begin{aligned} a_{11}x_1 + a_{12}x_2 \cdots + a_{1n}x_n &= b_1 \\ a_{21}x_1 + a_{22}x_2 \cdots + a_{2n}x_n &= b_2 \\ &\vdots \\ a_{l1}x_1 + a_{l2}x_2 \cdots + a_{ln}x_n &= b_l \end{aligned} \tag{2-52}$$

when this system of equations has a solution the system is said to be *consistent*. Otherwise the system is said to be *inconsistent*. For example the system

$$x_1 - x_2 = 4, \qquad 2x_1 - 2x_2 = 2 \tag{2-53}$$

is clearly inconsistent. Because there are no values of $x_1$ and $x_2$ that will simultaneously satisfy the above equations. A consistent system of equations may have just one solution or they may have infinitely many.

Our system of linear equations may be written in matrix form as

$$\begin{bmatrix} a_{11} & a_{12} & \cdots & a_{1n} \\ a_{21} & a_{22} & \cdots & a_{2n} \\ \vdots & \vdots & & \vdots \\ a_{m1} & a_{m2} & \cdots & a_{mn} \end{bmatrix} \begin{bmatrix} x_1 \\ x_2 \\ \vdots \\ x_n \end{bmatrix} = \begin{bmatrix} b_1 \\ b_2 \\ \vdots \\ b_m \end{bmatrix} \tag{2-54}$$

or

$$\mathbf{Ax} = \mathbf{b}$$

where $\mathbf{A} = [a_{ij}]$ and is called the coefficient matrix $\mathbf{x} = [x_1, x_2, \ldots, x_n]$ and $\mathbf{b} = [b_1, b_2, \ldots, b_m]$. We define the *augmented* matrix $[\mathbf{Ab}]$ by

$$[\mathbf{Ab}] = \begin{bmatrix} a_{11} & a_{12} & \cdots & a_{1n} & b_1 \\ a_{21} & a_{22} & \cdots & a_{2n} & b_2 \\ \vdots & \vdots & & \vdots & \vdots \\ a_{m1} & a_{m2} & \cdots & a_{mn} & b_m \end{bmatrix} \tag{2-55}$$

*Theorem* 2-5. A system of $m$ linear equations in $n$ unknowns is consistent if and only if the coefficient matrix and the augmented matrix have the same rank. Further, if in a consistent system of rank $r < n$, $n - r$ of the unknown are chosen so that the coefficient matrix of the remaining $r$ unknowns is of rank $r$. When these $n - r$ unknowns are assigned arbitrary values the $r$ unknowns are uniquely determined.

### 9.2 Nonhomogenous Equations

When a linear equation

$$a_1x_1 + a_2x_2 + \cdots + a_nx_n = b \tag{2-56}$$

is such that $b \neq 0$, it is called a nonhomogeneous equation. A system of equations

$$\mathbf{Ax} = \mathbf{b} \tag{2-57}$$

is nonhomogenous provided $\mathbf{b}$ is not a zero vector. A system of $n$ nonhomogenous equations in $n$ unknowns has a unique solution provided the rank of the coefficient matrix is $n$, that is if $\mathbf{A}$ is a nonsingular matrix. If we have a system of $n$ nonhomogenous equations in $n$ unknown written $\mathbf{Ax} = \mathbf{b}$, we can solve this system by means of *Cramer*'s rule. This is done by defining $\mathbf{A}_i$ for $i = 1, 2, \ldots, n$ as the matrix obtained from $\mathbf{A}$ by replacing its $i$th column by the column vector $\mathbf{b}$. The unique solution for the nonhomogenous equations is

$$x_1 = |\mathbf{A}_1| / |\mathbf{A}|, \qquad x_2 = |\mathbf{A}_2| / |\mathbf{A}|, \ldots, \qquad x_n = |\mathbf{A}_n| / |\mathbf{A}| \tag{2-58}$$

*Example.* Solve the system

$$\begin{aligned} 4x_1 + 2x_2 + 10x_3 + 2x_4 &= 10 \\ 2x_1 + 2x_2 - 6x_3 - 8x_4 &= -2 \\ 6x_1 + 12x_2 - 4x_3 + 2x_4 &= 16 \\ 4x_1 + 4x_2 + 4x_3 - 6x_4 &= 4 \end{aligned}$$

Using Cramer's rule

$$|\mathbf{A}| = \begin{vmatrix} 4 & 2 & 10 & 2 \\ 2 & 2 & -6 & -8 \\ 6 & 12 & -4 & 2 \\ 4 & 4 & 4 & -6 \end{vmatrix} = -240$$

$$|\mathbf{A}_1| = \begin{vmatrix} 10 & 2 & 10 & 2 \\ -2 & 2 & -6 & -8 \\ -16 & 12 & -4 & 2 \\ 4 & 4 & 4 & -6 \end{vmatrix} = -480$$

$$|\mathbf{A}_2| = \begin{vmatrix} 4 & 10 & 10 & 2 \\ 2 & -2 & -6 & -8 \\ 6 & 16 & -4 & 2 \\ 4 & 4 & 4 & -6 \end{vmatrix} = -48$$

$$|\mathbf{A}_3| = \begin{vmatrix} 4 & 2 & 10 & 2 \\ 2 & 2 & -2 & -8 \\ 6 & 12 & 16 & 2 \\ 4 & 4 & 4 & -6 \end{vmatrix} = 0$$

$$|\mathbf{A}_4| = \begin{vmatrix} 4 & 2 & 10 & 10 \\ 2 & 2 & -6 & -2 \\ 6 & 12 & -4 & 16 \\ 4 & 4 & 4 & 4 \end{vmatrix} = -192$$

so that $x_1 = -480/(-240)=2$, $x_2 = -48/(-240)=0.2$, $x_3=0/(-240) = 0$, $x_4 = -192/(-240) = 0.8$.

### 9.3 Homogenous Equations

A linear equation of the form

$$a_1x + a_2x_2 + \cdots + a_nx_n = 0 \tag{2-59}$$

is called a homogenous equation. A system of homogenous linear equations can be denoted by

$$\mathbf{Ax} = \mathbf{0} \quad \text{(zero column vector)} \tag{2-60}$$

Clearly the rank of **A** is always equal to the rank of the augmented matrix [**A0**]. This system is therefore always consistent and it is obvious that $\mathbf{x} = \mathbf{0}$ (i.e., $x_1 = x_2 = \cdots = x_n = 0$) is always a solution of this system. Such a solution is called the trivial solution. If the rank of **A** is $n$ then the trivial solution is the only solution. If on the other hand the rank of **A** is $r$ where $r < n$ then there exists a solution other than the trivial. In fact a necessary and sufficient condition for Eq. (2-60) to have a solution other than the trivial solution is that the rank of **A** be $r < n$. It also follows that for a system of $n$ homogenous equations in $n$ unknowns to have a solution other than the trivial, we must have $|\mathbf{A}| = 0$. Lastly, if the rank of $\mathbf{A}$ is $r < n$, the system has $n - r$ linearly independent solutions.

## 10. Quadratic Forms

Quadratic forms occur with considerable frequency in mathematical analysis as it relates to data fitting. In particular they occur in the statistical procedures of the analysis of variance.

### 10.1 Quadratic Forms

*Definition.* If $\mathbf{x}$ is an $n \times 1$ vector and its $i$th element is $x_i$, and if $\mathbf{A}$ is an $n \times n$ matrix where elements are $a_{ij}$ then the quadratic form is equal to

$$\mathbf{x}'\mathbf{A}\mathbf{x} = \sum_{i=1}^{n} \sum_{j=1}^{n} a_{ij} x_i x_j \tag{2-61}$$

The matrix $\mathbf{A}$ is called the matrix *associated* with the quadratic form.

*Example.* Consider the quadratic form given by

$$2x_1^2 + 4x_2^2 - 14x_3^2 - 8x_1x_2 + 16x_1x_3 \tag{2-62}$$

The matrix of this quadratic form can be written in a number of ways depending on how one handles the cross-product terms. For example for the term $-8x_1x_2$ we can choose the corresponding elements of the matrix $\mathbf{A}$ as $a_{12} = 0$ and $a_{21} = -8$ or we can have $a_{12} = 2$ and $a_{21} = -10$ or we may have $a_{21} = a_{12} = -4$. Since we can always choose $a_{ij} = a_{ji}$ for all $ij$ then it follows that the matrix of our quadratic form above is given by the symmetric matrix

$$\begin{bmatrix} 2 & -4 & 8 \\ -4 & 4 & 0 \\ 8 & 0 & -14 \end{bmatrix} \tag{2-63}$$

In particular this can be done for all quadratic forms. We shall consider only symmetric matrices in what follows. The rank of the quadratic form is defined as the rank of $\mathbf{A}$.

### 10.2 Positive Definite Forms

A quadratic form is said to be positive definite if and only if $\mathbf{x}'\mathbf{A}\mathbf{x} > 0$ for all vectors $\mathbf{x}$ where $\mathbf{x} \neq 0$. A quadratic form is positive semidefinite if and only if $\mathbf{x}'\mathbf{A}\mathbf{x} \geq 0$ for all $\mathbf{x}$ and $\mathbf{x}'\mathbf{A}\mathbf{x} = 0$ for some vector $\mathbf{x} \neq 0$. We also note that the matrix $\mathbf{A}$ of the quadratic form $\mathbf{x}'\mathbf{A}\mathbf{x}$ is positive definite (semidefinite) when the actual quadratic form is positive definite (semidefinite). A matrix

$$\mathbf{A} = \begin{bmatrix} a_{11} & a_{12} & \cdots & a_{1n} \\ a_{21} & a_{22} & \cdots & a_{2n} \\ \vdots & \vdots & & \vdots \\ a_{n1} & a_{n2} & \cdots & a_{nn} \end{bmatrix} \tag{2-64}$$

is positive definite if and only if

$$| a_{11} | > 0, \qquad \begin{vmatrix} a_{11} & a_{12} \\ a_{21} & a_{22} \end{vmatrix} > 0, \qquad \ldots, \qquad \begin{vmatrix} a_{11} & a_{12} & \cdots & a_{1n} \\ a_{21} & a_{22} & \cdots & a_{2n} \\ \vdots & \vdots & & \vdots \\ a_{n1} & a_{n2} & \cdots & a_{nn} \end{vmatrix} > 0 \tag{2-65}$$

To develop the theory of quadratic forms further, we describe some transformations in addition to the elementary transformations we described previously.

### 10.3 Linear Transformations

An equation of the form

$$\mathbf{y} = \mathbf{A}\mathbf{x} \tag{2-66}$$

where both $\mathbf{x}$ and $\mathbf{y}$ are vectors is a transformation of the vector $\mathbf{y}$ to the vector $\mathbf{x}$ by the matrix $\mathbf{A}$. A transformation is called linear if

$$\mathbf{A}(k\mathbf{x}_1 + h\mathbf{x}_2) = k\mathbf{A}\mathbf{x}_1 + h\mathbf{A}\mathbf{x}_2$$

where $k$ and $h$ are scalars.

The matrix $\mathbf{A}$ is called the matrix of the transformation and the transformation is called nonsingular if the matrix of the transformation is nonsingular. A nonsingular linear transformation carries linearly independent (dependent) vectors into linearly independent (dependent) vectors.

### 10.4 Orthogonal Matrices and Transformations

*Definition.* If $\mathbf{O}$ is an $n \times n$ matrix such that $\mathbf{O}'\mathbf{O} = \mathbf{I}$ then $\mathbf{O}$ is said to be an orthogonal matrix. The rows and columns of such a matrix are naturally orthogonal, that is to say if $\mathbf{g}_i$ and $\mathbf{g}_j$ are row vectors of an orthogonal matrix represented as

$$\mathbf{g}_i = [g_{i1}, g_{i2}, \cdots, g_{in}] \tag{2-67}$$

and

$$\mathbf{g}_j = [g_{j1}, g_{j2}, \cdots, g_{jn}] \tag{2-68}$$

then

$$\mathbf{g}_i\mathbf{g}_i = [g_{i1}g_{i1} + g_{i2}g_{i2} + \cdots + g_{in}g_{in}] = 1 \qquad \text{for all } i \tag{2-69}$$

and

$$\mathbf{g}_i\mathbf{g}_j = [g_{i1}g_{j1} + g_{i2}g_{j2} + \cdots + g_{in}g_{jn}] = 0 \qquad \text{for all } i \neq j \tag{2-70}$$

If $\mathbf{O}$ is an orthogonal matrix then $|\mathbf{O}| = \pm 1$. The reader can verify that the following matrix is orthogonal by determining that it has the properties given by Eq. (2-69) and (2-70) and that $|\mathbf{O}| = +1$ or $-1$.

$$\mathbf{O} = \begin{bmatrix} 1/\sqrt{6} & 1/\sqrt{2} & 1/\sqrt{3} \\ 2/\sqrt{6} & 0 & -1/\sqrt{3} \\ 1/\sqrt{6} & -1/\sqrt{2} & 1/\sqrt{3} \end{bmatrix}$$

A linear transformation is said to be orthogonal when the matrix of the transformation is orthogonal. This is shown as follows: if we consider the transformation $\mathbf{x} = \mathbf{F}\mathbf{y}$ then

$$\mathbf{x}'\mathbf{A}\mathbf{x} = (\mathbf{F}\mathbf{y})'\mathbf{A}(\mathbf{F}\mathbf{y}) = \mathbf{y}'\mathbf{F}'\mathbf{A}(\mathbf{F}\mathbf{y})$$

and it follows that by the transformation $\mathbf{x} = \mathbf{F}\mathbf{y}$ we have transformed the quadratic form $\mathbf{x}'\mathbf{A}\mathbf{x}$ into $\mathbf{y}'\mathbf{F}'\mathbf{A}\mathbf{F}\mathbf{y}$. Now, if $\mathbf{O}$ is an orthogonal matrix and if we make the transformation $\mathbf{x} = \mathbf{O}\mathbf{y}$ on $\mathbf{x}'\mathbf{x}$ then

$$\mathbf{x}'\mathbf{x} = \mathbf{x}'\mathbf{I}\mathbf{x} = \mathbf{y}'\mathbf{O}'\mathbf{I}\mathbf{O}\mathbf{y} = \mathbf{y}'\mathbf{O}'\mathbf{O}\mathbf{y} = \mathbf{y}'\mathbf{y} \tag{2-71}$$

We therefore see that an orthogonal matrix can transform the quadratic form $\mathbf{x}'\mathbf{x}$ into another quadratic form $\mathbf{y}'\mathbf{y}$.

To pursue the theory of quadratic forms further we have to define the characteristic equation of a matrix and its roots.

## 11. Eigenvalues and Eigenvectors

### 11.1 Transformations Giving Rise to Eigenvalue Problems

A vector $\mathbf{x}$ which can be transformed by a matrix $\mathbf{A}$ into $\lambda\mathbf{x}$ where $\lambda$ is a scalar is called an *invariant vector*. The equation for this transformation is

$$\mathbf{Ax} = \lambda\mathbf{x} \tag{2-72}$$

This equation can be written as

$$(\mathbf{A} - \lambda\mathbf{I})\mathbf{x} = 0 \tag{2-73}$$

which is a set of homogenous equations and possesses only the trivial solution $\mathbf{x} = 0$ unless the determinant of the coefficients is zero, i.e.

$$|\mathbf{A} - \lambda\mathbf{I}| = \begin{vmatrix} a_{11} - \lambda & a_{12} & \cdots & a_{1n} \\ a_{21} & a_{22} - \lambda & \cdots & a_{2n} \\ \vdots & \vdots & & \vdots \\ a_{n1} & a_{n2} & \cdots & a_{nn} - \lambda \end{vmatrix} = 0 \tag{2-74}$$

### 11.2 Characteristic Polynomial and Equation of a Matrix

Expanding the determinant (2-74) will yield a polynomial of degree $n$ in $\lambda$ written as

$$b_0 + b_1\lambda + b_2\lambda^2 + \cdots + b_n\lambda^n \tag{2-75}$$

where $b_0, b_1, \ldots, b_n$ are constants

*Definition.* The polynomial $f(\lambda) = |\mathbf{A} - \lambda\mathbf{I}|$ is called the *characteristic polynomial* of the matrix $\mathbf{A}$. The equation

$$f(\lambda) = |\mathbf{A} - \lambda\mathbf{I}| = 0 \tag{2-76}$$

is called the *characteristic equation* of the matrix $\mathbf{A}$, and the roots of $f(\lambda)$ are called *characteristic roots.*

### 11.3 Eigenvalues and Eigenvectors of Real Symmetric Matrices

If Eq. (2-76) has a nontrivial solution then at least one root is not equal to zero. Associated with each characteristic root is a *characteristic vector*. Characteristic roots are also called *latent roots* or *eigenvalues* and characteristic vectors are also called *eigenvectors.*

To illustrate this consider the matrix

$$\mathbf{A} = \begin{bmatrix} 2 & 2 & 1 \\ 1 & 3 & 1 \\ 1 & 2 & 2 \end{bmatrix} \tag{2-77}$$

its characteristic polynomial is

$$f(\lambda) = \begin{vmatrix} 2-\lambda & 2 & 1 \\ 1 & 3-\lambda & 1 \\ 1 & 2 & 2-\lambda \end{vmatrix} = -\lambda^3 + 7\lambda^2 - 11\lambda + 5 \tag{2-78}$$

Solution of the cubic equation $f(\lambda) = 0$ yields the roots $\lambda = -5$, and $\lambda_2 = \lambda_1 = -1$. For the case when $\lambda = -5$ we have to solve the following system of linear equations in order to determine the characteristic vector.

$$\begin{bmatrix} -3 & 2 & 1 \\ 1 & -2 & 1 \\ 1 & 2 & -3 \end{bmatrix} \begin{bmatrix} x_1 \\ x_2 \\ x_3 \end{bmatrix} = \begin{bmatrix} 0 \\ 0 \\ 0 \end{bmatrix} \tag{2-79}$$

One solution for the system of equations given by (2-79) is $x_1 = x_2 = x_3 = 1$. Consequently we can say that associated with the root $\lambda = -5$ is the characteristic vector $[1, 1, 1]'$. Further when the vector $[1, 1, 1]'$ is multiplied by a scalar $k$ we get other solutions to Eq. (2-79). Therefore all vectors of the form $[k, k, k]'$ are characteristic vectors or eigenvectors of $\mathbf{A}$.

For the case $\lambda_2 = \lambda_3 = 1$ we have to solve the following system of linear equations to obtain the characteristic vector

$$\begin{bmatrix} 1 & 2 & 1 \\ 1 & 2 & 1 \\ 1 & 2 & 1 \end{bmatrix} \begin{bmatrix} x_1 \\ x_2 \\ x_3 \end{bmatrix} = \begin{bmatrix} 0 \\ 0 \\ 0 \end{bmatrix} \tag{2-80}$$

This system of equations has two linearly independent solutions given by the vectors $\mathbf{x}_1 = [2, -1, 0]'$ and $\mathbf{x}_2 = [1, 0, -1]'$. In addition every vector $h\mathbf{x}_1 + k\mathbf{x}_2$ where $h$ and $k$ are scalars are characteristic vectors of the roots $\lambda_2 = \lambda_3 = -1$.

## 11.4 Theorems on Eigenvalues and Eigenvectors

*Theorem* 2-6. If $\lambda_1, \lambda_2, \ldots, \lambda_k$ are distinct (meaning nonequal) characteristic roots of a matrix $\mathbf{A}$ and if $\mathbf{x}_1, \mathbf{x}_2, \ldots, \mathbf{x}_k$ are the characteristic vectors associated with these roots then $\mathbf{x}_1, \mathbf{x}_2, \ldots, \mathbf{x}_k$ are linearly independent.

*Theorem* 2-7. If $\lambda_i$ is an $r$-fold [occurring $r$ times, in the above example for matrix (2-77) since $\lambda_2 = \lambda_3 = 1$, we had a 2-fold root] characteristic root of an $n \times n$ matrix $\mathbf{A}$ the rank of $|\mathbf{A} - \lambda\mathbf{I}|$ is not less than $n - r$ and there are linearly independent vectors that are characteristic vectors corresponding to that root.

*Theorem* 2-8. If $\mathbf{A}$ is $n \times n$ matrix of rank $r$ then at least $n - r$ of its characteristic roots are zero.

*Theorem* 2-9. If $\mathbf{O}$ is an orthogonal matrix then $\mathbf{A}$ and $\mathbf{OAO}'$ have the same characteristic roots.

*Theorem* 2-10. The characteristic roots of a symmetric matrix $\mathbf{A}$ are all real.

*Theorem* 2-11. The characteristic roots of a positive definite matrix $\mathbf{A}$ are all positive. For a positive semidefinite matrix the roots are either positive or zero.

*Theorem* 2-12. For every symmetric matrix $\mathbf{A}$ there exists an orthogonal matrix $\mathbf{O}$ such that

$$\mathbf{O}'\mathbf{AO} = \mathbf{D} \tag{2-81}$$

where $\mathbf{D}$ is diagonal. The diagonal elements of the matrix $\mathbf{D}$ are the characteristic roots of $\mathbf{A}$. It follows from the above theorem that any quadratic form $\mathbf{x}'\mathbf{Ax}$ can be reduced to a form having no cross terms, e.g.

$$d_1x_1^2 + d_2x_2^2 + d_3x_3^2 + \cdots + d_nx_n^2 \tag{2-82}$$

where $d_1, d_2, \ldots, d_n$ are constants which are the eigenvalues of the associated matrix $\mathbf{A}$. In addition it is true that if the matrix $\mathbf{A}$ is of rank $r < n$ then it can be reduced to the form

$$d_1x_1^2 + d_2x_2^2 + d_3x_3^2 + \cdots + d_rx_r^2 \tag{2-83}$$

## 12. Partitioned Matrices

### 12.1 Partitioned Matrices

In certain instances of matrix manipulation it is profitable to break up matrices into submatrices. This process is called partitioning. For example let us consider the matrices $\mathbf{A}$ and $\mathbf{B}$, partitioned by the dotted lines thus

$$\begin{bmatrix} b_{11} & b_{12} \\ b_{21} & b_{22} \\ \cdots & \cdots \\ b_{31} & b_{32} \end{bmatrix} \begin{bmatrix} a_{11} & \vdots & a_{12} & a_{13} & \vdots & a_{14} \\ a_{21} & \vdots & a_{22} & a_{23} & \vdots & a_{24} \end{bmatrix} = \mathbf{BA} \tag{2-84}$$

Multiplication of the matrices results in

$$\begin{bmatrix} b_{11}a_{11} + b_{12}b_{21} & \vdots & b_{11}a_{12} + b_{12}a_{22}b_{11}a_{13} + b_{12}a_{23} & \vdots & b_{11}a_{14} + b_{12}a_{24} \\ b_{21}a_{11} + b_{22}a_{21} & \vdots & b_{21}a_{12} + b_{22}a_{22}b_{11}a_{13} + b_{22}a_{23} & \vdots & b_{21}a_{14} + b_{22}a_{24} \\ \cdots\cdots & \vdots & \cdots\cdots\cdots\cdots & \vdots & \cdots\cdots \\ b_{31}a_{11} + b_{32}a_{21} & \vdots & b_{31}a_{13} + b_{32}a_{22}b_{31}a_{13} + b_{32}a_{23} & \vdots & b_{31}a_{14} + b_{32}a_{24} \end{bmatrix} \tag{2-85}$$

From the above results we note that

$$\begin{bmatrix} \mathbf{B}_1 \\ \mathbf{B}_2 \end{bmatrix} [\mathbf{A}_1 \quad \mathbf{A}_2 \quad \mathbf{A}_3] = \begin{bmatrix} \mathbf{B}_1\mathbf{A}_1 & \mathbf{B}_1\mathbf{A}_2 & \mathbf{B}_1\mathbf{A}_3 \\ \mathbf{B}_2\mathbf{A}_1 & \mathbf{B}_2\mathbf{A}_2 & \mathbf{B}_2\mathbf{A}_3 \end{bmatrix} \tag{2-86}$$

In general if we have a matrix $\mathbf{A}$ it can be partitioned into submatrices thus

$$\mathbf{A} = \begin{bmatrix} \mathbf{A}_{11} & \mathbf{A}_{12} \\ \mathbf{A}_{21} & \mathbf{A}_{22} \end{bmatrix} \tag{2-87}$$

where $\mathbf{A}$ is $m \times n$, $\mathbf{A}_{11}$ is $m_1 \times n_1$, $\mathbf{A}_{12}$ is $m_1 \times n_2$, $\mathbf{A}_{21}$ is $m_2 \times n_1$ $\mathbf{A}_{22}$ is $m_2 \times n_2$ and where we have $m = m_1 + m_2$ and $n = n_1 + n_2$. Further, if

$$\mathbf{C} = \begin{bmatrix} \mathbf{C}_{11} & \mathbf{C}_{12} \\ \mathbf{C}_{21} & \mathbf{C}_{22} \end{bmatrix} \tag{2-88}$$

is an $m \times n$ matrix and if

$$\mathbf{D} = \begin{bmatrix} \mathbf{D}_{11} & \mathbf{D}_{12} \\ \mathbf{D}_{21} & \mathbf{D}_{22} \end{bmatrix} \tag{2-89}$$

is an $n \times k$ matrix, then the resulting matrix on multiplication is

$$\mathbf{CD} = \begin{bmatrix} \mathbf{C}_{11}\mathbf{C}_{12} & \mathbf{D}_{11}\mathbf{D}_{12} & \mathbf{C}_{11}\mathbf{D}_{11} + \mathbf{C}_{12}\mathbf{D}_{21} & \mathbf{C}_{11}\mathbf{D}_{12} + \mathbf{C}_{12}\mathbf{D}_{22} \\ \mathbf{C}_{21}\mathbf{C}_{22} & \mathbf{D}_{21}\mathbf{D}_{22} & \mathbf{C}_{21}\mathbf{D}_{11} + \mathbf{C}_{22}\mathbf{D}_{21} & \mathbf{C}_{21}\mathbf{D}_{12} + \mathbf{C}_{22}\mathbf{D}_{22} \end{bmatrix} \tag{2-90}$$

### 12.2 Theorems on Partitioned Matrices

These theorems will be useful in the development of the statistical theory particularly the distribution of quadratic forms.

*Theorem* 2-13. If $\mathbf{A}$ is a positive definite symmetric matrix such that

$$\mathbf{A} = \begin{bmatrix} \mathbf{A}_{11} & \mathbf{A}_{12} \\ \mathbf{A}_{21} & \mathbf{A}_{22} \end{bmatrix} \tag{2-91}$$

and $\mathbf{B}$ its inverse such that

$$\mathbf{B} = \begin{bmatrix} \mathbf{B}_{11} & \mathbf{B}_{12} \\ \mathbf{B}_{21} & \mathbf{B}_{22} \end{bmatrix} \tag{2-92}$$

and $\mathbf{B}_{ii}$ and $\mathbf{A}_{ii}$ are each of dimension $m_i \times n_i$ and if we use the same notation for the rest of the submatrices then

$$| \mathbf{A}_{11}^{-1} | = | \mathbf{B}_{11} - \mathbf{B}_{12} \ \ \mathbf{B}_{22}^{-1}\mathbf{B}_{21} | \tag{2-93}$$

*Theorem* 2-14. If $\mathbf{A}$ is a square matrix so that

$$\mathbf{A} = \begin{vmatrix} \mathbf{A}_{11} & \mathbf{A}_{12} \\ \mathbf{A}_{21} & \mathbf{A}_{22} \end{vmatrix} \tag{2-94}$$

and if $\mathbf{A}_{22}$ is nonsingular then

$$| \mathbf{A} | = | \mathbf{A}_{22} | \ | \mathbf{A}_{11} - \mathbf{A}_{12}\mathbf{A}_{22}^{-1}\mathbf{A}_{21} | \tag{2-95}$$

## 13. Application to Coordinate Geometry and Bilinear Forms

### 13.1 Coordinate Transformations

The use of matrices and their associated concepts aid in computing and describing the transformations from one set of coordinates to another. We consider only the case of the plane (two-dimensional space), but the methods extend directly to higher dimensions.

#### 13.1.1 *Translation of Axes*

Let $[x_1, x_2]$ be the coordinates of a point $P$ relative to the usual $x_1$- and $x_2$-axes. Let us translate the axes parallel to themselves to the new position $O'y_1$ and $O'y_2$ and let the coordinates of the new origin $O'$ relative to the old axes be $[\xi_1, \xi_2]$. It is clear from Fig. 2-3 that the new coordinates $[y_1', y_2']$ of the point $P$ are connected with old coordinates by the following equations.

$$\begin{aligned} x_1 &= y_1' + \xi_1 \\ x_2 &= y_2' + \xi_2 \end{aligned} \tag{2-96}$$

In matrix form this translation can be written as

$$\begin{bmatrix} y_1 \\ y_2 \end{bmatrix} = \begin{bmatrix} x_1 \\ x_2 \end{bmatrix} - \begin{bmatrix} \xi_1 \\ \xi_2 \end{bmatrix} \tag{2-97}$$

and is called a *parallel translation* of axes.

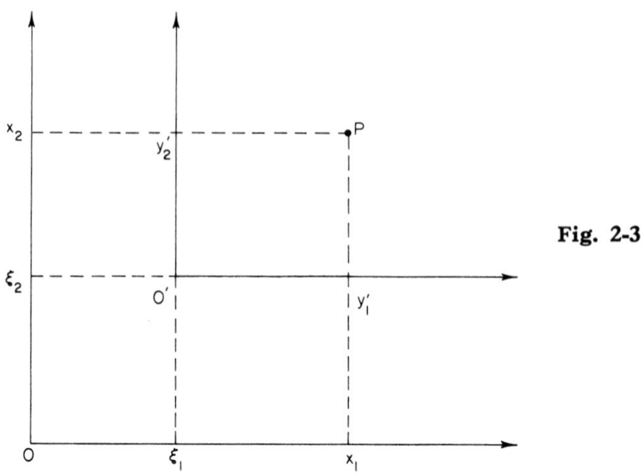


Fig. 2-3

13.1.2 *Rotation of Axes*

Suppose the coordinates of a point $P$ are $[x_1, x_2]$ and suppose that the coordinate system $Ox_1$ and $Ox_2$ is rotated to $Ox_1'$ and $Ox_2'$ by an angle $\phi$ (See Fig. 2-4). The new coordinates are connected to the old coordinates by the equations

$$x_1 = x_1' \cos\phi - x_2' \sin\phi, \qquad x_2 = x_1' \sin\phi + x_2' \cos\phi \tag{2-98}$$

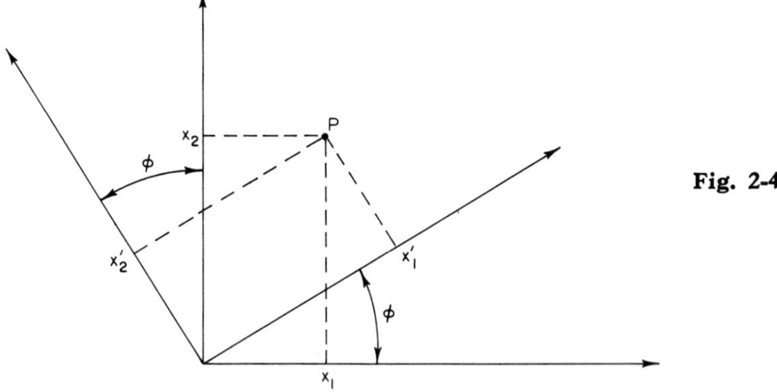


Fig. 2-4

In matrix notation we have

$$\begin{bmatrix} x_1 \\ x_2 \end{bmatrix} = \begin{bmatrix} \cos\phi & -\sin\phi \\ \sin\phi & \cos\phi \end{bmatrix} \begin{bmatrix} x_1' \\ x_2' \end{bmatrix} \tag{2-99}$$

or $\mathbf{x} = \mathbf{A}\mathbf{x}'$. Thus whether our transformation of coordinates is a translation or a rotation, matrices can be utilized to represent such transformations.

In 3 dimensions, rotation of coordinates by an angle $\phi$ leaving the $x_3$ axis fixed, results in the following relationship between the old and new coordinates.

$$x_1 = x_1' \cos \phi - x_2' \sin \phi, \qquad x_2 = x_1' \sin \phi + x_2' \cos \phi, \qquad x_3 = x_3' \tag{2-100}$$

In matrix form we have a similar result

$$\begin{bmatrix} x_1 \\ x_2 \\ x_3 \end{bmatrix} = \begin{bmatrix} \cos \phi & -\sin \phi & 0 \\ \sin \phi & \cos \phi & 0 \\ 0 & 0 & 1 \end{bmatrix} \begin{bmatrix} x_1' \\ x_2' \\ x_3' \end{bmatrix} \tag{2-101}$$

or again, $\mathbf{x} = \mathbf{A}\mathbf{x}'$. We can then rotate about the $x_2$ axis through an angle $\theta$ to obtain

$$\begin{bmatrix} x_1' \\ x_2' \\ x_3' \end{bmatrix} = \begin{bmatrix} \cos \theta & 0 & -\sin \theta \\ 0 & 1 & 0 \\ \sin \theta & 0 & \cos \theta \end{bmatrix} \begin{bmatrix} x_1'' \\ x_2'' \\ x_3'' \end{bmatrix} \tag{2-102}$$

or $\mathbf{x}' = \mathbf{B}\mathbf{x}''$. Finally we rotate about the $x_1$ axis through an angle $\beta$ to obtain

$$\begin{bmatrix} x_1'' \\ x_2'' \\ x_3'' \end{bmatrix} = \begin{bmatrix} 1 & 0 & 0 \\ 0 & \cos \beta & -\sin \beta \\ 0 & \sin \beta & \cos \beta \end{bmatrix} \begin{bmatrix} x_1''' \\ x_2''' \\ x_3''' \end{bmatrix} \tag{2-103}$$

or $\mathbf{x}'' = \mathbf{C}\mathbf{x}'''$.

The connection between the final $\mathbf{x}'''$ vector coordinates and the original $\mathbf{x}$ coordinates is found by direct substitution,

$$\mathbf{x} = \mathbf{A} \cdot \mathbf{B} \cdot \mathbf{C}\mathbf{x}''' \tag{2-104}$$

or $\mathbf{x} = \mathbf{D}\mathbf{x}'''$, and the new coefficient matrix $\mathbf{D}$ is the product of the coefficient matrices $\mathbf{A}$, $\mathbf{B}$, and $\mathbf{C}$ from each of the respective rotations.

These example applications illustrate how matrix multiplication can be interpreted as coordinate transformations.

### 13.2 The General Equation of Second Degree and Bilinear Forms

Of particular interest in Chapter 4 will be the general equation of second degree in two variables $x_1$ and $x_2$ given by

$$f(x_1, x_2) = Ax_1^2 + Bx_1x_2 + Cx_2^2 + Dx_1 + Ex_2 + F = 0 \tag{2-105}$$

We shall be interested in classifying geometrically what this function represents for different values of its coefficients $A$, $B$, $C$, $D$, $E$, and $F$.

Descartes, the discoverer of analytic geometry, studied this equation in its general form and gave its various geometric forms the name conic sections. These are, of course, the familiar ellipse (circle), parabola, hyperbola, and the degenerate case ($A = B = C = 0$) of straight lines. We shall see how these are identified.

We observe first that Eq. (2-105) is a quadratic form and may therefore be represented in matrix form as

$$F(x_1, x_2) = \mathbf{x}'\begin{bmatrix} A & B/2 \\ B/2 & C \end{bmatrix}\mathbf{x} + [D, E]\mathbf{x} + F \tag{2-106}$$

Clearly, the case $A = B = C = 0$ is simply a straight line and hence deserves no further comment. It is also clear that the geometric nature of (2-105) is completely determined by the nonzero character of $A$, $B$, and $C$. If we write (2-106) as

$$F(x_1, x_2) = \mathbf{x}'\mathbf{M}\mathbf{x} + [D, E]\mathbf{x} + F \tag{2-107}$$

then the geometric nature of (2-105) must be characterized by the properties of the coefficient matrix $\mathbf{M}$. We now examine these properties. The characteristic equation for the matrix $\mathbf{M}$ is

$$|\mathbf{M} - \lambda\mathbf{I}| = \begin{vmatrix} A - \lambda & B/2 \\ B/2 & C - \lambda \end{vmatrix} = 0$$

or expanding

$$\lambda^2 - (A + C)\lambda + AC - (B^2/4) = 0$$

Note that the constant term $[AC - (B^2/4)]$, is the determinant of $\mathbf{M}$ commonly called the discriminant function of Eq. (2-105). The eigenvalues of $\mathbf{M}$ are the solutions of Eq. (2-106) and are given by

$$\lambda_{1,2} = \frac{A + C}{2} \pm \left[\left(\frac{A + C}{2}\right)^2 + \left(\frac{B^2}{4} - AC\right)\right]^{1/2} \tag{2-108}$$

using the $+$ sign for one value and $-$ for the other. We can now classify the eigenvalues of $\mathbf{M}$ by means of the quantity $(B^2/4) - AC$. The quantity $[(A+C)/2]^2 + [(B^2/4) - AC]$ can be rewritten as $[(A - C)/2]^2 + (B^2/4)$ and is clearly always positive. Therefore only real values can occur in Eq. (2-108):

*Case* 1. When $(B^2/4) - AC < 0$ then $0 < (B^2/4) < AC$ which means $A$ and $C$ have the same sign. Since $0 < [(A + C)/2]^2 + [(B^2/4) - AC] < [(A + C)/2]^2$ both roots in Eq. (2-106) have the same sign (i.e., both are positive or both are negative). The two roots are equal if

$$[(A + C)/2]^2 = AC - (B^2/4).$$

This means that there is a coordinate transformation $\mathbf{V}$ as in Eq. (2-99) such that $\mathbf{x} = \mathbf{Vu}$, for which $\mathbf{M}$ is transformed to the diagonal form

$$\mathbf{V'MV} = \begin{bmatrix} \lambda_1 & 0 \\ 0 & \lambda_2 \end{bmatrix}$$

Equation (2-105) in these coordinates $u_1$ and $u_2$ becomes

$$F(u_1, u_2) = \lambda_1 u_1{}^2 + \lambda_2 u_2{}^2 + \alpha u_1 + \beta u_2 + F. \tag{2-109}$$

Clearly this is the equation of an ellipse, and is a circle if $\lambda_1 = \lambda_2$. Here $[\alpha, \beta] = [D, E]\mathbf{V}$.

*Case* 2. When $(B^2/4) - AC = 0$, then $\mathbf{M}$ has only one nonzero eigenvalue, namely $(A + C)$ and $\mathbf{M}$ is similar to the diagonal form

$$\begin{bmatrix} \lambda_1 & 0 \\ 0 & 0 \end{bmatrix}$$

again by means of the transformation

$$\mathbf{x} = \mathbf{Wu} \tag{2-110}$$

equation (2-105) then becomes

$$F(u_1, u_2) = \lambda_1 u_1{}^2 + \alpha' u_1 + \beta' u_2 + F \tag{2-111}$$

which is clearly a parabola.

*Case* 3. The final case is $(B^2/4) - AC > 0$ and (2-107) will have two roots of opposite signs. The matrix $\mathbf{M}$ is, in this case, similar to

$$\begin{bmatrix} \lambda_1 & 0 \\ 0 & \lambda_2 \end{bmatrix} \tag{2-112}$$

where $\lambda_1$ and $\lambda_2$ have opposite signs. In this case Eq. (2-105) takes the form

$$F(u_1, u_2) = \lambda_1 u_1{}^2 + \lambda_2 u_2{}^2 + \alpha'' u_1 + \beta'' u_2 + F \tag{2-113}$$

and is (because of the difference in signs of $\lambda_1$ and $\lambda_2$) a hyperbola. We could also use a different form,

$$\begin{bmatrix} 0 & \lambda_2 \\ \lambda_1 & 0 \end{bmatrix}$$

and obtain

$$F(u_1', u_2') = (\lambda_1 + \lambda_2)u_1'u_2' + \alpha'''u_1 + \beta'''u_2 + F \qquad (2\text{-}114)$$

which is also recognizable as a hyperbola.

What we have done in the preceeding discussion is go through the "rotation procedure" from high school algebra which removes the product term $Bx_1x_2$ in Eq. (2-105). The reader will recall that the angle of rotation is found by

$$\tan 2\theta = B/(A - C)$$

and the quantity $B^2 - 4AC$ is invariant under this rotation. We leave it as an exercise for the reader to verify that our method and the "rotation procedure" are the same.

We complete this discussion by putting Eq. (2-109) in its canonical form. We rewrite it by completing the squares in each variable $u_1$ and $u_2$ to obtain

$$\lambda_1[u_1 + (\alpha/2\lambda_1)]^2 + \lambda_2[u_2 + (\beta/2\lambda_2)]^2 + F - (\alpha^2/4\lambda_1) - (\beta^2/4\lambda_2) = 0 \qquad (2\text{-}115)$$

If we now make the parallel translation and scaling of variables

$$\boldsymbol{\nu} = \mathbf{u} + \boldsymbol{\xi}$$

where

$$\boldsymbol{\xi} = [(\alpha/2\lambda_1), (\beta/2\lambda_2)],$$

and defining

$$\psi = (\alpha^2/4\lambda_1) + (\beta^2/4\lambda_2) - F \geq 0$$

then if we set $a^2 = \psi\lambda_2$ and $b^2 = \psi\lambda_1$ we obtain by means of our transformations

$$(\nu_1^2/a^2) + (\nu_2^2/b^2) = 1 \qquad (2\text{-}116)$$

the canonical equation of an ellipse.

This procedure can be repeated for the parabola and hyperbola to obtain their canonical forms. We leave this as an exercise for the reader.

We return briefly to (2-105) and summarize our transformations for the case when $B^2 - 4AC < 0$ (i.e., the ellipse). Substituting

$$\mathbf{x} = \mathbf{V}\mathbf{u} \qquad \text{and} \qquad \mathbf{u} = \boldsymbol{\nu} - \boldsymbol{\xi}$$

we see that

$$\mathbf{x} = \mathbf{V}\boldsymbol{\nu} - \mathbf{V}\boldsymbol{\xi}$$

relates our canonical form (2-115) to the original form (2-105) and $\mathbf{V}$ is found from the eigenvectors of the matrix $\mathbf{M}$, $\boldsymbol{\xi}$ is found from the eigenvectors and eigenvalues of $\mathbf{M}$ and the constants $[D, E]$. In fact

$$\boldsymbol{\xi} = \frac{1}{2}\begin{bmatrix} 1/\lambda_2 & 0 \\ 0 & 1/\lambda_1 \end{bmatrix}\{[D, E]\mathbf{V}\}' \tag{2-117}$$

As is seen from this expression, everything is known once the eigenvalues of $\mathbf{M}$ and their corresponding eigenvectors are computed, (i.e., $\mathbf{V}$ is the matrix of eigenvectors).

### 13.3 Bilinear Forms

The concepts of Section (13.2) extend naturally to dimensions greater than two, with one notable exception. Instead of examining a discriminant function for the matrix $\mathbf{M}$ in

$$F(x_1, \ldots, x_7) = \mathbf{x}'\mathbf{M}\mathbf{x} + [a_1, \ldots, a_n]\mathbf{x} + b \tag{2-118}$$

where $\mathbf{x} = [x_1, \ldots, x_n]'$, we examine $\mathbf{M}$ for its positive semidefiniteness or other properties. Furthermore, only the hyperellipse (hypersphere) retains a geometric meaning in higher than two dimensions. The cases where $\mathbf{M}$ is positive definite or positive semidefinite are the ones of most interest. When $\mathbf{M}$ is positive definite we can transform it, exactly as in the case of two dimensions, to diagonal form which expresses (2-118) as

$$\lambda_1 u_1^2 + \cdots + \lambda_n u_n^2 + \alpha_1 u_1 + \cdots + a_n u_n + b = 0. \tag{2-119}$$

This is, of course, the hyperelliptic form mentioned above. We treat this further in Chapter 4.

In concluding this chapter it is pointed out that there are two minor topics not mentioned here that will aid in the elaboration of the subject. These are the differentiation of vectors and matrices, and they will be covered in the next chapter.

We have been quite brief in the treatment of some aspects of the subject presented here. Inasmuch as we do not present proofs of the theorems and present only the theorems that are immediately pertinent to our future work, this gives the appearance of disconnectedness in the presentation. We must point out however our primary interest is application, and consequently we feel that reference to theorems is sufficient. Proofs of all theorems are presented in the listed references.

## References

Aitken, A. C. (1937). "Determinants and Matrices." Oliver and Boyd, Edinburgh.

Bellman, R. (1960). "Introduction to Matrix Analysis." McGraw-Hill, New York.

Fuller, L. E. (1962). "Basic Matrix Theory." Prentice-Hall Englewood Cliffs, New Jersey.

Herstein, I. N. (1964). "Topics in Algebra." Ginn (Blaisdell), Boston, Massachusetts.

Hoffman, K. and Kunze, R. (1961). "Linear Algebra." Prentice-Hall, Engelwood Cliffs, New Jersey.

Noble, B. (1969). "Applied Linear Algebra." Prentice-Hall, Engelwood Cliffs, New Jersey.

Perlis, S. (1952). "Theory of Matrices." Addison-Wesley, Reading, Massachusetts.

*CHAPTER 3*

# ANALYSIS

This chapter is divided in two major parts. The first part deals with the main ideas in analysis which we shall call pure analysis. The second part deals with pertinent specific applications of the concepts developed in the first part. We have labeled the second part applied analysis.

Readers well versed in the calculus of functions of one or more variables may omit this chapter and proceed directly to Chapter 4. For other readers a first course of calculus amounting to the elements of differentiation and integration is probably required. As written, the subject matter may sound quite dry and abstract especially to those ultimately interested in applications. To those readers we recommend that on a first reading they focus attention on the ideas as they are presented in the diagrams and not pay a great deal of detailed attention to requirements of mathematical rigor.

# PURE ANALYSIS

## 1. Sets

### 1.1 Sets

The concept of a set is fundamental in mathematics. A collection of objects having some given property or properties is called a *set* or *class*. For example, the set of all infants of age two months or less, the set of all molecular biologists, or the set of all letters in the alphabet. An individual object in a set is a *member* or an *element* of the set and is written: the element $a$ in the set $A$, i.e.,

$a \in A$. The statement that $a = b$ means that the object $a$ and the object $b$ are identical. In particular if $A$ and $B$ are sets then $A = B$ if and only if every element of the set $A$ is also an element of the set $B$ and conversely, every element of the set $B$ is an element of the set $A$.

### 1.2 Subsets

If $A$ and $B$ are sets such that every element of $B$ is also an element of $A$ then $B$ is a subset of $A$ or $B$ is contained in $A$. This statement is written as $B \subset A$. Since $A$ is always a subset of itself, if $B \subset A$ and $A$ is not equal to $B$ then $B$ is a *proper* subset of $A$. For example consider the following sets:

$$A = (a, b, c, d, e), \qquad B = (a, b, c), \qquad C = (a, d, b),$$
$$D = (a, d, b, c, e), \qquad E = (e, f)$$

here $A = D$ and $B$ and $C$ are proper subsets of $A$ and $D$. Further $A$ and $D$ are subsets of each other and $E$ is *not* a subset of $A$, $B$, $C$, or $D$.

### 1.3 Null Sets and Universal Sets

*Definition.* We define the empty set as a set having no elements and it is denoted by $\varphi$. We observe that the empty set is a subset of every set. For any specific application of set theory discussed here, all sets will be considered as subsets of a *universal set*, (or space) $S$.

### 1.4 Unions, Intersections, and Complements

*Definition.* The *union* of two sets $A$ and $B$ is the set $C$ identified by the rule that $x$ is a member of $C$ if $x$ is a member of $A$ *or* $B$ *or both*. The union of $A$ and $B$ is denoted by $A \cup B$. In the previous examples of sets:

$$A \cup B = (a, b, c, d, e), \qquad A \cup E = (a, b, c, d, e, f),$$
$$B \cup C = (a, b, c, d), \qquad \text{etc.}$$

*Definition.* The *intersection* of two sets $A$ and $B$ is the set $D$ such that all members of $D$ belong to *both $A$ and $B$*. The intersection is written as $A \cap B$. Using the previous examples of sets:

$$A \cap D = (a, b, c, d, e), \qquad A \cap B = (a, b, c)$$
$$A \cap E = (a, b, c, d, e), \qquad B \cap C = (a, b), \qquad \text{etc.}$$

If the intersection of two sets contains no elements, that is, the null set $\phi$, then the two sets are called disjoint, or nonintersecting.

*Definition.* The relative complement of two sets $A$ and $B$ is the difference between $A$ and $B$ and is denoted by $A - B$, that is, the set of elements which belong to $A$ and not to $B$. Example: $A - B = (d, e)$.

*Definition.* The *absolute complement* of set $X$ is the set of elements in the space $S$ which do not belong to $X$ and is the difference between the set $X$ and the universal set. (i.e., $S - X = X^c$). Most set operations can be demonstrated by Venn diagrams as shown in Fig. 3-1. In what follows (see Section 1.7) the only sets we will generally be concerned with are sets of real numbers.

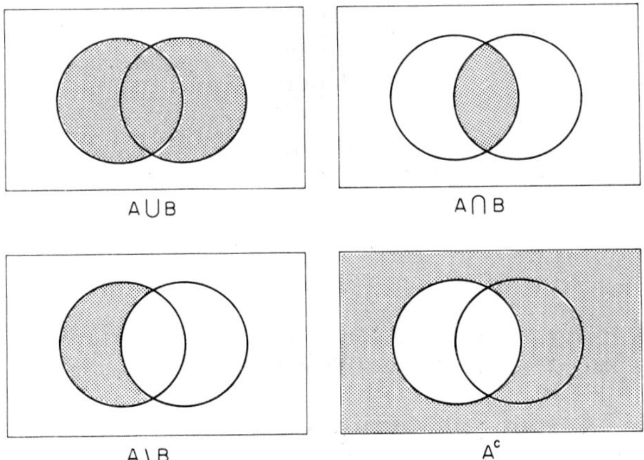


**Fig. 3-1.** Venn diagrams. In the diagram labeled $A^c$ the universal set is the rectangle in which $A$ and $B$ are enclosed and the shaded area is $A^c$.

### 1.5 Ordered Pairs

*Definition.* Given any objects $a$ and $b$ we can form their *ordered pair* $(a, b)$. Two ordered pairs $(a, b)$ and $(c, d)$ are equal if and only if $a = c$ and $b = d$. Whenever pairs are ordered, the representations $(a, b)$ and $(b, a)$ are different entities except when $a = b$.

### 1.6 Relationships

*Definition.* A relation $R$ is a set whose numbers are ordered pairs. We may indicate that $(x, y)$ is a member of a relation $R$ by saying $x$ is related to $y$ or $xRy$ or by writing $(x, y) \in R$.

*Definition.* The *domain* (Dom $R$) of a relation $R$ is the set of all first members of the ordered pair and the *range* (Rng $R$) is the set of all second members of the ordered pair.

*Definition.* The inverse of a relation $R$ is the relation $R^{-1}$ such that $(b, a) \in R^{-1}$ if and only if $(a, b) \in R$.

### 1.7 The Real Line, the Real Plane, Intervals, and Neighborhoods

As stated earlier one of the most important sets is the set of real numbers. All real numbers can be associated with points on a line which we call "the real line." See Fig. 3-2. This *universal set* or *space* in this case is denoted by $R^1$. If we consider the cartesian plane we note that the set of points making up the plane can be associated with a universal set of ordered pairs $(x, y)$. This set we denote by $R^2$. In what follows in Chapter 4 these ideas will be extended to higher dimensions and we will talk about Euclidean $m$-spaces denoted by $R^m$.

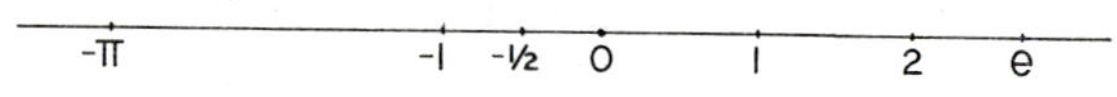


**Fig. 3-2.** The real line.

*Definition.* The set of all points between two numbers $a$ and $b$ on the real line is called an *interval.*

*Definition.* A set of points $x$ such that $a \leq x \leq b$ is called a *closed* interval and is denoted by $[a, b]$. The set of points $a < x < b$ is an *open* interval and is denoted by $(a, b)$. The symbol $x$ is a variable and represents any point in the set.

*Definition.* The set of all points such that $|x - a| < \delta$ where $\delta > 0$ is called a $\delta$ *neighborhood* of the point $a$. If $\delta = \frac{1}{2}$ and $a = 0$ then all points which are less than $\frac{1}{2}$ and greater than $-\frac{1}{2}$ are in the neighborhood of the point from zero. Further the set of all points $y = |x - a|$ such that $0 < |x - a| < \delta$ from which the point $x = a$ is excluded is called a *deleted* $\delta$ *neighborhood* of $a$.

*Definition.* A point $x$ is a *limit point* of a set $S$ if every deleted $\delta$ neighborhood of $x$ intersects the set $S$. That is if for every set $T$ whose points $y$ satisfy the relationship $0 < |y - x| < \delta$ for $\delta > 0$ is such that $T \cap S \neq \phi$ then $x$ is a limit point of $S$. For example the point $b$ is a limit point of the closed interval $[a, b]$.

With the previous definitions in mind, we note that a closed interval contains its limit points, whereas an open interval does not.

## 2. Functions and Their Properties

### 2.1 Functions

*Definition.* A function is a rule which establishes a relation $R$ between two sets of number. If for each value of a variable $x$ there corresponds one and only one value of a variable $y$ then we call $y$ a function of $x$ and we may write $y = f(x)$. (See Fig. 3-3a.) To be more explicit, we define a function as a relation $R$ such that if $(a, b) \in R$ and $(a, c) \in R$ then $b = c$. Therefore, what we call a function is *single valued* relation.

When the above does not hold one could say that $R$ is an $n$-valued relation if for some value of $x$ there are $n$ different values of $y$ but no more than $n$ values. Examples:

$$\text{(a)} \qquad y = \pm\sqrt{x}, \qquad x > 0$$

$$\text{(b)} \qquad \theta = \tan^{-1} y/x, \qquad x \neq 0$$

The relation (a) has two values for every value of $x > 0$, whereas the relation (b) has an infinite number of values each differing by multiples of $\pi$. In some texts, a "multivalued relation" is sometimes called a multivalued function.

There are several ways of looking at functions or explaining them and we list some of them as follows.

What a function $f(x)$ does is to *map* or send the set called the *domain* of $f(x)$ into the set called the *range* of $f(x)$. Each point of the domain must map into exactly one point of the range. However, *more* than one point of the domain may map into the same point of the range. A function is called *one-to-one* if whenever $f(a) = f(b)$ then $a = b$, or in terms of ordered pairs $(x, y) = (z, y)$ implies $x = z$.

With the letter $f$ symbolizing the function then $f(a)$ and $f(b)$ will denote the value of the function at the points $x = a$ and $x = b$ respectively. The set of values which $x$ can take is the *domain* of the function and the set of values which $y = f(x)$ can take is called the *range*; $x$ is called the *independent variable* and $y$ the *dependent variable.*

As examples of functions we may cite the following, if to each number in the interval $-1 \leqq x \leqq 1$ we associate a number $y$ given by $x^2$ then the correspondence between $x$ and $x^2$ defines a function $f$. Another example of a function is the absorption spectra of a compound. This spectrum is a function of wavelength. The optical rotatory dispersion of an optically active compound is also a function of wavelength. The graph of a function $y = f(x)$ can be defined by locating on rectangular coordinates all the points defined by the ordered pairs $(x, y)$ or $(x, f(x))$.

## 2.2 Inverse Functions

In general the inverse relation, $f^{-1}$ is not a function; however, if the inverse of a function is also a function then it must be true that no two ordered pairs can have the same second member with differing first members. This can only happen if and only if the function is one-to-one. Therefore we can say that if $f$ is one-to-one there exists function denoted by $f^{-1}$ which has the property that if $f$ sends $x$ to $y$ then $f^{-1}$ sends $y$ to $x$. Written symbolically, this is

$$f(f^{-1}(y)) = y, \qquad f^{-1}(f(x)) = x.$$

## 2.3 Bounds

Consider Fig. 3-3a on which are presented graphs of several functions. If there is a constant $M$ such that $f(x) \leq M$ in the interval $a \leq x \leq b$ then the function $f(x)$ is said to be *bounded from above* in the interval and $M$ is called an *upper bound*. Clearly, $M$ is one of an infinite number of upper bounds. In addition, if there exists a constant number such that $f(x) \geq m$ then $f(x)$ is said to be *bounded from below* in the interval $a \leq x \leq b$. Further, if $m \leq f(x) \leq M$ in an interval we call $f(x)$ a bounded function on the interval. The wiggly function shown in our Fig. 3-3a is bounded in the interval $a \leqq x \leqq b$. Similarly the function $f(x) = 4 + x$ is bounded from above in $-1 \leqq x \leqq 1$ because in this interval $f(x)$

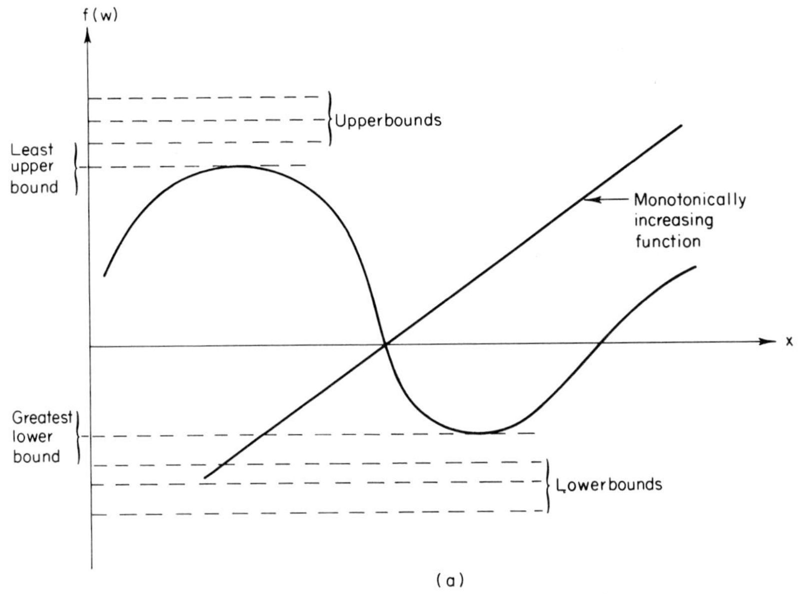


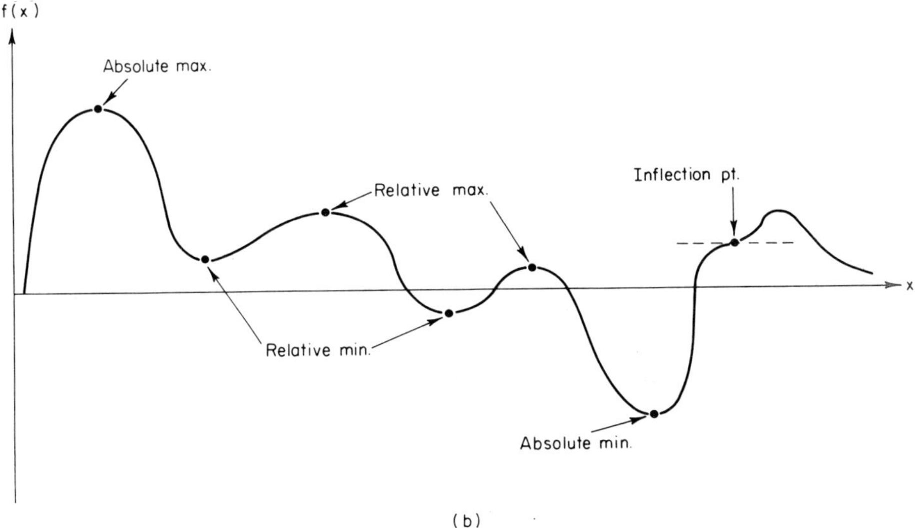


**Fig. 3-3.** (a) Functions and bounds. (b) Maxima and minima of functions.

cannot exceed 5 (or any number greater than 5) and in this interval $f(x)$ is also bounded from below because it cannot become smaller than 3 (or any number less than 3). It can be shown that if $f(x)$ has an upper bound (lower bound) it also has a *least upper bound* (*greatest lower bound*). The wiggly function in our figure exhibits those characteristics and both the greatest lower bound and least upper bound are shown in Fig. 3-3b.

A function is *monotonically increasing* in an interval if for any two points $x_1$ and $x_2$ in this interval where $x_1 \leqq x_2$, $f(x_1) \leqq f(x_2)$. In the event that strict inequality holds, that is to say if $f(x_1) < f(x_2)$ for $x_1 < x_2$, then the function is *strictly increasing*. Similarly, if $f(x_1) \geqq f(x_2)$ whenever $x_1 \geqq x_2$ then $f(x)$ is monotonically decreasing. Once again if $f(x_1) > f(x_2)$ for $x_1 < x_2$ then the function is *strictly decreasing*. Examples of increasing and decreasing functions in the indicated intervals are shown in Fig. 3-3b.

## 2.4 Maxima and Minima

If $x_0$ is a point in an interval $[a, b]$ such that $f(x_0) > f(x)$, $(f(x) > f(x_0))$ for all $x$ in that interval then $f(x)$ is said to assume an *absolute maximum* (*absolute minimum*) on this interval. If this is true only for all $x$ in a deleted $\delta$ neighborhood of $x_0$ where $\delta > 0$ then $f(x)$ is said to have a *relative maximum* or *minimum* at $x_0$. See Fig. 3-3b.

*Examples.* $y = x^2$ has minimum at $x = 0$, whereas $y = \cos x$ has maximum at $x = 0 \pm 2\pi, \pm 4\pi, \ldots, \pm 2n\pi$, where $n$ is an integer.

## 2.5 Limits of a Function

*Definition.* Suppose $f(x)$ is defined for all values of $x$ near $x = x_0$ except perhaps at the point $x = x_0$ [that is $f(x)$ is defined on a deleted $\delta$ neighborhood of $x_0$], then a number $l$ is defined as the limit of $f(x)$ as $x$ tends to or as $x$ approaches $x_0$ [written as $\lim_{x\to x_0} f(x) = l$] if, given any arbitrary positive $\varepsilon$ we can find a positive $\delta$ such that whenever $|x - x_0| < \delta$ then $|f(x) - l| < \varepsilon$. This means that as $x$ comes closer to $x_0$, $f(x)$ approaches the value $l$. Whenever a function has a limit, this limit is unique.

*Examples.* (a) $\lim_{x\to 1} [(x^2 - 1)/(x - 1)] = 2$. (b) $\lim_{x\to 0}[(\sin x)/x] = 1$. (c) To illustrate the rigorous definition of a limit consider the function of the real variable $x$

$$f(x) = x^2 + 2$$

This has

$$\lim_{x \to 2} f(x) = 6$$

Since $f(x)$ is defined in $0 \leq x \leq 4$ then

$$|f(x) - 6| = |x^2 - 4| = |x - 2||x + 2| < |x - 2||4 + 2|$$
$$= 6|x - 2|$$

because $x$ does not exceed 4 in the interval of interest. Taking $\delta \leqq \varepsilon/6$ we can write

$$|f(x) - 6| \leq 6|x - 2| \leq 6\,\delta \leq 6(\varepsilon/6) = \varepsilon$$

and this holds everywhere in the interval

$$0 < |x - 2| < \delta$$

## 2.6 Continuous Functions

*Definition.* Suppose $f(x)$ is defined for all values of $x$ in a $\delta$ neighborhood of $x_0$, then $f(x)$ is said to be continuous at $x = x_0$ if given any arbitrary $\varepsilon > 0$, we can find a $\delta > 0$ such that $|f(x) - f(x_0)| < \varepsilon$ whenever $|x - x_0| < \delta$, [this means that $\lim_{x \to x_0} f(x) = l$ exists], and such that $f(x_0) = l$. Continuity on an interval means a function $f(x)$ is continuous at each point in the interval. We may think of a function that is continuous on an interval as one whose graph is continuous in that its graphical representation can be drawn without lifting the pencil from the paper. Some examples of functions with discontinuities are (1) the function $f(x) = 1/x$ which is discontinuous at $x = 0$; (2) the function $f(x) = \tan x$ is discontinuous at the points $\pi/2$, $-\pi/2$, $3\pi/2$, . . . , etc.

## 2.7 Derivatives of Functions

If $f(x)$ is defined at any point $x_0$ in an interval $(a, b)$ the derivative of $f(x)$ at $x = x_0$ is defined as

$$f'(x_0) = \lim_{x \to x_0} \frac{f(x) - f(x_0)}{x - x_0} = \lim_{\Delta x \to 0} \frac{f(x_0 + \Delta x) - f(x_0)}{\Delta x}$$

When this limit exists the function is said to be differentiable at $x_0$. If a function has a derivative at all points in any given interval it is said to be differentiable on the interval.

## 2.8 Geometrical Consideration of a Derivative

Let $f(x)$ be the curve $ABCD$ as depicted in the Fig. 3-4. In this figure

$$\frac{CE}{BE} = \frac{f(x_0 + \Delta x) - f(x_0)}{\Delta x} = \tan\theta$$

Now as $\Delta x \to 0$ we obtain

$$\lim_{\Delta x \to 0} \frac{f(x_0 + \Delta x) - f(x_0)}{\Delta x} = \frac{SE}{BE} = \tan\alpha$$

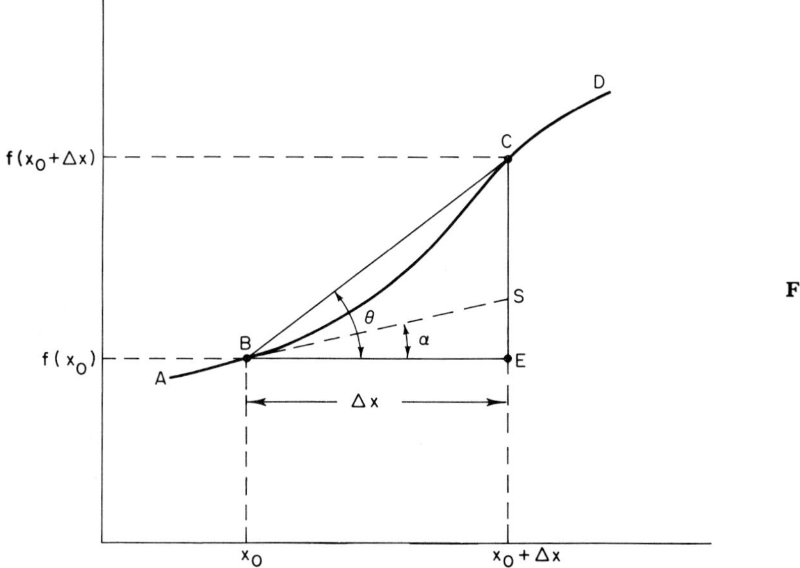


Fig. 3-4

Consequently the derivative at a point may be interpreted as the slope of the line tangent to the curve (function) at the point $x_0$.

## 2.9 Rules for Differentiation

If $f(x)$, $g(x)$, and $h(x)$ are differentiable then the following formulas are true.

(i) $\dfrac{d}{dx}[f(x) \pm g(x)] = f'(x) \pm g'(x)$

(ii) $\dfrac{d}{dx}[Cf(x)] = Cf'(x)$ where $C$ is constant

(iii) $\dfrac{d}{dx}[f(x)g(x)] = f(x)g'(x) + g(x)f'(x)$

(iv) $\dfrac{d}{dx}\dfrac{g(x)}{f(x)} = \dfrac{f(x)g'(x) - g(x)f'(x)}{[f(x)]^2}$

provided, of course, $f(x) \neq 0$.

To differentiate a function of a function, or a composite function, we set $y = f(v)$ and $v = g(x)$ then

(v) $dy/dx = f'(v)\, dv/dx = (df/dv)(dv/dx)$

This rule can be extended to several functions and for example if $y = f(v)$, $v = g(u)$ and $u = h(x)$ then

$$dy/dx = (dy/dv)(dv/du)(du/dx)$$

This operation is known as the chain rule for differentiation. Let $y = f(x)$ be a one-to-one function then as we have shown before $x = f^{-1}(y)$ is a function and we can establish the relationship

(vi) $\dfrac{dy}{dx} = \dfrac{1}{dx/dy}$

If $x = f(t)$ and $y = g(t)$ are functions differentiable with respect to $t$, then

(vii) $\dfrac{dy}{dx} = \dfrac{dy/dt}{dx/dt} = \dfrac{g'(t)}{f'(t)}$

The reader can verify the above rules by particular examples of differentiable functions.

## 2.10 Higher Order Derivatives

If $f(x)$ is differentiable in an interval and its derivative is given by $f'(x)$ and if, in addition, $f'(x)$ is also differentiable in an interval then $f(x)$ possesses a second derivative denoted by $f''(x)$. If this is also differentiable in the interval then it has a third derivative and so on as long as the succeeding derivatives are differentiable. The $n$th derivative is denoted by $f^n(x)$. For example consider the function

$$f(x) = x^3 + 3x^2 + 2x + \sin x$$

It has the following derivatives up to order four.

$$df(x)/dx = f'(x) = 3x^2 + 6x + 2 + \cos x$$
$$df'(x)/dx = f''(x) = d^2f(x)/dx^2 = 6x + 6 - \sin x$$
$$df''(x)/dx = f'''(x) = d^3f(x)/dx^3 = 6 - \cos x$$
$$df'''(x)/dx = f''''(x) = d^4f(x)/dx^4 = \sin x$$

This function, in fact, has an infinite number of continuous derivatives. One important application of higher order derivatives is the classification of relative maxima and minima or relative extrema of functions.

## 2.11 Maxima and Minima

Suppose that $x = x_0$ and $f(x)$ satisfies the condition

$$f'(x_0) = f''(x_0) = \cdots = f^{(2k-1)}(x_0) = 0$$

and $f^{(2k)}(x_0) \neq 0$ for some positive integer $k$ (usually 1), then $f(x)$ has a relative maximum at $x = x_0$ if $f^{(2k)}(x_0) < 0$, and $f(x)$ has a relative minimum at $x = x_0$ if $f^{(2k)}(x_0) > 0$. From the standpoint of finding those extrema we must first determine the point or points $x_0$ at which $f'(x) = 0$. Then we determine whether the function has a minimum or a maximum. The condition that $f'(x) = 0$ is such that we must be at a minimum, a maximum or a point of inflection. This can be interpreted geometrically because at this point the tangent to $f(x)$ at the point $x_0$ must parallel the $x$ axis. See Fig. 3-3b.

*Example.* Find the extreme values of

$$f(x) = 2x^3 - 15x^2 + 36x - 12$$

Taking the first derivative and equating to zero we find

$$f'(x) = 6x^2 - 30x + 36 = 0$$

the solutions of this equation are $x = 2$ and $x = 3$. Taking the point $x = 2$ we find that

$$f''(2) = 12(2) - 30 = -6$$

so that $f''(2) < 0$ and we conclude there is a local maximum at $x = 2$. At $x = 3$,

$$f''(3) = 12(3) - 30 = 6$$

so that $f''(3) > 0$ and the conclusion is, there is a minimum at $x = 3$. The reader can easily sketch the function

$$f(x) = 2x^3 - 15x^2 + 36x - 12$$

and verify these facts for himself. The reader will observe that these are only local minima and maxima since the function is unbounded on the real line.

## 2.12 Mean Value Theorems

These theorems enable us to relate the behavior of differentiable functions and their derivatives on intervals. In particular, Taylor's mean value theorem will play a powerful role in our subsequent work.

One of the theorems of the mean employed in differentiation is the following: If $f(x)$ is continuous on $[a, b]$ and differentiable in $(a, b)$ then there exists a point $\xi$ in $(a, b)$ such that

$$\frac{f(b) - f(a)}{b - a} = f'(\xi), \qquad a < \xi < b$$

An alternative form of writing this theorem is

$$f(b) = f(a) + f'(\xi)(b - a)$$

when $\xi$ is between $b$ and $a$. If the interval is small and $f'(a) \neq 0$, then we may write

$$f(b) - f(a) \simeq f'(a)(b - a)$$

which means that a small increment in the value of the function is approximately equal to the derivative of the function at the initial point multiplied by the distance from $a$ to $b$.

## 2.13 Taylor's Mean Value Theorem

If $f^n(x)$ is continuous on the interval $[a, b]$ and differentiable in $(a, b)$ then there exists a point $\phi$ in $(a, b)$ such that

$$f(b) = f(a) + f'(a)(b - a) + \frac{f''(a)(b - a)^2}{2!} + \cdots + \frac{f^n(a)(b - a)^n}{n!} + R_n(\phi)$$

where $R_n$ is the remainder and can be written in either of the following forms:

$$\text{Lagrange's form:}\quad R_n = f^{n+1}(\phi)(b-a)^{n+1}/n!, \qquad a < \phi < b$$

$$\text{Cauchy's form:}\quad R_n = f^{n+1}(\phi)(b-\phi)^{n}(b-a)/n!, \qquad a < \phi < b$$

This theorem is quite powerful, and it shows that if we know the value of a function at one end of an interval, $f(a)$ in this case, and the length of the interval then we can approximate the value of this function at any point in the interval by differentiation of the function at point $a$ as many times as possible and then apply Taylor's theorem.

*Example.* Find the Taylor's expansion of the function $f(x) = \sin x$ about the point $x = 0$. The reader can verify that this expansion is

$$\sin x = x - (x^3/3!) + (x^5/5!) - (x^7/7!) + \cdots + R_n(x)$$

since

$$f(0) = \sin 0 = 0, \qquad f'(0) = \cos 0 = 1,$$

$$f''(0) = -\sin 0 = 0, \qquad \text{and} \qquad f'''(0) = -\cos 0 = -1, \qquad \text{etc.}$$

## 2.14 The Definite Integral

The integral calculus is said to have been begun by the Greeks who were interested in computing the area of a circle. In general we ask what is the area bound by the curve $y = f(x)$, the $x$-axis and the ordinates erected at $x = a$ and $x = b$. If we subdivide the interval $a \leq x \leq b$ in Fig. 3-5 into $n$ arbitrary subintervals by placing the points $x_1, x_2, \ldots, x_{n-1}$ and choose arbitrary points $\xi_1, \xi_2, \ldots, \xi_{n-1}, \xi_n$ in the intervals $(a, x_1)$, $(x_1, x_2), \ldots, (x_{n-1}, b)$ respectively, then the sum is

$$S = f(\xi_1)(x_1 - a) + f(\xi_2)(x_2 - x_1) + \cdots + f(\xi_n)(b - x_{n-1})$$

also written as

$$S = \sum_{k=1}^{n} f(\xi_k)(x_k - x_{k-1}) = \sum_{k=1}^{n} f(\xi_k)\, \Delta x_k$$

where we have written $x_k - x_{k-1} = \Delta x_k$. This sum represents the area enclosed by all the rectangles in Fig. 3-5. As the number of the intervals $n$ tends to infinity, or as $\Delta x_k \to 0$, the sum approaches a limit which is equal to the area. This limit is represented by $\int_a^b f(x)\, dx$ and is called the definite integral of $f(x)$ between the limits of integration.

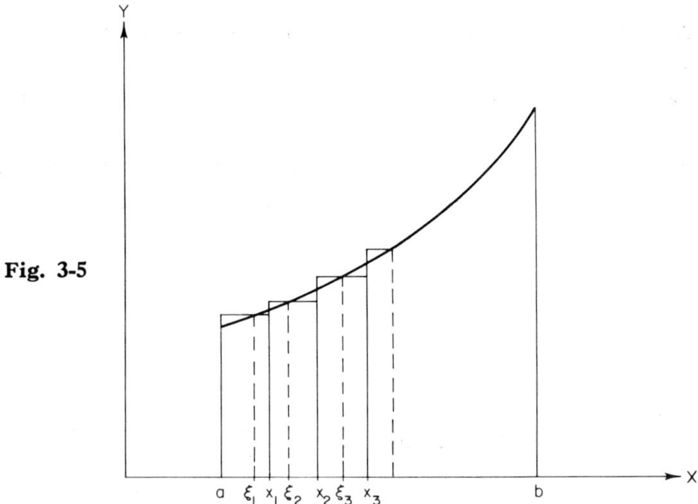


Fig. 3-5

When this limit exists $f(x)$ is said to be Riemann[†] integrable over $[a, b]$. Note that $f(x)$ must be bounded in the interval otherwise we cannot define the sum.

## 2.15 Properties of the Definite Integral

Addition and subtraction: If $f(x)$ and $g(x)$ are integrable in the interval $[a, b]$ we have

$$\int_a^b [f(x) \pm g(x)]\, dx = \int_a^b f(x)\, dx \pm \int_a^b g(x)\, dx$$

Multiplication by a constant: If $k$ is a constant we have

$$\int_a^b kf(x)\, dx = k \int_a^b f(x)\, dx$$

Dividing the interval: Here we have

$$\int_a^b f(x)\, dx = \int_a^c f(x)\, dx + \int_c^b f(x)\, dx$$

provided $f(x)$ is integrable in both the intervals $[a, c]$ and $[c, b]$. An

[†] Named after Bernard Riemann (1826–1866), the first person to precisely define the definite integral in that way.

important practical point to note in connection with this property of the definite integral is that whenever there is an interval where the data for a certain function cannot be obtained accurately one can omit that region from the analysis and integrate in the region where the data is good. For example if one is dealing with the absorption spectra for a certain compound which is opaque in a certain region as in Fig. 3-6.

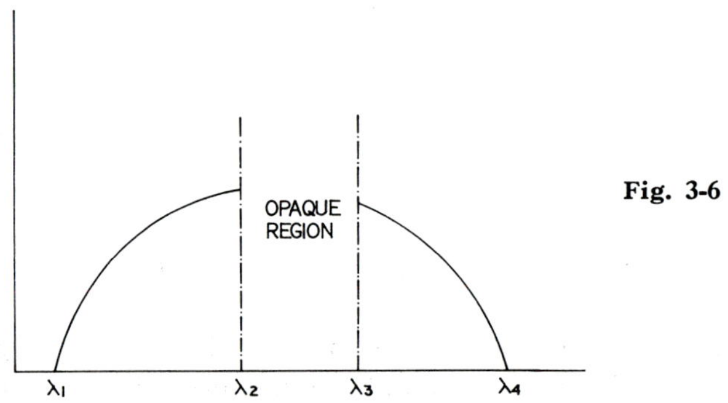


**Fig. 3-6**

If we desire to integrate the function presented in the interval $(\lambda_1, \lambda_4)$ then we do the following

$$\int_{\lambda_1}^{\lambda_2} f(\lambda)\, d\lambda + \int_{\lambda_4}^{\lambda_3} f(\lambda)\, d\lambda$$

thereby excluding the entire region $[\lambda_2, \lambda_3]$ for whatever analysis we happen to be conducting. Similarly for any other problem we might encounter, if data obtained in a certain region is unobtainable, then it can be omitted without affecting the analysis provided that the same region is also omitted for other functions required in the same analysis. We continue with the properties of the definite integral.

Reversing the limits of integration: Here we have

$$\int_a^b f(x)\, dx = -\int_b^a f(x)\, dx$$

When the function is bounded in an interval: If $a \leq x \leq b$ and $m \leq f(x) \leq M$ where $m$ and $M$ are constants then

$$m(b-a) \leq \int_a^b f(x)\, dx \leq M(b-a)$$

When two functions are unequal: If $a \leq x \leq b$ and $f(x) < g(x)$ in the interval $[a, b]$ then

$$\int_a^b f(x)\,dx < \int_a^b g(x)\,dx$$

The reader can verify all the above examples for some particular functions say $f(x) = x^2$ and $g(x) = x^3$ in the interval $[1, 2]$.

## 2.16 Indefinite Integrals

Given a continuous function $f(x)$, then any function $g(x)$ such that $g'(x) = f(x)$, is an indefinite integral or antiderivative of $f(x)$. If $g(x)$ is an indefinite integral of $f(x)$ then so is $g(x) + c$, where $c$ is an arbitrary constant because

$$d[g(x) + c]/dx = g'(x) = f(x)$$

For example if $g'(x) = x^3$ then $g(x) = (x^4/4) + c$ and is an antiderivative of $x^3$.

## 2.17 The Fundamental Theorem of Integral Calculus

If $f(x)$ is continuous in $[a, b]$ and $g(x)$ is a function such that $g'(x) = f(x)$ then

$$\int_a^b f(x)\,dx = g(b) - g(a)$$

We give an example of the use of this theorem. To compute $\int_1^2 x^3\,dx$ we have

$$g(x) = (x^4/4) + c$$

Now

$$g(2) = (2^4/4) + c \qquad \text{and} \qquad g(1) = (1^4/4) + c$$

so that

$$\int_1^2 x^3\,dx = 15/4$$

An indefinite integral can be expressed as definite integral with a variable upper limit by writing

$$\int^x f(u)\,du = \int_a^x f(u)\,du + c$$

It can be shown that

$$\frac{d}{dx}\int_a^x f(u)\, d(u) = f(x)$$

The fact that the derivative of a definite integral with variable limits is a continuously differentiable function is of some importance for our purposes for we will be attempting to fit data with functions which appear as

$$F(x) = \int_a^x f(u)\, du$$

## 2.18 Changing the Variable of Integration

Suppose we change the independent variable $x$ by the one-to-one monotone transformation $x = g(u)$. We then have in the indefinite integral $dx = g'(u)\, du$ and

$$\int^x f(x)\, dx + c = \int f(g(u))g'(u)\, du + c$$

*Example.* Suppose we wished to evaluate the integral

$$\int^x (x + 2) \sin(x^2 + 4x - 4)\, dx + c$$

If we put

$$(x^2 + 4x - 4) = u$$

then

$$du/dx = 2x + 4 \qquad \text{and} \qquad \tfrac{1}{2}(du/dx) = x + 2$$

or

$$\tfrac{1}{2}\, du = (x + 2)\, dx$$

The integral then becomes

$$\int f(x)\, dx + c = \tfrac{1}{2}\int^x \sin u\, du + c = -\tfrac{1}{2}(\cos u) + c$$

$$= -\tfrac{1}{2}\cos(x^2 + 4x - 4) + c$$

In the definite integral a change of integration variable produces the following:

$$\int_a^b f(x)\, dx = \int_{g^{-1}(a)}^{g^{-1}(b)} f(g(t))g'(t)\, dt.$$

A simple application of the above technique is the evaluation of

$$\int_{e}^{e^3} \frac{dx}{x(\ln x)^3}$$

If we change the variable by letting $y = \ln x$, then when $x = e$ then $y = 1$, and when $x = e^3$ then $y = 3$. We also have $dx/x = dy$ so that the integral becomes

$$\int_1^3 (dy/y^3) = [y^{-2}/-2]_1^3 = \tfrac{4}{9}$$

## 2.19 Integration by Parts

A valuable technique of integration is a consequence of the product rule for differentiation

$$d(fg)/dx = f(dg/dx) + g(df/dx)$$

where $f$ and $g$ are differentiable functions of $x$.
Rearranging and integrating the equation we obtain

$$\int g\frac{df}{dx} = \int \frac{d}{dx}(gf) - \int f\frac{dg}{dx} = gf - \int f\frac{dg}{dx}$$

Hence the integration by parts formula is derived as

$$\int f'(x)g(x)\,dx = f(x)g(x) - \int f(x)g'(x)\,dx.$$

## 2.20 Differentiation under the Integral Sign

Sometimes we will need to differentiate a function which is an integral. For example, if

$$\psi(\beta) = \int_{\alpha_1}^{\alpha_2} F(x, \beta)\,dx, \qquad a \le \beta \le b$$

where $\alpha_1$ and $\alpha_2$ may depend on $\beta$. They can be written as $\alpha_1(\beta)$ and $\alpha_2(\beta)$. Then we can write

$$d\psi(\beta)/d\beta = \int_{\alpha_1(\beta)}^{\alpha_2(\beta)} (\partial F/\partial\beta)\,dx + F(\alpha_2(\beta), \beta)(d\alpha_2(\beta)/d\beta)$$
$$- F(\alpha_1(\beta), \beta)(d\alpha_1(\beta)/d\beta)$$

for $a < \beta < b$, provided $F(x, \beta)$ and $\partial F/\partial \beta$ are continuous in both the variable $x$ and $\beta$ in a region on the $x\beta$ plane [remember $F(x, \beta)$ is defined on the $x\beta$ plane] including the regions $\alpha_1 \leq x \leq \alpha_2$ and $a \leq \beta \leq b$ and $\alpha_1(\beta)$ and $\alpha_2(\beta)$ are continuous and differentiable. In the event that $\alpha_1(\beta)$ and $\alpha_2(\beta)$ are constants then the last two terms vanish.

## 2.21 Improper Integrals

Improper integrals occur when the limits of a definite integral are infinite or when the function to be integrated is not bounded. These integrals occur in the theory of the Laplace and Fourier Transforms discussed later. We have considerable use for both the Laplace transforms and the Fourier transforms in the chapters on statistics and in certain instances in topics concerning molecular weight distribution functions. Consider the expression

$$\int_a^b f(x)\,dx$$

where any of the limits $a$ and $b$ may tend to either $+\infty$ or $-\infty$. When such is the case we have an *improper* integral. Further if $f(x)$ *is not bounded* in the interval, then the integral is also called *improper*. The points in the interval $a \leq x \leq b$, where $f(x)$ is not bounded are called *singular* points. Generally we are interested in the evaluation of the integral and consequently we look in the question of the existence of the integrals. If $f(x)$ is bounded and integrable in every finite interval $a \leq x \leq b$, then we may define

$$\int_a^\infty f(x)\,dx = \lim_{b\to\infty} \int_a^b f(x)\,dx$$

and we say the integral on the left *converges* or *diverges*, depending on whether or not the limit exists. If the integral diverges the limit has no useful meaning, and if it converges, we define its value to be the limit. The same thing can be said for the integral

$$\int_{-\infty}^b f(x)\,dx = \lim_{a\to-\infty} \int_a^b f(x)\,dx$$

When the integral is unbounded in an interval the integral cannot be evaluated in general. There exists special cases where the integral is unbounded only at the end points ofan interval so we evaluate these integrals by effectively eliminating this point or these points from consideration.

*Examples.* In the case where the integral $f(x)$ is unbounded consider

$$\text{(i)} \quad \int_0^1 (dx/x) = \lim_{x\to 0} [-lnx] = +\infty$$

This integral does not exist; whereas

$$\text{(ii)} \quad \int_0^1 (dx/\sqrt{x}) = \lim_{x\to 0} [2 - 2x^{1/2}] = 2$$

does exist. For the case of an infinite limit

$$\text{(iii)} \quad \int_0^\infty e^{-st}\, dt = \lim_{T\to\infty} \int_0^T e^{-st}\, dt$$

$$= \lim_{T\to\infty} \left[ -\frac{e^{-sT}}{s} + \frac{1}{s} \right]$$

and if $s > 0$ then we define

$$\int_0^\infty e^{-st}\, dt = 1/s$$

The above topics are essentially all we require in so far as functions of one variable. In what follows we extend that these ideas to functions of more than one variable and we will use functions of two or three variables as illustrative examples. From these examples the generalization to functions of several variables will be obvious. In as much as this is not a rigorous mathematical text, proofs, as before will be omitted.

## 3. Functions of Several Real Variables

This topic is probably the most important part of this chapter for ultimately our aim is the minimization of functions of several variables.

### 3.1 Functions of Two Variables

A function of two variables written $z = \phi(x, y)$ is defined as follows: For every pair of numbers $(x, y)$ a correspondence is set up between the set $z = \phi(x, y)$ and the given pair $(x, y)$. Stating it in another way, for each pair of independent variables $x$ and $y$ we can determine a value for $z = \phi(x, y)$. In such instances $z$ is called the dependent variable. This concept is easily extended to functions of several variables. An example of function of two variables is

$$z = \phi(x, y) = x^2 + y^2.$$

We might think of the absorption spectrum of DNA as a function of the two independent variables, wavelength and temperature and so on.

## 3.2 Neighborhoods and Regions in the Plane

In a manner similar to our definition of neighborhood on the real line, we can define a neighborhood of a point in the $xy$-plane.

*Definition.* A neighborhood of a point $(x_0, y_0)$ is defined as the set of all points $(x, y)$ such that $|y - y_0| < \delta$ and $|x - x_0| < \delta$ where $\delta > 0$. This neighborhood is a ***rectangular*** $\delta$ neighborhood of $(x_0, y_0)$. The set of all points $(x, y)$ such that

$$(x - x_0)^2 + (y - y_0)^2 < \delta^2$$

is a *circular* $\delta$ neighborhood of $(x_0\ y_0)$. One may also define a neighborhood of $(x_0, y_0)$ as the set of all $(x, y)$ such that

$$|x - x_0| + |y - y_0| < \delta.$$

This gives a diamond shaped neighborhood $(x_0, y_0)$. These three neighborhoods are illustrated in Fig. 3-7. These neighborhoods are called respectively squircle, circle, and dircle neighborhoods.

In contrast to the case for $R^1$, neighborhoods in $R^2$ are ***not uniquely*** defined in the sense that one has only circles, or squares, or diamonds as neighborhoods.

However each of these neighborhoods are equivalent to the others in the sense that given any one, there is one of the others that will fit inside it. This is evident from Fig. 3-7.

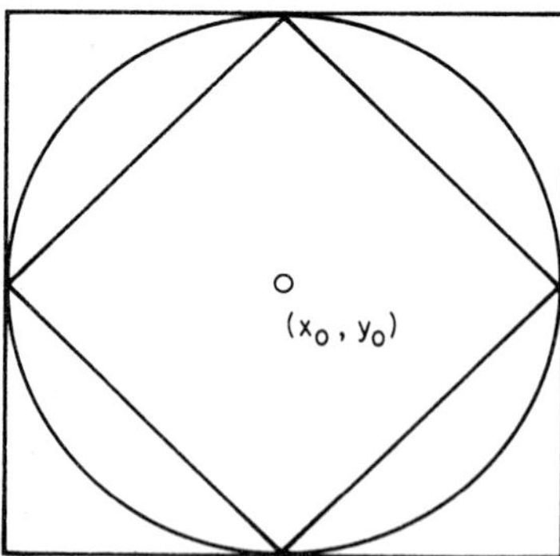


**Fig. 3-7**

A neighborhood of $(x_0, y_0)$ which excludes the point $(x_0, y_0)$ is a deleted $\delta$ neighborhood of $(x_0, y_0)$. In the same manner as we defined *intervals* on the real line we define *regions* in the $xy$-plane. Functions of two variables will be defined on regions in a similar manner as a function of a single variable was defined on an interval. An example of a region is the rectangular one shown in Fig. 3-8. This *closed* rectangular region with its boundary represents the set of points $a \leqq x \leqq b$, $c \leqq y \leqq d$. Without this boundary it represents the set of points $a < x < b, c < y < d$ and is called an *open* region.

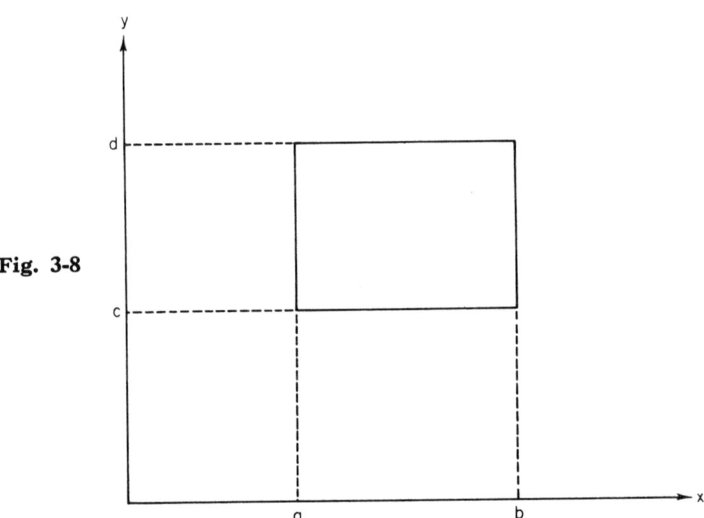

**Fig. 3-8**

*Definition.* A region is called *open* if every point within this region has a neighborhood which is entirely within the region.

### 3.3 Limit Points

Limit points for functions of two variables are defined similarly to limit points for a function of one variable.

*Definition.* A point $(x_0, y_0)$ is a limit point of a set $S$ if *every* deleted $\delta$ neighborhood of $(x_0, y_0)$ contains points of $S$.

*Definition.* A region is called *closed* if it contains all its limit points.

*Definition.* Limits of functions: If a function of two variables $\phi(x, y)$ is defined on a deleted $\delta$ neighborhood of $(x_0, y_0)$ [which means the function may *not* be defined at $(x_0, y_0)$], we may write

$$\lim_{\substack{x \to x_0 \\ y \to y_0}} \phi(x, y) = l$$

where $l$ is the limit of the function, and we say that this function has the limit $l$ if given any positive number $\varepsilon$, we can find some corresponding positive number $\delta$ such that $|\phi(x, y) - l| < \varepsilon$ whenever $0 < |x - x_0| < \delta$, $0 < |y - y_0| < \delta$.

## 3.4 Continuity

If $\phi(x, y)$ is defined at and on a neighborhood of $(x_0, y_0)$, then $\phi(x, y)$ is continuous at the point $(x_0, y_0)$, if given any positive number $\varepsilon$ we can find a $\delta$ such that $|\phi(x, y) - \phi(x_0, y_0)| < \varepsilon$ whenever $|x - x_0| < \delta$ and $|y - y_0| < \delta$. For a function $\phi(x, y)$ to be continuous at $(x_0, y_0)$ we must have the existence of $\lim \phi(x, y)$ as $x \to x_0$ and $y \to y_0$. The function must be defined at $(x_0, y_0)$ and

$$\lim_{\substack{x \to x_0 \\ y \to y_0}} \phi(x, y) = \phi(x_0, y_0)$$

For example the function

$$\phi(x, y) = \{1/(x + y)\}$$

is continuous everywhere except at the points $x = 0$, $y = 0$ and $x = -y$.

## 3.5 Partial Derivatives

The derivative of a function of several variables with respect to one of those variables keeping all the other variables constant is called the partial derivative of the function with respect to that particular variable. Partial derivatives of $f(x, y)$ with respect to $x$ can be denoted by several notations such as $\partial f/\partial x$ or $f_x$. Similarly $\partial f/\partial y$ and $f_y$ stands for differentiation with respect to $y$ keeping $x$ constant.

*Definition.* We can define the derivative of a function of three variables $(x, y, z)$ with respect to $x$ as follows

$$\frac{\partial f(x, y, z)}{\partial x} = \lim_{\Delta x \to 0} \frac{f(x + \Delta x, y, z) - f(x, y, z)}{\Delta x}$$

with respect to $y$

$$\frac{\partial f(x, y, z)}{\partial y} = \lim_{\Delta y \to 0} \frac{f(x, y + \Delta y, z) - f(x, y, z)}{\Delta y}$$

and lastly with respect to $z$

$$\frac{f(x, y, z)}{\partial z} = \lim_{\Delta z \to 0} \frac{f(x, y, z + \Delta z) - f(x, y, z)}{\Delta z}$$

## 3.6 Higher Order Partial Derivatives

In our explanation here we will not go beyond functions of three variables and their derivatives. The generalization to $n$-variables is obvious. Consider the case of two variables: we note that $\partial f(x, y)/\partial x$ and $\partial f(x, y)/\partial y$ are again functions of $x$ and $y$, then it may be possible to differentiate each of these functions again with respect to each of the variables $x$ and $y$. Such second derivatives are given by

$$\frac{\partial}{\partial x}\left(\frac{\partial f(x, y)}{\partial x}\right) = \frac{\partial^2 f(x, y)}{\partial x^2}$$

$$\frac{\partial}{\partial y}\left(\frac{f(x, y)}{\partial x}\right) = \frac{\partial^2 f(x, y)}{\partial y\, \partial x}$$

$$\frac{\partial}{\partial y}\left(\frac{\partial f(x, y)}{\partial y}\right) = \frac{\partial^2 f(x, y)}{\partial y^2}$$

and finally

$$\frac{\partial}{\partial x}\left(\frac{\partial f(x, y)}{\partial y}\right) = \frac{\partial^2 f(x, y)}{\partial x\, \partial y}$$

Still higher derivatives can be formed whenever the differentiation operation operation is valid. Such a higher order mixed partial derivative is

$$\frac{\partial}{\partial y}\left(\frac{\partial^2 f(x, y)}{\partial x\, \partial y}\right) = \frac{\partial^3 f(x, y)}{\partial y^2\, \partial x}$$

*Example.* Consider the function of two variables

$$f(x, y) = x^2 y + \exp(xy^2)$$

then the first derivatives are

$$\partial f/\partial x = 2xy + y^2 \exp(xy^2) \qquad \text{and} \qquad \partial f/\partial y = x^2 + 2xy \exp(xy^2)$$

The second derivatives are

$$\partial^2 f/\partial x^2 = 2x + y^4 \exp(xy^2)$$

$$\partial^2 f/\partial y^2 = 4x^2y^2 \exp(xy^2) + 2x \exp(xy^2)$$

$$\partial^2 f/\partial x\, \partial y = 2x + 2xy^3 \exp(xy^2) + 2y \exp(xy^2)$$

$$\partial^2 f/\partial y\, \partial x = 2x + 2xy^3 \exp(xy^2) + 2y \exp(xy^2)$$

In this example we note the equality of $\partial^2 f/\partial x\, \partial y = \partial^2 f/\partial y\, \partial x$ which is the case if the derivatives are continuous functions of $x$ and $y$.

## 3.7 Implicit Functions

The derivative of $y$ with respect to $x$ when $y$ is considered a function of $x$ defined implicitly by the relation $\phi(x, y) = c$ where $c$ is a constant, may be obtained indirectly. If $\phi(x, y)$ has continuous derivatives with respect to $x$ and $y$ and $\partial\phi(x, y)/\partial y \neq 0$ in some neighborhood of a point $(x, y)$ then

$$\frac{dy}{dx} = -\frac{\partial\phi(x, y)/\partial x}{\partial\phi(x, y)/\partial y}$$

*Example.* Given $\phi(x, y) = x^4y^4 + \sin y$, find $dy/dx$.

$$\partial\phi(x, y)/\partial x = 4x^3y^4, \qquad \partial\phi(x, y)/\partial y = 4x^4y^3 + \cos y$$

and

$$dy/dx = -\, 4x^3y^4/(4x^4y^3 + \cos y)$$

## 3.8 Partial Differentiation of Composite Functions

If $z = f(x, y)$ and $x = g(s, t)$ and $y = h(s, t)$ then

$$\frac{\partial z}{\partial s} = \frac{\partial z}{\partial x}\frac{\partial x}{\partial s} + \frac{\partial z}{\partial y}\frac{\partial y}{\partial s}, \qquad \frac{\partial z}{\partial t} = \frac{\partial z}{\partial x}\frac{\partial x}{\partial t} + \frac{\partial z}{\partial y}\frac{\partial y}{\partial t}$$

In general if $z = f_0(x_1, x_2, \ldots, x_n)$ and $x_1 = f_1(s_1, s_2, \ldots, s_p)$, $x_2 = f_2(s_1, s_2, \ldots, s_p)$ and so on such that $x_n = f_n(s_1, s_2, \ldots, s_p)$, then $z$ is ultimately a function of the variables $s_1, s_2, \ldots, s_p$, and

$$\frac{\partial z}{\partial s_k} = \frac{\partial z}{\partial x_1}\frac{\partial x_1}{\partial s_k} + \frac{\partial z}{\partial x_2}\frac{\partial x_2}{\partial s_k} + \cdots + \frac{\partial z}{\partial x_n}\frac{\partial x_n}{\partial s_k}$$

## 3.9 Total Derivatives

Suppose that $u = f(x(t), y(t))$ which is a function of two variables and suppose that $x(t)$ and $y(t)$ are differentiable functions of $t$. Then the total derivative of $u$ with respect to $t$ is given by

$$\frac{du}{dt} = \frac{\partial f}{\partial x}\frac{dx}{dt} + \frac{\partial f}{\partial y}\frac{dy}{dt}$$

Clearly this can be extended to functions of more than two variables.

*Example.* Given

$$u = \phi(x, y) = \sin(x/y), \qquad x(t) = t, \qquad y(t) = t^2$$

$$\frac{\partial \phi}{\partial x} = \frac{1}{y}\cos\frac{x}{y}, \qquad \frac{\partial \phi}{\partial y} = -\frac{x}{y^2}\cos\frac{x}{y}, \qquad \frac{dx}{dt} = 1, \qquad \frac{dy}{dt} = 2t$$

then

$$\frac{du}{dt} = \left(\frac{1}{t^2}\cos\frac{t}{t^2}\right) + \left(-\frac{t}{t^4}\cos\frac{t}{t^2}\right)(2t)$$

## 3.10 Taylor's Mean Value Theorem for Functions of Two or More Variables

If all the $n$th partial derivatives of $f(x, y)$ are continuous in a closed region and if the partial derivatives of order $(n + 1)$ exist in the open region (by open region we mean the closed region just mentioned without its boundaries) then we can expand the function $f(x, y)$ about the point $x = x_0$ and $y = y_0$ as follows

$$\begin{aligned} f(x_0 + \Delta x, y_0 + \Delta y) = f(x_0, y_0) &+ \left(\Delta x\frac{\partial}{\partial x} + \Delta y\frac{\partial}{\partial y}\right)f(x_0, y_0) \\ &+ \frac{1}{2!}\left(\Delta x\frac{\partial}{\partial x} + \Delta y\frac{\partial}{\partial y}\right)^2 f(x_0, y_0) + \cdots \\ &+ \frac{1}{n!}\left(\Delta x\frac{\partial}{\partial x} + \Delta y\frac{\partial}{\partial y}\right)^n f(x_0, y_0) + R_n \end{aligned}$$

where we have used the notation

$$\left(\Delta x\frac{\partial}{\partial x} + \Delta y\frac{\partial}{\partial y}\right)f(x_0, y_0)$$

and

$$\left(\Delta x\frac{\partial}{\partial x} + \Delta y\frac{\partial}{\partial y}\right)^2 f(x_0, y_0)$$

to denote respectively

$$\Delta x \frac{\partial f(x_0, y_0)}{\partial x} + \Delta y \frac{\partial f(x_0, y_0)}{\partial y}$$

and

$$(\Delta x)^2 \frac{\partial^2 f(x_0, y_0)}{\partial x^2} + 2(\Delta x)(\Delta y) \frac{\partial^2 f(x_0, y_0)}{\partial x\, \partial y} + (\Delta y)^2 \frac{\partial^2 f(x_0, y_0)}{\partial y^2}$$

All expressions of the form

$$\left(\Delta x \frac{\partial}{\partial x} + \Delta y \frac{\partial}{\partial y}\right)^n$$

are to be formally expanded by the binomial theorem. The expression for the remainder in the above expansion may be written as

$$R_n = \frac{1}{(n+1)!} \left(\Delta x \frac{\partial}{\partial x} + y \frac{\partial}{\partial y}\right)^{n+1} f(x_0 + \theta\, \Delta x, y_0 + \theta\, \Delta y)$$

where $\theta$ can take any value $0 < \theta < 1$.

The Taylor mean value theorem or expansion can be extended to functions of several variables. In attempting to minimize a function with respect to its parameters, it will happen frequently that when the parameters occur nonlinearly, a Taylor's expansion of the function we seek to minimize will be the starting point of our minimization techniques. Consequently, the importance of this section cannot be overemphasized. As an exercise the reader can expand $f(x, y) = \sin x + \cos y$ about the point $(x = 0, y = 0)$. The answer to this problem is

$$f(x, y) = 1 + x - (y^2/2!) - (x^3/3!) - (y^4/4!) + (x^5/5!) - (y^6/6!) - \cdots$$

## 3.11 Tangent Plane and Normal Line to a Surface

These concepts are given for the three dimensional cases only, and there should be no difficulty in extending them to $n$ dimensions. The equation of the tangent plane at point $(x_0, y_0, z_0)$ of a surface given by the equation $f(x, y, z) = 0$, may be written as

$$\left.\frac{\partial f}{\partial x}\right|_{(x_0, y_0, z_0)} (x - x_0) + \left.\frac{\partial f}{\partial y}\right|_{(x_0, y_0, z_0)} (y - y_0) + \left.\frac{\partial f}{\partial z}\right|_{(x_0, y_0, z_0)} (z - z_0) = 0$$

The equation of the normal line to the surface $f(x, y, z) = 0$ at $(x, y, z)$ is given by

$$\frac{x - x_0}{-\partial f/\partial x\,|_{(x_0,y_0,z_0)}} = \frac{y - y_0}{-\partial f/\partial y\,|_{(x_0,y_0,z_0)}} = \frac{z - z_0}{\partial f/\partial z\,|_{(x_0,y_0,z_0)}}$$

### 3.12 Directional Derivatives

If $f(x, y, z)$ is defined at a point $(x, y, z)$ on a space curve and if $f(x+\Delta x, y + \Delta y, z + \Delta z)$ is the value of the function at a neighboring point on the curve, and $\Delta s$ is the length of the arc to that neighboring point, the directional derivative along the curve at $(x, y, z)$ is defined as

$$\lim \frac{\Delta f}{\Delta s} = \lim_{\Delta s \to 0} \frac{f(x + \Delta x, y + \Delta y, z + \Delta z) - f(x, y, z)}{\Delta s}$$

and it is given by

$$\frac{\partial f}{\partial s} = \frac{\partial f}{\partial x}\frac{\partial x}{\partial s} + \frac{\partial f}{\partial y}\frac{\partial y}{\partial s} + \frac{\partial f}{\partial z}\frac{\partial z}{\partial s}$$

### 3.13 Maxima and Minima

The location of local maxima and minima of a function of several variables is of considerable importance to the future development of our subject. If $f(x_1, x_2, \ldots, x_n)$ is a function of $n$ variables and if all the partial derivatives $\partial f/\partial x_i$ are continuous in a region then $f$ attains local maxima or minima only at points where

$$\partial f/\partial x_1 = \partial f/\partial x_2 = \cdots = \partial f/\partial x_n = 0$$

The classification of these extrema and examples of this important subject are discussed in Section 4.2 when we present Jacobians and Hessians. See Chapter 4 for further detailed discussions.

# APPLIED ANALYSIS

## 4. Applications of Differentiation

The first topic we deal with here is the differentiation of vectors of matrices. This topic is of importance in the discussion and exposition of the methods of function minimization and is used in some of the statistical theory applied here.

### 4.1 Derivatives of Vectors and Matrices

Let $\mathbf{y}$ be $n \times 1$ vector with elements $y_i$ and let $\mathbf{a}$ be $n \times 1$ vector with elements $a_i$ and let $Z$ be a scalar such that $Z = \mathbf{y}'\mathbf{a} = \mathbf{a}'\mathbf{y}$. The derivative of $Z$ with respect to the vector $\mathbf{y}$ can be defined for our purposes as the $n \times 1$ vector whose $i$th element is $\partial Z/\partial y_i$, therefore we may write

$$\begin{bmatrix} \dfrac{\partial Z}{\partial y_1} \\ \dfrac{\partial Z}{\partial y_2} \\ \vdots \\ \dfrac{\partial Z}{\partial y_n} \end{bmatrix} = \begin{bmatrix} \dfrac{\partial \sum_{j=1}^{n} a_j y_i}{\partial y_1} \\ \dfrac{\partial \sum_{j=1}^{n} a_j y_i}{\partial y_2} \\ \vdots \\ \dfrac{\partial \sum_{j=1}^{n} a_j y_i}{\partial y_p} \end{bmatrix} = \mathbf{a} = \begin{bmatrix} a_1 \\ a_2 \\ \vdots \\ a_n \end{bmatrix}$$

Let $\mathbf{a}$ equal a $p \times 1$ vector, $\mathbf{b}$, a $q \times 1$ vector and $\mathbf{Y}$ a $p \times q$ matrix whose $ij$th element is equal to $y_{ij}$. Further, if we let

$$Z = \mathbf{a}'\mathbf{Y}\mathbf{b} = \sum_{m=1}^{q} \sum_{n=1}^{p} a_n y_{nm} b_m$$

then

$$\frac{\partial Z}{\partial y_{ij}} = \frac{\partial(\sum_{m=1}^{p} \sum_{n=1}^{p} a_n y_{nm} b_m)}{\partial y_{ij}} = a_i b_j$$

The differentiation of a quadratic form

$$Y = \mathbf{x}'\mathbf{A}\mathbf{x} = \sum_{i=1}^{p} \sum_{j=1}^{p} x_i x_j a_{ij}$$

is a special case of the above (here $p = q$) and can be performed in the following manner

$$\frac{\partial Y}{\partial a_{ij}} = \frac{\partial(\sum_{m=1}^{p} \sum_{n=1}^{p} x_m x_n a_{mn})}{\partial a_{ij}} = x_i x_j$$

when $i = j$, $\partial y/\partial a_{ij} = x_i^2$ and when $i \neq j$ then $\partial y/\partial a_{ij} = 2x_i x_j$ because $a_{ij} = a_{ji}$. Making use of these relations

$$\partial y/\partial \mathbf{A} = 2\mathbf{XX}' - \operatorname{diag}(\mathbf{XX}')$$

where $\operatorname{diag}(\mathbf{XX}')$ is a diagonal matrix whose elements are the diagonal elements of $\mathbf{XX}'$.

If we wish to differentiate the quadratic form with respect to the vector $\mathbf{x}$ we have

$$\frac{\partial}{\partial x_i}(\mathbf{x}'\mathbf{A}\mathbf{x}) = \frac{\partial y}{\partial x_i} = 2\sum_{n=1}^{p} x_n a_{in}$$

or

$$\partial \mathbf{y}/\partial \mathbf{x} = 2\mathbf{A}\mathbf{x}$$

All the above equations can be easily verified using 2-vectors and $2 \times 2$ matrices as examples.

## 4.2 Jacobians and Hessians

The Jacobian and Hessian are important matrices occuring in probability theory, statistics, and the theory of function minimization.

If $F(x, y)$ and $G(x, y)$ are differentiable in a region, the Jacobian of the function $F$ and $G$ with respect to $x$ and $y$ is defined as

$$\partial(F, G)/\partial(x, y) = \begin{bmatrix} \partial F/\partial x & \partial F/\partial y \\ \partial G/\partial x & \partial G/\partial y \end{bmatrix}$$

For the three functions having three variables $F(x, y, z)$, $G(x, y, z)$ and $H(x, y, z)$, the Jacobian of $F$, $G$, and $H$ with respect to $(x, y, z)$ is given by

$$\partial(F, G, H)/\partial(x, y, z) = \begin{bmatrix} \partial F/\partial x & \partial F/\partial y & \partial F/\partial z \\ \partial G/\partial x & \partial G/\partial y & \partial G/\partial z \\ \partial H/\partial x & \partial H/\partial y & \partial H/\partial z \end{bmatrix}$$

and so on for functions of several variables.

The Hessian $\mathbf{H}$ of a function $f(x_1, x_2, \ldots, x_n)$ is given by

$$\mathbf{H} = \begin{bmatrix} \partial^2 f/\partial x_1^2 & \partial^2 f/\partial x_1\,\partial x_2 & \cdots & \partial^2 f/\partial x_1\,\partial x_n \\ \partial^2 f/\partial x_2\,\partial x_1 & \partial^2 f/\partial x_2^2 & \cdots & \partial^2 f/\partial x_2\,\partial x_n \\ \vdots & & & \vdots \\ \partial^2 f/\partial x_n\,\partial x_1 & \cdots & \cdots & \partial^2 f/\partial x_n^2 \end{bmatrix}$$

## 4.3 Applications of the Hessian and Jacobian

The Hessian is used to classify extrema of functions of two or more variables. Specifically, if the function $f(x_1, \ldots, x_n)$ has continuous first and second partial derivatives with respect to all independent variables, then where

$$\partial f/\partial x_1 = \partial f/\partial x_2 = \cdots = \partial f/\partial x_n = 0$$

the function $f$ has a relative minimum if the Hessian $\mathbf{H}$ is positive definite and a relative maximum if $\mathbf{H}$ is positive definite.

Suppose we wish to find the extreme values of

$$G(x_1, x_2) = 2x_1^2 - 2x_1 + 2x_2^2 - 12x_2 + 14$$

Taking the first derivatives with respect to each variable we obtain

$$\partial G/\partial x_1 = 4x_1 - 2 = 0, \qquad \partial G/\partial x_2 = 4x_2 - 12 = 0$$

Solving those two equations we get $x_1 = \frac{1}{2}$ and $x_2 = 3$. Further differentiation and substitution of the appropriate values yields.

$$\partial^2 F/\partial x_1^2 = 4(\partial^2 F/\partial x_2^2) = 4(\partial^2 F/\partial x_1\, \partial x_2) = \partial^2 F/\partial x_2\, \partial x_1 = 0$$

which gives the following structure to Hessian of the function $G(x_1, x_2)$

$$\mathbf{H} = \begin{bmatrix} 4 & 0 \\ 0 & 4 \end{bmatrix}$$

and that is positive definite indicating a minimum at the points $x_1 = \frac{1}{2}$ and $x_2 = 3$.

Jacobians are used to transform problems using one set of coordinate variables to problems using another set of related coordinate variables. These transformations are generally employed because they facilitate the solution of a problem or because by putting it in another way the problem becomes more tractable when dealt with in terms of the second set of variables.

Suppose we have the function $f(x_1, x_2, \ldots, x_n)$ and suppose

$$x_i = \phi_i(y_1, y_2, \ldots, y_n), \qquad i = 1, 2, \ldots, n$$

To transform $f(x_1, x_2, \ldots, x_n)$ to a function $h(y_1, y_2, \ldots, y_n)$ we employ the following equation

$$h(y_1, y_2, \ldots, y_n) = f(\phi_1, \phi_2, \ldots, \phi_n)\,|\,\mathbf{J}\,|$$

where $\mathbf{J}$ is the Jacobian of the transformation (note absolute value).

*Example.*

$$f(x_1, x_2) = (1/\pi)\exp(-x_1^2 - x_2^2), \qquad x_1 = 4y_1 + y_2, \qquad x_2 = 2y_1 - 2y_2$$

According to the definition of the Jacobian we have

$$\mathrm{J} = \begin{vmatrix} 4 & 1 \\ 2 & -2 \end{vmatrix} = -6$$

and the new function $h(y_1, y_2)$ is given by

$$h(y_1 y_2) = (6/\pi) \exp[-(4y_1 + y_2)^2 + (2y_1 - 2y_2)^2]$$

## 5. Applications of Integration

### 5.1 Numerical Integration

Since most of the functions we deal with in this book are essentially experimentally obtained curves numerical methods of integration will be found to be most suitable. In these methods we subdivide the interval on which the function is defined say $[a, b]$ into $n$ equal parts of length $\Delta x = (b - a)/n$.

With this notation the following rules can be used.

#### 5.1.1. *The Regular Rule*

$$\int_a^b f(x)\, dx \simeq \Delta x[f(a) + f(a + \Delta x) + \cdots + f(a + (n - 1)\, \Delta x)]$$

or

$$\simeq \Delta x[f(a + \Delta x) + f(a + 2\, \Delta x) + \cdots + f(b)]$$

#### 5.1.2 *The Trapezoidal Rule*

$$\int_a^b f(x)\, dx \simeq (\Delta x/2)[f(a) + 2f(a + \Delta x) + \cdots + 2f(a + (n - 1)\, \Delta x) + f(b)]$$

#### 5.1.3 *Simpson's Rule*

$$\int_a^b f(x)\, dx \simeq (\Delta x/3)[f(a) + 4f(a + \Delta x) + 2f(a + 2\Delta x) + 4f(a + 3\Delta x) + 2f(a + 4\Delta x) + \cdots + 4f(a + (n - 1)\, \Delta x) + f(b)]$$

One way of handling the integration of a function when the interval is divided into unequal parts is to determine the upper and lower bounds of the function in the subintervals and to take their mean as the value of $y$ in that interval.

### 5.2 Quadrature

The basic task of quadrature methods is to find an approximate formula for a definite integral. For example, we wish to find a set of $A_i$ and $x_i$ to ensure as much accuracy as possible in the following formula

$$\int_a^b f(x)\,dx \simeq \sum_{i=1}^{n} A_i f(x_i)$$

The reason we want to do this depends on the nature of the problem. If $f(x)$ is an elementary function we can establish its indefinite integral in closed form. Then using the fundamental theorem of integral calculus we can compute $\int_a^b f(x)\,dx$ exactly, and there would be no quadrature problem to solve. If, on the other hand, as frequently occurs, $f(x)$ does not represent a primitive function which can be treated by elementary methods or $f(x)$ is a tabulated function, then elementary methods for evaluation are not applicable, in which case we must resort to other means. It is important that the reader realize that if a satisfactory set of $A_i$ and $x_i$ are to be found then $\int_a^b f(x)\,dx$ should be approximated $\sum_{i=1}^{n} A_i f(x_i)$ to a degree of accuracy which improves with increasing $n$ regardless of the nature of the function.

Actual applications of quadrature methods are in Chapter 12, and are applied to the evaluation of molecular weight distribution functions from sedimentation equilibrium, light scattering, and low-angle $X$-ray scattering.

We do not plan to show how the various quadrature formulae are derived. The interested reader should refer to Kopal (1955). What we point out here is that there are numerous quadrature schemes which specify various sets of $x_i$ and $A_i$. These $A_i$ and $x_i$ for any given quadrature scheme can be found tabulated in Kopal (1955). For the actual explanation of the use of quadrature procedures in specific applications we refer the reader to Chapter 12.

## 6. Miscellaneous Topics

### 6.1 The Gamma Function

*Definition.* The gamma function is defined by the integral

$$\Gamma(x) = \int_0^\infty t^{x-1} e^{-t}\,dt \qquad \text{for} \quad x > 0$$

One of the fundamental properties of the gamma function is

$$\Gamma(x+1) = x\Gamma(x).$$

When $x = 1$ we have

$$\Gamma(1) = \int_0^\infty e^{-t}\,dt = 1$$

For integer values of $x$, called the $n$, we note, using the above property of the $\Gamma$ function, that

$$\begin{aligned}\Gamma(n+1) &= n\Gamma(n) = n(n-1)\Gamma(n-1)\\ &= \cdots = n(n-1)(n-2)\cdots\Gamma(1)\end{aligned}$$

or

$$\Gamma(n+1) = n!$$

We also note that $0! = 1$.

## 6.2 The Beta Function

*Definition.* The beta function is defined by the integral

$$\beta(x, y) = \int_0^1 t^{x-1}(1-t)^{y-1}\,dt, \qquad x, y > 0$$

The relationship between a pair of gamma functions and the beta function is given by

$$\Gamma(x)\Gamma(y) = \beta(x, y)\Gamma(x+y)$$

so that

$$\beta(x, y) = \frac{\Gamma(x)\Gamma(y)}{\Gamma(x+y)} = \frac{\Gamma(y)\Gamma(x)}{\Gamma(y+x)} = \beta(y, x)$$

Showing that the beta function is symmetric in its variables.

Both the gamma and beta function occur in the functional form of certain distribution functions occuring in the statistical theory given in the next chapters.

## 6.3 Laplace Transforms

The Laplace transform of a function $f(x)$ is defined as

$$f(s) = L[f(x)] = \int_0^\infty e^{-sx}f(x)\,dx$$

when the integral exists. Some familiar Laplace transforms are the following: If $f(x) = a$ then

$$L[f(x)] = a/s \qquad \text{for} \qquad s > 0$$

If $f(x) = e^{ax}$ then

$$L[f(x)] = 1/(s - a) \qquad \text{for} \quad s > a$$

If $f(x) = x^n$, for every positive integer $n$ then

$$L[f(x)] = n!/s^{n+1}$$

The function $L^{-1}[f(x)]$ is called the inverse Laplace transform of $f(x)$.

## 6.4 Fourier Transforms

A Fourier transform pair of a function $F(x)$ is defined as

$$f(u) = \mathscr{F}[F(x)] = \int_{-\infty}^{+\infty} e^{-iux} F(x)\, dx$$

and

$$F(x) = (1/2\pi) \int_{-\infty}^{\infty} f(u) e^{iux}\, du$$

where $i = (-1)^{1/2}$.

Both the Laplace transforms and the Fourier transforms are improper integrals but they exist for certain classes of functions and can be evaluated by various methods. The Laplace transform will be used in the handling of a problem in the analysis of aggregative systems, and the Fourier transform as stated before is important in probability theory, and will be used in Chapter 5 in the definition of characteristic function. Ideally this chapter should also contain a section on differential equations and one on integral equations since we will encounter both to some extent in the text. In the interest of brevity, however, we refer the reader to an appropriate text.

## 6.5 Orthogonal Functions

This topic is a natural extension of orthogonal vectors and orthogonal matrices which we dealt with in Chapter 2. Suppose we have a set of functions $\phi_i(x)$ such that for $i = 1, 2, 3, \ldots$ the functions have the following properties

$$\int_a^b \phi_m(x)\phi_n(x)\,dx = 0, \qquad m \neq n$$

$$\int_a^b \{\phi_m(x)\}^2\,dx = \lambda_m > 0, \qquad m = n$$

These functions then form an orthogonal set on the interval [a, b] and each member of this set is orthogonal to every other member of the set on this interval. Suppose we let

$$\psi_m(x) = \varphi_m(x)/(\lambda_m)^{1/2}$$

we then obtain

$$\int_a^b \psi_m(x)\psi_n(x)\,dx = \begin{matrix} 0, & m \neq n \\ 1, & m = n \end{matrix}$$

The latter set of functions are called an orthonormal set and the process of transforming $\phi_m(x)$ to $\psi_m(x)$ is called *normalization.*

A number of orthogonal functions of interest are the trigonometric functions.

$$\cos x,\ \cos 2x,\ \ldots,\ \cos mx,\ \ldots$$

$$\sin x,\ \sin 2x,\ \ldots,\ \sin mx,\ \ldots$$

The Legendre polynomials $P_m(x)$ are also orthogonal functions and can be written as

$$\int_{-1}^{1} P_m(x)P_n(x)\,dx = \begin{matrix} 0, & m \neq n \\ 2/(2n+1), & m = n \end{matrix}$$

where

$$P_n(x) = \frac{1}{2^n n!}\,\frac{d^n}{dx^n}\,(x^2 - 1)$$

The Hermite polynomials maybe written as

$$\int_{-\infty}^{\infty} \exp(-x^2)H_m(x)H_n(x)\,dx = \begin{matrix} 0, & m \neq n \\ 2^n n!\sqrt{\pi}, & m = n \end{matrix}$$

where

$$H_n(x) = (-1)^n e^{x^2}\,\frac{d^n}{dx^n}\,[\exp(-x^2)]$$

The Laguerre polynomials are as follows

$$\int_0^{\infty} e^{-x}L_m(x)L_n(x)\,dx = \begin{matrix} 0, & m \neq n \\ n!, & m = n \end{matrix}$$

where

$$L_n(x) = e^x \frac{d}{dx^n} [e^{-x}x^n]$$

Two particular uses are made of orthogonal polynomials. First we use them in systems of Gaussian quadrature where we would like to approximate the definite integral of a function. Our desire to approximate a definite integral stems from the fact that in the analysis of aggregating systems by sedimentation equilibrium, light scattering (see Chapter 12), or other means we encounter such integrals.†

The second reason why we would like to consider orthogonal functions is that they give a diagonal structure to matrices arising in least square minimization problems.

**References**


Apostol, T. M. (1969). "Calculus." Ginn (Blaisdell), Boston, Massachusetts.

Bowman, F. and Gerard, F. A. (1967). "Higher Calculus." Cambridge Univ. Press, London and New York.

Burkill, J. C. and Burkill, H. (1970). "A Second Course in Mathematical Analysis." Cambridge Univ. Press, London and New York.

Haaser, N. B., La Salle, J. P., and Sullivan, J. A. (1964). "Intermediate Analysis," Vol. I and II. Ginn (Blaisdell), Boston, Massachusetts.

Kline, M. (1967). "Calculus," Parts I and II. Wiley, New York.

Kopal, Z. (1955). "Numerical Analysis," Wiley, New York.

Spiegel, M. (1963). "Advanced Calculus," Schaum's Outline Ser. McGraw-Hill, New York.


† In the solution of a Fredholm integral equation of the first kind.

*CHAPTER 4*

# MINIMIZATION OF FUNCTIONS

## 1. Introduction

Our discussion of the techniques of the minimization of functions of several variables will follow a compromise course. We follow a course which is not so theoretical that it would leave the biologist wondering why he is reading this book. We choose a presentation that hopefully will direct the reader to the core problem of how the evaluation of the parameters in which he is interested came about.

There are perhaps readers who may not care to read this entire chapter but who wish only to use its results. For this reader we have an applications section at the end of the chapter which indicates which techniques are favorable to the solution of a particular type of problem. In the various chapters on the biological topics we shall keep referring back to the applications section (Section 7).

In Chapter 1 we described some of the problems we were going to deal with, and it was pointed out that in most instances, we will be determining parameters which are subject to some constraints, (for instance that the parameters are all positive or that they lie within certain bounds). This chapter describes the methods and algorithms that enable us to make the desired computations.

The methods we describe determine the parameters of interest when the parameters occur both linearly and nonlinearly; consequently, most equations can be analyzed by these methods to yield a "best" set of parameters in the sense that some function of the experimental errors is minimized for these parameter values.

## 2. Functional Extrema

### 2.1 Functional Extrema

We consider first the extreme values of a function $f(x)$ of a single real variable. We have pointed out in Chapter 3 that for twice differentiable functions on some open interval or open set $S \subset R^1$, the points $x_i$ at which $f'(x_i) = 0$ are local maxima, minima, or points of inflection. These points were classified by the sign of $f''(x_i)$ being $< 0$, $> 0$, or $= 0$ respectively. This concept was extended to functions of several variables $f(x_1, \ldots, x_n)$ having partial derivatives on open sets $S \subset R^n$, and the necessary conditions for local extrema are

$$\partial f/\partial x_1 = \partial f/\partial x_2 = \cdots = \partial f/\partial x_n = 0 \tag{4-1}$$

where the Hessians $H(x_1, x_2, \ldots, x_n)$ are such that

$$H(x_1, \ldots, x_n)$$

is positive definite,

$$-H(x_1, \ldots, x_n) \tag{4-2}$$

is positive definite, or

$$H(x, \ldots, x_n)$$

is singular, at the points of interest.

We wish to point out here that these cases do not exhaust the mathematical possibilities, but do in fact represent the ones of particular interest.

In the case of a single variable, an extreme value of a function may occur if $f'(x_0)$ is undefined.

*Example.* The function

$$f(x) = 1 - x^\alpha \qquad \text{for} \quad 0 < \alpha < 1$$

has the extreme value 1 at $x = 0$. However

$$f'(x) = -\alpha x^{\alpha-1} = -\alpha/x^{1-\alpha}$$

and $f'(0)$ is undefined.

In *closed* sets (or intervals) extreme values can occur at the boundary points (end points).

*Example.* The function $f(x) = x$ defined on $0 \leqq x \leqq 1$ has the two extreme values $f(0) = 0$ and $f(1) = 1$ while $f'(x) = 1$, for $0 \leqq x \leqq 1$.

Similar examples can also be constructed for functions of several variables. The point to be made here is that the use of the differentiation test is a test for *local* extrema only. Other tests must be made if one seeks the *global* extrema (i.e., absolute maximum and absolute minimum of a given function).

## 2.2 An Example of Function Minimization

In the introduction we outlined to the reader what is involved in function minimization and curve fitting. There it was explained that some function of the experimental error was to be minimized. To illustrate more completely what is involved we consider the following: suppose we have the function sin $x$. What linear combination of the functions $x$ and $x^2$ will permit us to come as close as possible to sin $x$, over the interval from 0 to $\pi$? Stating it a little differently, what values of the two parameters $a$ and $b$ in the expression $ax + bx^2$ will come closest to sin $x$ on the interval $[0, \pi]$? Obviously it becomes important to define what, "as close as possible," means. There are an infinity of criteria that can be used to define "as close as possible." For example, we might choose to make minima of the following functions

$$\int_0^\pi |\sin x - ax + bx^2|\, dx \tag{4-3}$$

$$\int_0^\pi [\sin x - (ax + bx^2)]\, dx \tag{4-4}$$

$$\int_0^\pi [\sin x - (ax + bx^2)]^2\, dx \tag{4-5}$$

$$\int_0^\pi [\sin x - (ax + bx^2)]^3\, dx \tag{4-6}$$

$$\vdots$$

$$\int_0^\pi [\sin x - (ax + bx^2)]^p\, dx \qquad \text{for} \quad p \geq 1 \tag{4-7}$$

The reader, recalling the introduction, may see the relationship of this example to the problems mentioned there. If one considers $\sin x$ as the actual response, $x$ and $x^2$ as the approximating functions, then the error will correspond to the difference, namely, $\sin x - ax + bx^2$. The expressions depicted above are obviously some function of the experimental error and our objective is to minimize an appropriate one of these functions.

In the above expressions we choose to *integrate* over the interval $[0, \pi]$ because effectively we want *all* the values of $ax + bx^2$ in the interval to come "close" to *all* the values of $\sin x$ in this interval. If we just wanted to make $\sin x - ax + bx^2$ as small as possible at only one point $x_1$ in the interval, we would have no difficulty in making this quantity zero. This is done by setting either $b = 0$ or $a = 0$ and solving the expression $\sin x_1 - ax_1 + bx_1^2 = 0$. In the first instance, we find that $a$ becomes

$$a = (\sin x_1)/x_1$$

and $b = 0$. In the second instance we find

$$-b = (\sin x_1)/x_1^2$$

and $a = 0$. It is also easy to make $\sin x - (ax + bx^2)$ zero at two distinct points in the interval, say $x_1$ and $x_2$. This is done by solving the simultaneous equations

$$\sin x_1 = ax_1 + bx_1^2, \qquad \sin x_2 = ax_2 + bx_2^2 \tag{4-8}$$

for $a$ and $b$. However, in general, what we want to do, is, to find an $a$ and $b$ so that a particular one of the quantities shown in the preceeding expressions is as small as possible over the *entire* interval, not just at a particular finite set of points. This is the reason for integrating over the entire interval $[0, \pi]$.

If we minimize

$$\int_0^\pi [\sin x - (ax + bx^2)]\, dx \tag{4-9}$$

then we are minimizing the *area* between the curve $\sin x$ and the curve $ax + bx^2$. Minimization of

$$\int_0^\pi [\sin x - (ax + bx^2)]^2\, dx \tag{4-10}$$

is the familiar *least squares* approximation and we shall use it as an example to demonstrate function minimization.

To compute the parameters $a$ and $b$ for the least squares approximation of $\sin x$ by the polynomial $ax + bx^2$ over the interval $[0, \pi]$, we proceed as follows. First we note that the expression in Eq. (4-5) is a function of $a$ and $b$. Consequently we can write

$$F(a, b) = \int_0^\pi [\sin x - (ax + bx^2)]^2 \, dx \tag{4-11}$$

As in Chapter 3, we find the relative extreme points when

$$\partial F(a, b)/\partial a = \partial F(a, b)/\partial b = 0 \tag{4-12}$$

Carrying out the differentiations as described in Chapter 3, we find

$$\begin{aligned} \frac{\partial F}{\partial a} &= \int_0^\pi \frac{\partial}{\partial a} [(\sin x) - (ax + bx^2)]^2 \, dx \\ &= -2 \int_0^\pi x[(\sin x) - (ax + bx^2)] \, dx = 0 \end{aligned} \tag{4-13}$$

$$\begin{aligned} \frac{\partial F}{\partial b} &= \int_0^\pi \frac{\partial}{\partial b} [(\sin x) - (ax^2 + bx^2)]^2 \, dx \\ &= -2 \int_0^\pi x^2[(\sin x) - (ax + bx^2)] \, dx = 0 \end{aligned} \tag{4-14}$$

We have then

$$a \int_0^\pi x^2 \, dx + b \int_0^\pi x^3 \, dx = \int_0^\pi x \sin x \, dx \tag{4-15}$$

and

$$a \int_0^\pi x^3 \, dx + b \int_0^\pi x^4 \, dx = \int_0^\pi x^2 \sin x \, dx \tag{4-16}$$

On carrying out the required integrations we obtain the following simultaneous equations for $a$ and $b$

$$(\pi^3 a/3) + (\pi^4 b/4) = \pi \tag{4-17}$$

and

$$(\pi^4 a/4) + (\pi^5 b/5) = \pi^2 - 4 \tag{4-18}$$

Solving for $a$ and $b$ we find $a_0 \simeq -0.4$ and $b_0 \simeq 1.25$. We leave it as an exercise for the reader to plot the curve $\sin x$ in the interval $[0, \pi]$ and the polynomial $-0.4x + 1.25x^2$ in the interval $[0, \pi]$ and see how closely they come to each other. We have not, as yet, classified this extreme point of $F(a, b)$. However, it is clear by inspection of expression (4-11) that $F(a, b)$ is unbounded as a function of $a$ and $b$. The reader can see this by allowing $a$ or $b$ to be arbitrarily large. The point $(a_0, b_0)$ must therefore be a local minimum of $F(a, b)$. The reader can verify this by constructing the Hessian of $F(a, b)$ and evaluating it at $(a_0, b_0)$. We also leave it as an exercise for the reader to find another $a$ and $b$ using trial and error or any other means and see if he or she can bring the two curves $\sin x$ and the polynomial $ax + bx^2$ "closer" than the choice made in our worked example. Readers attempting the latter exercise must, of course, define the sense in which the term "closer" is used.

In general the more parameters one uses the better one fits the data, or function. If instead of finding the best least squares approximation of the polynomial $ax + bx^2$ to $\sin x$ in $[0, \pi]$, we sought the least squares approximation by the polynomial $ax + bx^2 + cx^3$ to $\sin x$ in $[0, \pi]$, then we would find, on carrying out the computations and plotting the results, that $ax + bx^2 + cx^3$ comes *closer* in the least squares sense to $\sin x$ than does $ax + bx^2$. This fact leads us to a powerful theorem concerning polynomial approximation due to Weierstrass, which we state without proof.

### 2.3 Weierstrass Approximation Theorem

Let $f(x)$ be a continuous (bounded) function defined on the interval $[a, b]$. Then given any $\varepsilon > 0$, we can find a polynomial $p_n(x)$ of sufficiently high degree such that

$$|f(x) - p_n(x)| \leqq \varepsilon \qquad \text{for all} \quad a \leq x \leq b$$

We say that $f(x)$ is *uniformly* approximated by $p_n(x)$ on $[a, b]$ whenever this holds.

### 2.4 Constrained Extrema (Lagrangian Multipliers)

A problem that occurs with some frequency is that of finding the extrema of a function $f(x_1, x_2, \ldots, x_n)$ subject to the constraining relation, $\Phi(x_1, \ldots, x_n) = C$, $C$ a constant. Geometrically this might mean finding the maximum or minimum of a function $f$ on some surface $\Phi$. A later example will illustrate this interpretation. As we have seen, the local extrema of $f(x_1, \ldots, x_n)$ occur at points where

$$f_{x_1} = f_{x_2} = \cdots = f_{x_n} = 0 \tag{4-19}$$

where $f_{x_1}$ indicates the derivative of $f(x_1, x_2, \ldots, x_n)$ with respect to $x_1$. We must select from these extrema the ones (if any) which satisfy the constraining relation $\Phi(x_1, \ldots, x_n) = C$.

The system of equations

$$f_{x_1} = f_{x_2} = f_{x_n} = \Phi(x_1, \ldots, x_n) - C = 0 \tag{4-20}$$

is an *overdetermined* system since we have $n + 1$ conditions to satisfy and only $n$ unknowns to find. We relieve this overdeterminancy by introducing another variable, or parameter $\lambda$. We do this by constructing a new function of $n + 1$ variables

$$F(x_1, x_2, \ldots, x_n, \lambda) = f(x_1, x_2, \ldots, x_n) + \lambda\Phi(x_1, \ldots, x_n) \tag{4-21}$$

where $\lambda$ is a parameter to be determined. A word of motivation is in order here. Suppose that an appropriate value for $\lambda$ is known, then the new function is

$$F(x_1, \ldots, x_n, \lambda) = f(x_1, \ldots, x_n) + \lambda C \tag{4-22}$$

That is, the values of $F(x_1, \ldots, x_n, \lambda)$, for a fixed $\lambda$, differ from those of $f(x_1, \ldots, x_n)$ by only a constant value. Since the points of $R^n$ at which the extrema of a function $f$ occur, are not affected by adding a constant to $f$, the function $F(x_1, \ldots, x_n, \lambda)$ and the function $f(x_1, \ldots, x_n)$ must have extreme points in common (for a fixed $\lambda$). The specific extreme points that we seek are those which satisfy the constraining relation $\Phi(x_1, \ldots, x_n) = C$. The extreme values of $F(x_1, \ldots, x_n, \lambda)$ we seek must satisfy

$$F_{x_1} = F_{x_2} = \cdots = F_{x_n} = 0 \tag{4-23}$$

and

$$\Phi(x_1, \ldots, x_n) = C \tag{4-24}$$

We note that $\Phi(x_1, x_2, \ldots, x_n) = F_\lambda$ and therefore our constrained extrema are given exactly by the extrema of the function $F$ in $R^{n+1}$ space with the independent variables $(x_1, x_2, \ldots, x_n, \lambda)$. The simultaneous system of $n + 1$ equations (usually nonlinear)

$$F_{x_1} = F_{x_2} = \cdots = F_{x_n} = F_\lambda = 0 \tag{4-25}$$

determine the desired local extrema, and the parameter $\lambda$ corresponding to each extreme point of $f(x_1, \ldots, x_n)$ which satisfies $\Phi(x_1, \ldots, x_n) = C$ are exactly the extrema of $F(x_1, \ldots, x_n, \lambda)$. This method is due to

Lagrange (1736–1813) and the parameter $\lambda$ is called a Lagrangian multiplier.

When there is more than one constraint, the process extends quite easily by introducing a Lagrangian multiplier $\lambda_i$ for each constraint $\Phi_i$. We then find the extrema of

$$F(x_1, \ldots, x_n, \lambda_1, \lambda_2, \ldots, \lambda_k) = f(x_1, \ldots, x_n) + \lambda_1 \Phi_1 + \cdots + \lambda_k \Phi_k \tag{4-26}$$

just as in the case for a single parameter. The simultaneous equations

$$F_{x_1} = F_{x_2} = \cdots = F_{x_n} = F_{\lambda_1} = \cdots = F_{\lambda_k} = 0 \tag{4-27}$$

then determine the desired extrema and the $k$ Lagrangian multipliers at each extreme point. We caution the reader that solutions of this system are, except for simple cases, quite difficult to obtain by elementary methods. This is because of the nonlinear character of the resulting equations and the difficulty of the handling simultaneous systems of nonlinear algebraic equations.

To illustrate finding extreme values of a function subject to constraints we take a familiar example from coordinate geometry. Suppose we wish to find the extreme values of the function

$$f(x_1, x_2, x_3) = x_1^2 + x_2^2 + x_3^2 \tag{4-28}$$

subject to the constraints

$$(x_1^2/4) + (x_2^2/5) + (x_3^2/25) = 1, \qquad \text{and} \quad x_1 + x_2 - x_3 = 0 \tag{4-29}$$

In order to do this we form the auxiliary function

$$\begin{aligned} F(x_1, x_2, x_3, \lambda_1, \lambda_2) = f(x_1, x_2, x_3) &+ \lambda_1[(x_1^2/4) + (x_2^2/5) + (x_3^2/25) \\ &- 1] + \lambda_2(x_1 + x_2 - x_3) \end{aligned} \tag{4-30}$$

Differentiating and equating to zero

$$\begin{aligned} \partial F/\partial x_1 &= 2x_1 + (\lambda_1 x_1/2) + \lambda_2 = 0 \\ \partial F/\partial x_2 &= 2x_2 + (2\lambda_1 x_2/5) + \lambda_2 = 0 \\ \partial F/\partial x_3 &= 2x_3 + (2\lambda_1 x_3/25) - \lambda_2 = 0 \\ \partial F/\partial \lambda_1 &= (x_1^2/4) + (x_2^2/5) - (x_3^2/25) + 1 = 0 \\ \partial F/\partial \lambda_2 &= x_1 + x_2 - x_3 = 0 \end{aligned} \tag{4-31}$$

Solving for $x_1$, $x_2$, and $x_3$ in the first three of the above equations we obtain

$$x_1 = -2\lambda_2(\lambda_1 + 4), \qquad x_2 = -5\lambda_2(2\lambda_1 + 10), \qquad x_3 = 25\lambda_2/(2\lambda_1 + 50)$$

Using the fifth relation $x_1 + x_2 - x_3 = 0$ of Eq. (4-31) above then dividing out $\lambda_2 \neq 0$, we have

$$\frac{2}{\lambda_1 + 4} + \frac{5}{2\lambda_1 + 10} + \frac{25}{2\lambda_1 + 50} = 0 \qquad (4\text{-}33)$$

This yields a quadratic equation in $\lambda_1$,

$$17\lambda_1^2 + 245\lambda_1 + 750 = 0 \qquad (4\text{-}34)$$

whose solutions are $\lambda_1 = -10$ and $\lambda_1 = -75/17$. When $\lambda_1 = -10$ then $x_1 = \lambda_2/3$, $x_2 = \lambda_2/2$, and $x_3 = 5\lambda_2/6$. Substituting these values in the final equation, that is in

$$(x_1^2/4) + (x_2^2/5) + (x_3^2/25) = 1$$

yields a quadratic in $\lambda_2$ whose solutions are the values $\pm 6(5/19)^{1/2}$. Since we have determined $x_1$, $x_2$ and $x_3$ in terms of $\lambda_2$ our critical points are:

$$\begin{aligned} x_1 &= 2(5/19)^{1/2}, & x_2 &= 3(5/19)^{1/2}, & x_3 &= 5(5/19)^{1/2} \\ x_1 &= -2(5/19)^{1/2}, & x_2 &= -3(5/19)^{1/2}, & x_3 &= -5(5/19)^{1/2} \end{aligned} \qquad (4\text{-}35)$$

The value of the function $f(x_1, x_2, x_3)$ at each of these two critical points is 10.

Looking at the other root of $\lambda_1$, namely $\lambda_1 = -75/17$, and using the values of $x_1$, $x_2$, $x_3$ in terms of $\lambda_2$ and substituting those values in the fourth relation

$$(x_1^2/4) + (x_2^2/5) + (x_3^2/25) = 1$$

we obtain another quadratic in $\lambda_2$. The solutions are

$$\lambda_2 = \pm 140[17/(646)^{1/2}].$$

These values give us the following critical points:

$$\begin{aligned} x_1 &= 40/(646)^{1/2}, & x_2 &= -35/(646)^{1/2}, & x_3 &= 5/(646)^{1/2} \\ x_1 &= -40/(646)^{1/2}, & x_2 &= 35/(646)^{1/2}, & x_3 &= -5/(646)^{1/2} \end{aligned} \qquad (4\text{-}36)$$

When either of these sets of values are substituted in the original function $f(x_1, x_2, x_3)$, they yield a value of 75/17. Therefore, the minimum is found at the points given by Eq. (4-36) and the maximum is found at the points given by Eq. (4-35). Note that there is more than one maximum and minimum.

Minimization problems subject to constraints occur quite frequently in this kind of work. Unfortunately the method of the Lagrangian multipliers does not constrain the variables so that they are all positive, a condition we will frequently require in this text.

## 3. Least Squares

### 3.1 Least Squares Straight Line

We introduce the mathematical concepts of least squares linear regression as background for the more complex applications of the least squares method to the fitting of nonlinear models. The topic is best introduced through a simple example of fitting straight lines to data.

The equation of a straight line is given by

$$y = \alpha_1 + \alpha_2 x \tag{4-37}$$

or in the notation we shall adopt for our models

$$f(x) = \alpha_1 f_1(x) + \alpha_2 f_2(x) \tag{4-38}$$

where $y = f(x)$, $1 = f_1(x)$, and $x = f_2(x)$. For specific data points $(x_j, y_j)$ we set $y_j = f(x_j)$ and $x_j = f_2(x_j)$. If we incorporate experimental error in Eq. (4-38), we write it as

$$f(x) = \alpha_1 + \alpha_2 f_2(x) + \delta \tag{4-39}$$

Equation (4-39) evaluated at a typical data point is written as

$$f(x_j) = \alpha_1 + \alpha_2 f_2(x_j) + \delta_j \tag{4-40}$$

For $m$ data points Eq. (4-40) can be written $m$ times with $j = 1, 2, \ldots, m$. Our objective is to determine $\alpha_1$ and $\alpha_2$ such that $\sum_{j=1}^{m} \delta_j^2$ is a minimum. Using $m$ equations such as (4-40), we can write $\sum_{j=1}^{m} \delta_j^2$ as follows

$$\sum_{j=1}^{m} \delta_j^2 = \sum_{j=1}^{m} [f(x_j) - \alpha_1 - \alpha_2 f_2(x_j)]^2 \tag{4-41}$$

To obtain the minimum of the above expression with respect to the parameters $\alpha_1$ and $\alpha_2$ we differentiate $\sum_{j=1}^{m} \delta_j^2$ with respect to $\alpha_1$ and $\alpha_2$ which are considered as variables, and equate them to zero. We digress for a moment to clarify a point.

The observant reader might well inquire why we have called $\alpha_1$ and $\alpha_2$ *variables*, since they are in fact the constants or the parameters of the model. The $\alpha$ are treated as variables in order to minimize the function $\sum_{j=1}^{m} \delta_j^2$. When this minimum is found, the values of these parameters which produce this minimum are the constants or the parameters the investigator seeks. We have explained this to some extent in the introduction. To explain this further, suppose one knew nothing about the analytical techniques of function minimization and you were asked to minimize a function such as shown in Eq. (4-41). Since the $f(x_j)$ are known and the $f_2(x_j)$ are also known the only unknowns in that equation are the $\alpha$. Therefore you choose a set of $\alpha$ and substitute them in Eq. (4-41). This procedure will enable you to calculate a number. You then choose another set of $\alpha$ substitute them in Eq. (4-41) and see if the number you obtain this time is smaller than the first number. Repeating this procedure, you arrive at the smallest number you can find. When that happens you say you have minimized the function denoted by (4-41) by trial and error, and you have done so by repeatedly varying your set of $\alpha$. This is the reason we call the $\alpha$ variables; we might remark, incidentally, that the method of trial and error has been used extensively by a number of authors, in the biochemical literature, but we shall not give a detailed bibliography. Clearly the methods used here are a great deal more sophisticated than any crude trial and error methods.

Returning to the least squares problem we differentiate Eq. (4-41) with respect to $\alpha_1$ and $\alpha_2$ and equate to zero.

$$\begin{aligned} \frac{\partial \sum_{j=1}^{m} \delta_j^2}{\partial \alpha_1} &= -2 \sum_{j=1}^{m} [f(x_j) - \alpha_1 - \alpha_2 f_2(x_j)] = 0 \\ \frac{\partial \sum_{j=1}^{m} \delta_j^2}{\partial \alpha_2} &= -2 \sum_{j=1}^{m} [f(x_j) - \alpha_1 - \alpha_2 f_2(x_j)] f_2(x_j) = 0 \end{aligned} \tag{4-42}$$

or

$$\begin{aligned} \sum_{j=1}^{m} f(x_j) &= \alpha_1 \sum_{j=1}^{m} 1 + \alpha_2 \sum_{j=1}^{m} f_2(x_j) \\ \sum_{j=1}^{m} f(x_j) f_2(x_j) &= \alpha_1 \sum_{j=1}^{m} f_2(x_j) + \alpha_2 \sum_{j=1}^{m} f_2^2(x_j) \end{aligned} \tag{4-43}$$

The above equations are simultaneous linear equations in the unknowns $\alpha_1$ and $\alpha_2$. In matrix notation these can be written as

$$\begin{bmatrix} \sum_{j=1}^{m} 1 & \sum_{j=1}^{m} f_1(x_j) f_2(x_j) \\ \sum_{j=1}^{m} f_2(x_j) & \sum_{j=1}^{m} f_2^2(x_j) \end{bmatrix} \begin{bmatrix} \alpha_1 \\ \alpha_2 \end{bmatrix} = \begin{bmatrix} \sum_{j=1}^{m} f(x_j) \\ \sum_{j=1}^{m} f(x_j) f_2(x_j) \end{bmatrix}$$

Note that $\sum_{j=1}^{m} 1 = m$. Solving the above equations for $\alpha_1$ and $\alpha_2$ yields the least squares straight line estimate for $\alpha_1$ and $\alpha_2$.

*Example.* Find the least square estimate of $\alpha_1$ and $\alpha_2$ for the set of data points in Table 4-1.

**Table 4-1**

| $y_j = f(x_j)$ | $x_j = f_2(x_j)$ |
|---|---|
| 54 | 18 |
| 64 | 26 |
| 54 | 28 |
| 62 | 34 |
| 68 | 36 |
| 70 | 42 |
| 76 | 48 |
| 66 | 52 |
| 76 | 54 |
| 74 | 60 |

For this set of numbers we find that

$$\sum_{j=1}^{m} f_2(x_j) = 398, \qquad \sum_{j=1}^{m} f(x_j) = 664$$

$$\sum_{j=1}^{m} f_2^2(x_j) = 17{,}524, \qquad \sum_{j=1}^{m} f(x_j) f_2(x_j) = 27{,}268$$

The matrix equation becomes

$$\begin{bmatrix} 10 & 398 \\ 398 & 17{,}524 \end{bmatrix} \begin{bmatrix} \alpha_1 \\ \alpha_2 \end{bmatrix} = \begin{bmatrix} 664 \\ 27{,}268 \end{bmatrix}$$

On solving the simultaneous equations for $\alpha_1$ and $\alpha_2$ we find that $\alpha_1 = 46.5$ and $\alpha_2 = 0.4994$.

### 3.2 Least Squares in General

The statement of the linear problem in general goes as follows: It is required to approximate a given response function of a system, $f(x)$, in terms of the functions $f_1(x), f_2(x), \ldots, f_n(x)$ by determining the parameters $\alpha_1, \alpha_2, \ldots, \alpha_n$ in the following equations:

$$
\begin{aligned}
f(x_1) &= \sum_{i=1}^{n} \alpha_i f_i(x_1) + \delta_1 \\
f(x_2) &= \sum_{i=1}^{n} \alpha_i f_i(x_2) + \delta_2 \\
&\vdots \\
f(x_j) &= \sum_{i=1}^{n} \alpha_i f_i(x_j) + \delta_j \\
&\vdots \\
f(x_m) &= \sum_{i=1}^{n} \alpha_i f_i(x_2) + \delta_m
\end{aligned}
\tag{4-44}
$$

such that $\sum_{j=1}^{m} \delta_j^2$ is a minimum. Equations (4-44) may be expressed in matrix form as follows:

$$\mathbf{Y} = \mathbf{X}\boldsymbol{\alpha} + \mathbf{d} \tag{4-45}$$

where the matrices $\mathbf{Y}$ and $\mathbf{d}$ are column vectors with $m$ elements, the matrix $\mathbf{X}$ is an $m \times n$ matrix, and $\boldsymbol{\alpha}$ is a column vector with $n$ elements, which may be written as

$$
\mathbf{Y} = \begin{bmatrix} f(x_1) \\ f(x_2) \\ \vdots \\ f(x_j) \\ \vdots \\ f(x_m) \end{bmatrix}, \quad
\boldsymbol{\alpha} = \begin{bmatrix} \alpha_1 \\ \alpha_2 \\ \vdots \\ \alpha_n \end{bmatrix}, \quad
\mathbf{X} = \begin{bmatrix} f_1(x_1) & f_1(x_2) & \cdots & f_1(x_n) \\ f_2(x_1) & f_2(x_2) & \cdots & f_2(x_n) \\ \vdots & \vdots & & \vdots \\ f_m(x_1) & f_m(x_2) & \cdots & f_m(x_n) \end{bmatrix}, \quad
\mathbf{d} = \begin{bmatrix} \delta_1 \\ \delta_2 \\ \vdots \\ \delta_m \end{bmatrix}
\tag{4-46}
$$

When using the least squares approximation we have the coefficients $\alpha_1, \alpha_2, \ldots, \alpha_n$ which minimize

$$\sum_{j=1}^{m} \delta_j^2 = \mathbf{d}'\mathbf{d} \tag{4-47}$$

If we define

$$\phi(\boldsymbol{\alpha}, x_j) = \sum_{i=1}^{n} \alpha_i f_i(x_j) \tag{4-48}$$

then essentially we are trying to minimize the function

$$F(\boldsymbol{\alpha}, x) = \sum_{j=1}^{m} \delta_j^2 = \sum_{j=1}^{m} [f(x_j) - \phi(\boldsymbol{\alpha}, x_j)]^2 \tag{4-49}$$

with respect to variables $\alpha_1, \alpha_2, \ldots, \alpha_n$.

The function $F(\boldsymbol{\alpha}, x)$ will have extreme points at

$$\partial F(\boldsymbol{\alpha}, x)/\partial \alpha_i = 0 \qquad \text{for} \quad i = 1, 2, \ldots, n \tag{4-50}$$

Expanding the quantity in square brackets in Eq. (4-49) we obtain

$$\partial\left\{\sum_{j=1}^{m} [f^2(x_j) - 2f(x_j)\phi(\boldsymbol{\alpha}, x_j) + \phi^2(\boldsymbol{\alpha}, x_j)]\right\}\Big/\partial \alpha_i = 0 \text{ for } i = 1, 2, \ldots, n. \tag{4-51}$$

Carrying out the differentiation for any $i$ we find the first term yields

$$\partial\left[\sum_{j=1}^{m} f^2(x_j)\right]\Big/\partial \alpha_i = 0 \tag{4-52}$$

The second term gives

$$\partial\left[\sum_{j=1}^{m} f(x_j)\phi(\boldsymbol{\alpha}, x_j)\right]\Big/\partial \alpha_i = 2 \sum_{j=1}^{m} f(x_j) f_i(x_j) \tag{4-53}$$

and the last term yields

$$\partial\left[\sum_{j=1}^{m} \phi^2(\boldsymbol{\alpha}, x_j)\right]\Big/\partial \alpha_i = 2 \sum_{k=1}^{n} \alpha_k \sum_{j=1}^{m} f_i(x_j) f_k(x_j) \tag{4-54}$$

For any $i$ therefore, we have

$$\partial F(\boldsymbol{\alpha}, x)/\partial \alpha_i = 0 = 0 - 2 \sum_{j=1}^{m} f(x_j) f_i(x_j) + 2 \sum_{k=1}^{n} \alpha_i \sum_{j=1}^{m} f_i(x) f_j(x) \tag{4-55}$$

or

$$\sum_{=1}^{m} f(x_j) f_i(x_j) = \sum_{k=1}^{n} \alpha_k \sum_{j=1}^{m} f_i(x_j) f_j(x_j) \qquad \text{for} \quad i = 1, 2, \ldots, n \tag{4-56}$$

Equations (4-56) are called the normal equations and are linear in the undetermined parameters $\alpha_i$. They can be solved by the techniques for linear equations shown in Chapter 2.

In matrix notation Eqs. (4-56) become

$$\begin{bmatrix} \sum_{j=1}^{m} f_1^2(x_j) & \sum_{j=1}^{m} f_1(x_j)f_2(x_j) & \cdots & \sum_{j=1}^{m} f_1(x_j)f_n(x_j) \\ \sum_{j=1}^{m} f_1(x_j)f_i(x_j) & \sum_{j=1}^{m} f_2^2(x_j) & \cdots & \sum_{j=1}^{m} f_2(x_j)f_n(x_j) \\ \vdots & \vdots & & \vdots \\ \sum_{j=1}^{m} f_n(x_j)f_1(x_j) & \sum_{j=1}^{m} f_n(x_j)f_2(x_j) & \cdots & \sum_{j=1}^{m} f_n^2(x_j) \end{bmatrix} \begin{bmatrix} \alpha_1 \\ \alpha_2 \\ \vdots \\ \alpha_n \end{bmatrix}$$

$$= \begin{bmatrix} \sum_{j=1}^{m} f(x_j)f_1(x_j) \\ \sum_{j=1}^{m} f(x_j)f_2(x_j) \\ \vdots \\ \sum_{j=1}^{m} f(x_j)f_n(x_j) \end{bmatrix} \tag{4-57}$$

or

$$\mathbf{X'X\alpha} = \mathbf{YX'} \tag{4-58}$$

The conditions for the existence of the solution for the above linear equations (4-58) were discussed in Chapter 2. If the matrix on the left hand side of Eq. (4-57) is singular we will encounter difficulties in the solution. Singularity will occur when one of the $f_i(x)$ is linearly dependent on the rest. When the $f_i(x)$ are linearly independent then the solution vector $\boldsymbol{\alpha}$ is unique.

When the matrix is singular or nearly singular we have a system that is *ill-conditioned.*

### 3.3 Ill-Conditioning

The singularity of the matrix of coefficients is closely associated with ill-conditioning. To see what this means, suppose we are considering two systems of simultaneous equations where the coefficients differ by very little; for example consider the system given by

$$\alpha_1 - \alpha_2 = 1, \qquad \alpha_1 - 1.00001\alpha_2 = 0 \tag{4-59}$$

and another system given by

$$\alpha_1' - \alpha_2' = 1, \qquad \alpha_1' - 0.9999\alpha_2' = 0 \tag{4-60}$$

The solution of the first system of equations is $\alpha_1 = 100{,}001$ and $\alpha_2 = 100{,}000$ while the solution of the second system of equations is $\alpha_1' = -99.999$ and $\alpha_2' = -100{,}000$. These two vastly differing solutions point out effects of the ill-conditioning of the problem. It is readily observed that the coefficients of the system differ by very little, but the solutions are quite different. Ill-conditioning is present whenever very small differences in the coefficients yield large differences in the solution.

In general, we consider the system

$$\alpha_1 + \alpha_2 = 1, \qquad \alpha_1 + \alpha_2(1 + \varepsilon) = 1 + p \tag{4-61}$$

and the solution

$$\alpha_1 = 1 - (p/\varepsilon) \qquad \text{and} \qquad \alpha_2 = p/\varepsilon \tag{4-62}$$

We have no solution when $\varepsilon = 0$, and an infinity of solutions when $p = \varepsilon = 0$ (see Chapter 2 on linear equations). However when $\varepsilon \neq 0$ the solution is very sensitive to small changes in $\varepsilon$ in a neighborhood of $\varepsilon = 0$. The ill-conditioning observed in Eqs. (4-59)–(4-61) is reflected in the near *singularity* of the matrix of coefficients. This concept is clarified in a more extensive development in which we emphasize the geometric interpretations.

### 3.4 Geometric Considerations of Least Squares and the Geometric Effects of Ill-Conditioning

We again consider the fitting of straight lines, or in higher dimensions, a hyperplane, by the least squares criteria for goodness of fit (approximation). That is, we are considering the minimization of the function

$$F(\boldsymbol{\alpha}, \mathbf{x}) = [\mathbf{y} - \mathbf{Y}]'[\mathbf{y} - \mathbf{Y}] \tag{4-63}$$

where $\mathbf{Y}$ is our model, and may be equated to $\phi(\boldsymbol{\alpha}, \mathbf{x})$ of (4-48).

Given $m$ observation vectors of the form $(y, x_1, \ldots, x_n)$ where $y$ is the value of the response, and $x_i$, $i = 1, \ldots, n$, are the values of the associated independent (controlled) variables, the $f_i(x)$ of Eq. (4-44); consider once again the linear model

$$Y_j = \sum_{i=2}^{n} \alpha_i (x_{i-1})_j + \alpha_1 \qquad (j = 1, 2, \ldots, n) \tag{4-64}$$

where $Y_j$ denotes the calculated (or "expected") value or the function

$\phi(\boldsymbol{\alpha}, \mathbf{x})$ of Eq. (4-48) or our model. Here again we wish to minimize the quantity in (4-63) given

$$F(\boldsymbol{\alpha}, \mathbf{x}) = \sum_{j=1}^{m} (y_j - Y_j)^2 \qquad \text{where} \quad \boldsymbol{\alpha} = [\alpha_1, \ldots, \alpha_n]' \tag{4-65}$$

We shall examine only the case $n = 2$ in detail. The procedure is exactly the same for all finite dimensions. Substituting $Y_j = \alpha_2 x_j + \alpha_1$ into Eq.(4-65) and expanding

$$F(\alpha_1, \alpha_2) = B\alpha_1^2 + C\alpha_1\alpha_2 + D\alpha_2^2 + E\alpha_1 + F\alpha_2 + G \tag{4-66}$$

where

$$B = \sum_{j=1}^{m} x_j^2, \qquad C = 2\sum_{j=1}^{m} x_j, \qquad D = \sum_{j=1}^{m} 1$$

$$E = -2\sum_{j=1}^{m} x_j y_j, \qquad F = -2\sum_{j=1}^{m} y_j, \qquad G = \sum_{j=1}^{m} y_j^2$$

We note that Eq. (4-66) is a quadratic form and therefore we may write Eq. (4-66) as the matrix equation

$$F(\alpha_1, \alpha_2) = \boldsymbol{\alpha}'\mathbf{A}\boldsymbol{\alpha} + [E, F]\boldsymbol{\alpha} + [G, 0]\,[1, 0], \tag{4-67}$$

where

$$\mathbf{A} = \begin{bmatrix} B & C/2 \\ C/2 & D \end{bmatrix}.$$

What is the curve defined by $F(\alpha_1, \alpha_2) = K$, where $K$ is an arbitrary *positive* constant? Since Eq. (4-66) is a general equation of second degree (see Chapter 2) in the parameters $(\alpha_1, \alpha_2)$, its geometric character is revealed by the discriminant function

$$\Delta = C^2 - 4\mathrm{BD} \tag{4-68}$$

Using the definitions appended to Eq. (4-66), it can be shown that

$$\Delta = -2\sum_{i=1}^{m}\sum_{j=1}^{m} (x_i - x_j)^2 \tag{4-69}$$

Thus $\Delta < 0$ for any meaningful case (i.e., if $m > 1$ and $x_i \neq x_j$ for at least one $j$). Therefore $F(\alpha_1, \alpha_2) = K$ represents a family of *ellipses* in the $(\alpha_1, \alpha_2)$ parameter space. To represent Eq. (4-66) so that it has the standard elliptic form, the original coordinates are rotated so as to remove the cross product term, $C\alpha_1\alpha_2$. Removing this term in this manner corresponds to diagonalizing the matrix $\mathbf{A}$ in Eq. (4-67). Since $\mathbf{A}$ is a positive definite symmetric matrix, it is similar to a diagonal matrix of all positive elements, namely $\mathbf{\Lambda} = (\lambda_{ii})$, which are the eigenvalues for $\mathbf{A}$ (see Chapter 2). $\mathbf{A}$ can therefore be transformed to $\mathbf{\Lambda}$ by the relation

$$\mathbf{V}'\mathbf{A}\mathbf{V} = \mathbf{\Lambda} \tag{4-70}$$

Where $\mathbf{V}\mathbf{V}' = \mathbf{I}$ and the columns of the matrix $\mathbf{V}$ are proportional to the linearly independent eigenvectors corresponding to each of the eigenvalues $\lambda_{11}$ and $\lambda_{22}$ of $\mathbf{A}$. A linear transformation that removes the cross-product term in Eq. (4-66) can now be defined as

$$\mathbf{u} = \mathbf{V}'\boldsymbol{\alpha}, \tag{4-71}$$

where $\mathbf{u} = [u_1, u_2]'$, $\boldsymbol{\alpha} = [\alpha_1, \alpha_2]'$, $u_1$ and $u_2$ are the new (rotated) coordinate variables. With this transformation, Eq. (4-66) becomes

$$F(u, v) = \mathbf{u}'\mathbf{V}'\mathbf{A}\mathbf{V}\mathbf{u} + [E, F]\mathbf{V}\mathbf{u} + [G, 0]\,[1, 0] \tag{4-72}$$

since

$$\mathbf{V}'\mathbf{A}\mathbf{V} = \mathbf{\Lambda} = \begin{bmatrix} \lambda_{11} & 0 \\ 0 & \lambda_{22} \end{bmatrix}$$

and setting $[E, F]\mathbf{V} = [\beta, \gamma]$, Eq. (4-72) becomes

$$F(u_1, u_2) = \lambda_{11}u_1^2 + \lambda_{22}u_2^2 + \beta u_1 + \gamma u_2 + G \tag{4-73}$$

This equation may be written as the standard equation for an ellipse as

$$\Phi(u_1, u_2) = \lambda_{11}[u_1 + (\beta/2\lambda_{11})]^2 + \lambda_{22}[u_2 + (\gamma/2\lambda_{22})]^2 + G - (\beta^2/4\lambda_{11}) - (\gamma^2/4\lambda_{22}) \tag{4-74}$$

Our principal interest is the set of parameters $\boldsymbol{\alpha}$ that minimize the right-hand side of Eq. (4-65), or (for $n = 2$) its equivalent in the linear case, Eq. (4-66). Since $\lambda_{11}$, $\lambda_{22}$, and $G$ are *positive* (Chapter 2), this unique minimum is obtained when $u_1 = -\beta/2\lambda_{11}$ and $u_2 = -\gamma/2\lambda_{22}$. These values are transformed back to the $(\alpha_1, \alpha_2)$ coordinates by

$$\begin{bmatrix} \alpha_1 \\ \alpha_2 \end{bmatrix} = \mathbf{V}\begin{bmatrix} u_1 \\ u_2 \end{bmatrix}, \qquad \boldsymbol{\alpha} = \mathbf{V}\mathbf{u} \tag{4-75}$$

These values clearly denote the *center* of the family of ellipses (4-66) (i.e., the least squares solution). To verify that (4-75) does indeed give the least squares solution of Eq. (4-66), we use the substitutions

$$\begin{bmatrix} -\beta/2\lambda_{11} \\ -\gamma/2\lambda_{22} \end{bmatrix} = -\tfrac{1}{2}\mathbf{\Lambda}^{-1}\begin{bmatrix} \beta \\ \gamma \end{bmatrix} = -\tfrac{1}{2}\mathbf{\Lambda}^{-1}\mathbf{V}\begin{bmatrix} E \\ F \end{bmatrix} \tag{4-76}$$

Thus

$$\begin{bmatrix} \alpha_1{}^c \\ \alpha_2{}^c \end{bmatrix} = -\tfrac{1}{2}\mathbf{V}\mathbf{\Lambda}^{-1}\mathbf{V}\begin{bmatrix} E \\ F \end{bmatrix} = -\tfrac{1}{2}\mathbf{A}^{-1}\begin{bmatrix} E \\ F \end{bmatrix} \tag{4-77}$$

Taking the partial derivatives of Eq. (4-66) with respect to each of the parameters and setting them equal to zero simultaneously, we obtain

$$\begin{aligned} \alpha_1 B + \alpha_2(C/2) &= -E/2 \\ \alpha_1(C/2) + \alpha_2 D &= -F/2 \end{aligned} \tag{4-78}$$

which rewritten in matrix form yields

$$\begin{bmatrix} \alpha_1 \\ \alpha_2 \end{bmatrix} = -\tfrac{1}{2}\mathbf{A}^{-1}\begin{bmatrix} E \\ F \end{bmatrix} \tag{4-79}$$

This proves that Eq. (4-77) is, in fact, the least squares solution for Eq. (4-66). Therefore the complete solution of the *linear* least squares problem can be obtained by means of the eigenvalues and eigenvectors for the matrix **A**, and the vector $[E, F]'$. The minimum sum of squares is given by

$$F(\alpha_1, \alpha_2) = G - \tfrac{1}{4}[E, F]\mathbf{A}^{-1}[E, F]' \tag{4-80}$$

The use of the eigenvalues and eigenvectors has other distinct advantages. We have seen that in the eigenvector coordinate system, the cross-product terms do not appear. This means, geometrically, that the eigenvector coordinate directions $(u_1, u_2)$ are parallel to each of the principal axes of the family of ellipses. Each ellipse represents a constant value of the sum of squares [i.e., Eq. (4-66) set equal to a value $K$]. The relationship of the eigenvalues and the lengths of the principal axes of the family of ellipses can be obtained from Eq. (4-74). See Fig. 4-1 for the complete geometry.

Let $K$ be a fixed positive constant identifying a specific nontrivial ellipse. Then Eq. (4-74) can be written as

$$\lambda_{11}(u_1 + \beta/2\lambda_{11})^2 + \lambda_{22}(u_2 + \gamma/2\lambda_{22})^2 = \theta \tag{4-81}$$

where

$$\theta = K + (\beta^2/2\lambda_{11}) + (\gamma^2/2\lambda_{22}) - G$$

Suppose we translate the origin to the *center* of the ellipses by letting $\chi_1 = u_1 + (\beta/2\lambda_{11})$ and $\chi_2 = u_2 + (\gamma/2\lambda_{22})$. Then Eq. (4-74) becomes $\lambda_{11}\chi_1{}^2 + \lambda_{22}\chi_2{}^2 = \theta$, the canonical equation for an ellipse. The length of the principal axis in a particular direction is found by setting all

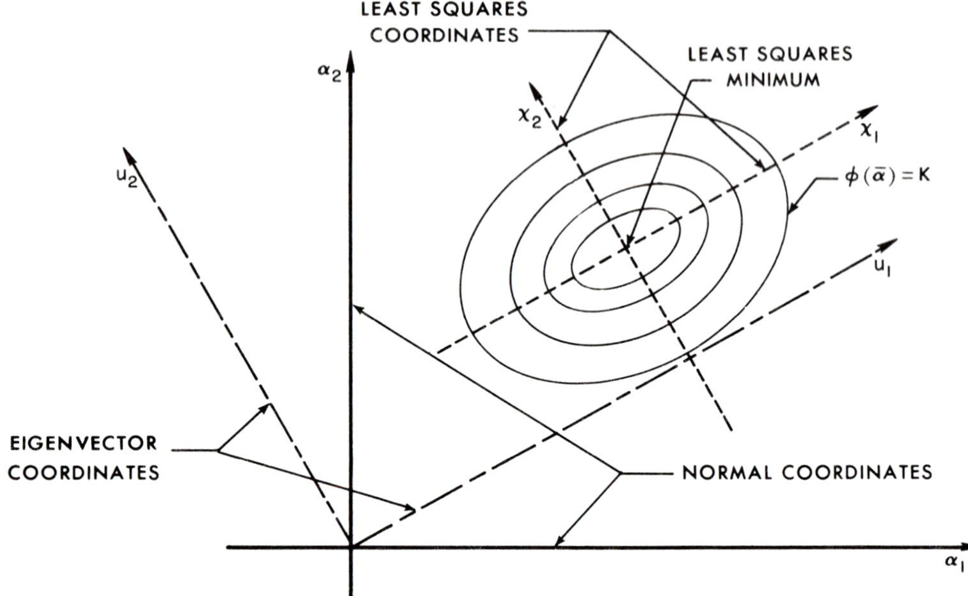


**Fig. 4-1.** Linear least squares geometry ($N = 2$). An overbar in the figure corresponds to boldface in the text (Fletcher and Shrager, 1968).

variables not in that direction equal to zero. The general formula for the principal axis is therefore

$$l_i = (\theta/\lambda_{ii})^{1/2} \tag{4-82}$$

This formula shows that the ellipse is elongated most in the direction of the smallest eigenvalue, (i.e., the largest principal axis is in the direction of the eigenvector corresponding to the smallest eigenvalue). The property of extreme elongation with very small eigenvalues shows that the relationship between the value of the sum of squares and a least squares solution can be a weak one. Consider in Fig. 4-2 a particular ellipse and points $X$ and $Y$ on such an ellipse. Point $Y$ is clearly *much closer* to the least squares solution at 0 than is point $X$; however, the *sum of squares* is *exactly* the same for both points. This weakening of the goodness of fit relationship is particularly evident whenever the matrix **A** is ill-conditioned. Ill-conditioning occurs when we have a very small eigenvalue, illustrated in Fig. 4-2. The occurrence of a small eigenvalue may therefore *neutralize* the usefulness of the sum of squares as a goodness of fit criteria.

Experimentally, ill-conditioning is an indicator of faulty data, an inappropriate model, or both. By faulty data we mean measurements that are not completely independent of each other or that contain excessive errors. An inappropriate model will have either too many parameters, or all the model's parameters cannot be resolved independently with the particular set of data. The ill-conditioning can sometimes be removed by either extending the range of the data or by simplifying the model being used so that it contains less parameters.

The vector $[E, F]$ also has special geometric meaning for the linear model. Recall that $F(\boldsymbol{\alpha}) = K$ is, in general, a surface of constant sum of squares in the parameter space. The gradient vector normal to this surface in general is

$$\boldsymbol{\omega} = [\partial F/\partial \alpha_1, \partial F/\partial \alpha_2, \ldots, \partial F/\partial \alpha_n]' \tag{4-83}$$

or in the two parameter case

$$\boldsymbol{\omega} = [2B\alpha_1 + C\alpha_2 + E, 2D\alpha_2 + C\alpha_1 + F]'$$

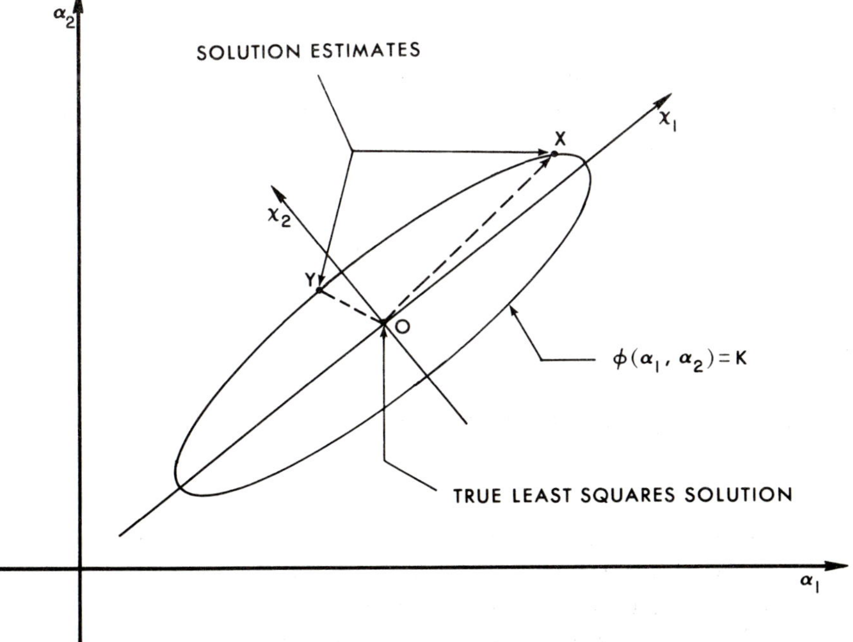


**Fig. 4-2.** Relationship of sum of squares to true solution point (ill-conditioning). (Fletcher and Shrager, 1968).

If $\alpha_1 = 0$ and $\alpha_2 = 0$ are taken as the initial estimates then $\boldsymbol{\omega} = [E, F]'$. With this notation, Eq. (4-83) becomes (in the general case)

$$\boldsymbol{\alpha} = -\tfrac{1}{2}\,\mathbf{A}\boldsymbol{\omega} \tag{4-84}$$

where it is understood that the gradient vector $\boldsymbol{\omega}$ is calculated at the origin $\boldsymbol{\alpha} = 0$.

For purposes of machine computation, it is not usually convenient to compute directly the matrix **A**. Note that if

$$F_j = \alpha_2 x_j + \alpha_1$$

is the model and we define a matrix **P** as

$$(p_{ij}) = \frac{\partial F_j}{\partial \alpha_i} = \begin{bmatrix} x_1 & 1 \\ x_2 & 1 \\ \vdots & \vdots \\ x_m & 1 \end{bmatrix}$$

then

$$\mathbf{P}'\mathbf{P} = \mathbf{A} \tag{4-85}$$

The fact that **A** is a symmetric positive definite matrix is then clear, and one may replace **A** by **P′P** in all the preceeding equations. We can also express the gradient in terms of **P** by letting

$$\mathbf{y} = \begin{bmatrix} y_1 \\ \vdots \\ y_m \end{bmatrix}$$

then

$$[E, F] = -2\mathbf{P}'\mathbf{y} \tag{4-86}$$

Using Eq. (4-85) and (4-86) in Eq. (4-84) we obtain the solution vector as

$$\boldsymbol{\alpha} = (\mathbf{P}'\mathbf{P})^{-1}\mathbf{P}'\mathbf{y} \tag{4-87}$$

This convenient form gives the *linear* least squares solution completely in terms of the data arrays, and matrix operations are the only mathematical computations required. In the following discussions we shall refer to **P′y** simply as the gradient vector, and we shall extend Eq. (4-87) to nonlinear models. The geometric meaning of Eq. (4-87) is that the gradient vector to the ellipse passing through the origin is transformed

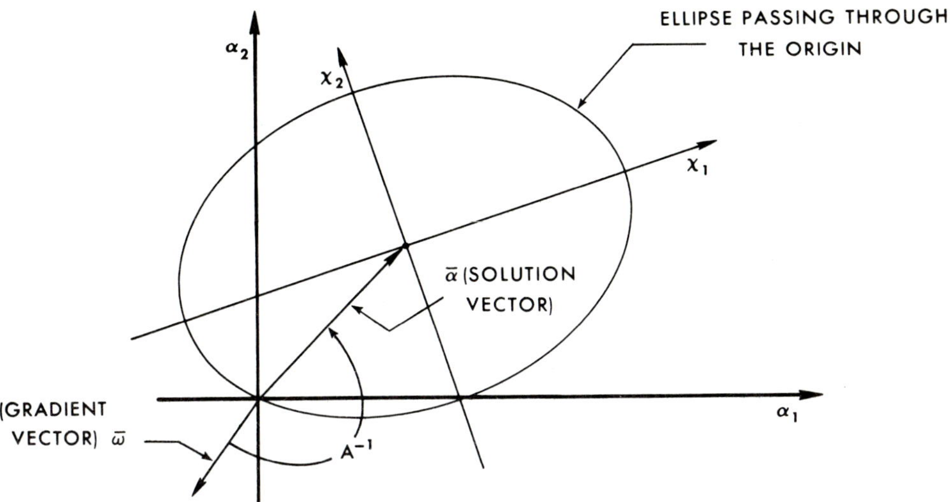


**Fig. 4-3.** A geometric interpretation of Gauss–Newton (Fletcher and Shrager, 1968).

by the matrix $\mathbf{A}^{-1}$ to the solution vector $\boldsymbol{\alpha}$ which is the least squares solution of Eq. (4-65). See Fig. 4-3.

### 3.5 Weighted Least Squares

When there is unequal experimental error in data being used for a given least squares problem the use of a *weighted* least squares method is in order.

When we have observations of unequal accuracy we employ weights. The weights used in the least squares analysis are inversely proportional to the variance of the measurements, or the square of the standard deviation. The weight assigned to the *j*th observation $w_j$ is written as

$$w_j = 1/(\sigma_j^2) \tag{4-88}$$

where $\sigma_j^2$ is the variance of the *j*th observation and the weights are generally normalized so that

$$\sum_{j=1}^{m} w_j = m \tag{4-89}$$

For the weighted least squares the quantity to be minimized with respect to the parameters $\alpha_1, \alpha_2, \ldots, \alpha_n$ is

$$\sum_{j=1}^{m} w_j[f(x_j) - \phi(\boldsymbol{\alpha}, x_j)]^2 \tag{4-90}$$

On carrying out the necessary differentiation we obtain the following normal equations

$$\sum_{j=1}^{m} w_j f(x_j) f_i(x_j) = \sum_{k=1}^{n} \alpha_k \sum_{j=1}^{m} w_j f_i(x_j) f_k(x_j) \qquad (i = 1, 2, \ldots, n) \qquad (4\text{-}91)$$

Once again Eq. (4-91) is a set of $n$ linear equations in the undetermined parameters and are solved by the usual techniques. Writing these equations in matrix form we have

$$\mathbf{X}'\mathbf{W}\mathbf{X}\boldsymbol{\alpha} = \mathbf{X}'\mathbf{W}\mathbf{Y} \qquad (4\text{-}92)$$

where the matrix $\mathbf{W}$ may be represented as

$$\mathbf{W} = \begin{bmatrix} w_1 & & & 0 \\ & w_2 & & \\ & & \ddots & \\ 0 & & & w_m \end{bmatrix} \qquad (4\text{-}93)$$

### 3.6 Least Squares Subject to Linear Constraints

As explained in the introduction, constraints on the parameters $\alpha_1$, $\alpha_2, \ldots, \alpha_n$ occur in certain problems. The type of linear constraint which the least squares approximation is capable of dealing with is one where we have

$$\alpha_1^* + \alpha_2^* + \cdots + \alpha_n^* = k \qquad (4\text{-}94)$$

where $k$ is some constant and $\alpha_1^*, \alpha_2^*, \ldots, \alpha_n^*$ are the parameters subject to the restriction imposed by Eq. (4-94) (as opposed to $\alpha_1, \alpha_2, \ldots, \alpha_n$ which are the least squares solution found when no restrictions are imposed).

It must be noted that the linear restrictions shown here do not restrict the parameters regarding their sign (i.e., do not restrict them to be either wholly positive or wholly negative).

Whenever we are finding the extreme value of a function subject to constraints we must resort to the technique of Lagrangian multipliers.

To solve our problem we minimize the quantity

$$F(\boldsymbol{\alpha}^*, x) = \sum_{j=1}^{m} [f(x_j) - \phi(\boldsymbol{\alpha}^*, x_j)]^2 + 2\lambda(\alpha_1^* + \alpha_2^* + \cdots \alpha_n^*) \qquad (4\text{-}95)$$

where

$$\phi(\boldsymbol{\alpha}^*, x_j) = \sum_{i=1}^{n} \alpha_i^* f_i(x_j)$$

and $\lambda$ is the Lagrangian multiplier. The factor 2 in front of that quantity is put there for convenience and does not affect the argument. To obtain a minimum we must have

$$\partial F(\boldsymbol{\alpha}^*, x)/\partial \alpha_i^* = 0 \qquad \text{for all} \quad i = 1, 2, \ldots, n \tag{4-96}$$

This leads to the following normal equation

$$\alpha_1^* \sum_{j=1}^{m} f_i(x_j)f_1(x_j) + \alpha_2^* \sum_{j=1}^{m} f_i(x_j)f_2(x_j) + \cdots + \alpha_n^* \sum_{j=1}^{m} f_i(x_j)f_n(x_j)$$
$$= \sum_{j=1}^{m} f(x_j)f_i(x_j) - \lambda \qquad \text{for all} \quad i = 1, 2, \ldots, n \tag{4-97}$$

In matrix notation Eq. (4-97) can be written as

$$\mathbf{X'X}\boldsymbol{\alpha}^* = \mathbf{X'Y} - \boldsymbol{\lambda} \tag{4-98}$$

where $\boldsymbol{\lambda}$ is a column vector of identical elements $\lambda$

Equating the matrix $\mathbf{X'X}$ to $\boldsymbol{\varkappa}$ we can write a typical matrix element of $\boldsymbol{\varkappa}$ as

$$\varkappa_{ih} = \sum_{j=1}^{m} f_i(x_j)f_h(x_j) \tag{4-99}$$

We will also call a matrix of elements of $\boldsymbol{\varkappa}^{-1}$, $\varkappa^{ih}$. To solve for the $\boldsymbol{\alpha}^*$ we have the following matrix equation

$$\boldsymbol{\alpha}^* = \boldsymbol{\varkappa}^{-1}\mathbf{X'Y} - \boldsymbol{\varkappa}^{-1}\boldsymbol{\lambda}. \tag{4-100}$$

since $\boldsymbol{\varkappa}^{-1}\mathbf{X'Y} = \boldsymbol{\alpha}$ from Eq. (4-58), the unconstrained least squares solution, then Eq. (4-100) becomes

$$\boldsymbol{\alpha}^* = \boldsymbol{\alpha} - \boldsymbol{\varkappa}^{-1}\boldsymbol{\lambda} \tag{4-101}$$

If we can determine the Lagrangian multiplier $\boldsymbol{\lambda}$ then we can solve for the $\boldsymbol{\alpha}^*$ and determine the solution for the constrained least squares norm. To solve for $\boldsymbol{\lambda}$ we must note that we have the constraint that $\sum_{i=1}^{n} \alpha_i^* = k$. Using this fact as well as Eq. (4-101) provides a solution for $\boldsymbol{\lambda}$ thus

$$\sum_{i=1}^{n} \alpha_i^* = \sum_{i=1}^{n} \alpha_i - \lambda \sum_{i=1}^{n} \sum_{j=1}^{n} \varkappa^{ij} \tag{4-102}$$

In Eq. (4-102) if we replace $\sum_{i=1}^{n} \alpha^*$ by $k$ then all the quantities are known except $\lambda$ and we solve for that. Having obtained that value of $\lambda$ we substitute it in Eq. (4-103) and solve for the $\alpha_i^*$ individually.

### 3.7 Least Squares Approximations with Special Functions

Least squares approximation with special functions will occur quite frequently during the course of the work of a molecular biologist. The functions that one generally uses are polynomials, trigonometric functions, or other orthogonal functions. The special functions are used in data processing and data smoothing of various kinds. The specific uses that these functions have been put to are mostly in the domain of concentration-dependent aggregating systems, and in the processing of osmotic coefficient data from sedimentation equilibrium. The potential applications of such data fitting methods to all kinds of problems in molecular biology and chemistry are quite numerous. In Chapter 12 we point out some applications to aggregating systems.

#### 3.7.1 *Least Squares with Polynomials*

In this case we seek to fit our reponse function to a polynomial which may be written as

$$y = f(x) \cong \alpha_0 + \alpha_1 x^1 + \alpha_2 x^2 + \cdots + \alpha_n x^n \tag{4-103}$$

Consequently our functions $f_i(x)$ have the following form

$$f_0(x) = 1, \qquad f_1(x) = x, \qquad f_2(x) = x^2, \ldots, f_n(x) = x^n \tag{4-104}$$

In general, polynomial fits are used for smoothing data, particularly when the data requires further integration or differentiation. Needless to say both the weighted and the constrained least squares methods can also be used to fit polynomials to data.

A direct least squares polynomial solution, despite the fact that it in theory defines the parameters uniquely, gives rise to computational problems in extracting these parameters. This is due to the structure of the matrices involved.

The normal equations in matrix notation for such a case with data points $j = 1, 2, \ldots, m$ can be written

$$\begin{bmatrix} m & \sum_{j=1}^{m} x_j & \sum_{j=1}^{m} x_1{}^2 & \cdots & \sum_{j=1}^{m} x_i{}^n \\ \sum_{j=1}^{m} x_j & \sum_{j=1}^{m} x_j{}^2 & & \cdots & \sum_{j=1}^{m} x_j^{n+1} \\ \vdots & & & & \vdots \\ \sum_{j=1}^{m} x_j{}^n & & & \cdots & \sum_{j=1}^{m} x_j^{2n} \end{bmatrix} \begin{bmatrix} \alpha_1 \\ \alpha_2 \\ \vdots \\ \alpha_n \end{bmatrix} = \begin{bmatrix} \sum_{j=1}^{m} y_j \\ \sum_{j=1}^{m} x_i y_j \\ \vdots \\ \sum_{j=1}^{m} x_j{}^n y_j \end{bmatrix} \tag{4-105}$$

The $\mathbf{X}$ matrix has the form

$$\mathbf{X} = \begin{bmatrix} 1 & x_1 & x_1^2 & \cdots & x_1^n \\ 1 & x_2 & x_2^2 & \cdots & x_2^n \\ \vdots & \vdots & \vdots & & \vdots \\ 1 & x_m & x_m^2 & \cdots & x_m^n \end{bmatrix} \tag{4-106}$$

It can be shown (Hamming, 1962) that the determinant of $\mathbf{X}$ behaves like the Hilbert determinant except for a constant multiplier. The Hilbert determinant of order $n$ (which may be taken as the order of the polynomial) has the value

$$H_n = \frac{[1!\,2!\,3!\cdots(n-1)!]^3}{n!\,(n+1)!\cdots(2n-1)!} \tag{4-107}$$

As $n$ increases it is clear that $H_n$ tends rapidly to zero and as a guide (particularly to the programmers) to its size we have Table 4-2. The smallness of this determinant as $n$ increases demonstrates that ill-conditioning becomes a problem as one increases the order (degree) of the

**Table 4-2**

| $n$ | $H_n$ |
|---|---|
| 1 | 1 |
| 2 | $8.3 \times 10^{-2}$ |
| 3 | $4.6 \times 10^{-4}$ |
| 4 | $1.7 \times 10^{-7}$ |
| 5 | $3.7 \times 10^{-12}$ |
| 6 | $5.4 \times 10^{-18}$ |
| 7 | $4.8 \times 10^{-25}$ |
| 8 | $2.7 \times 10^{-33}$ |
| 9 | $9.7 \times 10^{-43}$ |

polynomial approximation. To overcome this handicap one uses orthogonal polynomials. These give rise to diagonal $\mathbf{X}'\mathbf{X}$ matrices, thus greatly facilitating machine computation.

Another practical point which must be noted is one frequently mentioned throughout this volume. Suppose an investigator obtains a least squares solution of a response function and obtains answers which do not seem to be consistent with what is felt to be the true physical situation.

It is possible that he has used a set of ill-conditioned equations in obtaining his solution.

If it is recalled that the set of functions which are used to achieve the solution are (or must be) linearly independent, then this set of functions may be orthogonalized as in Chapter 2. Consequently, if unrealistic situations arise in a least squares computation one might try orthogonalizing the approximating functions. Orthogonalization will also help in constructing solutions with functions which are *nearly* linearly independent.

### 3.7.2 *Fitting with Orthogonal Polynomials*

Suppose one has values of a function $f(x)$ in a given interval $[a, b]$, and it is desired to approximate this function by means of a set of orthogonal functions. We set

$$f(x) \cong \alpha_0 + \alpha_1 f_1(x) + \alpha_2 f_2(x) + \cdots + \alpha_n f_n(x) \tag{4-107a}$$

where

$$\int_a^b f_i(x) f_j(x)\, dx \begin{array}{ll} = 0 & \text{if } i \neq j \\ = 1 & \text{if } i = j \end{array}$$

Then because the $\mathbf{X'X}$ matrix is diagonal we have

$$\int_a^b f(x) f_j(x)\, dx = \alpha_j \qquad \text{for} \quad j = 0, 1, 2, \ldots, n$$

It can be shown that these coefficients are the least squares solution of the minimization problem

$$F(\alpha_1, \ldots, \alpha_n) = \int_a^b [f(x) - \sum_{i=0}^{n} \alpha_n f_n(x)]^2\, dx$$

whenever the $f_n(x)$ are orthogonal. We turn now to algorithms which carry out these ideas in computation.

In this case we suppose that the $m$ data points we are using are equally spaced. Here the response $f(x_j)$ will be represented as follows

$$\begin{aligned} f(x_j) = \beta_0 + \beta_1 P_1\{j - [(m+1)/2]\} + \beta_2 P_2\{j - [(m+1)/2]\} + \cdots \\ + \beta_n P_n\{j - [(m+1)/2]\} + \delta_j \qquad \text{for all} \quad j = 1, 2, \ldots, m \end{aligned} \tag{4-108}$$

where the $\beta_i$ are the constants to be determined and where

$$P_t\{j - [(m+1)/2]\}$$

is a $t$th degree polynomial in the variable $\{j - [(m + 1)/2]\}$. For the time being the $P_t\{j - [(m + 1)/2]\}$ are not orthogonal. To orthogonalize the $P_t$ or more specifically to orthogonalize the matrix in which they ultimately occur we write

$$P_t = P_t\{j - [(m + 1)/2]\} = a_{0t} + a_{1t}\{j - [(m + 1)/2]\} + a_{2t}\{j - [(m + 1)/2]\}^2 + \cdots + a_{tt}\{j - [(m + 1)/2]\}^t \quad (4\text{-}109)$$

Here $a_{0t}, a_{1t}, \ldots, a_{tt}$ are constants to be determined to satisfy the orthogonality conditions or to make the matrix $\mathbf{Q'Q}$ diagonal (see Eq. 4-112).

The representation of $P_t$ by Eq. (4-109) may appear to the reader as being quite arbitrary. The reader will recall, however, that we generally wish to represent the response in terms of polynomials up to a certain power. If the reader bears that in mind then he will note that there is a certain fixed number of $a_{it}$ depending on $t$ which are required so that all quantities in (4-109) are known. These quantities can be found in special tables and are required to construct the matrix $\mathbf{Q}$.

For the estimation of $\boldsymbol{\beta}$ in Eq. (4-108) we have by the least squares theory the following matrix equation

$$\boldsymbol{\beta} = (\mathbf{Q'Q})^{-1}\mathbf{Q'Y} \quad (4\text{-}110)$$

where the matrix $\mathbf{Q}$ is $n \times m$ and can be written as

$$\mathbf{Q} = \begin{bmatrix} 1 & P_1\left(1 - \frac{m+1}{2}\right) & \cdots & P_n\left(1 - \frac{m+1}{2}\right) \\ 1 & P_1\left(2 - \frac{m+1}{2}\right) & \cdots & P_n\left(2 - \frac{m+1}{2}\right) \\ \vdots & \vdots & \cdots & \vdots \\ 1 & P_1\left(m - \frac{m+1}{2}\right) & \cdots & P_n\left(m - \frac{m+1}{2}\right) \end{bmatrix} \quad (4\text{-}111)$$

and

$$\mathbf{Q'Q} = \begin{bmatrix} n & \Sigma P_1 & \Sigma P_2 & \cdots & \Sigma P_n \\ \Sigma P_1 & \Sigma P_1^2 & & \cdots & \Sigma P_1 P_n \\ \vdots & \vdots & & \cdots & \vdots \\ \Sigma P_n & \Sigma P_n P_1 & & \cdots & \Sigma P_n^2 \end{bmatrix} \quad (4\text{-}112)$$

where the notation employed for the matrix elements of the matrix $\mathbf{Q'Q}$

is such that

$$\Sigma P_r'P_s = \sum_{j=1}^{n} P_r\left(j - \frac{m+1}{2}\right)P_s\left(j - \frac{m+1}{2}\right) \tag{4-113}$$

further

$$\mathbf{Q'Y} = \begin{bmatrix} \Sigma y_j \\ \Sigma y_j P_1 \\ \Sigma y_j P_2 \\ \vdots \\ \Sigma y_j P_n \end{bmatrix} \tag{4-114}$$

where the summations in the matrices (4-112) and (4-114) are taken from $j = 1$ to $j = m$. The coefficients $a_{tt}$ in Eq. (4-109) are read from the tables. Those coefficients have been tabulated for $n$ up to 104 by Anderson and Houseman (1941). Once the vector $\boldsymbol{\beta}$ is evaluated with the help of the $a_{tt}$ then Eq. (4-100) reproduces $f(x)$ in terms of orthogonal polynomials which can be easily differentiated or integrated according to how we further use the data.

### 3.7.3 *Least Squares Fits with Fourier (Trigonometric) Series*

In this case we represent $f(x_j)$ in terms of trigonometric series as follows

$$f(x_j) = (a_0/2) + \sum_{k=1}^{j} \{a_k \cos[2\pi/(2j+1)]kf(x_j) + b_k \sin[2\pi/(2j+1)]kf(x_j)\} \tag{4-115}$$

in the above equation $a_0$, $a_k$ and $b_k$ are constants to be determined. The number of data points that one should use are $2(j+1)$ and our objective is to minimize the quantity

$$R = \sum_{x_j=0}^{2j} [f(x_j) - T_m\{f(x_j)\}]^2 \tag{4-116}$$

where $T_m\{f(x_j)\}$ is a truncated trigonometric series such that $M < j$. We can write $T_m\{f(x_j)\}$ as follows

$$T_m\{f(x_j)\} = (a_0/2) + \sum_{k=1}^{M} a_k \cos[2\pi/(2j+1)]kf(x_j) + b_k \sin[2\pi/(2j+1)]kf(x_j) \tag{4-117}$$

The minimum of the quantity $R$ can be written

$$R_{\min} = [(2j+1)/2] \sum_{k=m+1}^{j} (a_k^2 + b_k^2) \tag{4-118}$$

The coefficients $a_k$ and $b_k$ are obtained by making use of the orthogonality property of trigonometric functions and can be written as

$$a_k = [2/(2j+1)] \sum_{f(x_j)=0}^{2j} f(x_j) \cos[2\pi/(2j+1)]f(x_j) \qquad (k = 0, 1, \ldots, j)$$

and (4-119)

$$b_k = [2/(2j+1)] \sum_{f(x_j)=0}^{2j} f(x_j) \sin[2\pi/(2j+1)]f(x_j) \qquad (k = 1, 2, \ldots, j)$$

The equations of this subsection establish the representation of $f(x_j)$ by means of a trigonometric series which, as in the case of the polynomials and orthogonal polynomials, can be differentiated or integrated according to our requirements.

Trigonometric functions are used to smooth data particularly if such data is periodic. They have been used by Jeffrey and Coates (1966) to process sedimentation equilibrium data.

## 4. Minimax Curve Fitting or Chebyshev Minimization

### 4.1 Elementary Considerations

This rather specialized topic has not been exploited extensively in applications to biology and molecular biology. We treat this topic here, under the heading of the linear problem because we shall show its application only in the linear case. Unlike the least squares methods which, as we show in Chapters 5 and 6, have an extensive parallel statistical treatment, Chebyshev minimization does not give rise to known statistical theories, and consequently, has not attained the popularity of the least squares methods.

*Essentially, what the Chebyshev minimax method does is to minimize the maximum error or the maximum difference between the response function and the functions by which we seek to approximate that response function.* For example, suppose we wish to fit a straight line through a set of data $(y_j, x_j)$ where $j = 1, 2, \ldots, m$, Our straight line model is given by $y_j = \alpha_1 + \alpha_2 x_j$. Assuming, as is common in a practical situation, that the set of experimental data will not all fall on the straight line, we draw a straight line through two of the experimental points. Let the equation of this straight line be $\alpha_1^{(1)} + \alpha_2^{(1)} x$. Since all the data are not on this straight line, let the amount by which this straight line misses the $j$th data point be $h_j = \alpha_1^{(1)} + \alpha_2^{(1)} x_j - y_j$. Also, let $H$ be the largest of the set of $|h_j|$.

The Chebyshev or min-max approximation for this problem, is the particular straight line for which $H$ has the smallest value. That is, the straight line that minimized the maximum difference between the response function and the approximating function or functions.

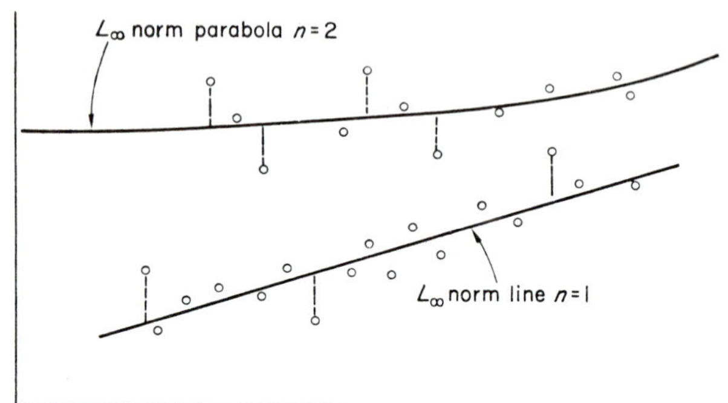


**Fig. 4-4.** Approximation in the $L_\infty$-norm for two sets of data by a straight line $n = 1$ and a parabola $n = 2$.

Since a picture is worth a thousand words, Fig. 4-4 shows a min-max straight line passing through a set of data. In this figure we note that our line has the following characteristics:

a. It "misses" three data points emphasized in the figure by equal quantities.
b. Those three data points are further than any other point from the straight line.
c. If we move along the $x$ axis, each one of those three data points is on opposite sides of the one following it in relation to our straight line.

To pursue this matter of illustration further, suppose our set of data points $(y_j, x_j)$ for $j = 1, 2, \ldots, m$ is to be fitted by the min-max parabola given by $\alpha_1 + \alpha_2 x + a_3 x^2$. Figure 4-4 shows the behavior of such a parabola. What has been said of the min-max straight line above can also be said of the parabola, except that instead of using the phrase "three data points," we must use the phrase "four data points," and so on for higher order polynomials.

## 4.2 Properties of Chebyshev Minimization

All this leads us to the consideration of the properties of min-max polynomials in general, and they may be stated as follows without proof:

1. The min-max approximation exists and is unique.

2. The equal error property is the identifying attribute of the min-max polynomial and can be rigorously stated by the Chebyshev equioscillation theorem.

*Chebyshev Equioscillation Theorem.* Let $f(x)$ (our response function in general) be a member of the set of continuous functions defined on the interval $[a, b]$ and let $p(x)$ be the min-max polynomial approximation of degree $n$ to $f(x)$. Further let

$$p_n = \max_{a \le x \le b} |f(x) - p(x)| \qquad \text{and} \qquad \varepsilon(x) = f(x) - p(x)$$

There are at least $n + 2$ points $a \leqq x_1 < x_2 < \cdots < x_{n+2} \leqq b$ where $\varepsilon(x)$ assumes values $\pm p_n$ and with alternating signs thus

$$\begin{aligned} \varepsilon(x_i) &= \pm p_n & (i &= 1, 2, \ldots, n + 2) \\ \varepsilon(x_i) &= -\varepsilon(x_{i+1}) & (i &= 1, 2, \ldots, n + 1) \end{aligned}$$

What this theorem says is essentially illustrated in Fig. 4-4 for $n = 1$ and $n = 2$.

A min-max fit can also be obtained by a set of arbitrary known functions, and is not restricted to polynomials. These functions, however, must be linearly independent (Rice, 1964).

### 4.3 Computation of the Chebyshev Approximation with an Arbitrary Set of Functions (the Exchange Algorithm)

If we bear in mind that polynomials are really a special case of the arbitrary function then we won't have any difficulty in applying the exchange algorithm to the problem. Suppose we have a set of $m$ data points $f(x_j)$, where $j = 1, 2, \ldots, m$, representing our response function, and we wish to express this function in terms of a linear combination of functions $f_1(x), f_2(x), \ldots, f_n(x)$, which can be written as $\phi(\alpha, x) = \sum_{i=1}^{n} \cdot \alpha_i f_i(x)$, and we wish to determine the set $[\alpha_1, \alpha_2, \ldots, \alpha_n]$ which minimizes the maximum error. We note that $m$ must be greater than $n + 1$. Choosing any arbitrary set of $n + 1$ data points from $m$ data points in the interval in question, we can write the following $n + 1$ equations

$$\begin{aligned} f(x_1) - \alpha_1 f_1(x_1) + \alpha_2 f_2(x_1) + \cdots + \alpha_n f_n(x_1) &= l \\ f(x_2) - \alpha_1 f_1(x_2) + \alpha_2 f_2(x_2) + \cdots + \alpha_n f_n(x_2) &= -l \\ &\vdots \\ f(x_n) - \alpha_1 f_1(x_n) + \alpha_2 f_2(x_n) + \cdots + \alpha_n f_n(x_n) &= l \\ f(x_{n+1}) - \alpha_1 f_1(x_{n+1}) + \alpha_2 f_2(x_{n+1}) + \cdots + \alpha_n f_n(x_{n+1}) &= -l \end{aligned} \tag{4-120}$$

where $l$ is a number to be determined. The Eqs. (4-120) form a set of $n + 1$ simultaneous linear equations which are solved for the parameters $\alpha_1, \alpha_2, \ldots, \alpha_n$ and $l$, by the usual methods. The vector $\boldsymbol{\alpha} = [\alpha_1, \alpha_2, \ldots, \alpha_n]$ and $l$ obtained as the solution of (4-120) may or may not be the solution of the problem. To determine whether or not this set $\boldsymbol{\alpha}$ and $l$ are the solution we now test *all* data points *not* included in the initial $n + 1$ set by means of the following procedure. Let $f(x_k)$ be any data point *not included* in the initial set. If

$$| f(x_k) - \alpha_1 f_1(x_k) + \alpha_2 f_2(x_k) + \cdots + \alpha_n f_n(x_k) | \leq l \quad \text{(4-120a)}$$

for all $x_k$, then the vector $\boldsymbol{\alpha}$ provides the parameters for the Chebyshev approximation of $\phi(\alpha, x)$ to $f(x)$. If on the other hand, for any $f(x_k)$, we have the following

$$| f(x_k) - \alpha_1 f_1(x_k) + \alpha_2 f_2(x_k) + \cdots + \alpha_n f_n(x_n) | > l \quad \text{(4-121)}$$

then this specific $f(x_k)$ is exchanged with any one of $n + 1$ initial data points and the $n + 1$ simultaneous equations (4-120) are now solved again to determine a new vector $\boldsymbol{\alpha}$ and $l$. The same testing procedure is now applied to all data points not included in the present computation until the condition of Eq. (4-120a) is satisfied for all data points. At this stage we have determined the min-max approximation. Intuitively, it may appear that the exchange algorithm is cumbersome and takes a long time to converge. In practice, however, this is not the case and proof of convergence has been shown (Scheid, 1968).

## 5. Linear Programming

Before developing the concepts of linear programming and some other related aspects of function minimization, it will be necessary to extend some of the notions developed in Chapters 2 and 3.

### 5.1 Euclidean $m$-Spaces

In Chapter 2 we noted that a 2-vector represents a point in the plane or in 2-dimensional space. Extending that concept to higher dimensions, we note that a 3-vector will represent a point in 3-dimensional space. When the vectors have more than three components, they represent points in $m$-dimensional space or $m$-space, and the set of all points in $m$-space is denoted by $R^m$. From elementary considerations of analytic geometry, the distance $d_2$ between two points or vectors in the plane $(x_1, y_1)$ and $(x_2, y_2)$ is given by

$$d_2 = [(x_1 - y_1)^2 + (x_2 - y_2)^2]^{1/2} \tag{4-122}$$

and the distance $d_m$ between any two points or vectors in $m$-space is by analogy

$$d_m = \left(\sum_{i=1}^{m} (x_i - y_i)^2\right)^{1/2} \tag{4-123}$$

*Definition.* The Euclidean norm or length of an $m$-vector is defined by

$$\| x \| = (x_1^2 + x_2^2 + \cdots + x_m^2)^{1/2} \tag{4-124}$$

and is seen to be the distance from the origin to the terminal point of the $m$-vector.

## 5.2 Bounded Set

A subset $X$ of $R^m$ is bounded if and only if there exists a real number $M > 0$ such that $\| x \| < M$ for every $x \in X$. For example, in $R^2$ the subset $X$ in Fig. 4-5 consists of all points in the shaded area bounded by a circle which is centered at the origin and is of radius $M$.

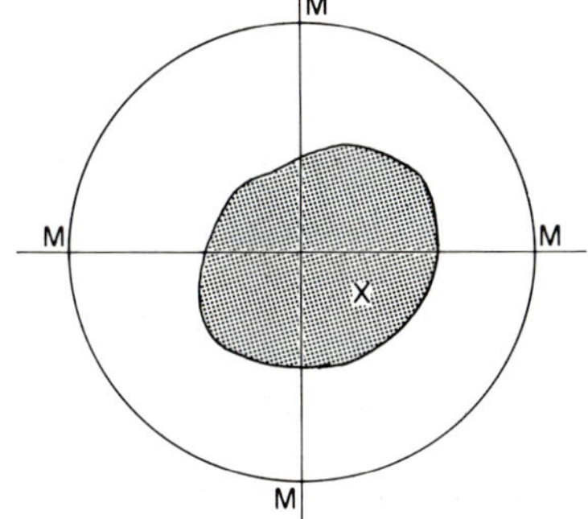


**Fig. 4-5.** Bounded set in $R^2$.

## 5.3 Line Segments and Convex Sets

The line segment joining the points $\mathbf{P}_1(x_1, x_2, \ldots, x_m)$ and $\mathbf{P}_2(y_1, y_2, \ldots, y_m)$ in $R^m$ is denoted by $\overline{\mathbf{P}_1\mathbf{P}_2}$ and is given by the set of points

$$t\mathbf{P}_1 + (1 - t)\mathbf{P}_2 = [tx_1 + (1 - t)y_1, tx_2 + (1 - t)y_2, \ldots, tx_m + (1 - t)y_m] \tag{4-125}$$

where $0 \leq t \leq 1$. Now a subset $X$ of $R^m$ is convex if and only if every line segment joining any two points $\mathbf{P}_1$ and $\mathbf{P}_2$ in the subset is contained in the subset. This implies $\overline{\mathbf{P}_1\mathbf{P}_2} \in X$.

An example of convex sets in the plane is shown in Fig. 4-6. In this figure we also illustrate an important theorem: that the intersection of two convex sets is convex, diagram (d). Diagram (c) of Fig. 4-6 shows a nonconvex set.

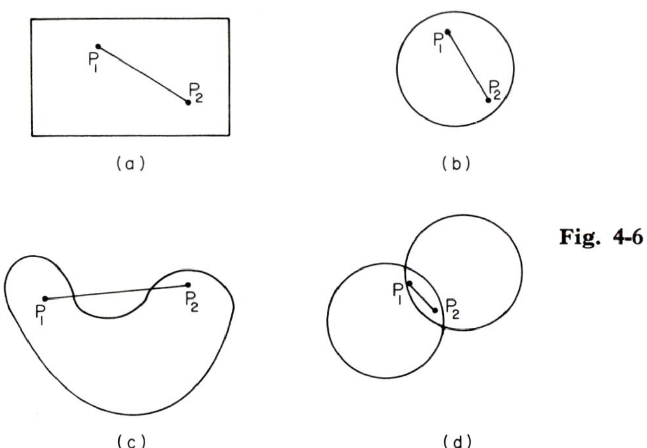


Fig. 4-6

## 5.4 Linear Functions over Convex Sets

*Definition.* A function $f$ defined over $R^m$ is said to be linear in the variables $(x_1, \ldots, x_m)$ if it is of the form

$$f = f(x_1, x_2, \ldots, x_m) = c_1x_1 + c_2x_2 + \cdots + c_mx_m \tag{4-126}$$

where $c_1, c_2, \ldots, c_m$ are constants.

*Theorem.* Let $f$ be a linear function defined over a line segment $\overline{\mathbf{P}_1\mathbf{P}_2}$. Then for every point $\mathbf{A} \in \overline{\mathbf{P}_1\mathbf{P}_2}$ either

$$f(\mathbf{P}_1) \leq f(\mathbf{A}) \leq f(\mathbf{P}_2) \tag{4-127}$$

or

$$f(\mathbf{P}_2) \leq f(\mathbf{A}) \leq f(\mathbf{P}_1) \tag{4-128}$$

An illustration of the above theorem in one dimension can readily be shown. Suppose the linear function is $4x_1$. If $\mathbf{P}_1 = 4$ and $\mathbf{P}_2 = 8$, then as we move on the real line from 4 to 8, we find that for any intermediate point **A**, $f(\mathbf{A})$ increases. On the other hand, if $\mathbf{P}_1 = -4$ and $\mathbf{P}_2 = -8$, we find that for any intermediate point that $f(\mathbf{A})$ decreases.

## 5.5 Planes and Hyperplanes

Consider the linear equation

$$c_1x_1 + c_2x_2 + \cdots + c_mx_m = b \tag{4-129}$$

where $x_1, x_2, \ldots, x_m$ are unknowns and where $c_1, c_2, \ldots, c_m$ and $b$ are constants such that $c_i \neq 0$. In vector notation Eq. (4-129) can be written as

$$\mathbf{c}' \cdot \mathbf{x} = b \tag{4-130}$$

The set of all points in $R^m$ which satisfy Eq. (4-130) are called hyperplanes. In $R^2$ the hyperplanes are straight lines and in $R^3$ the hyperplanes are planes. The equation $\mathbf{c}' \cdot \mathbf{x} = b$ partitions the set $R^m$ into 3 subsets, those being the points on the hyperplane satisfying the equation, points such that $\mathbf{c}' \cdot \mathbf{x} < b$, and points such that $\mathbf{c}' \cdot \mathbf{x} > b$. This is illustrated in $R^2$ space in Fig. 4-7.

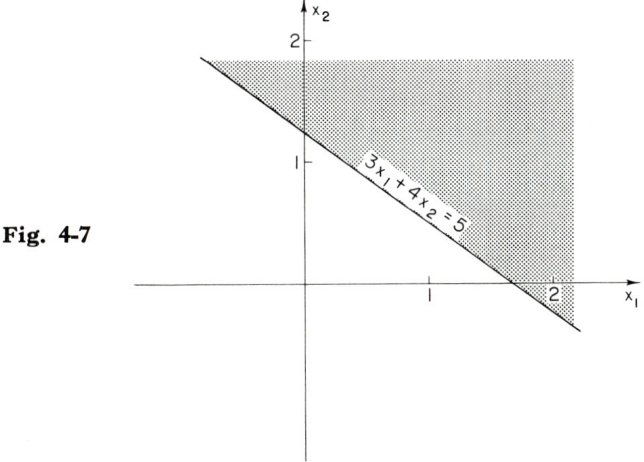

**Fig. 4-7**

Suppose the equation is

$$3x_1 + 4x_2 = 5 \tag{4-131}$$

then the shaded area represents the set of points in $R^2$ satisfying $3x_1 + 4x_2 < 5$ and is called a half space. The straight line satisfies Eq.

(4-131) and the unshaded area satisfies $3x_1 + 4x_2 > 5$ and is also called a half space. Two things can be noted about Eq. (4-131) and half spaces. First, the half spaces are convex, and second, all values satisfying $\mathbf{c}' \cdot \mathbf{x} > b$ and $\mathbf{c}' \cdot \mathbf{x} < b$ lie in the respective half spaces.

## 5.6 Polygonal Convex Sets, Polyhedral Convex Sets, and Extreme Points

The set $X$ which forms the intersection of noncomplimentary half spaces in $R^2$ is a *polygonal convex set*. This set is convex because it is formed by the intersection of half spaces which are convex. A point $\mathbf{P} \in X$ is an extreme point if $\mathbf{P}$ is the intersection of two bounding straight lines. The above ideas are easily extended to higher dimensions, and we say the set $X$ that is the intersection of half spaces in $R^m$ is a *polyhedral convex set*. A point $\mathbf{P} \in X$ is an extreme point if it is the intersection of $m$ bounding hyperplanes.

*Theorem.* The solution of a system of linear inequalities given by $\mathbf{Ax} \leq \mathbf{b}$ is a polyhedral convex set.

We now illustrate some of the above remarks for the two-dimensional case

$$\alpha_1 \leq -1, \qquad \alpha_2 \leq 1, \qquad \alpha_1 + 2\alpha_2 \leq 4, \qquad \alpha_1 - \alpha_2 \leq 4 \qquad (4\text{-}132)$$

The above inequalities form the bounds of a polygon (see Fig. 4-8). The extreme points are given by $\mathbf{A} = (1, -1)$, $\mathbf{B} = (3, -1)$, $\mathbf{C} = (4, 0)$,

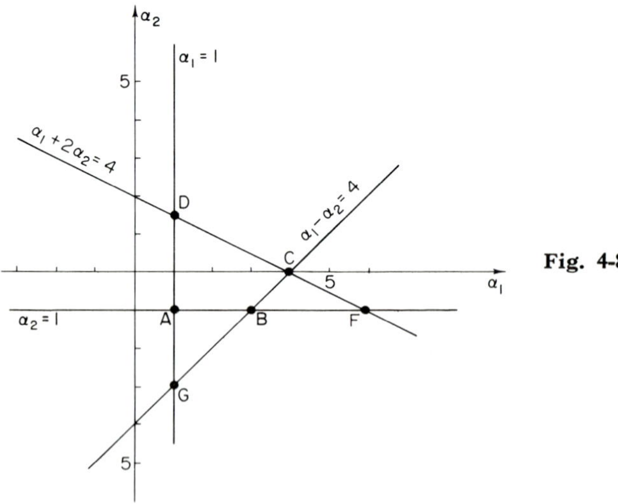


Fig. 4-8

and $\mathbf{D} = (1, \frac{3}{2})$. All the points in the shaded area are consistent with the solutions of the inequalities. The reader will note that there are other intersections between the lines, namely **G** and **F**, but these points are not consistent with the solutions of all the inequalities, as the reader can readily verify. Not all linear inequalities have solutions. For example, the set of inequalities

$$x_1 + 2x_2 \leq -5, \qquad x_1 - x_2 \leq 0, \qquad x_2 \leq -1 \tag{4-133}$$

does not form the bounds of a convex set and therefore, it has no extreme points. Such a set of equations is said to be inconsistent. The reader can readily verify that there are no two values of $x_1$ and $x_2$ that can simultaneously satisfy the inequalities in (4-133).

## 5.7 Linear Functions on Polyhedral Convex Sets

Before proceeding to explain linear programming, one last point needs to be demonstrated. When we seek the maximum or the minimum of a linear function given by $f = \mathbf{c}' \cdot \mathbf{x}$, where the unknowns are subject to the inequalities of the form $\mathbf{Ax} \leq \mathbf{b}$, we are in effect seeking the optimum of the above function over a polyhedral convex set.

*Theorem.* The maximum (and/or) minimum of a linear function defined over a bounded polyhedral convex set is attained at an extreme point.

We consider the geometrical implications of this theorem. Suppose that the linear function is

$$f = 2x_1 + 5x_2 \tag{4-134}$$

defined on the convex polygon given by

$$x_1 + 2x_2 \leq 4, \qquad x_1 - x_2 \leq 4, \qquad x_1 \leq -1, \qquad x_2 \leq 1 \tag{4-135}$$

Figure 4-9 shows the family of parallel lines represented by $2x_1 + 5x_2 = C$, where $C$ is some constant different from line to line. From the diagram we note that the family of lines intersects the convex polygon *first* at point **A**. We also note that the family of lines with large $f$ intersects the convex polygon first at the point **D**. From these considerations we note that the function $2x_1 + 5x_2$ will attain its minimum at **A** and its maximum at **D** or at two extreme points of the convex set $X$.

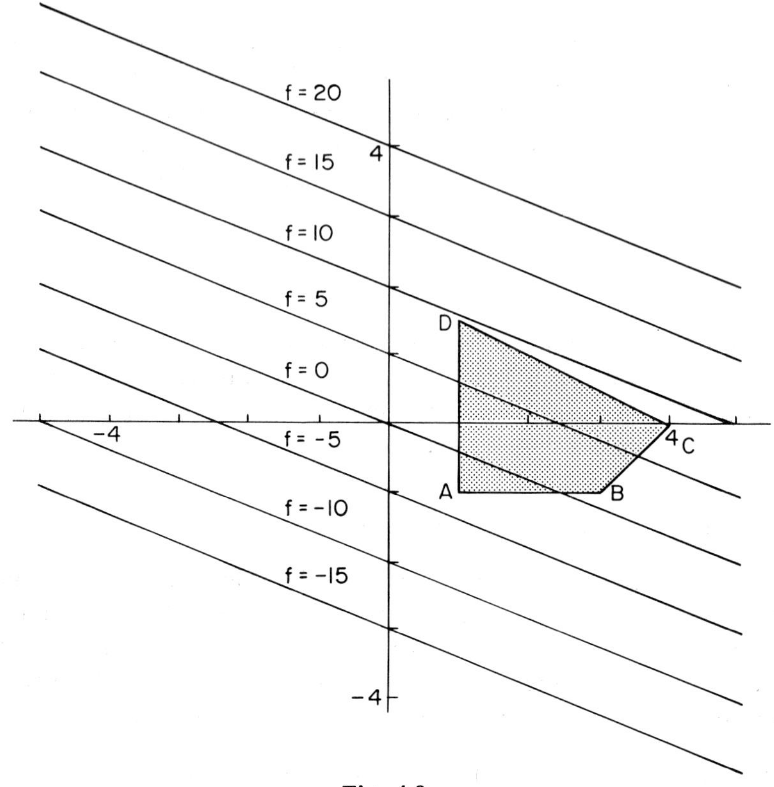


Fig. 4-9

## 5.8 Linear Programming

### 5.8.1 *Statement of Problem*

In general, a linear programming problem requires that we minimize or maximize a linear function, also called an objective function which can be written as

$$H(\alpha) = c_1\alpha_1 + c_2\alpha_2 + \cdots + c_n\alpha_n \tag{4-136}$$

where $c_1, c_2, \ldots, c_n$ are constants. This function will be maximized or minimized subject to the constraints of the following form which can be written in our notation

$$\sum_{k=1}^{n} \alpha_k f_k(x_i) \leq f(x_i) = y_i \tag{4-137}$$

and

$$\alpha_j \geq 0 \tag{4-138}$$

where we have $i = 1, 2, \ldots, m$ and $j = 1, \ldots, n$. In matrix notation we can write the following

$$H(\boldsymbol{\alpha}) = \mathbf{c}'\boldsymbol{\alpha} = \text{minimum} \qquad (\mathbf{X}\boldsymbol{\alpha} \leq \mathbf{Y}, \quad 0 \leq \boldsymbol{\alpha}) \tag{4-139}$$

The above means we have $n + m$ conditions of constraint. Any set of values $[\alpha_1, \alpha_2, \ldots, \alpha_n]$ which satisfy the constraints given by the above equations is called a feasible point (in $n$-dimensional space). An extreme feasible point is one which satisfies the equation when at least $m$ of the constraints become equalities. The fundamental theorem of linear programming is that the minimum occurs at an extreme feasible point.

### 5.8.2 *Slack Variables and Their Relationship to Experimental Error*

To elaborate the linear programming method and to show its connection to our work, we introduce the slack variables $\alpha_{n+1}, \alpha_{n+2}, \ldots, \alpha_{n+m}$ which convert the above inequalities to equalities as follows

$$\begin{aligned} f(x_1) &= \alpha_1 f_1(x_1) + \alpha_2 f_2(x_1) + \alpha_3 f_3(x_1) + \cdots + \alpha_n f_n(x_1) + \alpha_{n+1} \\ f(x_2) &= \alpha_1 f_1(x_2) + \alpha_2 f_2(x_2) + \alpha_3 f_3(x_2) + \cdots + \alpha_n f_n(x_2) + \alpha_{n+2} \\ &\vdots \\ f(x_m) &= \alpha_1 f_1(x_m) + \alpha_2 f_2(x_m) + \alpha_3 f_3(x_m) + \cdots + \alpha_n f_n(x_m) + \alpha_{n+m} \end{aligned} \tag{4-140}$$

The observant reader will immediately realize that if we consider the slack variables $\alpha_{n+1}, \alpha_{n+2}, \ldots, \alpha_{n+m}$ as our $\delta$ or experimental error, and if we equate the constants $c_1, c_2, \ldots, c_n$ to zero in Eq. (4-136) and $c_{n+1}, c_{n+2}, \ldots, c_{n+m}$ to unity then what we seek to minimize is the following modified objective function

$$H(\boldsymbol{\alpha}) = |\, \alpha_{n+1} + \alpha_{n+2} + \cdots + \alpha_{n+m} \,| \tag{4-141}$$

which is equivalent to the sum of the experimental error. We note that the linear programming problem with the slack variables $\alpha_{n+1}, \alpha_{n+2}, \ldots, \alpha_{n+m}$ allows the experimental error to have only one sign, that is $\alpha_{n+1} \geq 0$ for all $i$. This requirement is too restrictive because we would like the experimental error to have both signs. Consequently, we modify our equations such that

$$\begin{aligned} &\alpha_1 f_1(x_1) + \alpha_2 f_2(x_1) + \cdots + \alpha_n f_n(x_1) + \alpha_{n+1} - \alpha_{n+m+1} = f(x_1) \\ &\alpha_1 f_1(x_2) + \alpha_2 f_2(x_2) + \cdots + \alpha_n f_n(x_2) + \alpha_{n+2} - \alpha_{n+m+2} = f(x_2) \\ &\vdots \\ &\alpha_1 f_1(x_m) + \alpha_2 f_2(x_m) + \cdots + \alpha_n f_n(x_m) + \alpha_{n+m} - \alpha_{n+2m} = f(x_m) \end{aligned} \tag{4-142}$$

and our objective function is now given by

$$H(\boldsymbol{\alpha}) = | \alpha_{n+1} + \alpha_{n+2} + \cdots + \alpha_{n+m} + \alpha_{n+m+1} + \cdots + \alpha_{n+2m} | \quad (4\text{-}143)$$

and it is the function that we seek to minimize.

To arrive at the minimum of this function we use a simplex method, which essentially starts us at an extreme feasible point. By exchanging this with other extreme points, we determine the one that is the solution of the problem. After noting the duality theorem we shall demonstrate by example a simple minimization problem using the simplex method.

### 5.8.3 *The Duality Theorem in Linear Programming*

Linear programming is not confined to problems of minimization; problems of maximization can also be handled by it. In fact, the duality theorem which we state shows the close connection.

The two linear programming problems with objective function

$$\mathbf{C}' \cdot \boldsymbol{\alpha} = \text{minimum} \qquad \boldsymbol{\beta}' \cdot \mathbf{b} = \text{maximum}$$

and constraints

$$\boldsymbol{\alpha} \geq 0, \qquad \mathbf{A}\boldsymbol{\alpha} \geq b, \qquad \boldsymbol{\beta} \geq 0, \qquad \boldsymbol{\beta}'\mathbf{A} \leq \mathbf{C}'$$

are called dual problems and the duality theorem states: It one problem has a solution, then so does the other; further, the point where $\mathbf{C}'\boldsymbol{\alpha}$ has a minimum is the *same* as the point where $\boldsymbol{\beta}' \cdot \mathbf{b}$ has a maximum. Here $\mathbf{A}$ is the matrix of our known functions.

As we indicate earlier, the only change in making the computations in one problem or the other is the change in the indicators.

### 5.8.4 *The Simplex Algorithm, Examples of Linear Programming*

Minimize the objective function $H(\boldsymbol{\alpha}) = \alpha_1 - \alpha_2$ subject to the constraints

$$\alpha_1 \geq 0, \qquad \alpha_2 \geq 0, \qquad -\alpha_1 + 2\alpha_2 \leq 2, \qquad \alpha_1 + \alpha_2 \leq 4, \qquad \alpha_1 \leq 3 \quad (4\text{-}144)$$

Since we have 5 conditions of constraint, we will have 5 variables altogether. We therefore introduce the slack variables $\alpha_3$, $\alpha_4$, and $\alpha_5$. Our initial equations with the added slack become

$$-\alpha_1 + 2\alpha_2 + \alpha_3 = 2, \qquad \alpha_1 + \alpha_2 + \alpha_4 = 4, \qquad \alpha_1 + \alpha_5 = 3 \quad (4\text{-}145)$$

Using these equations we make up what is called a simplex tableau or an initial simplex tableau since it is the first one. This is constructed from (a) the coefficients of the variables, (b) the value of the objective function at an extreme feasible point, (c) the coefficients of the variables in the objective function.

An extreme feasible point (a possible solution) for this problem is $\alpha_1 = \alpha_2 = 0$, making $\alpha_3 = 2$, $\alpha_4 = 4$, $\alpha_5 = 3$. The variables which satisfy all the conditions of constraint given above and are not equal to zero are called the *basic variables*. In our case $\alpha_3$, $\alpha_4$, and $\alpha_5$ are the basic variables. The initial tableau may be constructed as follows:

| | $R_1$ | $R_2$ | $R_3$ | $R_4$ | $R_5$ | | |
|---|---|---|---|---|---|---|---|
| $R_3$ | −1 | ② | 1 | 0 | 0 | 2 | |
| $R_4$ | 1 | 1 | 0 | 1 | 0 | 4 | |
| $R_5$ | 1 | 0 | 0 | 0 | 1 | 3 | |
| | −1 | 1 | 0 | 0 | 0 | 0 | ← objective function value |

indicators

The construction of the last row involves steps (b) and (c) above. The value of the objective function at the point we selected is put in the last column, since $\alpha_1 = \alpha_2 = 0$, and $H(\boldsymbol{\alpha}) = 0$. The rest of the last row is made up of *indicators* which are the negatives of the coefficients of the variables occurring in the objective function. Since the coefficient of $\alpha_1$ is 1, the entry in the first column is −1 and the coefficient of $\alpha_2$ is −1, so the entry in the second column is 1 and so on. The rows are labelled with the numbers of the basic variables. One of the central ideas of passing from one simplex tableau to another is to change the values of basic variables until the minimum is reached.

In general, the initial simplex tableau of the following set of equations

$$\begin{aligned} &\alpha_1 x_{11} + \alpha_2 x_{12} + \cdots + \alpha_n x_{1n} + \alpha_{n+1} &&= b_1 \\ &\alpha_1 x_{21} + \alpha_2 x_{22} + \cdots + \alpha_n x_{2m} + \cdots + \alpha_{n+2} &&= b_2 \\ &\vdots \\ &\alpha_1 x_{k1} + \alpha_2 x_{k2} + \cdots + \alpha_n x_{kn} + \cdots + \alpha_{n+k} &&= b_k \end{aligned} \tag{4-146}$$

and objective function

$$H(\alpha) = c_1\alpha_1 + c_2\alpha_2 + \cdots + c_n\alpha_n \tag{4-147}$$

can be constructed as follows:

$$
\begin{array}{lcccccccc|c}
 & R_1 & R_2 & & R_n & R_{n+1} & & & R_{n+k} & \\
R_{n+1} & x_{11} & x_{12} & \cdots & x_{1n} & 1 & 0 & \cdots & 0 & b_1 \\
R_{n+2} & x_{21} & x_{22} & \cdots & x_{2n} & 0 & 1 & \cdots & 0 & b_2 \\
\vdots & & & & & & & & & \\
R_{n+k} & x_{k1} & x_{k2} & \cdots & x_{kn} & 0 & 0 & \cdots & 1 & b_k \\
\hline
 & -c_1 & -c_2 & \cdots & -c_n & 0 & 0 & & 0 & 0 \\
 & \multicolumn{4}{c}{\underbrace{\hphantom{-c_1 \quad -c_2 \quad \cdots \quad -c_n}}_{\text{indicators}}} & & & & &
\end{array}
$$

*provided* the origin is an extreme feasible point, i.e. $\alpha_1 = \alpha_2 = \cdots = \alpha_n = 0$.

When the origin is not an extreme feasible point, certain other measures have to be taken to construct the initial tableau and these will be pointed out in the final example.

Returning to our problem, the second step in the determination of the new simplex tableau is the selection of the *pivot element*. This is done by selecting (in a minimization problem) a column with a positive indicator. In a maximization problem one chooses a column with a negative indicator. Since we are minimizing the objective function we have only one choice as to the selection of columns from our initial simplex tableau. This is the second column. If there had been more than one positive indicator, then the column having the *greater* positive indicator will be chosen. The pivot element is found by (a) dividing the element in the last column by the corresponding element in the pivot column, (b) performing the above only for *positive* elements in the pivot column, and (c) selecting the smallest element.

In our example, since $2/2 < 4/1$, then the pivot or pivot element is 2. Having found the pivot element, we proceed to construct a new simplex tableau. By dividing all elements of the row containing the pivot by the value of the pivot we obtain the first row in the second tableau. This is

$$-\tfrac{1}{2} \quad 1 \quad \tfrac{1}{2} \quad 0 \quad 0 \quad 1$$

The other new rows (for example, the *i*th row) are obtained by adding an appropriate multiple of the row *just obtained* to the element of the *i*th row. The appropriate multiple is chosen such that the other elements in the pivot column become zero. In our case, in order for the element in the pivot column in the second row to become zero, the appropriate multiple is $-1$. This makes the second row of the second tableau element by element

$$(-1)(-\tfrac{1}{2}) + 1 \quad (-1)(1) + 1 \quad (-1)(\tfrac{1}{2}) + 0 \quad (-1)(0) + 1 \quad (-1)(0) + 0 \quad (-1)(1) + 4$$

or

$$\tfrac{3}{2} \quad 0 \quad -\tfrac{1}{2} \quad 1 \quad 0 \quad 3$$

For the third row the appropriate multiple is zero, so that row will be unchanged. The details of this operation element by element are given below for illustration.

$$(0)(-\tfrac{1}{2}) + 1 \quad (0)(1) + 0 \quad (0)(\tfrac{1}{2}) + 0 \quad (0)(0) + 1 \quad (0)(0) + 0 \quad (0)(1) + 3$$

or

$$1 \quad 0 \quad 0 \quad 1 \quad 0 \quad 3$$

For the 4th row the appropriate multiple is $(-1)$, giving element by element

$$(-1)(-\tfrac{1}{2}) - 1 \quad (-1)(1) + 1 \quad (-1)(\tfrac{1}{2}) + 0 \quad (-1)(0) + 0 \quad (-1)(0) + 0 \quad (-1)(1) + 0$$

or

$$-\tfrac{1}{2} \quad 0 \quad -\tfrac{1}{2} \quad 0 \quad 0 \quad -1$$

The entire new simplex tableau becomes

| | $R_1$ | $R_2$ | $R_3$ | $R_4$ | $R_5$ | | |
|---|---|---|---|---|---|---|---|
| $R_2$ | $-\frac{1}{2}$ | 1 | $\frac{1}{2}$ | 0 | 0 | 1 | ← solution point, the other solution point is in the $R_1$ row and is equal to zero |
| $R_4$ | $\frac{1}{2}$ | 0 | $-\frac{1}{2}$ | 1 | 0 | 3 | |
| $R_5$ | 1 | 0 | 0 | 0 | 1 | 3 | |
| | $-\frac{1}{2}$ | 0 | $-\frac{1}{2}$ | 0 | 0 | $-1$ | ← objective function value |

(The $R$ labels have been inserted in the second tableaux to indicate a change from the first tableaux. This change is such that the basic variable 3 has been exchanged for the basic variable 2).

Since there are no more positive indicators, the problem is solved. The value of the objective function is $-1$ and it occurs at the point $\alpha_1 = 0$,

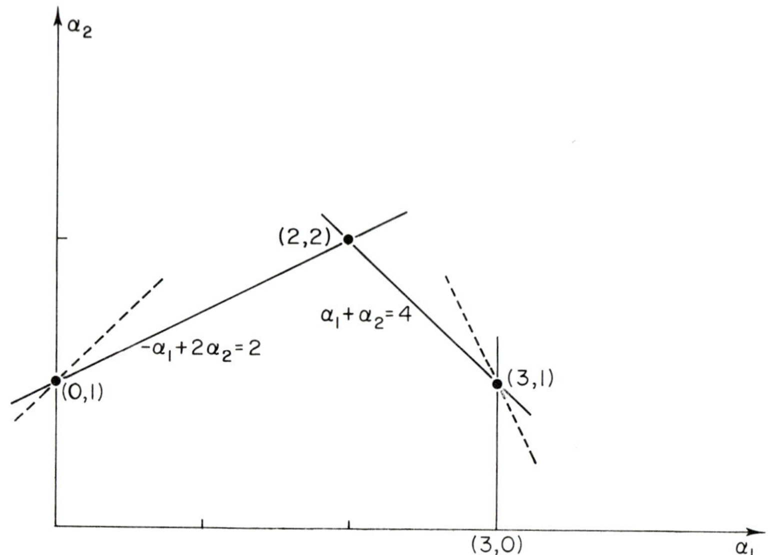


**Fig. 4-10.** Geometry of simplex solution of the linear programming problem.

$\alpha_2 = 1$. Figure 4-10 shows the geometry of the problem. The simplex method for a minimization problem can be summed up as follows: (1) construct the initial simplex tableau, (2) select a column containing a positive indicator, (3) find the pivot in that column, and (4) calculate a new tableau and repeat the process until no other positive indicators are found.

Another example will suffice to demonstrate the method. Our objective function in this case is $H(\boldsymbol{\alpha}) = -3\alpha_1 - \alpha_2$, subject to the constraints

$$-\alpha_1 + 2\alpha_2 \leq 2, \qquad \alpha_1 + \alpha_2 \leq 4, \qquad \alpha_1 \leq 3$$

As before the origin is an extreme feasible point. (See next worked-out example where an extreme feasible point is not obvious.) The initial tableau becomes

| | $R_1$ | $R_2$ | $R_3$ | $R_4$ | $R_5$ | |
|---|---|---|---|---|---|---|
| $R_3$ | $-1$ | 2 | 1 | 0 | 0 | 2 |
| $R_4$ | 1 | 1 | 0 | 1 | 0 | 4 |
| $R_5$ | ① | 0 | 0 | 0 | 1 | 3 |
| | 3 | 1 | 0 | 0 | 0 | 0 |

The pivot element can be chosen from either $R_1$ or $R_2$ since the indicators of those columns are positive. Selecting from $R_1$, the element that becomes the pivot is the one circled, since 3/1 is less than 4/1. The *third* row of the second tableau becomes

$$1 \quad 0 \quad 0 \quad 0 \quad 1 \quad 3$$

The other rows are constructed as follows. For the first row we have, element by element,

$$(1)(1) + (-1) \quad (1)(0) + 2 \quad (1)(0) + 1 \quad (1)(0) + 0 \quad (1)(1) + (0) \quad (3)(1) + 2$$

or

$$0 \quad 2 \quad 1 \quad 0 \quad 1 \quad 5$$

The second row element by element is

$$(-1)(1) + \quad (-1)(0) + 1 \quad (-1)(0) + 0 \quad (-1)(0) + 1 \quad (-1)(1) + 0 \quad (-1)(3) + 4$$

or

$$0 \quad 1 \quad 0 \quad 1 \quad -1 \quad 1$$

The fourth row element by element becomes

$$(-3)(1) + 3 \quad (-3)(0) + 1 \quad (-3)(0) + 0 \quad (-3)(0) + 0 \quad (-3)(1) + 1 \quad (-3)(3) + 0$$

or

$$0 \quad 1 \quad 0 \quad 0 \quad 2 \quad -9$$

So that the new tableau becomes

| | $R_1$ | $R_2$ | $R_3$ | $R_4$ | $R_5$ | |
|---|---|---|---|---|---|---|
| $R_3$ | 0 | 2 | 1 | 0 | 1 | 5 |
| $R_4$ | 0 | (1) | 0 | 4 | −1 | 1 |
| $R_1$ | 1 | 0 | 0 | 0 | 1 | 3 |
| | 0 | 1 | 0 | 0 | −2 | −9 |

The pivot element must now be selected from $R_2$ because it has the only positive indicator. The pivot element here is now circled and the choice of this element is clear, since $3/1 > 5/2 > 1/1$. Carrying out the computations makes our new tableau

| | $R_1$ | $R_2$ | $R_3$ | $R_4$ | $R_5$ | | |
|---|---|---|---|---|---|---|---|
| $R_3$ | 0 | 0 | 1 | $-2$ | 3 | 3 | |
| $R_2$ | 0 | 1 | 0 | 1 | $-1$ | 1 | ← } solution points |
| $R_1$ | 1 | 0 | 0 | 0 | 1 | 3 | ← } solution points |
| | 0 | 0 | 0 | $-1$ | $-1$ | $-10$ | ← objective function |

and since this has no positive indicators, then we have reached the minimum of the function, namely $-10$ at the point $\alpha_1 = 3$ and $\alpha_2 = 1$.

In the last example we will show how we can overcome some complications. In certain instances it is not clear where we have a feasible point. For example, suppose we want to find the minimum of a function

$$H(\boldsymbol{\alpha}) = 2\alpha_1 + 4\alpha_2 + 3\alpha_3 \tag{4-148}$$

subject to the constraints

$$\alpha_1 - \alpha_2 - \alpha_3 \leq 1, \qquad -2\alpha_1 - \alpha_2 \leq -1 \tag{4-149}$$

The origin is not an extreme feasible point. This is readily determined from the following equations

$$\alpha_1 - \alpha_2 - \alpha_3 - \alpha_4 = 1, \qquad -2\alpha_1 - \alpha_2 + \alpha_5 = -1 \tag{4-150}$$

At the origin, $\alpha_1 = \alpha_2 = \alpha_3 = 0$ which makes $\alpha_5 = -1$, which is inconsistent with the constraints that all $\alpha_i \geq 0$.

To overcome this problem we introduce the slack variable $\alpha_6$ so that the second of our equations becomes

$$-2\alpha_1 - \alpha_2 + \alpha_5 - \alpha_6 = -1 \tag{4-151}$$

and we formulate another objective function

$$H = 2\alpha_1 + 4\alpha_2 + 3\alpha_3 + M\alpha_6 \tag{4-152}$$

Here it must be noted that when we change the problem in this manner, $\alpha_6$ must ultimately equal zero so that the objective function is not changed. This means that $M$ has to be a very large positive number. As it stands, Eq. (4-151) is not adequate and $-1$ must be changed to 1 which can be arranged by changing signs. Accordingly, Eq. (4-151) becomes

$$2\alpha_1 + \alpha_2 - \alpha_5 + \alpha_6 = 1 \tag{4-153}$$

We are now ready to start solving our problem.

While it is generally the case that the indicators are the coefficients of the objective function with the sign changed, this will only be true if the origin is an extreme feasible point. When the origin is not an extreme feasible point, a general formula for computing the indicators, labelled as $h_j$ is given below

$$h_j = \sum_{i=1}^{m} x_{ij} c_{n+j} - c_j \tag{4-154}$$

Using this equation gives us our complete initial tableaux

| | $R_1$ | $R_2$ | $R_3$ | $R_4$ | $R_5$ | $R_6$ | |
|---|---|---|---|---|---|---|---|
| $R_4$ | 1 | −1 | −1 | 1 | 0 | 0 | 1 |
| $R_6$ | ② | 1 | 0 | 0 | −1 | 1 | 1 |
| | $2M - 2$ | $M - 4$ | −3 | 0 | $-M$ | 0 | $M$ |

Since $M$ is a very large number, we can choose the elements from $R_1$ or $R_2$. Choosing the circled element from $R_1$ we have the following tableaux

| | $R_1$ | $R_2$ | $R_3$ | $R_4$ | $R_5$ | | |
|---|---|---|---|---|---|---|---|
| $R_4$ | 0 | $-\frac{3}{2}$ | −1 | 1 | $\frac{1}{2}$ | $\frac{1}{2}$ | |
| $R_1$ | 1 | $\frac{1}{2}$ | 0 | 0 | $-\frac{1}{2}$ | $\frac{1}{2}$ | ← solution point |
| | 0 | −3 | −3 | 0 | −1 | 1 | ← objective function value |

where $R_6$ has been dropped because it is of no further interest. Since all the indicators are negative, we have reached the minimum and this minimum is equal to 1. Further, from the last column, this minimum is to be found at the point $\alpha_1 = \frac{1}{2}$, $\alpha_2 = 0$, and $\alpha_3 = 0$.

## 6. Determination of Parameters in Nonlinear Models

### 6.1 Introduction and Plan of Procedure

As we noted in Chapter 1 there are a large number of equations in chemistry, biology, and, in particular, molecular biology in which the parameters occur nonlinearly. For example, we require nonlinear methods to determine parameters of the Gaussian bands in a complex circular dichroism curve, to decompose an optical rotatory dispersion curve into its component circular dichroic bands, to determine the parameters (relaxation times) in rapid reactions, to solve problems of enzyme kinetics and protein ligand equilibria, and for many other similar applications.

Frequently we encounter the problem of fitting a response function by a linear sum of exponentials. We give a bit of extra attention to this kind of analysis because of its frequency of occurrence. However, it must be emphasized that the analysis of a response function in terms of a sum of exponentials falls under the general heading of fitting nonlinear models.

Our procedure in what follows will be first, to describe the notation we use; second, we artificially divide the problem of nonlinear minimization into minimization with constraints and minimization without constraints. Most problems we encounter require constrained minimization of some type. A large number of authors have extended the methods of unconstrained minimization to obtain solutions to problems with constraints. Under the headings of contrained minimization we shall discuss these methods and point out how some authors have incorporated the constraints. Under the heading of constrained minimization we do not treat any specific methods. We only state the problem and delineate the conditions under which solutions can be obtained. The unconstrained nonlinear methods we describe here are the following: methods based on the linearization by a Taylor series (Gauss–Newton); the Newton–Raphson method, the method of steepest descent; a method due to Marquardt based on Levenberg's principle; and a search method due to Nelder and Mead (1965), which is based primarily on a method advocated by Spendley *et al.* (1962).

We describe only the methods mentioned above in detail; however, we note that there is a fairly complete bibliography on the subject up to 1965 in Chapter 10 of Draper and Smith (1966). For later references and descriptions of algorithms we refer the reader to Pierre (1969). While we discuss which method to use under given circumstance later in this chapter we point out the following: It is the task of mathematicians in striving for utmost generality to create the function or functions for which some or all of these methods will fail. We, of course, shall not pursue such esoteric practices here because the experimentalist will already have some functions to be fitted to his data. We rather hope that at least one of the methods given here will be appropriate for his task. It must be remembered that if *only one* method determines the required parameters for the investigator then that is all that is required.

## 6.2 General Description of Nonlinear Models

In this section we describe how we represent our nonlinear models. A general functional equation may be written as

$$y = f(x_1, x_2, \ldots, x_k, \alpha_1, \alpha_2, \ldots, \alpha_n) + \delta \qquad (4\text{-}155)$$

where $y$ in this case is the actual response function and is written as such in order to avoid confusion in the notation and $f(x_1, x_2, \ldots, x_k, \alpha_1, \alpha_2, \ldots, \alpha_n)$ is the function (generally nonlinear in the parameters $\alpha_1, \alpha_2, \ldots, \alpha_n$) by which we seek to characterize the response function. If we have $m$ measurements of the response function of the form

$$(y_j, x_{1j}, x_{2j}, x_{3j}, \ldots, x_{kj}) \qquad (4\text{-}156)$$

for $j = 1, 2, \ldots, m$, then we can write Eq. (4-156) $m$ times as follows

$$\begin{aligned} y_1 &= f(x_{11}, x_{21}, \ldots, x_{k1}, \alpha_1, \alpha_2, \ldots, \alpha_n) + \delta_1 \\ y_2 &= f(x_{12}, x_{22}, \ldots, x_{k2}, \alpha_1, \alpha_2, \ldots, \alpha_n) + \delta_2 \\ &\vdots \\ y_m &= f(x_{1m}, x_{2m}, \ldots, x_{km}, \alpha_1, \alpha_2, \ldots, \alpha_n) + \delta_m \end{aligned} \qquad (4\text{-}157)$$

Equations (4-157) may be written compactly as

$$y_j = f(\mathbf{x}_j, \boldsymbol{\alpha}) + \delta_j \qquad (j = 1, 2, \ldots, m) \qquad (4\text{-}158)$$

For simplicity and to demonstrate how the various symbols correspond to actual quantities, suppose we have a response function represented as a sum of 3 exponentials thus

$$y = \sum_{i=1}^{3} \alpha_i \exp(-t/\varrho_i) \qquad (4\text{-}159)$$

In this case we have six parameters to determine namely $\alpha_1, \alpha_2, \alpha_3$, and $\varrho_1, \varrho_2, \varrho_3$. These are the specific quantities we have labeled above as $\alpha_1, \alpha_2, \ldots, \alpha_6$. In the above case we have only the one variable $t$ which will be labeled as $x_1$. Consequently, if we have $m$ measurements for a response function such as the one shown in Eq. (4-159) we can write the following $m$ equations in the notation of Eq. (4-157) as follow

$$\begin{aligned}
y_1 &= \alpha_1 \exp(-x_{11}/\alpha_4) + \alpha_2 \exp(-x_{11}/\alpha_5) + \alpha_3 \exp(-x_{11}/\alpha_6) \\
y_2 &= \alpha_1 \exp(-x_{12}/\alpha_4) + \alpha_2 \exp(-x_{12}/\alpha_5) + \alpha_3 \exp(-x_{12}/\alpha_6) \\
&\vdots \\
y_m &= \alpha_1 \exp(x_{1m}/\alpha_4) \quad + \alpha_2 \exp(-x_{1m}/\alpha_5) + \alpha_3 \exp(-x_{1m}/\alpha_6)
\end{aligned} \tag{4-160}$$

Another example might be the representation of a response function, say the measured circular dichroism, which we want to analyze in terms of two Gaussian components (see Chapter 1). Thus we write

$$\begin{aligned}
y_1 &= \alpha_1 \exp[-(x_{11} - \alpha_2)^3/\alpha_3^2] + \alpha_4 \exp[-(x_{11} - \alpha_5)^2/\alpha_6^2] \\
y_2 &= \alpha_1 \exp[-(x_{12} - \alpha_2)^2/\alpha_3^2] + \alpha_4 \exp[-(x_{12} - \alpha_5)^2/\alpha_6^2] \\
&\vdots \\
y_m &= \alpha_1 \exp[-(x_{1m} - \alpha_2)^2/\alpha_3^2] + \alpha_4 \exp[-(x_{1m} - \alpha_5)^2/\alpha_6^2]
\end{aligned} \tag{4-161}$$

In this case $y_j$ is the measured circular dichroism at wavelength $j$ and $\alpha_1$, $\alpha_2, \ldots, \alpha_6$ are the parameters to be determined. Once again we have only one independent variable $x_1$ and, as before, $x_{1j}$ is the $j$th wavelength. With these illustrative examples we can now proceed to discuss specific methods.

## 6.3 Unconstrained Minimization

### 6.3.1 *The Newton–Raphson Method*

What we seek to minimize using the Newton–Raphson method is $\sum_{j=1}^{m} \delta_j^2$,[†] which is easily recognized from Eq. (4-157) as a function of the response function, the variables $x_1, x_2, \ldots, x_k$, and the parameters $\alpha_1, \alpha_2, \ldots, \alpha_n$. Let us write this function in general as $F(\mathbf{y}, \mathbf{X}, \boldsymbol{\alpha})$.

This minimization technique essentially approximates the quantity to be minimized by expanding in a Taylor series. We then trucate this expansion after the term involving second derivative and begin an iterative process. This iterative process will require a "first guess" of the answer

[†] In function minimization, in general, we need not necessarily minimize $\sum_{j=1}^{m} \delta_j^2$ (i.e., the sum of squares). The suitability of minimizing the sums of squares as a goodness of fit criteria results from the fact that one can apply in parallel, an extensive statistical theory (see Chapters 5 and 6), and the mathematical formalisms are much easier to handle.

which would be improved by the procedure we describe. We start our iterations at our initial guess vector $\boldsymbol{\alpha}^{(0)} = [\alpha_1^{(0)}, \alpha_2^{(2)}, \ldots, \alpha_n^{(0)}]$ where the superscript denotes the iteration number, and we assume that the required minimum is at $\boldsymbol{\alpha} = [\alpha_1^0 + \varepsilon_1, \alpha_2^0 + \varepsilon_2, \ldots, \alpha_n^0 + \varepsilon_n]$ then we can find the minimum at

$$[\partial F(\mathbf{y}, \mathbf{X}, \boldsymbol{\alpha})/\partial \alpha_i]_{\boldsymbol{\alpha}=\boldsymbol{\alpha}^{(0)}+\boldsymbol{\varepsilon}} = 0 \qquad (i = 1, 2, \ldots, n) \tag{4-162}$$

Expanding Eq. (4-162) in a Taylor series (see Chapter 3) about $\boldsymbol{\alpha}^{(0)}$ we have for the left-hand side of that equation for a given $i$,

$$[\partial F(\mathbf{y}, \mathbf{X}, \boldsymbol{\alpha})/\partial \alpha_i]_{\boldsymbol{\alpha}=\boldsymbol{\alpha}^{(0)}} + \sum_{k=1}^{n} [\partial^2 F(\mathbf{y}, \mathbf{X}, \boldsymbol{\alpha})/\partial \alpha_i \, \partial \alpha_k]_{\boldsymbol{\alpha}=\boldsymbol{\alpha}^{(0)}} \varepsilon_i + O(\varepsilon^2) \tag{4-163}$$

Define the following elements of the vector $\boldsymbol{\omega}$ and the matrix $\boldsymbol{\Omega}$

$$\omega_i = [\partial F(\mathbf{y}, \mathbf{X}, \boldsymbol{\alpha})/\partial \alpha_i]_{\boldsymbol{\alpha}=\boldsymbol{\alpha}^{(0)}} \tag{4-164}$$

and

$$\Omega_{ik} = [\partial F(\mathbf{y}, \mathbf{X}, \boldsymbol{\alpha})/\partial \alpha_i \, \partial \alpha_k]_{\boldsymbol{\alpha}=\boldsymbol{\alpha}^{(0)}} \tag{4-165}$$

so that Eq. (4-162) becomes

$$\boldsymbol{\varepsilon} = -\boldsymbol{\Omega}^{-1}[\boldsymbol{\omega} + O(\varepsilon^2)] \tag{4-166}$$

Neglecting $O(\varepsilon^2)$ Eq. (4-166) becomes

$$\boldsymbol{\varepsilon} = -\boldsymbol{\Omega}^{-1}\boldsymbol{\omega} \tag{4-167}$$

This is a set of linear simultaneous equations in the unknowns $\varepsilon_1, \varepsilon_2, \ldots, \varepsilon_n$ which are the quantities that must be added to our initial guesses in order to obtain the starting vector for our next iteration $\boldsymbol{\alpha}^{(1)} = [\alpha^{(1)}, \alpha_2^{(1)}, \ldots, \alpha_n^{(1)}]$. The Newton–Raphson method will have converged to a solution when

$$\boldsymbol{\alpha}^{(n+1)} = \boldsymbol{\alpha}^{(n)} - \boldsymbol{\Omega}^{-1}\boldsymbol{\omega}^{(n)} \tag{4-168}$$

We observe that

$$\frac{\partial^3 F(\mathbf{y}, \mathbf{X}, \boldsymbol{\alpha})}{\partial \alpha_i^3} = \frac{\partial^3 F(\mathbf{y}, \mathbf{X}, \boldsymbol{\alpha})}{\partial \alpha_i^2 \, \partial \alpha_k} = \frac{\partial^3 F(\mathbf{y}, \mathbf{X}, \boldsymbol{\alpha})}{\partial \alpha_i \, \partial \alpha_k^2} = \frac{\partial^3 F(\mathbf{y}, \mathbf{X}, \boldsymbol{\alpha})}{\partial \alpha_k^3} = 0 \tag{4-169}$$

for the models in which the parameters occur linearly. The significance of (4-169) is that the Newton–Raphson method for a linear problem must converge to the linear least squares solution on the first iteration. Consequently, this method is quite useless if the linear least squares solution does not yield the answer we require. If the reader wishes to work an example using the Newton–Raphson method the following steps are recommended:

1. Select initial guess $\boldsymbol{\alpha}^{(0)}$.
2. Differentiate function with respect to the parameters for each parameter holding the others constant. This yields the elements of $\boldsymbol{\omega}$ which are evaluated at the initial guess or at $\boldsymbol{\alpha}^{(0)}$.
3. Differentiate the function a second time with respect to the parameter. These derivatives evaluated at the initial guess form the elements of the $\boldsymbol{\Omega}$ matrix.
4. Solve the simultaneous equations (4-167) for $\boldsymbol{\varepsilon}$ to obtain the correction for the initial guess.
5. Restart the procedure with corrected vector.

The advantages and disadvantages of this method have been enumerated by Powell (1967). From a good initial guess convergence is quadratic and rapid. On the other hand, a bad initial guess may delay convergence. Another difficulty in the Newton–Raphson method is that $\boldsymbol{\Omega}$ may be both locally or globally singular or very nearly so. To speed up convergence in the nonsingular cases we can use the following iterative formula

$$\boldsymbol{\alpha}^{(n+1)} = \boldsymbol{\alpha}^{(n)} - \Pi \boldsymbol{\Omega}^{-1} \boldsymbol{\omega}^{(n)} \tag{4-170}$$

where $\Pi$ is a parameter to be determined. To avoid the use of the second derivative, the method of steepest descent is an iterative technique that can be used.

### 6.3.2 *The Method of Steepest Descent*

This is another iterative technique and is capable of handling both the linear and nonlinear problem. Once again we minimize the quantity $F(\mathbf{y}, \mathbf{X}, \boldsymbol{\alpha})$. The idea of the method of steepest descent is to move from the starting point of our iterations along the vector $\boldsymbol{\omega}$ whose components are $[\omega_1, \omega_2, \ldots, \omega_n]$ given by Eq. (4-164). Our iterations will be given by

$$\boldsymbol{\alpha}^{(n+1)} = \boldsymbol{\alpha}^{(n)} - \Pi \boldsymbol{\omega}^{(n)} \tag{4-171}$$

where $\Pi$ is a constant so chosen to make $F(\mathbf{y}, \mathbf{X}, \boldsymbol{\alpha})$ a minimum and is determined in a one dimensional minimization problem. The superscripts are the iteration numbers as before. To illustrate how to determine $\Pi$ in a one dimensional problem we recall that our original function $F(\mathbf{y}, \mathbf{X},$

$\boldsymbol{\alpha}$) is a function of $\Pi$. By setting

$$dF(\mathbf{y}, \mathbf{X}, \boldsymbol{\alpha})/d\Pi = 0 \tag{4-172}$$

we get an equation which is solved to obtain $\Pi$. This value of $\Pi$ is now introduced in Eq. (4-171) to obtain $\boldsymbol{\alpha}^{(1)}$ which enables us to compute $\boldsymbol{\omega}^{(1)}$. The entire process is repeated until we see little difference between the $n$th iteration $\boldsymbol{\alpha}^{(n)}$ and $(n + 1)$th iteration in which case the method has converged sufficiently close to a solution.

*Example.* Use the method of steepest descent to minimize $F(\alpha_1, \alpha_2)$ where

$$F(\alpha_1, \alpha_2) = [\alpha_1 - \sin(\alpha_1 + \alpha_2)]^2 + [\alpha_2 - \cos(\alpha_1 - \alpha_2)]^2$$

The method of steepest descent begins with an initial point$(\alpha_1^{(0)}, \alpha_2^{(0)})$. At this point the value of the function decreases most rapidly in the direction of the vector $\boldsymbol{\omega}$ given by

$$\boldsymbol{\omega} = [-\partial F/\partial \alpha_1, -\partial F/\partial \alpha_2]_{\boldsymbol{\alpha}=\boldsymbol{\alpha}^{(0)}}$$

Choosing $\boldsymbol{\alpha}^{(0)} = [0.5, 0.5]$ and noting that

$$\tfrac{1}{2}(\partial F/\partial \alpha_1) = [\alpha_1 - \sin(\alpha_1 + \alpha_2)][1 - \cos(\alpha_1 + \alpha_2)] + [\alpha_2 - \cos(\alpha_1 - \alpha_2)][\sin(\alpha_1 - \alpha_2)]$$

$$\tfrac{1}{2}(\partial F/\partial \alpha_2) = [\alpha_1 - \sin(\alpha_1 + \alpha_2)][-\cos(\alpha_1 + \alpha_2)] + [\alpha_2 - \cos(\alpha_1 - \alpha_2)][1 - \sin(\alpha_1 - \alpha_2)]$$

we find that $-\boldsymbol{\omega}^{(0)} = [0.3, 0.6]$. This establishes the direction where $F(\alpha_1, \alpha_2)$ is decreasing most rapidly at $(\alpha_1^{(0)}, \alpha_2^{(0)})$. The question now is how far in the direction of this vector does the function keep decreasing. This brings us to our algorithm, and

$$\alpha_1^{(1)} = 0.5 + 0.3\Pi, \qquad \alpha_1^{(2)} = 0.5 + 0.6\Pi$$

since $\Pi$ is a constant. We can make

$$\alpha_1^{(1)} = 0.5 + \Pi \qquad \alpha_2^{(1)} = 0.5 + 2\Pi$$

We now have to find the minimum of $F(0.5 + \Pi, 0.5 + 2\Pi)$ with respect to $\Pi$. This can be treated as a one variable minimization problem in $\Pi$ so that we set

$$\partial F/\partial \Pi = 0$$

and attempt to find the value of $\Pi$ which produces this minimum. The reader can verify that $\Pi \cong 0.3$ so that

$$\alpha_1^{(1)} = 0.8 \qquad \text{and} \quad \alpha_2^{(1)} = 1.1$$

These values become the starting points of the next iteration which is

$$-\boldsymbol{\omega}^{(1)} = [0.5, \ -0.25]$$

and

$$\alpha_1^{(2)} = 0.8 + 0.5\Pi, \qquad \alpha_2^{(2)} = 1.1 - 0.25\Pi$$

or

$$\alpha_1^{(2)} = 0.8 + 2\Pi, \qquad \alpha_2^{(2)} = 1.1 - \Pi$$

The value of $\Pi$ that gives the minimum of $F(0.8 + 2\Pi, 1.1 - \Pi)$ is then found and the process is repeated. The reader should continue the process as an exercise. The complete process is summarized in the table. The correct minimum is at $\alpha_1 = 1$, $\alpha_2 = 1$, and there $F = 0$.

| | | | | | | | | | |
|---|---|---|---|---|---|---|---|---|---|
| $\alpha_1^{(0)}$ | 0.5 | $\alpha_1^{(1)}$ | 0.8 | $\alpha_1^{(2)}$ | 0.94 | $\alpha_1^{(3)}$ | 0.936 | $\alpha_1^{(4)}$ | 0.934 |
| $\alpha_2^{(0)}$ | 0.5 | $\alpha_2^{(1)}$ | 1.1 | $\alpha_2^{(2)}$ | 1.006 | $\alpha_2^{(3)}$ | 1.002 | $\alpha_2^{(4)}$ | 0.998 |
| $F_0$ | 0.36 | $F_1$ | 0.04 | $F_2$ | 0.00013 | $F_3$ | 0.00025 | $F_4$ | 0.00002 |

The method of steepest descent may not in general converge rapidly. An algorithm to speed up the rate of convergence was given by Davidon (1959). This algorithm modifies the iterations as follows:

$$\boldsymbol{\alpha}^{(n+1)} = \boldsymbol{\alpha}^{(n)} - \Pi \mathbf{H}^{(n)} \boldsymbol{\omega}^{(n)} \tag{4-173}$$

where $\mathbf{H}^{(n)}$ is a positive definite matrix that changes with each successive iteration. It is chosen in such a way that

$$\boldsymbol{\alpha}^{(n+1)} - \boldsymbol{\alpha}^{(n)} = \mathbf{H}^{(n+1)} \boldsymbol{\omega}^{(n+1)} - \boldsymbol{\omega}^{(n)} \tag{4-174}$$

For the first iteration, however, we start with an arbitrary $\mathbf{H}^{(0)}$.

The relationship of Davidon's method to the Newton–Raphson procedure has been demonstrated by Fletcher and Powell (1963) who have proved that $\mathbf{H}^{(n+1)}$ always remains positive definite and that it converges to $\boldsymbol{\Omega}^{-1}$ evaluated at the minimum.

In actual application the following procedure is recommended and should be used for all iterative techniques. If the parameters are constrained such that $\alpha_i \geq 0$ for all $i$, or in any matter whatsoever, then we accept as our solution the last iteration which yields a solution consistent with those constraints. Further, since the starting point of all our iterative procedures are arbitrary then there is no reason why we cannot replace any parameter by what we feel to be an appropriate value whenever an iteration yields an inadmissible value of such a parameter. We recommend *several* starting points for all iterative techniques and the criteria of termination will depend on the judgment of the investigator.

The method of steepest descent is quite prone to converge to local minima which are not the global minima of the function. Consequently it is wise to start the iterative procedure from several different "initial guess vectors" to guard against the possibility of converging to a false minimum. The importance of good starting estimates cannot be overemphasized.

It is pertinent to point out, however, from the standpoint of our analyses (particularly when the parameters are constrained, to be all positive, for instance) that the mathematical global minimum may give rise to physically unacceptable solutions or solutions which are inconsistent with the constraints on the parameters. In such cases it is incumbent on the investigator to accept the solution located at a "false" minimum but which is consistent with the physical situation.

### 6.3.3 *The Gauss–Newton Procedure*

The theory and procedure of linear regression, largely due to Gauss (1777–1855), when formally extended to nonlinear models by Newton's method of local linearization is known as the Gauss–Newton procedure. It was shown in Section 3.4 that the linear least squares problem always has a solution whenever the matrix $\mathbf{P}'\mathbf{P}$ is well-conditioned. Moreover, this solution is unique. This can be stated, in part, because the matrix $\mathbf{P}$ is independent of the initial parameter estimates, so that whenever $(\mathbf{P}'\mathbf{P})$ is invertible the solution

$$\boldsymbol{\alpha} = (\mathbf{P}'\mathbf{P})^{-1}\mathbf{P}'\mathbf{y} \tag{4-175}$$

is unique. For nonlinear problems this is no longer the case. In fact whenever $\mathbf{y} = F(\boldsymbol{\alpha}, x)$ is a nonlinear function of the parameters $\boldsymbol{\alpha} = (\alpha_1, \alpha_2, \ldots, \alpha_n)$ then $\mathbf{P}$ is *always* a function of the parameters. Thus one cannot even initiate the procedure unless *initial estimates* for the parameters are available. That is, the matrix $\mathbf{P}$ cannot be computed unless initial values are supplied for the model's parameters. We therefore *linearize* the model in a neighborhood of some initial parameter estimate $\boldsymbol{\alpha}^0$ and construct a procedure that will, hopefully, improve the estimate $\boldsymbol{\alpha}^0$ by iteration. We assume the model can be approximated (in a neighborhood of $\boldsymbol{\alpha}^0$) by the Taylor's expansion

$$\mathbf{y} = F(\boldsymbol{\alpha}, x) \simeq F(\boldsymbol{\alpha}^{(1)}, x) + \sum_{j=1}^{n} [\partial F(\boldsymbol{\alpha}^0, x)/\partial \alpha_j]\varepsilon_j \tag{4-176}$$

where $\varepsilon_j = (\alpha_j^1 - \alpha_j^0)$. Then $\mathbf{y}$ is *linear* in the *unknown* parameters $\varepsilon_j$ and the *initial* estimate for these *linear* parameters is always $\boldsymbol{\varepsilon} = \mathbf{0}$. We redefine our "data" vector now as $\mathbf{y}^* = \mathbf{y} - F(\boldsymbol{\alpha}, x)$ and the "linearized" solution becomes

$$\boldsymbol{\varepsilon} = (\mathbf{PP})^{-1}\mathbf{Py}^*. \tag{4-177}$$

New parameter estimates are computed according to

$$\boldsymbol{\alpha}^{(1)} = \boldsymbol{\alpha}^{(0)} + \boldsymbol{\varepsilon} \tag{4-178}$$

and the entire process repeated. In this manner, a sequence of parameter values is constructed $\boldsymbol{\alpha}^{(0)}, \boldsymbol{\alpha}^{(1)}, \boldsymbol{\alpha}^{(2)}, \ldots, \boldsymbol{\alpha}^{(n)}$ which hopefully converges to a least squares minimum for the $\mathbf{y} = F(\boldsymbol{\alpha}, x)$. The following discussion will treat some modifications of the mathematical methodology and algorithms which are based on this concept.

### 6.3.4 *Marquardt's Method Using Levenberg's Principle*

In the previous section Eqs. (4-176)–(4-178) establish a procedure for computing a sequence of parameter values which may approximate the solution of a least squares problem. There are two fundamental mathematical difficulties in carrying out this procedure:

1. The matrix $\mathbf{A} = (\mathbf{P'P})$ may be ill-conditioned; thus, its inversion may not be computationally meaningful.
2. For nonlinear models the sequence of parameter values $\boldsymbol{\alpha}^{(0)}, \boldsymbol{\alpha}^{(1)}, \boldsymbol{\alpha}^{(2)}, \ldots, \boldsymbol{\alpha}^{(n)}$ need not converge to a least squares solution for any initial estimate $\boldsymbol{\alpha}^{(0)}$.

Levenberg (1955) proposed a set of alternatives to circumvent these difficulties:

1. Define a diagonal matrix $\mathbf{D} = (d_{ij})$ such that

$$\begin{aligned} d_{ij} &= a_{ij}^{1/2} \qquad (i = j) \\ &= 0 \qquad (i \neq j) \end{aligned} \tag{4-179}$$

where $(a_{ij}) = \mathbf{A}$.

2. Choose a positive scalar $k$ and let $\mathbf{g} = \mathbf{P}'\mathbf{y}^*$ then set

$$\boldsymbol{\varepsilon} = (\mathbf{A} + k\mathbf{D}^2)^{-1}\mathbf{g}. \tag{4-180}$$

Note that $(\mathbf{A} + k\mathbf{D}^2)^{-1}$ always exists for $k > 0$, and the conditioning can be controlled by the parameter $k$.

We are now concerned with the choice of $k$. Let $\mathbf{B} = \mathbf{D}^{-1}\mathbf{A}\mathbf{D}^{-1}$ then

$$\begin{aligned} b_{ij} &= 1 \qquad (i = j) \\ -1 \leq b_{ij} &\leq 1 \qquad (i \neq j) \end{aligned} \tag{4-181}$$

The matrix $\mathbf{B}$ is commonly called the "correlation matrix," a useful but not strictly accurate description. By substituting $\mathbf{DBD}$ for $\mathbf{A}$ in Eq. (4-180) one obtains

$$\boldsymbol{\varepsilon} = \mathbf{D}^{-1}(\mathbf{B} + k\mathbf{I})^{-1}\mathbf{D}^{-1}\mathbf{g}. \tag{4-182}$$

Eq. (4-182) is the form of Levenberg's principle used by Marquardt (1963). In order to find a natural choice for $k$ we examine the eigenvalues and eigenvectors of the "normalized" matrix $\mathbf{B}$. Let $\boldsymbol{\Lambda}$ and $\mathbf{V}$ represent the matrix of eigenvalues and eigenvectors respectively. That is, $\mathbf{BV} = \mathbf{V\Lambda}$ where $(\lambda_{ii})$ is diagonal, $\lambda_{ii} \geq \lambda_{i+1,i+1}$. Then (4-182) may be rewritten as

$$\boldsymbol{\varepsilon} = \mathbf{D}^{-1}\mathbf{V}(\boldsymbol{\Lambda} + k\mathbf{I})^{-1}\mathbf{V}^{\mathrm{T}}\mathbf{D}^{-1}\mathbf{g}. \tag{4-183}$$

Recalling that the eigenvectors are a basis set for the parameter space, one may interpret Eq. (4-183) geometrically as a series of transformations. The vector $\mathbf{g}$ is normalized to $\mathbf{g}^* = \mathbf{D}^{-1}\mathbf{g}$; rotated to the eigenvector coordinate system by $\mathbf{g}^{**} = \mathbf{V}\mathbf{g}^*$; transformated to a normalized parameter correction vector $\boldsymbol{\delta} = (\boldsymbol{\Lambda} + \boldsymbol{\varepsilon}\mathbf{I})^{-1}\mathbf{g}^{**}$; rotated back to the original coordinate system by $\boldsymbol{\delta}^* = \mathbf{V}\boldsymbol{\delta}$; and finally expressed in unnormalized form as $\boldsymbol{\delta}^{**} = \mathbf{D}^{-1}\boldsymbol{\delta}^*$.

The elements of $\boldsymbol{\Lambda}$ are such that $0 \leq \lambda_{ij} < n$ and $\sum_{j=1}^{m} \lambda_{jj} = n$. A conditioning problem for $\mathbf{B}$ and therefore for $\mathbf{A}$ is indicated whenever any $\lambda_{jj}$ is very small when compared to 1.0. A natural choice for $k$ is then a value which *relieves* the *ill-conditioning produced by the smallest eigenvalue in* $\boldsymbol{\Lambda}$. For purposes of computation one sets a lower limit for eigenvalues in a given problem. When this lower limit is exceeded by an eigenvalue one corrects the entire diagonal (all eigenvalues) by the amount required to bring up the smallest eigenvalue to the lower limit. This procedure may, of course, produce *biased* results, but it often serves to relieve local singularity. That is, a particular location on the nonlinear least squares surface may produce temporary ill-conditioning of the matrix **A**. This correction procedure forces the search path in the parameter space to move away from such locations even when ill-conditioning is present.

The problem of convergence of the sequence of parameter values is more difficult to handle. When the problem is linear and well-conditioned, we have seen that the sequence always converges in one step. That is, the correction vector $\boldsymbol{\varepsilon}$ gives the correct set of parameter values with only a single iteration. In the nonlinear case, the least squares surface may be distorted so that the entire correction vector may not produce an optimum improvement in the sum of squares. One would expect, however, that some vector in the parameter space near the Newton correction would produce an optimum improvement in the sum of squares. We shall show how to generate a search for this vector using the parameter $k$. We use the initial parameter estimate as a local origin and the Gauss–Newton correction as the outer limit of search. For simplicity, we examine this concept geometrically in the orthonormal eigenvector coordinate system.

### 6.3.5 *An Optimum Search Path in the Parameter Space*

Let $x_i$, $i = 1, \ldots, n$, be the normalized orthogonal coordinate variables; then we show the following result: The locus of tangency of the family of spheres centered at the initial parameter estimate and the linearized family of ellipsoids centered at the parameter's corrected value is a curve in the parameter space generated by single variable.

Let $\alpha_i^{(0)}$ be the $i$th component of the current normalized initial parameter estimate and $\alpha_i^{(0)} + \delta_i$ the $i$th normalized component of the linearized parameter estimate and $\alpha^{(0)} + \delta_i$ the $i$th normalized component of the linearized parameter correction, each expressed in the eigenvector coordinate system (see Section 3.4). Then

$$S = \sum_{i=1}^{n} x_i^2 \tag{4-184}$$

is the family of spheres centered at the origin (initial estimate), and

$$E = \sum_{i=1}^{n} \lambda_{ii}(x_i - \delta_i)^2 \tag{4-185}$$

is the family of ellipses centered at the corrected value. At a point of tangency the direction numbers of the two surfaces are proportional.

That is, for a fixed constant $k$,

$$\frac{\partial E/\partial x_i}{\partial S/\partial x_i} = k \qquad \text{for each} \quad i = 1, 2, \ldots, n \tag{4-186}$$

Therefore

$$\lambda_{ii}(x_i - \delta_i)/x_i = k \tag{4-187}$$

and solving for the coordinate variable,

$$x_i = [\lambda_{ii}/(\lambda_{ii} + k)]\, \delta_i \qquad (i = 1, \ldots, n) \tag{4-188}$$

Thus it is clear that the parameter $k$ generates a path $[x_1(k), x_2(k), \ldots, x_n(k)]$, which moves from the origin $\boldsymbol{\alpha}^{(0)}$, $k = +\infty$, to the point $\boldsymbol{\alpha}^{(0)} + \boldsymbol{\varepsilon}$, $k = 0$, in the parameter space. This path is the locus of the tangencies of the hyperspheres about the estimate $\boldsymbol{\alpha}^{(0)}$ to the hyperellipses about the correction $\boldsymbol{\alpha}^{(0)} + \boldsymbol{\varepsilon}$.

Searching parametrically along the path just described is equivalent to choosing that $\mathbf{x}(k)$ which minimizes $F(\boldsymbol{\alpha})$ on and within a sphere of radius $\| \mathbf{x}(k) \|$. Examining trial values along the path is in effect testing the limits of the locally linear assumption given by Eq. (4-176). Thus it provides an optimum search path in the parameter space for reducing the sum of squares with the *Gauss–Newton procedure*. If we write Eq. (4-182) in vector matrix form, we obtain

$$\mathbf{x}(k) = (\boldsymbol{\Lambda} + k\mathbf{I})^{-1}\boldsymbol{\Lambda}\boldsymbol{\varepsilon}_n$$

where the subscript $n$ denotes a normalized variable. Since the normalized correction vector is related to the vector $\mathbf{g}$ by the transformation $\boldsymbol{\varepsilon}_n = \boldsymbol{\Lambda}^{-1}\mathbf{V}'\mathbf{D}^{-1}\mathbf{g}$ then $\boldsymbol{\Lambda}\boldsymbol{\varepsilon}_n = \mathbf{V}'\mathbf{D}^{-1}\mathbf{g}$ and

$$\mathbf{x}(k) = (\boldsymbol{\Lambda} + k\mathbf{I})^{-1}\mathbf{V}'\mathbf{D}^{-1}\mathbf{g} \tag{4-189}$$

By comparing this equation to Eq. (4-182), one finds that each trial correction vector along the search path can be generated as a function of $k$ according to

$$\boldsymbol{\delta}(k) = \mathbf{D}^{-1}\mathbf{V}(\boldsymbol{\Lambda} + k\mathbf{I})^{-1}\mathbf{V}'\mathbf{D}^{-1}\mathbf{g} \tag{4-190}$$

We have therefore arrived at a search path which not only *relieves* the problem of *ill-conditioning*, but generates the *optimum* linear corrections for least squares problems.

It is instructive to compare this method with the well known steepest descent and unmodified Gauss–Newton correction procedures. We show

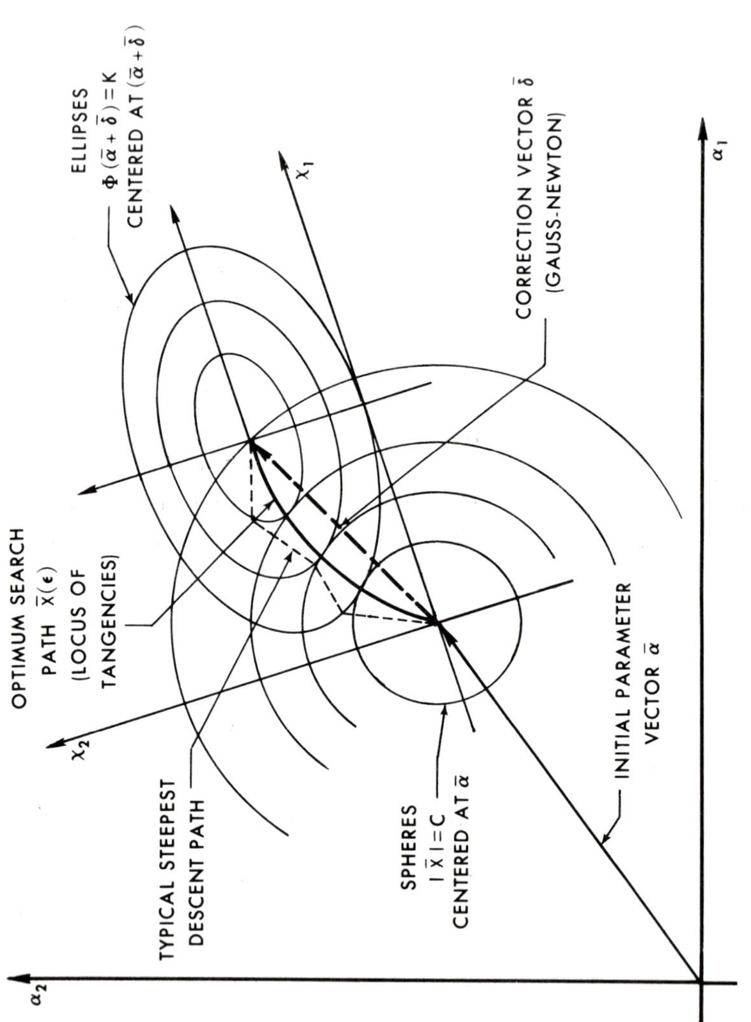


**Fig. 4-11.** Optimum search path in parameter space (Fletcher and Shrager, 1968).

that the direction of the search path at the origin ($k = +\infty$) is that of the antigradient $-\mathbf{g}$, and the correction vector reduces to the Gauss–Newton correction for $k = 0$. Since

$$\partial x_i(k)/\partial k = -\lambda_{ii}\,\delta_{in}/(\lambda_{ii} + k)^2 \tag{4-191}$$

then

$$\lim_{k\to\infty} [\partial x_i(k)/\partial k] = -\lambda_{ii}\varepsilon_{in} \tag{4-192}$$

or in vector form

$$d\mathbf{x}(k)/dk\,|_{k=+\infty} = -\mathbf{V}'\mathbf{D}^{-1}\mathbf{g}$$

Thus

$$d\,\boldsymbol{\delta}(k)/dk\,|_{k=+\infty} = -\mathbf{g} \tag{4-193}$$

The steepest descent method is known to be a very stable but slow method for reducing the sum of squares. We have thus shown that as $k$ increases the method used here approaches the steepest descent method for computing least squares solutions. On the other hand, for $k = 0$, we obtain the Gauss–Newton correction given previously in Eq. (4-177). This is easily seen by setting $k = 0$ in Eq. (4-189). The direction cosines of the path in this case are:

$$\partial x_i/\partial k\,|_{k=0} = \lambda_{ii}\varepsilon_{in}/(\lambda_{ii})^2$$

or unnormalized

$$\partial \varepsilon_i/\partial k\,|_{k=0} = -g_i/(\lambda_{ii})^2 \tag{4-194}$$

It is thus clear from Eqs. (4-194) how disastrous the effect of a small eigenvalue can be. The search path is greatly distorted by the very small divisor and the process may converge very slowly, if at all.

It is now clear that this method can avoid both the disadvantages of the gradient method (i.e., slow convergence), and of the Gauss–Newton method (i.e., ill-conditioning) while it retains the desirable properties of each (i.e., stability and rapid convergence). The correction vector $\varepsilon(k)$ is a monotone decreasing function of $k$, that is, increasing $k$ moves the correction vector $\boldsymbol{\varepsilon}$ along the optimum path from the Gauss–Newton correction *back* to the origin (starting estimate) (see Fig. 4-11). Using the sum of squares as a criterion for increasing $k$ one simply increases $k$ monotonically as long as the sum of squares decreases. In this way convergence of the sequence of iterates is assured.

In summary, we have a stable mathematical method for solving nonlinear least squares problems. The method avoids the problem of ill-conditioning, and whenever good initial estimates for the nonlinear parameters are provided, it converges to the local minimum.

### 6.3.6 *The Nelder and Mead Method*

This method is essentially a search method. Our explanation will also give some idea of what is involved in minimization problems. Suppose we wish to find the minimum of a function of one parameter $\alpha_1$. In the notation of this section this would be equivalent to finding the minimum of the function $F(\mathbf{y}, \mathbf{X}, \alpha_1)$, where $\alpha_1$ is the only unknown.

It is clear that by inserting different values of $\alpha_1$ in $F(\mathbf{y}, \mathbf{X}, \alpha_1)$ we can compute the value of this function at each value of $\alpha_1$ since all other quantities are known. We can keep doing that until we arrive at the value of $\alpha_1$ which gives us the minimum value of $F(\mathbf{y}, \mathbf{X}, \alpha_1)$. This is equivalent to moving along the $\alpha_1$ axis in Fig. 4-12 until we arrive at the point $\alpha_{1s_{\min}}$. Now if our search procedure is to evaluate $F(\mathbf{y}, \mathbf{X}, \alpha_1)$ at every point on the $\alpha_1$ axis, we would take an infinite amount of time before we would arrive at the minimum, since there are an infinity of points on the $\alpha_1$ axis. We must, therefore, move in finite steps.

An efficient search method should tell us how to move or what steps to take. For instance, in Fig. 4-12 if we move along the $\alpha_1$ axis in the

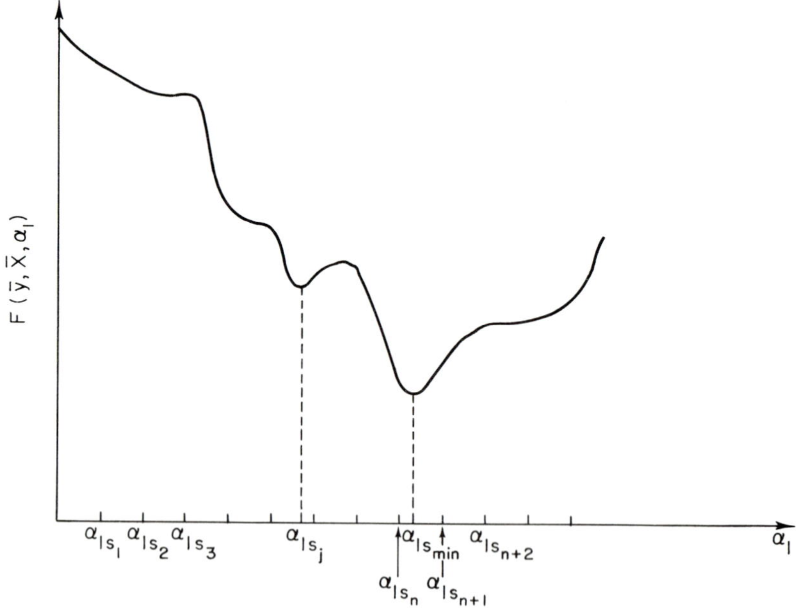


**Fig. 4-12.** Search method for determination of minimum of function $F(\mathbf{y}, \mathbf{X}, \alpha_1)$.

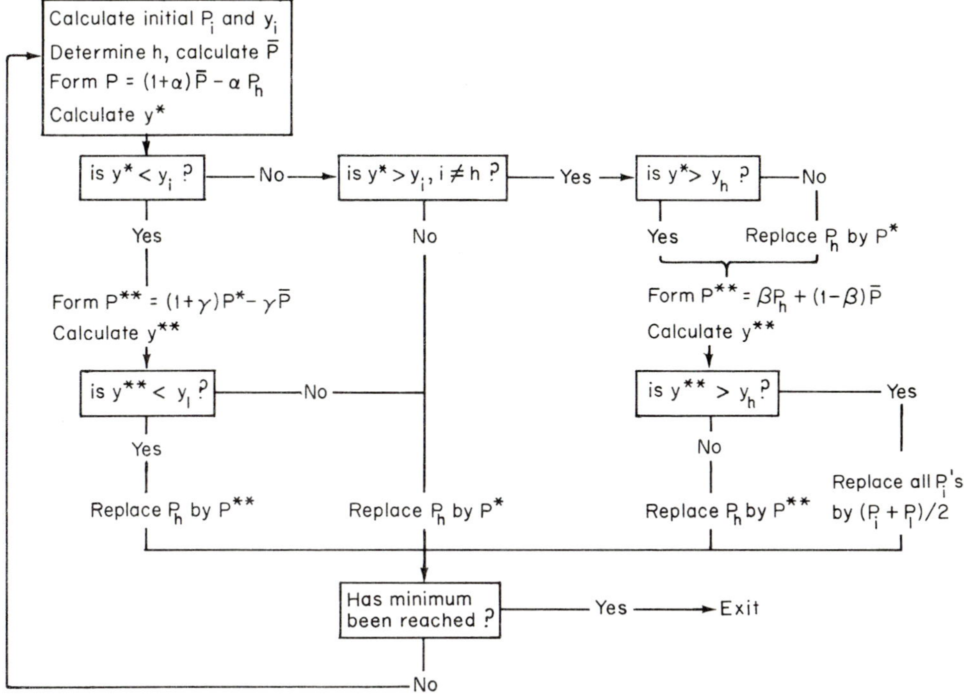


**Fig. 4-13.** Flow diagram of the Nelder and Mead simplex method. For simplicity in this flow diagram the $y_i$ is the value of the function $F(\mathbf{y}, \mathbf{X}, \alpha)$ at $P_i$ and similarily for other subscripts the $y$. The definition of symbols are as follows. $P_i$ point at $i$th vertex of simplex, $y_i$ value of function $F(\mathbf{y}, \mathbf{X}, \alpha)$ at point $P_i$, $y_h$ is the maximum value of function $F(\mathbf{y}, \mathbf{X}, \alpha)$ and is evaluated at $P_h$, $\bar{P}$ is the centroid of the points with $i \neq h$. The reflection of $P_h$ is $P^*$ and is defined by the relation $P^* = (1 + \alpha)\bar{P} - \alpha P_h$, where $\alpha$ is a positive constant, such that $[P^*\bar{P}] = \alpha[P\bar{P}]$, where the square brackets denote the distance between the two points in question. The value of the function at $P^*$ is $y^*$. The operation of expansion of $P^*$ to $P^{**}$ is defined by $P^{**} = \gamma P^* + (1 - \gamma)\bar{P}$, where $\gamma$ is greater than unity and is the ratio of $[P^{**}\bar{P}]$ to $[P^*\bar{P}]$. The last operation defined in this flow diagram is a contraction and is defined by $P^{**} = \beta P + (1 - \beta)\bar{P}$, where $\beta$ lies between **0** and **1** is the ratio between the distance $[P^{**}\bar{P}]$ to $[P\bar{P}]$. Lastly $P_e$ is the point where the function has the lowest value. To visualize the procedure in this flow diagram it is best to consider your simplex as a triangle (a two-dimensional simplex) covering the $xy$ plane and the function values defined on the $z$ axis.

steps indicated $\alpha_{1s_1}, \alpha_{1s_2}, \ldots, \alpha_{1s_{n+1}}, \alpha_{1s_n}, \alpha_{1s_{n+1}}$ we find that at $\alpha_{1s_n}$ the value of the function is smaller than any of its previous values $\alpha_{1s_1}$, $\alpha_{1s_2}, \ldots$. However, at $\alpha_{1s_{n+1}}$ the value of the function is now greater than its value at $\alpha_{1s_n}$ so that somewhere between $\alpha_{1s_n}$ and $\alpha_{1s_{n+1}}$ on the $\alpha_1$ axis, we may expected to find a minimum of the function. It is easy to see that if we had taken different steps, we might have found our minimum at $\alpha_{1s_j}$; but, as one can see from the figure, that is not really the minimum of the function. Efficient search techniques are designed to arrive as rapidly as possible at the minima.

The method of Nelder and Mead can minimize a function of several variables. This is done by evaluating the function at the vertices of a regular simplex. A regular simplex in $n$-dimensional space has $(n+1)$ vertices and edges of equal length. In one dimension this is a straight line, in two dimension it is an equilateral triangle and in three dimensions it is a regular tetrahedron and so on. In the search procedure of Nelder and Mead the vertex at which the function has the highest value is found and replaced by another point which is its reflextion in the hyperplane of the simplex. Following this procedure it is found that the simplex "contracts" to a minimum. The search procedure is best outlined by the flow diagram given in Fig. 4-13. Each operation is defined in the figure legend. Nelder and Mead also supply a procedure for the estimation of the Hessian matrix in the neighborhood of the minimum. The inverse of the Hessian matrix under certain conditions is the variance–covariance matrix of the estimated parameters.

## 6.4 Constrained Minimization

A great deal of what is said subsequently can be followed with greater ease if Section 4 has been carefully read.

### 6.4.1 *Convex Functions*

A function $F(\boldsymbol{\alpha})$, $\boldsymbol{\alpha}' = [\alpha_1, \ldots, \alpha_n]$, defined on $R^m$ is called convex on the set $S$ if for any two points $\boldsymbol{\alpha}^1$ and $\boldsymbol{\alpha}^2$ of $S$ we have

$$F(t\boldsymbol{\alpha}^1 + (1-t)\boldsymbol{\alpha}^2) \leq tF(\boldsymbol{\alpha}^1) + (1-t)F(\boldsymbol{\alpha}^2) \qquad (0 < t < 1) \qquad (4\text{-}195)$$

The function is strictly convex if $\leq$ can be replaced by $<$. When the convex functions are differentiable functions, we have

1. $$F(\boldsymbol{\alpha}^2) - F(\boldsymbol{\alpha}^1) \geq (\boldsymbol{\alpha}^2 - \boldsymbol{\alpha}^1)\left[\frac{\partial F}{\partial \boldsymbol{\alpha}}\right]_{\boldsymbol{\alpha}^1} \qquad (4\text{-}196)$$

for all $\boldsymbol{\alpha}^1$ and $\boldsymbol{\alpha}^2$.

2. The Hessian of the function is positive semidefinite (positive definite) if the function is convex (strictly convex).

3. For any matrix **A** which is positive semidefinite (positive definite) the function

$$P(\boldsymbol{\alpha}) = \boldsymbol{\alpha}'\mathbf{A}\boldsymbol{\alpha} \tag{4-197}$$

is convex (strictly convex).

### 6.4.2 *Statement of the Problem and Definitions*

Let $F(\boldsymbol{\alpha})$ and $f_j(\boldsymbol{\alpha})$, $j = 1, 2, \ldots, k$, be convex functions of $n$ variables $\alpha_1, \alpha_2, \ldots, \alpha_n$. We seek to minimize $F(\boldsymbol{\alpha})$ subject to the constraints that

$$f_j(\boldsymbol{\alpha}) \leq 0 \qquad (j = 1, 2, \ldots, k)$$

and

$$\alpha_1 \geq 0, \qquad \alpha_2 \geq 0, \qquad \ldots, \qquad \alpha_n \geq 0 \tag{4-198}$$

As in linear programming $F(\boldsymbol{\alpha})$ is called the objective function. In two dimensions, the search for the function minimum will be conducted in the domain shown in Fig. 4-14. In this figure we note that $f_j(\boldsymbol{\alpha})$ are

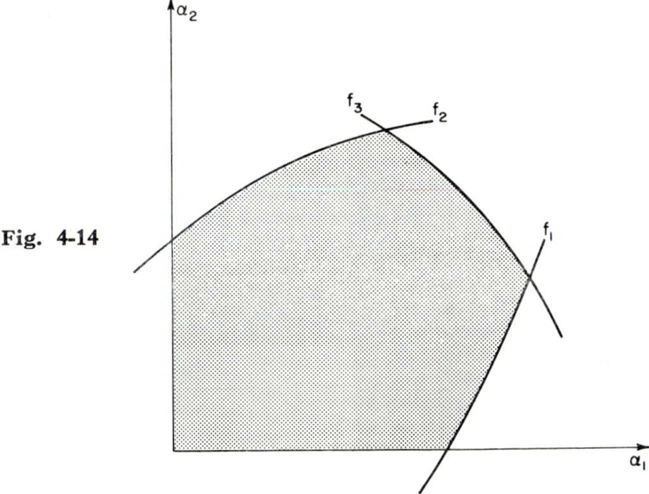


Fig. 4-14

not straight lines but curves. A point $(\alpha_1, \alpha_2)$ satisfying all the constraints is a feasible point. The set of all feasible points is a feasible domain $D$. Since the $f_j(\boldsymbol{\alpha})$ are convex, the domain $D$ is also convex. Two things must be noted. First, when the domain is empty then there is no solution, and second in order to find a minimum for $f(\boldsymbol{\alpha})$, it must be bounded on $D$. Due to these conditions we can show that if $D$ is bounded and not empty, at least one solution exists.

### 6.4.3 *Conditions of Solution. The Kuhn–Tucker Theorem and Its Implications*

As indicated earlier the problem of constrained minimization requires the introduction of the Lagrangian multipliers. As we stated the problem, we have $k$ constraints and we introduce the vector $\boldsymbol{\lambda}' = [\lambda_1, \lambda_2, \ldots, \lambda_k]$ of Lagrangian multipliers to form the function

$$G(\boldsymbol{\alpha}, \boldsymbol{\lambda}) = F(\boldsymbol{\alpha}) + \sum_{j=1}^{k} \lambda_j f_j(\boldsymbol{\alpha}) \qquad (4\text{-}199)$$

According to the Kuhn–Tucker theorem, a vector $\hat{\boldsymbol{\alpha}}$ is a solution to the problem stated in Section 6-4.2 if and only if there exists a vector $\hat{\boldsymbol{\lambda}}$ such that the following conditions are satisfied:

$$\hat{\boldsymbol{\alpha}} \geq 0, \qquad \hat{\boldsymbol{\lambda}} \geq 0, \qquad G(\hat{\boldsymbol{\alpha}}, \boldsymbol{\lambda}) \leq G(\boldsymbol{\alpha}, \boldsymbol{\lambda}) \leq G(\boldsymbol{\alpha}, \hat{\boldsymbol{\lambda}}) \qquad (4\text{-}200)$$

for all $\boldsymbol{\alpha} \geq 0$ and $\boldsymbol{\lambda} \geq 0$.

If $F(\boldsymbol{\alpha})$ and the $f_j(\boldsymbol{\alpha})$ are differentiable functions, then it can be shown that the above conditions are equivalent to

$$\begin{aligned} &[\partial G/\partial \alpha_i]\,|_{\hat{\boldsymbol{\alpha}},\hat{\boldsymbol{\lambda}}} \geq 0, \qquad \hat{\alpha}_i[\partial G/\partial \alpha_i]\,|_{\hat{\boldsymbol{\alpha}},\hat{\boldsymbol{\lambda}}} \geq 0 \qquad (\alpha_i \geq 0) \\ &[\partial G/\partial \lambda_j]\,|_{\hat{\boldsymbol{\alpha}},\hat{\boldsymbol{\lambda}}} \leq 0, \qquad \hat{\lambda}_j[\partial G/\partial \lambda_j]\,|_{\hat{\boldsymbol{\alpha}},\hat{\boldsymbol{\lambda}}} = 0 \qquad (\hat{\lambda}_j \geq 0) \end{aligned} \qquad (4\text{-}201)$$

In attacking a problem of constrained minimization, not all the conditions in the above equations are imposed. If we relax some conditions, we obtain the following results: If a constraint $\alpha_i \geq 0$ is missing, then the three conditions imposed by Eq. (4-200) are replaced by

$$[\partial G/\partial \alpha]\,|_{\hat{\boldsymbol{\alpha}},\hat{\boldsymbol{\lambda}}} = 0 \qquad (4\text{-}202)$$

In the case where $f_j(\alpha)$ is linear, then the conditions of Eqs. (4-200) are simplified to

$$[\partial G/\partial \lambda_j]\,|_{\hat{\boldsymbol{\alpha}},\hat{\boldsymbol{\lambda}}} = 0 \qquad (4\text{-}203)$$

The above conditions only tell us when we have arrived at the required minimum. In order to get to that minimum, the production of an efficient algorithm (a subject of current intensive investigation) is required. For some of these algorithms, we refer the reader to Beale (1955), Wolfe (1960, 1962), Shrager (1970) Kunzi and Krelle (1962), Rosen (1960, 1961), and Zoutendijk (1957).

## 6.5 Fitting a Response Function by a Series of Exponentials (Special Methods)

Two things cannot be overemphasized in discussing this topic, first, its importance, and second, that it is possible to handle it by the nonlinear techniques previously described. An equation such as (4-204) shows the type of approximation that we require for the response function. As before, we seek to determine the parameters $A_0, A_1, \ldots, A_n, \varrho_1, \varrho_2, \ldots, \varrho_n$. Familiar semilogarithmic plots have been used to analyze such data Prony's algebraic method which we now describe.

### 6.5.1 *Prony's Method*

Strictly speaking this is not a technique of function minimization, but is a solution for the unknowns $A_0, A_1, \ldots, A_n, \varrho_1, \varrho_2, \ldots, \varrho_n$ in an equation of the form

$$f(x_j) = A_0 e^{\varrho_0 x_j} + A_1 e^{\varrho_1 x_j} + \cdots + A_{n-1} e^{\varrho_{n-1} x_j} \tag{4-204}$$

where the data are *equally spaced* values of $x_j$, $i = 1, 2, \ldots, m$. It is evident that if one wishes to use Prony's method, one must have extremely precise data since we are solving an exact system of equations. Consequently, unless such precise data is available, Prony's method will be quite unsuitable.

As for the method itself Prony noted that each $\exp(\varrho_i x)$ for $i = 1, 2, \ldots, n-1$ is the solution an $n$th order difference equation† with constant coefficients and whose characteristic roots are

$$S = e^{\varrho_i} \tag{4-205}$$

Consequently, the function $f(x)$ must also satisfy this difference equation,

† A difference equation is one involving finite differences. To demonstrate a finite difference suppose we are given a discrete function at a finite set of arguments $x_1, x_2, \ldots, x_n$, and the value of the function at those arguments is $y_1, y_2, \ldots, y_n$. Suppose further that the arguments are equally spaced so that for any $x_i$, $x_{i+1} - x_i = n$. The first differences of $y_i$ values are given by $\Delta y_i = y_{i+1} - y_i$. The second difference is

$$\Delta^2 y_i = \Delta(\Delta y_i) = y_{i+2} - 2y_i + y_i$$

and so on for other differences. An example of a difference equation would be

$$\Delta^2 y_i + 2\,\Delta y_i + y_i = 0$$

and a solution of a difference equation is a sequence of $y_i$ values which obey the difference equation for a set of consecutive integers $i$.

which may be written as,

$$f(j) + C_1 f(j+1) + \cdots + C_n f(j+n) = 0 \qquad (j = 0, 1, 2, \ldots, m) \tag{4-206}$$

where $C_1, C_2, \ldots, C_n$ are unknown constants. We know that Eq. (4-206) is a simultaneous linear equation in the unknowns $C_1, C_2, \ldots, C_n$, consequently when $m = n$ we can solve Eq. (4-206) to obtain the coefficients $C_1, C_2, \ldots, C_n$. Alternatively when $m > n$ we can use linear least squares to obtain a solution for $C_1, C_2, \ldots, C_n$ and in this respect, we are using function minimization. Once the set of $C_1, C_2, \ldots, C_n$ are determined we replace them in the polynomial equation

$$S^n + C_1 S^{n-1} + C_2 S^{n-2} + \cdots + C_n = 0 \tag{4-207}$$

This equation is solved for the $n$ roots of $S$ from Eq. (4-205) we obtain the $\varrho_i$. The known values of the $\varrho_i$ are then substituted into the original Eq. (4-206) which provide $m$ equations in $n$ unknowns $A_1, A_2, \ldots, A_n$ which can be solved by the least squares method.

### 6.5.2 *Other Methods and References*

Several additional methods, graphical and other types, have been proposed for the determination of the parameters. Discussion of various aspects of specific problems in the use of sums of exponentials can be found in Berman and Schoenfeld (1956).

### 6.5.3 *Use of Nonlinear Techniques for Fitting Exponentials and a Word of Caution*

All the nonlinear techniques described in this chapter can, in principle, be applied to determine the parameters occurring in an equation which is the sum of exponentials. For investigators fitting a response function by a sum of exponentials, a word of caution is necessary. While the methods described herein can in principle determine the parameters, in actual practice there is some difficulty. This is illustrated by a simple example given by Hamming (1962). Suppose we wish to distinguish between a single term such as $Ae^{-\varrho x}$ and two terms such as $(A/2)e^{-(\varrho+\varepsilon)x} + (A/2)e^{-(\varrho-\varepsilon)x}$. We note that

$$(A/2)e^{-(\varrho+\varepsilon)x} + (A/2)e^{-(\varrho-\varepsilon)x} = A[e^{-\varepsilon x} + e^{-\varepsilon x}/2]e^{-\varrho x} \tag{4-208}$$

The quantity in bracket is the hyperbolic cosine of $\varepsilon x$ and has a series representation of

$$\cosh(\varepsilon x) = 1 + (\varepsilon^2 x^2/2) + (\varepsilon^4 x^4/4) + \cdots \tag{4-209}$$

Consequently the difference between $Ae^{-\varrho x}$ and the sum in expression (4-208) will depend on $\varepsilon^2$, and it is only when $x$ is large can we hope to detect this difference. However, at large $x$ the quantity $e^{-\varrho x}$ is small, thus neutralizing our ability to detect the difference.

What this example shows is that several sums of exponentials may be consistent with the same set of data, and the question of uniqueness of representation is left open in many applications.

## 7. Application

This section is written particularly for those who have no desire to go through the mathematics. In the previous section we have taken the trouble to explain the mathematics not only for the satisfaction of those people who like to know how everything works, but also in the belief that "It is rather unlikely that great physical insight will arise in the mind of a professional coder (programmer); if insights are to arise, and they are what we most want, then it follows that the man (or woman) with the problem must comprehend and follow the computing" Hamming (1962). Stated differently, "The purpose of computation is insight, not numbers."

Undoubtedly there are those who will not agree with the above and who feel that only good fits and parameters are required and never mind the details, computations, and algebra. We have for those people the next section which indicates which method should be used where.

We assume you have come to the stage where you gathered the experimental data and have an equation which represents a model with parameters and you wish to determine the parameters of this model.

### 7.1 Linear Models

If the parameters occur linearly then there are three questions you should ask. (a) Are the observations of equal accuracy? (b) Are the parameters to be constrained so that their sum should equal some number regardless of the sign of each individual parameter? (c) Are the parameters to be constrained so that they are all positive?

In addition to the least squares methods and linear programming mentioned in Table 4-1 the method of steepest descent and the Nelder and Mead method can be used to solve the problem. In the case of the method of steepest descent we accept as our answer the last iteration which yields only positive coefficients. In the case of the Nelder and Mead method we can restrict the coefficients to positive values only as outlined by those authors. Finally for the linear methods we note that the least squares methods, the Nelder and Mead method, and the method of steepest descent all minimize the sum of the squares of the error. In contrast, linear programming methods minimize the sum of the absolute errors.

**Table 4-1**

| Case | Question (a) | (b) | (c) | Method to use |
|---|---|---|---|---|
| 1 | Yes | No | No | Unconstrained, unweighted least squares method |
| 2 | No | No | No | Weighted least squares method with each weighted according to its standard error |
| 3 | Yes | Yes | No | Constrained least squares method |
| 4 | No | Yes | No | A weighted constrained least squares method |
| 5 | Yes | No | Yes | Use unweighted unconstrained least squares method first. It that fails to give meaningful answers then use linear programming. The least squares method is used first since it is much more advantageous from a point of view of the statistics probably needed subsequently. |
| 6 | No | No | Yes | Use weighted least squares method first. If that fails to give meaningful answers linear programming also can be used with weights. |

## 7.2 Nonlinear Models

If the parameters occur nonlinearly then the methods described for such purposes may be used. Offhand one cannot recommend one method over another. The method must work rapidly (to save computer time) and yield physically meaningful parameters and the confidence intervals these parameters. What nonlinear technique to use will largely depend on the problem. Since all these methods have not been tried with all of the problems we are essentially at the starting point of our evaluation of these methods with respect to their ability to solve our problems. Later more can be said about the ability of each method in relation to any specific problem.

One final note to the investigator. At your computer center you will probably find various versions of the methods described herein all "canned" and ready for your use. All you have to do in such instances is insert your raw data and carefully follow the instruction on the program.

### References

Anderson, R. L., and Houseman, E. E. (1941). *Iowa State Coll. Eng. Exp. Stat. Bull.* No. 297.

Beale, E. M. L. (1955). *J. Roy. Stat. Soc.* **17B**, 173.

Berman, M., (1965). *In* "Computers in Biomedical Research," (R. W. Stacy and B. D. Waxman, eds.), Vol. 2. Academic Press, New York.

Berman, M., and Schoenfeld, R. (1956). *J. Appl. Phys.* **27**, 1361.

Davidon, W. C. (1959). *At. Energy Comm. Dev. Rept.* An1-5990.

Draper, N. R., and Smith, H. (1966). "Applied Regression Analysis." Wiley, New York.

Fletcher, J. E., and Powell, M. J. D. (1963). *Comput. J.* **6**, 163.

Fletcher, J. E., and Shrager, R. I. (1968). *Tech. Rep.* 1, Division of Computer Research and Technology, National Institutes of Health, Bethesda, Maryland.

Forsythe, G. E., and Guolb, G. H. (1965). Maximizing a second degree polynomial on the unit sphere, *Tech. Rept.* CS 16. Comput. Sci. Dept., Standford Univ., Standford, California.

Gass, S. I. (1958). "Linear Programming." McGraw-Hill, New York.

Hamming, R. W. (1962). "Numerical Methods for Scientists and Engineers." McGraw-Hill, New York.

Indritz, J. (1963). "Methods in Analysis," pp. 1–127. Macmillan, New York.

Jeffrey, P. D., and Coates, H. H. (1966). *Biochemistry* **5**, 489.

Kunzi, H. O., Tzschach, H. G., and Zehnder, C. A. (1968). "Numerical Methods of Mathematical Optimization." Academic Press, New York.

Kunzi, H. P., and Krelle, W. (1966). "Nonlinear Programming." Ginn (Blaisdell), Boston, Massachusetts.

Lanczos, C. (1965). "Applied Analysis," pp. 167–170. Prentice-Hall, Englewood Cliffs, New Jersey.

Levenberg, K. (1955). *Quart. Appl. Math.* 2, 164–168.

Marquardt, D. W. (1963). *SIAM Rev.* 2, 431–441.

Meeter, D. A. (1966). *SIAM J. Appl. Math.* **14**, 1176–1179.

Nelder, J. A., and Mead, R. (1965). *Computer J.* **7**, 308.

Pierre, D. A. (1969). "Optimization Theory with Application." Wiley, New York.

Powell, M. J. D. (1967). "Numerical Analysis," (J. Walsh, ed.). Thompson, London.

Powell, M. J. D. (1965). *Computer J.* **7**, 303.

Ralston, A. (1965). "A First Course in Numerical Analysis," pp. 496, 516. McGraw-Hill, New York. 496, pp. 516.

Rice, J. R. (1964). "The Approximation of Functions," Addison-Wesley, Reading. Massachusetts.

Rosen, J. B. (1960). *SIAM J. Appl. Math.* **8**, 181.

Rosen, J. B. (1961). *SIAM J. Appl. Math.* **9**, 514.

Scheid, F. (1968). "Numerical Analysis." McGraw-Hill, New York.

Shrager, R. I. (1970). *J. Ass. Comput. Mach.* **17**, 446.

Spendley, W., Hext, G. R., and Himsworth, R. F. (1962). *Technometrics* **4**, 441–461.

Wolfe, R. (1960). Recent Development in nonlinear programming, *Rand Corp. Rept.* p. 2028.

Wolfe, R. (1962). *Oper. Res.* **10**, 438.

Zoutendijk, G. (1957). Studies in nonlinear programming. *Koninklije/Shell Laboratories Rept.* Amsterdam.

*CHAPTER 5*

# STATISTICS I

## 1. Introduction

The need for extensive statistical analysis in most of the work that shall be mentioned later will be evident any time one wishes to determine how well or how badly any parameter has been estimated. Enzyme kinetics is a good example of what is meant. This particular field was selected for special consideration fundamentally because most reports of kinetic parameters are given as numerical values, with no confidence intervals on the given values leaving one to wonder at the extent of the parameters' statistical reliability. Further, it is not an infrequent occurrence that the same kinetic parameters reported by two or three different laboratories differ by an order of magnitude and in some instances by two orders of magnitude. The appalling thing about all this is that little seems to be thought of these differences, even though arguments about metabolic pathways, control mechanisms, and reasonable *in vivo* estimates of the concentration of the metabolites depend on reasonable estimates of the kinetic parameters. Now if the parameters can differ by so much from laboratory to laboratory it seems that a pertinent question to ask is what is the statistical reliability of those parameters. If we assume that all investigators did exactly the same thing in the measurement of a certain kinetic parameter, then we must conclude that if the result reported from two or more laboratories was imagined to come from one laboratory the result would be quite horrendous to contemplate.

We conducted some statistical analyses by taking some data from steady state kinetics. To these data we added or substracted random errors comparable to what might be expected to occur in actual experiments. We then determined $K_m$ and $V_{max}$ using Lineweaver and Burk plots to see the effect of such errors on the computed parameters. We found that such perturbations can cause quite a bit of spread among the computed parameters leading us to wonder about the statistical reliability of results in steady state enzyme kinetics. As a matter of fact, it has been stated that even using linear least squares to compute $K_m$ and $V_{max}$ biochemists find biased results from unbiased observations (Colquhoun, 1971). According to Colquhoun using nonlinear least squares to compute $V_{max}$ and $K_m$ instead of the linear Lineweaver and Burk plot would be a much better way of computing the parameters. What is essential to point out here is that by not using statistical methods we can obtain biased results from essentially unbiased data. It is also important to note the great variation in results among investigators purportedly measuring the same thing.

This is not to criticize any particular investigator or work; we merely wish to pose the following question: Why is it the case that the statistical reliability of certain parameters is not considered to be important when it is readily conceded that there could be great variability in the reports of those parameters? We feel that it is in the interest of meaningful progress that this equestion be asked, not only in relation to the field of steady state enzyme kinetics but in every other endeavor of biology, biochemistry, and molecular biology where such variability occurs. In short, how can we hope to argue quantitative questions if we have no idea of the extent of statistical reliability of other people's data?

In the hope that the above paragraphs have emphasized the necessity of statistics we proceed with the explanation of statistical concepts.

## 2. Sample Space and Elementary Consideration of Probability Distributions

Before we take up the subject of statistics we will outline our mode of presentation of this subject. This is done to facilitate certain omissions the reader might want to make.

A certain number of notions and concepts must be understood before we can get into the heart of the matter that interests the readers most. Concepts such as the mean of a set of numbers, their standard deviation, variances, some elementary notions of probability, etc., are fairly easily understood. The important concept we would like to familiarize the reader with is the concept of distributions or probability distributions. This subject is explained in an elementary fashion in Section 2.1 on sample spaces and taken up once again in Section 2.3 on probability. The importance of probability distributions cannot be overemphasized because the parameters we estimate have certain probability distributions, and most statistically interesting quantities have well-documented probability distributions. Once one understands the meaning of a probability distribution the motivation for the statements of most of the theorems mentioned in this chapter and Chapter 6 becomes clearer. In most of those theorems we are merely trying to establish the probability distribution of certain quantities of importance. If a reader is willing to accept our assertion that a certain quantity has the distribution we claim it has, this chapter and Chapter 6 can be read far more rapidly than would otherwise be possible.

### 2.1 The Sample Space

There is no question that one of the best ways of obtaining knowledge is to observe the outcome of experiments. The set of all possible outcomes of an experiment is the *sample space.* The outcome of a single experiment is a *sample point.* For example, if you are a crap shooter (thrower of a pair of dice) the possible outcome of your "experiment" (your *sample space*) consists of thirty-six points. This is easily determined in the following manner: suppose one of your dice was red and the other blue, there is only *one* way you can make a two and that is by having an ace on the red dice and an ace on the blue dice. There are *two* ways of making a three and that is with an ace on the blue dice and deuce on the red dice, or an ace on the red dice and a deuce on the blue dice. A little more reflection and the reader will determine that there are three ways to make a four, four ways to make a five, five ways to make a six, six ways to make a seven, five ways to make an eight, four ways to make a nine, three ways to make a ten, two ways to make an eleven, and one way to make a twelve. Consequently our sample space consists of $1 + 2 + 3 + 4 + 5 + 6 + 5 + 4 + 3 + 2 + 1 = 36$ points.

Another example of a sample space might be the number of oranges produced by a tree. Every point which corresponds to an integer on one half of the real line (the positive half) is a sample point and all the positive integers are in the sample space. The fact that an orange tree cannot bear a billion oranges does not matter; the sample space must have points corresponding to every conceivable outcome, however unlikely.

### 2.2 Random Variable

One of the most important concepts in statistics is that of a random variable. A random variable is any function defined on a sample space. In the case of the crap game the possible outcomes (call them $x$) can equal any one of the numbers 2, 3, 4, 5, 6, 7, 8, 9, 10, 11, 12. Here $x$ (the outcome) can be considered a random variable; so can any function of $x$ such as $x^2$, $x^3$, $\log x$, and so on.

### 2.3 A Probability Distribution

To illustrate a probability distribution rather early in the game we can use above example from the game of craps. Since the sample space consists of 36 points the probability that we can throw a two is 1/36 or there is one chance in 36 of making a two. Using the above reasoning we can form Table 5-1.

**Table 5-1**

| Outcome of pair of dice, $x$ (point in sample space) | Probability associated with sample point |
|---|---|
| 2 | 1/36 |
| 3 | 2/36 |
| 4 | 3/36 |
| 5 | 4/36 |
| 6 | 5/36 |
| 7 | 6/36 |
| 8 | 5/36 |
| 9 | 4/36 |
| 10 | 3/36 |
| 11 | 2/36 |
| 12 | 1/36 |

In functional notation we could have written the right column of Table 5-1 as follows: $f(2), f(3), \ldots, f(12)$. This suggests that we can plot $f(x)$ where $x = 2, 3, \ldots, 12$ as shown in Fig. 5-1.

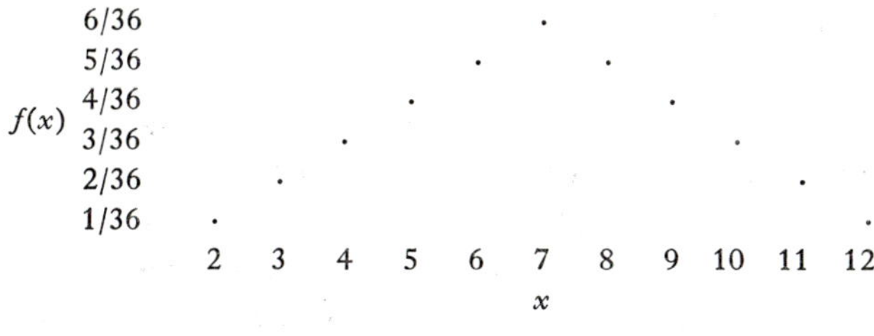


**Fig. 5-1**

The function $f(x)$, in Fig. 5-1 is called a *discrete probability function.* A number of things may be noted about this probability distribution. The sum of the separate probabilities is equal to unity, e.g., $\sum_{x=2}^{12} f(x) = 1$. The most likely outcome for a throw of a pair of dice is seven, since a seven is six times as likely as a two or a twelve and twice as likely as a four or a ten, and so on.

It is not difficult to see that the concepts of discrete probability distributions can be extended to *continuous distributions* if we suppose that $x$ can assume any of a set of continuous values. We depict a continuous probability distribution in Fig. 5-2. We sometimes call $f(x)$ in this figure a *probability density function* or a *density function.* When this density function is given we say a *continuous probability distribution* for $x$ has been defined. A variable like $x$ which possess a probability distribution is called a *variate.*

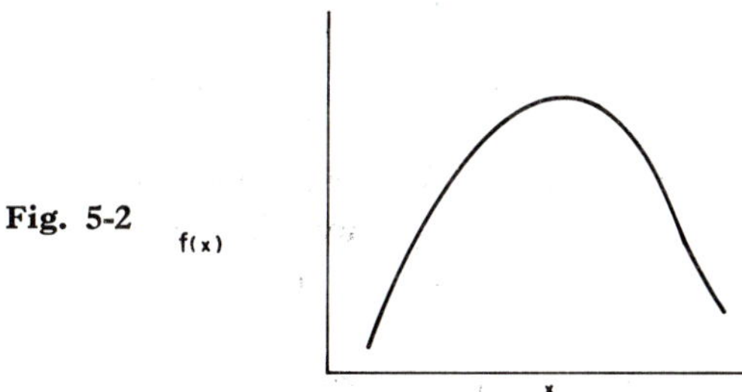


**Fig. 5-2**

## 2.4 Relationship of a Probability Distribution to the Outcome of an Experiment

As we stated earlier all estimated parameters will have probability distributions. This statement must now be explained further. Suppose that we are concerned with a single parameter which we shall call $\alpha$. Strictly speaking the parameter $\alpha$ has a single and definite but unknown value. This value of $\alpha$ can be determined only in a *perfect* experiment, that is to say, in an experiment where there is absolutely no experimental error. It is clear, however, that such an ideal is unattainable and experimental errors inevitably creep in. Consequently the parameter $\alpha$ estimated in an experiment will have a probability distribution and not a single value. When we say experiment here we mean a series of measurements which either yield several values of the parameter directly or a series of measurements of the response of a system which are analyzed in terms of a model containing the parameter as discussed in Chapter 1.

Suppose we estimate $\alpha$ from two experiments each of which gives a probability distribution $L(\alpha)$ for the answer as a function of $\alpha$. We depict those functions in Fig. 5-3. In this figure two things must be explained; first, the meaning of one of those distributions; and second, the difference between them. With respect to the meaning we note that the highest probability for the parameter $\alpha$ in the first experiment is $\alpha^*$. The prob-

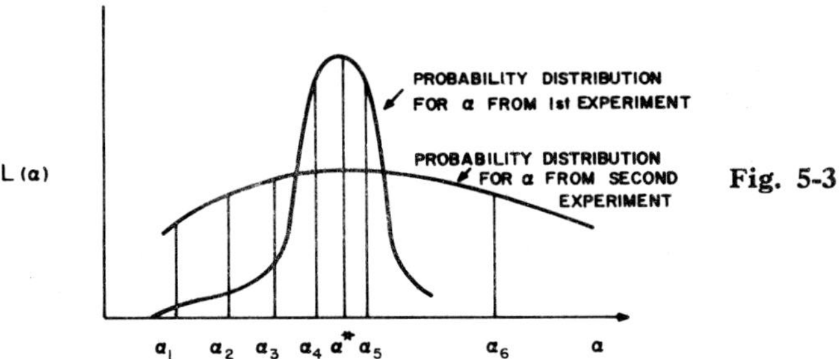


Fig. 5-3

ability at the value of the parameter $\alpha = \alpha_5$ is much smaller than the probability at $\alpha = \alpha^*$. That is to say, $L(\alpha^*) > L(\alpha_5)$. However, we note that $\alpha^*$ is not too different from $\alpha_5$. We also note that in the first experiment the probability $L(\alpha_1)$ is exceedingly low; in fact for all practical purposes there is almost no possibility that the first experiment will determine $\alpha$ to equal $\alpha_1$.

On the other hand, the second experiment shows that while $\alpha = \alpha^*$ gives us the maximum probability, the probability at $\alpha = \alpha^*$ and at $\alpha = \alpha_6$ is not too different, that is $L(\alpha^*)$ is almost equal to $L(\alpha_6)$. We note, however, that $\alpha^*$ is at a respectable distance from $\alpha_6$. This means that two quite different values of $\alpha$ are almost equally probable in their representation of the value of $\alpha$.

From this explanation we hope it is clear to the reader that the first experiment yielded a much better determination of $\alpha$ than the second experiment. A perfect experiment would have given us a single value of

$\alpha$ with perfect certainty or the probability distribution would have been the Dirac delta function.

The above reasoning illustrates the importance of probability distributions for once we know the probability distribution of a given parameter we can tell with what degree of precision this parameter has been estimated.

## 3. Frequency Distributions

### 3.1 Discrete Frequency Distributions

Consider now a sample space and conduct some experiments to determine the number of times the various points in the space occur as a result of your experiments. Suppose you determined the melting temperature $T_m$ for some DNA sample twenty times or suppose you measured the optical rotatory dispersion of poly-$L$-glutamate. Considering the first case the outcome of your experiments for $T_m$ is given by Table 5-2.

**Table 5-2**

| $f_1$ | $f_2$ | $f_3$ | $f_4$ | $f_5$ | $x_1$ | $x_2$ | $x_3$ | $x_4$ | $x_5$ |
|---|---|---|---|---|---|---|---|---|---|
| 2 | 4 | 6 | 5 | 3 | 55 | 55.5 | 56 | 56.5 | 57 |

In this experiment a temperature of 55° is measured two times and temperatures of 55.5°, 56°, 56.5°, and 57° are measured 4, 6, 5, and 3 times respectively.

The array of numbers in Table 5-2 is a *frequency distribution* of melting temperature. The sum of the frequencies $N$ is given by

$$\sum_{i=1}^{5} f_i = 2 + 4 + 6 + 5 + 3 = 20 = N$$

and is the number of times the experiment was conducted.

### 3.2 Mean

The *mean* of this frequency distribution is the arithmetic mean of the 20 trial values and is denoted by

$$\bar{x} = (1/N)(f_1x_1 + f_2x_2 + \cdots + f_5x_5) = (1/N) \sum_{i=1}^{5} f_ix_i$$

In the above case this is

$$[(2\times 55) + (4\times 55.5) + (6\times 56) + (5\times 56.5) + (3\times 57)]/20 = 56.075.$$

In general the mean of a set of numbers $x_1, x_2, \ldots, x_n$ is given by

$$\bar{x} = (1/n) \sum_{i=1}^{n} x_i \tag{5-1}$$

### 3.3 Mean Deviation

The mean deviation is a measure of spread of the numbers. For $N$ numbers $x_1, x_2, \ldots, x_i, \ldots, x_N$ it is defined as

$$\lambda = \sum_{i=1}^{N} [| x_i - \bar{x} |/N] \tag{5-2}$$

### 3.4 Standard Deviation and Variance

The *standard deviation* $v$ of a set of $N$ numbers $x_1, x_2, \ldots, x_i, \ldots, x_N$ is given by

$$v = \left[ \sum_{i=1}^{N} (x_i - \bar{x})^2/N \right]^{1/2} \tag{5-3}$$

The *variance* of a set of data is simply $v^2$. The standard deviation of the mean and the variance represent measures of data scatter. The variance in particular is of some import in our future work. When we have observations of unequal accuracy (that is some with greater spreads than others) it would be a mistake to give equal weight to all observations. Accordingly it would be a good idea in our computations to weight the data. The weights we put on data will be inversely proportional to the variance of those data. In what follows we list the variance of various functions.

i. *Variance of the sum or difference of two variables.* Given the variance of a variable $x$ and the variance of another variable $y$, it can be shown that

$$\begin{aligned} \operatorname{var}(x \pm y) &= \sum_{i=1}^{N} \frac{(x_i - \bar{x})^2}{N-1} + \sum_{i=1}^{N} \frac{(y_i - \bar{y})}{N-1} \pm 2 \sum_{i=1}^{N} \frac{(x_i - \bar{x})(y_i - \bar{y})}{N-1} \\ &= \operatorname{var}(x) + \operatorname{var}(y) \pm 2 \operatorname{cov}(x, y) \end{aligned}$$

where the covariance (cov) is defined by the last term in the above equation.

ii. *Multiplying a variable by a constant, a,* shows that

$$\mathrm{var}(ax) = a^2\,\mathrm{var}(x)$$

iii. *Adding a constant, a,* shows that

$$\mathrm{var}(a + x) = \mathrm{var}(x)$$

iv. *Approximate variance of any function* $f(x_1, x_2, \ldots, x_n)$ *of the uncorrelated variables* $x_1, x_2, \ldots, x_n$ is given by

$$\mathrm{var}(f) \cong \left(\frac{\partial f}{\partial x_1}\right)^2_{x=\bar{x}} \times \mathrm{var}(x_1) + \left(\frac{\partial f}{\partial x_2}\right)^2_{x=\bar{x}} \mathrm{var}(x_2) + \cdots + \left(\frac{\partial f}{\partial x_n}\right)^2_{x=\bar{x}} \mathrm{var}(x_n)$$

Here as elsewhere (see below) uncorrelated means that the covariance of any two variables $x_i, x_j$ is equal to zero, $\mathrm{cov}(x_i, x_j) = 0$. The derivatives in the above expression are evaluated at the mean. One notable example of the above result is the variance of a variable $(1/x)$ which occurs in enzyme kinetics) given by

$$\mathrm{var}\left(\frac{1}{x}\right) = \left[\frac{d(1/x)}{dx}\right]^2_{x=\bar{x}} \mathrm{var}(x) = \frac{\mathrm{var}(x)}{\bar{x}^4}$$

This shows that when the variance of $x$ is constant the variance of $1/x$ is proportional to the fourth power of $x$. Consequently, when we do not weight enzyme kinetic data and use Lineweaver and Burk plots we can expect very bad results.

## 3.5 Moments

The $r$th *moment* of a set of data $x_1, x_2, \ldots, x_i, \ldots, x_N$ is defined by

$$\bar{x}_r = (x_1^r + x_2^r + \cdots + x_N^r)/N = \left(\sum_{i=1}^{N} x_i^r\right)/N \tag{5-4}$$

From Eq. (5-4) it is clear that if $r = 1$ we are computing the arithmetic mean of the data [see Eq. (5-1)].

The $r$th moment about the mean $\bar{x}$ of the above set of data is defined by

$$m_r = \left[\sum_{i=1}^{N} (x_i - \bar{x})^r\right]/N \tag{5-5}$$

This definition leads to the result that when $r = 1$ then $m_1 = 0$ and when $r = 2$ then $m_2 = v^2$ or the variance.

The $r$th moment about any origin or number $A$ is defined by $m_r'$ where

$$m_r' = \left[ \sum_{i=1}^{N} (x_i - A)^r \right] / N \tag{5-6}$$

### 3.6 Continuous Frequency Distributions

The frequency distribution of the melting temperature $T_m$ of any DNA sample shown in Table 5-2 is a discrete one and so is the distribution in our crap game. This is to be compared with continuous distribution, one in which the random variables (the melting temperature in the case of DNA) take every value between certain limits. The total frequency in any interval of $[a, b]$ (for definition of interval see Chapter 3) is therefore infinite and so is the frequency in any subinterval of $[a, b]$. We shall pay attention to the *relative frequency* in the infinitesimal interval lying between $x - \frac{1}{2}\,dx$ and $x + \frac{1}{2}\,dx$ where $dx$ is small. The relative frequency in that interval can be expressed by $f(x)\,dx$ where $f(x)$ is a continuous function of $x$ and is called the *relative frequency density.* The continuous curve

$$y = f(x) \tag{5-7}$$

is the relative frequency curve for the distribution. In what follows we will call $f(x)$ or similar functions, *frequency distribution functions,* or simply *distribution functions.* The relative frequency in the infinitesimal interval $x - \frac{1}{2}\,dx$ to $x + \frac{1}{2}\,dx$ is represented by the area under the

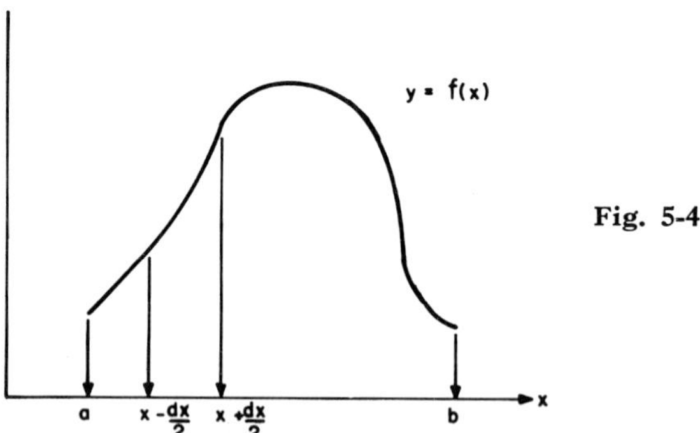


Fig. 5-4

curve between the ordinates of that interval, Fig. 5-4. In the following discussion of distributions the sum of all relative frequencies is unity. Thus we can write

$$\int_a^b f(x)\,dx = 1 \tag{5-8}$$

## 3.7 Cumulative Distribution Function

From an intuitive point of view the concept of frequency distribution functions appears to be well suited for statistics. From the point of view of rigorous mathematical theory, however, a more general mathematical function is considered. This function $F(x)$, called the *cumulative distribution function*, is always continuous on the right and we require that $F(x)$ shall be zero at the lower point of the range, which may be $-\infty$. Next we require that $F(x)$ should equal a constant $N$ at the upper end of the range, which may be $+\infty$. Lastly $F(x)$ shall not decrease at any point in the range. When $N = 1$ the cumulative distribution function is said to be normalized. In Fig. 5-5 we show the graph of a uniform frequency distribution function defined by

$$f(x) = 1/2a \qquad \text{for} \quad -a \leq x \leq a$$

and

$$f(x) = 0 \qquad \text{otherwise}$$

together with the graph of its cumulative distribution function. The reader will note on this graph the properties we prescribed for $F(x)$.

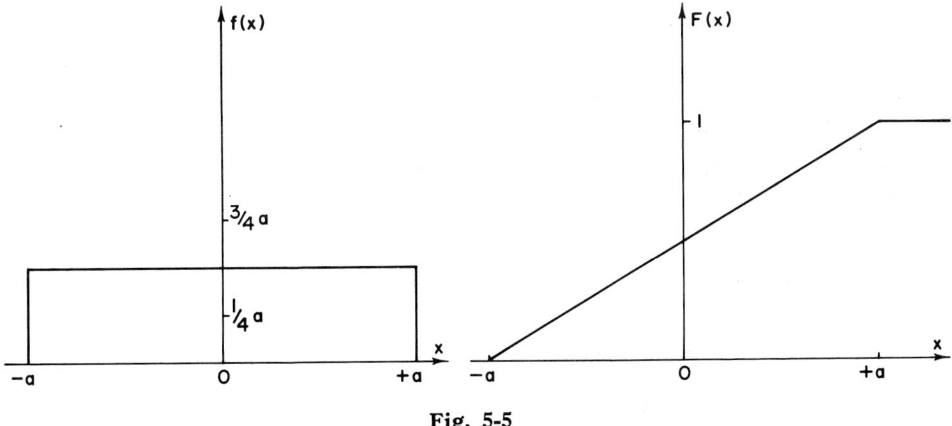


**Fig. 5-5**

For the purposes of this book when we speak of distribution functions, whether this be a frequency distribution function or a probability distribution function, we do not mean the cumulative distribution function.

### 3.8 Moments of Continuous Frequency Functions

In the same manner as we defined moments of order $r$ for discrete distributions we can define them as follows for continuous distributions for the $r$th moment about zero.

$$\bar{x}_r = \int_a^b x^r f(x)\, dx \tag{5-9}$$

so that when $r = 1$ we obtain the mean $\bar{x}$ written as

$$\bar{x} = \int_a^b x f(x)\, dx \tag{5-10}$$

The $r$th moment about the mean is given by

$$m_r = \int_a^b (x - \bar{x})^r f(x)\, dx \tag{5-11}$$

and the $r$th moment about any origin $A$ is

$$m_r' = \int_a^b (x - A)^r f(x)\, dx \tag{5-12}$$

Means, standard deviations, and moments characterize frequency distributions whether continuous or discrete and are valuable quantities in the description of distributions. The arithmetic of the computations of those quantities as observed in the preceding equations is quite simple. Apart from the use of these quantities for reporting experimental data in molecular biology, these concepts will also be used in the theory of sedimentation equilibrium in our attempts to determine molecular weight distribution functions.

## 4. Probability

### 4.1 General Considerations

Probability enters quite naturally in the discussion of statistics since the axiomatic basis of the science of statistics is founded on mathematical probability. There is a great deal of logical difficulty involved in the definition of probability and there have been many definitions of probability. Intuitively, some of our notions of probability are based on the observation that when a given set of circumstances, say the flipping of a symmetrical coin, has associated with it a fairly constant proportion for one result, e.g., that the coin will turn up heads, we are generally interested in the invariant part of that set of circumstances, or the constant proportion of the result of the event of interest. When we consider the toss of a symmetrical coin we would expect an equal chance for the outcome to be heads or tails. Now a logician might ask what is an equal chance and we might answer that the event "heads" and the event "tails" are equally likely. The logician might then ask what is "equally likely" and does it not mean the same thing as "equally probable" and this being the case, are we not defining probability by what is probable and thus using circular reasoning.

Conceding the logician's point of view, we shall nevertheless go on to define probability assuming that we and the reader recognize the same kind of "equal likeliness" in the sense that we recognize that in a perfectly symmetrical coin a head is equally likely as a tail and that an ace on a perfectly symmetrical die is equally likely to turn up as deuce and so on.

### 4.2 Empirical Definition of Probability

With the above stipulations in mind we state that, when an event $E$ can happen in $m$ times out of $n$ possible equally likely times then the probability of occurrence of this event is denoted by

$$p = P(E) = m/n \tag{5-13}$$

The probability of monoccurrence of the event $q$ is given by

$$q = P(\sim E) = 1 - p = 1 - P(E) \tag{5-14}$$

### 4.3 Dependent and Independent Events. Conditional Probability

If $E_1$ and $E_2$ are two events, the probability that $E_2$ occurs given that $E_1$ has occurred is denoted by $P(E_2 \text{ given } E_1)$. The quantity $P(E_2 \text{ given } E_1)$ is called the conditional probability of $E_2$ given $E_1$ and it is clear that this quantity may or may not depend on the occurrence of $E_1$. For instance it is possible that $E_2$ will not occur if $E_1$ has occurred in which case $P(E_2 \text{ given } E_1) = 0$, and it is also possible that $E_2$ has got absolutely nothing to do with $E_1$. If the occurrence of $E_2$ depends in some way on the occurrence of $E_1$, then $E_1$ and $E_2$ are dependent events; if on the other hand the occurrence of $E_1$ has no bearing on the occurrence of $E_2$ then the events are independent and $P(E_2 \text{ given } E_1) = P(E_2)$.

To illustrate some of the above suppose we wish to compute the probability of obtaining two events such as two consecutive elevens in a crap game. Obviously, these two events are independent of each other and the probability is $(1/18) \times (1/18) = 1/324$. Similarly the probability of throwing three consecutive sevens is $(1/6) \times (1/6) \times (1/6) = 1/216$. On the other hand, if we have a purse containing two newly minted dimes and three old dimes and wish to calculate the probability of pulling out the new dimes in succession we can proceed as follows. The probability that we draw a new dime the first time is $2/(2 + 3) = 2/5$. Now, the probability of drawing a new dime on the second draw is clearly going to depend on whether or not we drew a new dime the first time. If we did then this probability of drawing a new dime the second time is $1/(1 + 3) = 1/4$; on the other hand, if we had not drawn a new dime the first time the probability of drawing a new dime the second time will be $2/(2 + 2) = 1/2$. From the rules stated above the probability of drawing the two new dimes in *succession* is

$$P(E_1E_2) = P(E_1)P(E_2 \text{ given } E_1) = (2/5) \times (1/4) = 1/10$$

### 4.3.1 *Combined Events*

We denote the probability that both $E_1$ and $E_2$ occur by $P(E_1E_2)$ and for such a probability we have the following equation:

$$P(E_1E_2) = P(E_1)\, P(E_2 \text{ given } E_1) \tag{5-15}$$

When $E_1$ and $E_2$ are independent then

$$P(E_1E_2) = P(E_1)\, P(E_2) \tag{5-16}$$

The same reasoning follows for three events where we have

$$P(E_1E_2E_3) = P(E_1)P(E_2 \text{ given } E_1)P(E_3 \text{ given } E_2 \text{ and } E_1)$$

but if these are independent events we obtain

$$P(E_1E_2E_3) = P(E_1)P(E_2)P(E_3) \tag{5-17}$$

### 4.3.2 *Mutually Exclusive Events*

Two or more events are mutually exclusive if the occurrence of one of them precludes the occurrence of all the others, so that $P(E_1E_2) = 0$ when $E_1$ and $E_2$ are mutually exclusive. If $P(E_1 + E_2)$ denotes the probability that either $E_1$ or $E_2$ or both occur we then have

$$P(E_1 + E_2) = P(E_1) + P(E_2) - P(E_1E_2) \tag{5-18}$$

and for mutually exclusive events

$$P(E_1 + E_2) = P(E_1) + P(E_2) \tag{5-19}$$

since $P(E_1E_2) = 0$.

As an illustration of mutually exclusive events if $E_1$ is the event of drawing an ace from a deck of cards and if $E_2$ is the event of drawing a deuce from a deck of cards, then the probability of drawing either an ace or a deuce is

$$P(E_1 + E_2) = P(E_1) + P(E_2) = 2/13$$

These events are mutually exclusive we cannot, in a single draw, draw both an ace or a deuce. On the other hand, if event $E_1$ denoted the drawing of a deuce and if event $E_2$ denoted the drawing of a deuce of diamonds then

$$P(E_1 + E_2) = (1/13) + (1/4) - (1/52) = 4/13$$

because the drawing of a deuce and a deuce of diamonds are not mutually exclusive since we can draw the deuce of diamonds.

## 4.4 Discrete and Continuous Probability Distribution

A discrete probability distributions has been defined before and we repeat its definition here. In this distribution a variable $x$ can take a number of discrete values $x_1, x_2, \ldots, x.$ with probabilitics $p_1, p_2, \ldots, p_r$ where $\sum_{i=1}^{r} p_i = 1 \cdots$. Going back to our crap game example we have the following discrete probability distribution (from Table 5-1).

| $x =$ 2, | 3, | 4, | 5, | 6, | 7, | 8, | 9, | 10, | 11, | 12 |
|---|---|---|---|---|---|---|---|---|---|---|
| $p =$ 1/36, | 2/36, | 3/36, | 4/36, | 5/36, | 6/36, | 5/36, | 4/36, | 3/36, | 2/36, | 1/36 |

The analogy between the above distribution and the relative frequency distribution we have been discussing is clear when we let probabilities replace relative frequencies. We can also extend this analogy to continuous distributions. Consequently, as in the case of frequency distributions, the variable may take any value between certain limits $a$ and $b$, The continuous function $\phi(x)$ (Fig. 5-6) which denotes the probability at any $x$ is called the probability density function or simply the probability density. The area under the curve $\phi(x)$ from $x = \alpha$ to $x = \beta$ represents the probability that the value $x$ will fail within the interval $[\alpha, \beta]$. The

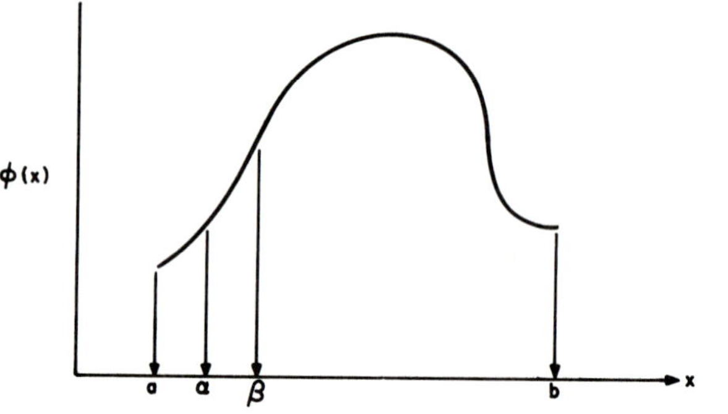


Fig. 5-6

total area under the curve is equal to unity; that is to say

$$\int_a^b \phi(x)\, dx = 1 \tag{5-20}$$

When $\phi(x)$ is a constant for all values of $n$ then we have a uniform or rectangular distribution and the value of this constant will be $1/(b - a)$ because of Eq. (5-20). Closely related to the idea of probability distribution is mathematical expectations which we consider next.

## 4.5 Mathematical Expectation

Mathematical expectation can be defined for both the discrete and continuous cases. An elementary idea of expectation can be given by the statement that if the probability that one will receive a sum of money $S$ is $p$ then the expectation is given by $pS$. For the discrete case if $n$ can take the values $x_1, x_2, \ldots, x_N$ with probabilities $p_1, p_2, \ldots, p_N$ respectively the mathematical expectation of $x$ denoted by $E(x)$ is defined simply as

$$E(x) = \sum_{i=1}^{N} p_i x_i \tag{5-21}$$

For our continuous probability distribution in Fig. 5-6 the expected value of $x$ is given by

$$E(x) = \int_a^b x\phi(x)\, dx \tag{5-22}$$

When we are talking simply of the variable $x$ then the expected value of $x$ is also the mean. This is in contrast to defining the expected value of a function $g(x)$ which is given by

$$E[g(x)] = \sum g(x)\phi(x) \tag{5-23}$$

for the discrete case and

$$E[g(x)] = \int_a^b g(x)\phi(x)\,dx \qquad (5\text{-}24)$$

for the continuous case.

## 4.6 Moments of Probability Distributions

For probability distributions, the $r$th moment of the distribution $\phi(x)$ about a particular value is the $r$th *moment* of the probability distribution about this value. Consequently the $r$th moment about $x = A$ is given by

$$\mu_r' = \int_a^b (x - A)^r \phi(x)\,dx \qquad (5\text{-}25)$$

With the above definition we notice that the first moment about $x = 0$ is simply the expected value of $x$. The $r$th moment about the expectation value of $x$ is given by

$$\int_a^b [x - E(x)]^r \phi(x)\,dx \qquad (5\text{-}26)$$

The variance of $x$ is the second moment about the expected value or the mean of $x$ and is given by

$$\operatorname{var} x = \int_a^b [x - E(x)]^2 \phi(x)\,dx \qquad (5\text{-}27)$$

## 4.7 Calculation of Complex Probability: Combinatorial Analysis

When one considers problems of probability it is often the case that one should be able to count the various possible ways in which events can occur. The subject of combinatorial analysis facilitates this counting.

A fundamental principle of combinatorial analysis is that if an event can happen in any one of $n_1$ ways and if another event can happen in any one of $n_2$ ways then the number of ways both events can happen is $n_1 n_2$.

### 4.7.1 *Permutations*

A permutation has been previously defined in Chapter 2 when we were considering the subject of determinants, however, we define it here once again for convenience.

Permutations pay particular attention to the order of arrangements, and the number of permutations of $n$ objects taken $r$ at a time is denoted by ${}_nP_r$ where

$$ {}_nP_r = n!/(n-r)! \qquad (5\text{-}28) $$

Thus the number of permutations of the letters $ABC$ taken one at a time is $3!/(3-1)!$ and these permutations are simply $A$, $B$, and $C$. The number of permutations of $A$, $B$, $C$ taken two at a time are $3!/(3-2)!=6$ and those are $AB$, $BA$, $AC$, $CA$, $BC$, and $CB$. Lastly the number of permutations of $ABC$ taken three at a time are $3!/(3-3)! = 6$ and those are $ABC$, $CAB$, $ACB$, $CBA$, $BAC$, and $BCA$. This last result suggests that the number of permutations of $n$ objects taken $n$ at a time is $n!$. Lastly the number of permutations of $n$ objects consisting of several groups in which we have $n_1$ alike, $n_2$ alike etc. is given by

$$ n!/n_1!n_2!\cdots \qquad \text{where } n_1 + n_2 + \cdots = n \qquad (5\text{-}29) $$

### 4.7.2 *Combinations*

The subject of combination pays no attention to the order of arrangement; thus the number of combinations of $n$ objects taken $r$ at a time is denoted by

$$ {}_nC_r = n!/r!(n-r)! \qquad (5\text{-}30) $$

Thus the number of combinations of the letters $ABC$ taken one at a time is three or $A$, $B$, and $C$. The number of combinations of the letters $ABC$ taken two at a time is three and those are $AB$, $AC$, and $BC$, and we note that while $AB$ and $BA$ are different permutations of the letter $A$ they are the same combination because no attention is paid to order. The uses of combinatorial analysis are obvious in the formulation of frequency distributions, and we will have to refer to it whenever we wish to compute the probability of complicated events.

## 4.8 Some Theorems on Probability and Probability Distributions

Here we state without proof three interesting theorems on the probability of variates. The theorems effectively set limits to the values these variates can assume.

### 4.8.1 *Chebyshev's Theorem*

Chebyshev's theorem states that in a random choice of a value of a variate whose standard deviation is $v$, the probability that the value chosen will differ from the mean by more than $\lambda v$ does not exceed $1/\lambda^2$ where $\lambda$ is a given positive number.

### 4.8.2 *Bernoulli's Theorem*

The next theorem we state is due to Bernoulli. In this theorem we are concerned with the number of successes $m$ obtained in $n$ independent trials and what happens to the ratio $(m/n)$ as $n$ increases indefinitely. If the probability of success is given by $p$, Bernoulli proved that given any positive number $\varepsilon$, however small, the probability of $|(m/n) - p|$ exceeding $\varepsilon$ tends to zero as $n$ tends to infinity.

*Bernoulli's Theorem.* Let $\varepsilon$ and $\delta$ be two given positive numbers however small and let $m$ be the number of successes in $n$ independent trials, in which the probability of success is $p$. Then the probability that inequality

$$|(m/n) - p| < \varepsilon \tag{5-31}$$

will hold is greater than 1-$\delta$, provided $n$ is greater than a certain $N$ depending on $\varepsilon$ and $\delta$.

### 4.8.3 *The Central Limit Theorem*

The last theorem we state is of considerable importance and is the one which makes the normal distribution, which we use extensively, so unique in statistical analysis.

*Central Limit Theorem.* Given a set of $n$ independent random variates $x_1, x_2, \ldots, x_i, \ldots, x_n$ each with an *arbitrary* probability distribution function with finite mean $\mu_i$ and second moment $v_i^2$, consider the sum $y = \sum_{i=1}^n \alpha_i x_i$ where the $\alpha_i$ are constants. The distribution of $y$ approaches the normal distribution with mean $\mu = \sum_{i=1}^n \alpha_i \mu_i$ and variance $v^2 = \sum_{i=1}^n \alpha_i^2 v_i^2$ as $n$ tends to infinity.

The importance of this theorem can be appreciated when we realize that during the course of an experiment there are many independent sources of error. The distribution of each of the error components may or may not be normal, however, regardless of how the individual errors are distributed the central limit theorem tells us that the total sum of those errors is approximately normally distributed.

In the first chapter we denoted the experimental error by $\delta$. Later in Chapter 6 when we estimate the parameters of a model we shall assume that $\delta$ is normally distributed. Since we do not known the actual distribution of $\delta$ this assumption appears to be quite drastic. However, in any experiment it is generally true that $\delta$ is made up of a sum of several sources of experimental error. Consequently, by virtue of the central limit theorem, when we assume that $\delta$ is normally distributed we do not seriously compromise ourselves.

## 4.9 Moment Generating Functions

Associated with probability density functions are *moment generating functions*. These functions will be very useful in proving a number of theorems about probability distributions and have general applications to some of the topics we deal with in the subsequent application sections. Let $\phi(x)$ be the probability density function. Then the expected value of $e^{tx}$ as a function of a dummy variable $t$ is given by

$$M(t) = \int e^{tx}\phi(x)\,dx \tag{5-32}$$

where the integration is taken over all $x$. By expanding the exponentional and integrating term by term we obtain

$$M(t) = 1 + \mu_1' t + \mu_2' t^2/2! + \mu_3' t^3/3! + \cdots \tag{5-33}$$

where

$$\mu_r' = \int x^r \phi(x)\,dr \tag{5-34}$$

and that is the moment of order $r$ about the origin. Here $M(t)$ is called the moment generating function about the origin. Similarly the moment generating function about $x = A$ is defined as the expected value of $e^{t(x-A)}$. This is given by

$$M_A(t) = \int e^{t(x-A)}\phi(x)\,dA \tag{5-35}$$

Now the coefficients of $t^r/r!$ is the moment of order $r$ about $x = A$. We also note the following identity relating $M_0(t)$ the moment generation function about $x = 0$ to the moment generating function about $x = A$

$$M_A(t) = e^{-At}M_0(t)$$

For a discrete distributions the moment generating function about $x = A$ can be written as

$$M_A(t) = \sum_i p_i \exp[t(x_i - A)] \tag{5-36}$$

where $p_i$ is the probability of the variable $x_i$. Tho moment generating function of the sum of *two independent variables*[†] $x$ and $y$ is the product of their moment generating function. Thus the moment generating function of two variables $x$ and $y$ can be written as

$$E(e^{t(x+y)}) = E(e^{tx}e^{ty}) \tag{5-37}$$

This can be expanded as the following product.

$$(1 + \mu_1't + \mu_2't^2/2! + \cdots)(1 + m_1't + m_2't^2/2! + \cdots) \tag{5-38}$$

where

$$m_r' = \int y^r\phi(y)\,dy \tag{5-39}$$

On multiplication we find that the coefficient of $t$ is equal to $\mu_1' + m_1'$. Consequently the mean of the sum of two variates is the sum of their means. Next we note that the coefficient of $t^2$ is $\mu_2' + 2\mu_1'm_1' + m_2'$. Therefore the second moment of $x + y$ about their means

$$\begin{aligned} \mu_2' + 2\mu_1'm_1' + m_2' - (\mu_1' + m_1')^2 &= (\mu_2' - \mu'^2) + (m_2' - m'^2) \\ &= \operatorname{var}(x) + \operatorname{var}(y) \end{aligned} \tag{5-40}$$

is equal to the sum or difference of the variances of the independent variates, a result previously noted.

Equation (5-33) can be recognized as a Taylor series (see Chapter 3) in which the coefficients are the moments of the distribution function. Inasmuch as a Taylor expansion is unique if we know the moment generating function of a probability distribution we can obtain the $n$th moment of the distribution by evaluating the $n$th derivative of the moment

† Independence plays a large part in statistical analysis and will be returned to again. In the meantime we consider two variables $x$ and $y$ independent if the probability of either of them will have a certain value is not influenced by the value given to the other. This is independence in the statistical or probabilistic sense. We can state this symbolically by saying that the probability that the value of $x$ falls in the interval $dx$ is independent of $y$ and is given by $\phi(x)$ a function $x$ only. Similarly, the probability that $y$ lies in the interval $dy$ is independent of $x$ and is given by $\phi(y)$. Statistical independence also means that $\operatorname{cov}(x, y) = 0$. This idea of independence is in contrast to functional independence where, given one variable, the other is not uniquely determined. In other words, $x$ and $y$ are functionally dependent when there exists a functional relation $F(x, y)$ such that $F(x, y) = 0$. When no such relationship exists $x$ and $y$ are functionally independent. In Chapters 5 and 6 independence will mean independence in the statistical or probabilistic sense.

generating function at $t = 0$. Accordingly *a distribution is completely specified by a complete set of its moments.* If during the course of experiments we determine the moments of an unknown population (e.g., in a sedimentation equilibrium experiment we obtain some moments of the molecular weight distribution function), then we can attempt to construct the distribution from its moments. For actual use of some of those ideas the reader is referred to Chapter 12 where we use the Chebyshev–Markov inequalities to help in determining molecular weight distributions.

## 5. Standard Distributions with a Single Variable

The probability distribution of interest will be considered next. The binomial distribution, the Poisson distribution and most importantly the normal distribution and the distributions derived from it will be dealt with. Sketching the curves representing those distributions is one way of obtaining facility with them (this exercise is recommended to the reader). The curves are found in various statistical texts. Alternatively statistical tables could be used to construct the curves.

### 5.1 The Binomial Distribution

If the probability that an event will happen in a single trial is $p$, then $q = 1 - p$ is the probability that the event will not happen. The probability that the event will happen exactly $x$ times in $N$ trials is given by

$$p(x) = [N!/x!(N - x)!]p^x q^{N-x} \tag{5-41}$$

Eq. (5-41) represents the binomial distribution. The reader may wish to construct $p(x)$ for $x$ ranging from 0 to 10 for a fair coin, that is, for a coin where $p = q = 0.5$, what are the chances $[p(x)]$ of obtaining no heads, 1 head, to 10 heads? Some of the properties of the binomial distribution are the following:

$$\text{mean} = Np, \qquad \text{variance} = Npq, \qquad \text{standard deviation} = (Npq)^{1/2} \tag{5-42}$$

### 5.2 The Poisson Distribution

Another distribution of interest due to Poisson has the following equation for the probability of $x$

$$p(x) = (\lambda^x e^{-\lambda})/x! \tag{5-43}$$

In the above $\lambda$ is a given constant. The properties of the Poisson distribution are as follows. Its mean is equal to $\lambda$, its variance is also $\lambda$ and its standard deviation is $\lambda^{1/2}$.

The Poisson distribution is related to the binomial distribution in the following manner: When $q$ is close to unity, indicating an unlikely event, the binomial distribution approximates the Poisson distribution.

## 5.3 The Normal Distribution

This distribution is most important in all subsequent work. The normal distribution is an ideal some distributions reach and other distributions approximate under certain circumstances. The equation for the normal distribution is

$$p(x) = [1/(2\pi v^2)^{1/2}] \exp[-(x - \mu)^2/2v^2] \tag{5-44}$$

where $\mu$ is the mean of the distribution and $v$ the standard deviation. As for all distributions we shall consider that

$$\int_{-\infty}^{\infty} [1/(2\pi v)^{1/2}] \exp[-(x - \mu)^2/2v^2]\, dx = 1 \tag{5-45}$$

The probability that a point $x$ lies between $a$ and $b$ is given by

$$\int_{a}^{b} [1/(2\pi v^2)^{1/2}] \exp[-(x - \mu)^2/2v^2]\, dx \tag{5-46}$$

By making a linear transformation among the variables and setting $z = (x - \mu)/v$, Eq. (5-44) now assumes its *standard form* and we obtain the following equation for the probability of $z$

$$p(z) = (1/2\pi)^{1/2} e^{-z^2/2} \tag{5-47}$$

Here $z$ is said to be a normally distributed variable with zero mean and a variance of one. A graph of a standardized normal distribution is shown in Fig. 5-7. A plot of $p(z)$ vs. $z$ shows that 68.27%, 95.45% and 99.73% of the area lie between $z = -1$ and $+1$, $z = -2$ and $+2$, and $z = -3$ and $+3$ respectively. The general properties of the normal distribution are as follows, mean $= \mu$, variance $= v^2$, standard deviation $= v$.

Both the Poisson distribution and the binomial distribution are related to the normal distribution. When $p$ and $q$ are not close to zero and $N$ is large the binomial distribution closely approximates the normal distribu-

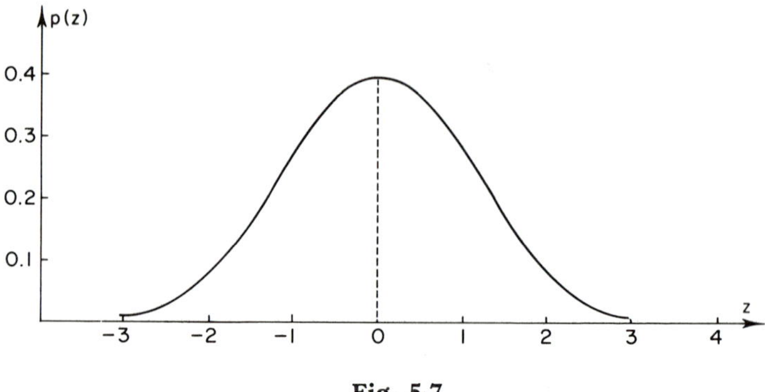


**Fig. 5-7**

tion with a standardized variable given by

$$z = (x - Np)/(Npq)^{1/2} \tag{5-48}$$

This approximation is quite good if *both* $Np$ and $Nq$ exceed 5. The relationship between the Poisson distribution and a normal distribution with a standardized variable $(x - \lambda)/\lambda^{1/2}$ is readily noted if we let $\lambda$ tend to infinity.

## 5.4 Distributions Derived from the Normal Distribution

The normal distribution is of fundamental importance and from it we derive several important distributions. Of these we introduce first the chi-square distribution, written $\chi^2$.

### 5.4.1 *Chi-Square Distribution*

If $X_1, X_2, \ldots, X_n$ are normally and independently distributed with zero mean and a variance of unity then the quantity $\varrho$ given by

$$\varrho = \sum_{i=1}^{n} X_i^2 \tag{5-49}$$

is distributed as chi-square with $n$ degrees of freedom and is designated by $\chi^2(n)$. The functional form of chi-square is

$$f(\varrho) = [(\varrho)^{(n-2)/2}e^{-\varrho/2}]/2^{n/2}\Gamma(n/2) \quad (0 < \varrho < \infty) \tag{5-50}$$

where $\Gamma$ represents the gamma function defined in Chapter 3. If a quantity $X$ is distributed as $\chi^2(n)$ and another quantity $Z$ is distributed as $\chi^2(m)$, then $X + Z$ is distributed as $\chi^2(m + n)$. This property of the chi-squared distribution is called the reproductive property and can be extended to any finite number of variables with chi-square distributions.

### 5.4.2 *Snedecor's F-Distribution*

If $X$ is distributed as $\chi^2(n)$ and $Z$ is distributed as $\chi^2(m)$ and if $X$ and $Z$ are independent then the quantity

$$\varrho = (n/m)(Z/X) \tag{5-51}$$

is distributed as Snedecor's $F$-distribution with $m$ and $n$ degrees of freedom and is denoted by $F(m, n)$. The frequency function of the $F$ distribution is given by

$$f(\varrho) = \frac{\Gamma[(m+n)/2][m/n]^{m/2}\varrho^{(m-2)/2}}{\Gamma(m/2)\Gamma(n/2)[1+(m\varrho/n)]^{(m+n)/2}} \qquad (0 < \varrho < \infty) \tag{5-52}$$

### 5.4.3 *Student's Distribution*

This distribution was discovered by Gosset who used the nom de plume Student. If $X$ is distributed normally with zero mean and variance $v^2$, $Z^2/v^2$ is distributed as chi-square with $n$ degrees of freedom, and $X$ and $Z$ are independent, then $\varrho = X\sqrt{n}/Z$ is distributed as Student's $t$-distribution with $n$ degrees of freedom. The frequency function for $\varrho$ is given by

$$f(\varrho) = \Gamma[(n+1)/2]/[(n\pi)^{1/2}\Gamma(n/2)(1+\varrho^2/n)^{(n+1)/2}] \qquad (-\infty < \varrho < \infty) \tag{5-53}$$

This formal definition may not convey much to some readers about the Student's $t$-distribution as it is called. Actually for future work we use the $t$-distribution to test whether some parameters (the slope in a least squares straight line fit, for example) are equal to certain values. Another way of looking at the $t$-distribution is to consider the distribution of the random variable

$$t = (x - \mu)/s(x)$$

where $x$ is a normally distributed variable with a true mean $\mu$ and unknown variance, and $s(x)$ is an estimate of the standard deviation

$$s(x) = \left[\sum_{i=1}^{N} (x_i - \bar{x})^2/N\right]^{1/2}$$

In this case $t$ is distributed as the Student $t$-distribution with $N - 1$ degrees of freedom. As the number of degrees of freedom increases, the

$t$-distribution approximates the normal distribution. The $t$-distribution as stated earlier is used to test hypotheses about quantities which have $t$-distribution (see sections on hypothesis testing in Chapter 6).

## 6. Bivariate Distributions

### 6.1 Introductions and Definitions

Before proceeding with the subject of distributions having more than one variable, in particular the multivariate normal distribution, we give some idea of the two variate distribution for two main reasons. First, we wish to simplify the discussion of the multivariate normal distribution and second, we want to introduce some ideas about *correlation and regression.*

As an example of a bivariate distribution suppose we consider the height and weight of a number of men. The $i$th man has a height denoted by $x_i$ and a weight denoted by $y_i$. If we plot all the points $(x_i, y_i)$ in the

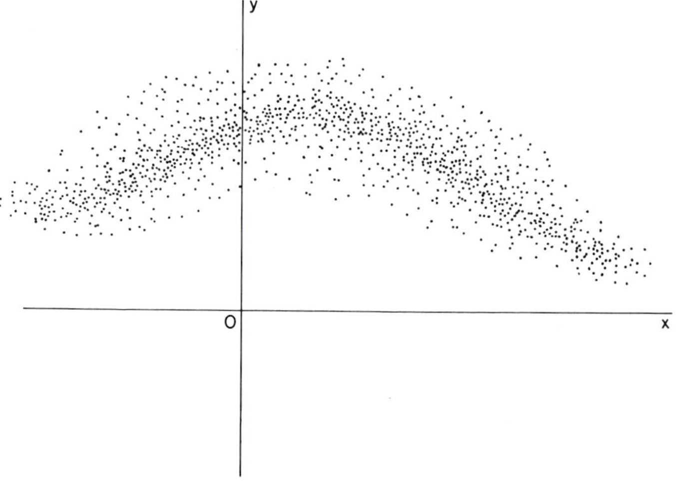


**Fig. 5-8**

plane we will get a diagram where there seems to be some relationship between the $y$ values and the $x$ values as shown in Fig. 5-8. This we can observe despite the scatter of points and it is no surprise, since, in general, the greater the height the greater the weight. A relationship of the form

$$y = f(x)$$

can be established for our diagram where the locus of $f(x)$ will go through the densest portion of our diagram. This functional relationship is called the *regression* of $y$ on $x$. This sort of reasoning can be extended to several variables where, for example we may write the regression of $y$ on the variables $\alpha$, $\beta$, $\gamma$, $\delta$, $\varepsilon$, . . . as follows:

$$y = \alpha_1 f(\beta) + \alpha_2(\gamma) + \alpha_3 f(\delta) + \alpha_4 f(\varepsilon) + \cdots \tag{5-54}$$

Such instances of *multiple regression* occur frequently in biology. The coefficients $\alpha_i$ are generally parameters to be estimated and, as we shall observe later, part of our statistical problem is to estimate those parameters.

Returning to our bivariate distribution we can consider that the density of dots in the rectangular region denoted by all points

$$x - (dx/2) < x < x + (dx/2), \qquad y - (dy/2) < y < y + (dy/2)$$

to constitute a bivariate frequency. Further, in the same manner as we extended considerations of frequency distributions to probability distributions we can consider the function $\phi(x, y)$ to represent a *bivariate probability distribution function.* The features of such distributions are quite analogous to those of univariate distributions. For instance such distributions have the property that

$$\iint \phi(x, y)\, dx\, dy = 1 \tag{5-55}$$

where the integration is carried over the region in the plane where the distribution is defined. The mean of this distribution $(\bar{x}, \bar{y})$ is given by the following two equations

$$\bar{x} = \iint x\phi(x, y)\, dy\, dx, \qquad \bar{y} = \iint y\phi(x, y)\, dy\, dx \tag{5-56}$$

As an example of the general equation for higher moments of bivariate distributions, the moment of order $r$ with respect to $x$ and the moment or order $s$ with respect $y$ is given by

$$\mu_{rs} = \iint \phi(x, y) x^r y^s\, dy\, dx \tag{5-57}$$

Other properties of interest are given by the same form of equation.

### 6.2 The Bivariate Normal Distribution

The distribution function for a *bivariate normal distribution* is given by

$$\phi(x, y) = \frac{1}{2\pi v_x v_y (1 - \varrho^2)^{1/2}} \exp - \left\{ \left[ \left( \frac{x - \mu_x}{v_x} \right)^2 - 2\varrho \left( \frac{x - \mu_x}{v_x} \right) \left( \frac{y - \mu_y}{v_y} \right) + \left( \frac{y - \mu_y}{v_y} \right)^2 \right] \Big/ 2(1 - \varrho^2) \right\} \tag{5-58}$$

where $\mu_x$ is the mean of the variable $x$ and similary $\mu_y$ is the mean of $y$ and where $v_x$ is the square root of the variance of the distribution $x$ and $v_y$ is the square root of the variance of the distribution in $y$ (both are defined in Eqs. (5-60) and (5-61). The parameter $\varrho$ is called the *correlation coefficient.* The correlation coefficient between the two variables $x$ and $y$ is also given by

$$\varrho = \operatorname{cov}(x, y)/(v_x^2 v_y^2)^{1/2} \tag{5-59}$$

where

$$\operatorname{cov}(x, y) = \int_{-\infty}^{\infty} \int_{-\infty}^{\infty} [y - E(y)][x - E(x)] \phi(x, y)\, dx\, dy \tag{5-60}$$

and

$$\begin{aligned} v_y^2 &= \int_{-\infty}^{\infty} [y - E(y)]^2 \phi(x, y)\, dy \\ v_x^2 &= \int_{-\infty}^{\infty} [x - E(x)]^2 \phi(x, y)\, dx \end{aligned} \tag{5-61}$$

To obtain an estimate of $\varrho$ (call the estimate $r$) for a finite sample (say $m$) of the form $(x_1, y_1), (x_2, y_2), \ldots, (x_m, y_m)$ we write

$$r = \sum_{j=1}^{m} (x_j - \bar{x})(y_j - \bar{y}) \Big/ \left\{ \left[ \sum_{j=1}^{m} (x_j - \bar{x})^2 \right] \left[ \sum_{j=1}^{m} (y_j - \bar{y})^2 \right] \right\}^{1/2} \tag{5-62}$$

The quantity $r$ is called the *sample correlation coefficient.* One can show that both $r$ and $\varrho$ can take any value between and including $-1$ to 1. If $\varrho = 1$ then $x$ and $y$ are perfectly positively correlated and if $\varrho = -1$ then $x$ and $y$ are perfectly negatively correlated and when $\varrho = 0$ the variables are uncorrelated. When $\varrho = \pm 1$ the possible points $(x, y)$ lie on straight lines with positive and negative slopes respectively.

In Eq. (4-58) it will be observed that when we put $\varrho = 0$ the probability distribution $\phi(x, y)$ will become a simple product of two normal distributions in $x$ and $y$ and that the variates are independent.

## 6.3 Standard Form of the Bivariate Normal Distribution

In Eq. (5-58) if we let

$$u = (x - \mu_x)/v_x \qquad \text{and} \qquad \nu = (y - \mu_y)/v_y \tag{5-63}$$

then our probability distribution $\phi(x, y)$ is transformed to the *standard form* given by (5-64)

$$\phi(u, \nu) = [1/2\pi(1 - \varrho^2)^{1/2}] \exp[-(u^2 - 2\varrho u\nu + \nu^2)/2(1 - \varrho^2)]\, du\, d\nu \tag{5-64}$$

The analogy of the above reasoning to the standardization of univariate distributions is clear.

Considerations of distributions having more than one variable leads to the definition of *marginal distributions* and *conditional distributions.* For the bivariate distribution the marginal distribution of $x$, for example, is by definition

$$\phi^*(x) = \int_{-\infty}^{\infty} \phi(x, y)\, dy \tag{5-65}$$

*Example.* If

$$\phi(x, y) = e^{-(x+y)} \qquad \text{for} \quad 0 < x < \infty; \quad 0 < y < \infty$$

then the marginal distribution of $x$, $\phi^*(x)$ is

$$\phi^*(x) = \int_0^{\infty} e^{-(x+y)}\, dy = e^{-x} \qquad \text{for} \quad 0 < x < \infty$$

It can be shown by carrying out the integration after making a change of variable using the first of Eqs. (5-63) that

$$\phi^*(x) = \frac{1}{(2\pi)^{1/2} v_x} \exp\left[-\left(\frac{x - \mu_x}{v_x}\right)^2 \Big/ 2\right] \tag{5-66}$$

or that the marginal distribution of $x$ is a univariate normal distribution. Similary one can show that

$$\phi^*(y) = \int_{-\infty}^{\infty} \phi(x, y)\, dx = \frac{1}{(2\pi)^{1/2} v_y} \exp\left[-\left(\frac{y - \mu_y}{v_y}\right)^2 \Big/ 2\right] \tag{5-67}$$

The conditional distribution of $x$ for a given value of $y$ is

$$\phi(x/y) = \phi(x, y)/\phi^*(y) \tag{5-68}$$

*Example.* If

$$\phi(x, y) = e^{-(x+y)} \qquad \text{for} \quad 0 < x < \infty; \quad 0 < y < \infty$$

then the conditional distribution of $x$ given $y$

$$\phi(x/y) = e^{-(x+y)}/e^{-y} = e^{-x} \qquad \text{for} \quad 0 < x < \infty$$

The conditional distribution of $x$ given $y$ for the bivariate normal distribution is

$$\phi(x|y) = \frac{1}{(2\pi)^{1/2} v_x (1 - \varrho^2)^{1/2}} \exp\left\{-\left[\frac{1}{2v_x^2(1 - \varrho^2)}\right] \times \left[x - \mu_x - \varrho \frac{v_x}{v_y}(y - \mu_y)\right]^2\right\} \tag{5-69}$$

which is a univariate normal distribution with mean $\mu_x + \varrho(v_x/v_y) \times (y - \mu_y)$ and variance $v_x^2(1 - \varrho^2)$.

A regression line of $x$ on $y$ is defined by the conditional distribution of the bivariate normal distribution. Such regression lines for a given $y$ have mean and variance indicated by the quantities just noted.

A relationship between correlation and regression in conditional bivariate distribution can be observed when we have a set of data $(x_1, y_1)$, $\ldots, (x_m, y_m)$ and we postulate a model of the form $y = \alpha_0 + \alpha_1 x + \delta$ to fit the data. Here $\alpha_0$ and $\alpha_1$ are regression coefficients, or parameters to be estimated, and $\delta$ is the error. If we estimate the regression coefficients $\alpha_0$ and $\alpha_1$ by minimizing the sum of the squares of the errors, then this estimate of the regression coefficient $\alpha_1$, call it $\hat{\alpha}_1$, can be expressed in terms of $r$ the sample correlation coefficient and the data as follows

$$\hat{\alpha}_1 = \left[\sum_{j=1}^{m} (y_j - \bar{y})^2 \Big/ \sum_{j=1}^{m} (x_j - \bar{x})^2\right]^{1/2} r$$

and $\hat{\alpha}_1$ is a measure of $r$ scaled by the ratio of the spread of the $x_i$ divided by the spread of the $y_i$.

Since matrix notation will be used throughout in the description of multivariate normal distribution an illustration of this kind of notation for bivariate distributions is extremely helpful. For the bivariate distribution we denote the two variables by $X_1$ and $X_2$ and their means by $\mu_1$ and $\mu_2$. From this point and for the rest of the chapter the mean of the variates in multivariate distribution will be given by $\mu_1, \mu_2, \ldots$, and $\boldsymbol{\mu}$ shall denote the vector that represents the array of the means. For a bivariate distribution $\boldsymbol{\mu} = [\mu_1, \mu_2]$. The variances of $X_1$ and $X_2$ will be denoted by $v_{11}$ and $v_{22}$. Instead of the correlation $\varrho$ the covariance will be used, which from Eq. (5-59) is given by $\varrho(v_{11}v_{22})^{1/2}$, and this quantity will be denoted by $v_{12} = v_{21}$.

The *variance–covariance matrix* or simply the *covariance matrix* for the bivariate distribution will then be written as

$$\mathbf{V} = \begin{bmatrix} v_{11} & v_{12} \\ v_{21} & v_{22} \end{bmatrix} \tag{5-70}$$

For simplicity in notation we shall denote a multivariate normal distribution by the symbol $N(\boldsymbol{\mu}, \mathbf{V})$ where $\boldsymbol{\mu}$ is a vector whose components are the expected values of the variables and $\mathbf{V}$ is the covariance matrix. In the case of the bivariate distribution $\boldsymbol{\mu} = [\mu_1, \mu_2]$ as indicated above and $\mathbf{V}$ is given by the matrix shown in (5-70). For the bivariate distribu-bution the determinant of the matrix $|\mathbf{V}|$ is given by Eq. (5-71)

$$|\mathbf{V}| = v_{11}v_{22} - v_{12}v_{21} = v_{11}v_{22}(1 - \varrho^2) \tag{5-71}$$

using the previous notation in terms of $x$ and $y$. The inverse of this matrix is

$$\mathbf{V}^{-1} = \begin{bmatrix} v_{11}/|\mathbf{V}| & -v_{12}/|\mathbf{V}| \\ -v_{21}/|\mathbf{V}| & v_{22}/|\mathbf{V}| \end{bmatrix}$$

$$= \begin{bmatrix} v^{11} = 1/v_{11}(1 - \varrho^2) & v^{12} = -\varrho/(v_{11}v_{22})^{1/2}(1 - \varrho^2) \\ v^{21} = -\varrho/(v_{11}v_{22})^{1/2}(1 - \varrho^2) & v^{22} = 1/v_{22}(1 - \varrho^2) \end{bmatrix} \tag{5-72}$$

where $v^{ij}$ are the elements of the inverse matrix as depicted by (5-72). The determinant of the inverse matrix is

$$|\mathbf{V}^{-1}| = 1/v_{11}v_{22}(1 - \varrho^2)$$

A glance at Eq. (5-72) shows that the matrix elements of the inverse of the covariance matrix can be written as the coefficients of the exponential term of expression (5-58). In our new notation that means that the exponential can be written as

$$-\tfrac{1}{2}[v^{11}(x_1 - \mu_1)^2 + 2v^{12}(x_1 - \mu_1)(x_2 - \mu_2) + v^{22}(x_2 - \mu_2)^2] \tag{5-73}$$

The expression (5-73) is called the quadratic form of this bivariate distribution. The constant term in the distribution $1/2\pi[v_{11}v_{22}(1 - \varrho^2)]^{1/2}$ can be written as $1/2\pi(|\mathbf{V}|)^{1/2}$ so the distribution function may be written as

$$\phi(x_1, x_2) = [1/2\pi(|\mathbf{V}|)^{1/2}] \exp\left\{-\frac{1}{2}\left[\sum_{i=1}^{2} \sum_{j=1}^{2} v^{ij}(x_1 - \mu_i)(x_j - \mu_j)\right]\right\} \tag{5-74}$$

In what follows we continue with the subject of distributions having several variables proceeding ultimately to the multivariate normal distributions. It may be helpful to some readers to leave this section for a later reading and to go on to subject of statistical inference.

## 7. Multivariate Distributions

### 7.1 Introduction and Definitions

The ideas about distributions of one and two variables can be extended several variables. For instance, when we say the variables $X_1, X_2, X_3$ are jointly distributed as $f(X_1, X_2, X_3)$ we mean that the probability of finding $X_1$ in the interval $[a, b]$ and the probability of finding $X_2$ in the interval $[c, d]$ and $X_3$ in $[e, f]$ is given by

$$\int_e^f \int_c^d \int_a^b f(X_1, X_2, X_3)\, dX_1\, dX_2\, dX_3 \tag{5-75}$$

The above considerations can of course be generalized to several variables and infinite intervals. Thus, we can say that the variables $X_1, X_2, \ldots, X_p$ have a joint frequency function provided

$$f(X_1, X_2, \ldots, X_p) \geq 0 \qquad (-\infty < X_i < \infty; \quad i = 1, 2, \ldots, p) \tag{5-76}$$

and

$$\int_{-\infty}^{\infty} \int_{-\infty}^{\infty} \cdots \int_{-\infty}^{\infty} f(X_1, X_2, \ldots, X_p)\, dX_1\, dX_2 \cdots dX_p = 1 \tag{5-77}$$

In discussing multivariate distributions we also consider marginal distributions and conditional distributions.

Considering first the marginal distribution, if the variables $X_1, X_2, \ldots, X_p$ have the joint frequency function $f(X_1, X_2, \ldots, X_p)$, then the marginal distribution of $X_1, X_2, \ldots, X_r$, where $r < p$, is given by

$$f^*(X_1, X_2, \ldots, X_r) = \int_{-\infty}^{\infty} \int_{-\infty}^{\infty} \cdots \int_{-\infty}^{\infty} f(X_1, X_2, \ldots, X_r)\, dX_{r+1}\, dX_{r+2} \cdots dX_p \tag{5-78}$$

This means that the marginal distribution of the variables $X_1, X_2, \ldots, X_r$ is found by integrating over the variables not included in the subset $r$.

The conditional distributions of the same subset of variables $X_1, X_2, \ldots, X_r$ is obtained when the variables $X_{r+1}, X_{r+2}, \ldots, X_p$ not included in the subset are given certain values. This is given by

$$\begin{aligned} f(X_1, X_2, \ldots, X_r \mid X_{r+1}, X_{r+2}, \ldots, X_p) \\ = f(X_1, X_2, \ldots, X_r)/f^*(X_{r+1}, X_{r+2}, \ldots, X_p) \end{aligned} \tag{5-79}$$

The relationship between marginal distributions, conditional distributions, and statistical independence or independence in the probability sense is delineated by the following statement. Two random variables $X_1$ and $X_2$ are statistically independent [i.e., $\text{cov}(X_1, X_2) = 0$] if and only if the conditional distribution of $X_1$ given $X_2$ is equal to the marginal distribution of $X_1$. This can be written as we have seen before for two variables as

$$f(X_1 \mid X_2) = f^*(X_1) \tag{5-80}$$

The extension to several random variables is obvious and we say that the random variables $X_1, X_2, \ldots, X_p$ are jointly statistically independent if and only if the joint frequency function equals the product of the marginal distributions. Statistical independence between several random variables implies that the following equation must hold

$$f(X_1, X_2, \ldots, X_p) = f^*(X_1)f^*(X_2) \cdots f^*(X_p) \tag{5-81}$$

and all covariances are equal to zero.

## 7.2 Multivariate Normal Distribution

As stated previously, the most important multivariate distribution is the multivariate normal distribution because frequently the set of parameters we estimate have a multivariate normal distribution. Therefore, it is highly desirable that we should have some idea of the properties of such a distribution.

### 7.2.1 *General Properties*

If $X_1, X_2, \ldots, X_r$ are random variables represented by the vector $\mathbf{X} = [X_1, X_2, \ldots, X_r]$. We define the $r$-variate multivariate normal frequency distribution by

$$f(\mathbf{X}) = K \exp[-\tfrac{1}{2}(\mathbf{X} - \boldsymbol{\mu})'\mathbf{R}(\mathbf{X} - \boldsymbol{\mu})] \tag{5-82}$$

where $K$ is a constant, $\mathbf{R}$ a positive definite matrix and $\boldsymbol{\mu}$ is equal to the expected value of the vector $\mathbf{X}$, which can be written as $\mathbf{E}(\mathbf{X}) = \boldsymbol{\mu}$.

With respect to the mean of a single variable it can be shown that

$$E(X_i) = K \int_{-\infty}^{\infty} \int_{-\infty}^{\infty} \cdots \int_{-\infty}^{\infty} X_i \times \exp[-\tfrac{1}{2}(\mathbf{X} - \boldsymbol{\mu})'\mathbf{R}(\mathbf{X} - \boldsymbol{\mu})]\, dX_1 \cdots dX_r = \mu_i \qquad (5\text{-}83)$$

The value of $K$ in the $r$-variate normal distribution given by (5-82) can be shown to be $|\,\mathbf{R}\,|^{1/2}/(2\pi)^{r/2}$.

In the same manner as we transformed the single variate and bivariate normal distributions to their standard form we can transform the multivariate normal distribution to a standard form by putting

$$X_i - \mu_i = Z_i \qquad (i = 1, 2, \ldots, r) \qquad (5\text{-}84)$$

Now, since $dX_i/dZ_i = 0$ if $i \neq j$ and $dX_i/dZ_i = 1$ otherwise, the Jacobian (see Chapter 3) of the transformation is unity and we find that

$$\int_{-\infty}^{\infty} \int_{\infty}^{\infty} \cdots \int_{-\infty}^{\infty} K \exp[-\ \tfrac{1}{2}(\mathbf{X} - \boldsymbol{\mu})'\mathbf{R}(\mathbf{X} - \boldsymbol{\mu})]\, dX_1\, dX_2 \cdots dX_r$$

can be transformed to

$$\int_{-\infty}^{\infty} \int_{-\infty}^{\infty} \cdots \int_{-\infty}^{\infty} K \exp[-(\tfrac{1}{2}\mathbf{Z}'\mathbf{R}\mathbf{Z})]\, dZ_1\, dZ_2 \cdots dZ_r \qquad (5\text{-}85)$$

In the above multivariate normal distribution, whether in the general form given by Eq. (5-82) or in the standard form set with $\mathbf{Z}$ as the vector of the variables, we call $(\mathbf{X} - \boldsymbol{\mu})'\mathbf{R}(\mathbf{X} - \boldsymbol{\mu})$ and $\mathbf{Z}'\mathbf{R}\mathbf{Z}$ the quadratic forms of the multivariate normal distribution. The positive definite matrix $\mathbf{R}$ is the inverse of the variance–covariance matrix $\mathbf{V}$.

In an $r$-variate normal distribution there are $r$ variances and $\frac{1}{2}(r^2 - r)$ covariances, one for each combination of a pair of random variables such as $X_i$ and $X_j$, where $i \neq j$. The $r \times r$ matrix, with diagonal elements the variances of $X_i$ and with $ij$th elements the covariances of $X_i$ and $X_j$, can be written as

$$E\{[\mathbf{X} - \mathbf{E}(\mathbf{X})][\mathbf{X} - \mathbf{E}(\mathbf{X})]'\} = E[(\mathbf{X} - \boldsymbol{\mu})(\mathbf{X} - \boldsymbol{\mu})'] = \mathbf{V} \qquad (5\text{-}86)$$

Further, one can prove that the positive definite matrix $\mathbf{R}$ occurring in the frequency function shown in Eq. (5-82) is related to $\mathbf{V}$ by the equation

$$\mathbf{R}^{-1} = \mathbf{V} \qquad (5\text{-}87)$$

and the resulting distribution has the form [compare Eq. (5-74) for the bivariate]

$$f(\mathbf{X}) = [1/(2\pi)^{r/2}|\mathbf{V}|]^{1/2} \exp\left[-\frac{1}{2}\left(\sum_{i=1}^{r}\sum_{j=1}^{r} v^{ij}X_iX_j\right)\right]$$

### 7.2.2 *Properties of Marginal Distributions of Multivariate Normal Distributions*

In what follows we state some properties of marginal distributions of multivariate normal distributions.

We begin by noting that a marginal distribution of any one random variable in an $r$-variate normal distribution is a simple normal distribution of a single variable

$$f^*(X_1) = [1/(2\pi)^{1/2}v_{11}] \exp\{-[(X_1 - \mu_1)/2v_{x_1}^2]\} \qquad (-\infty < X_1 < \infty) \tag{5-88}$$

Next, one can show that the joint marginal distribution of a subset $l$ of an $r$-variate normal distribution is an $l$-variate normal distribution.

### 7.3.3 *Linear Functions of Normal Variables*

We define a linear function of normal variates as a linear combination of a set of variates $X_1, X_2, \ldots, X_r$ each having a univariate normal distribution. It is important to point out, for such linear combinations, that if $X_1, X_2, \ldots, X_r$ are each distributed as single variate normal distribution then a linear combination of the $X_i$ may or may not be normally distributed. On the other hand if $X_1, X_2, \ldots, X_r$ have a multivariate normal then a linear combination of the $X_i$ is normally distributed. In particular, it can be shown that if the components of the vector $\mathbf{X}$ have a multivariate normal distribution with mean $\boldsymbol{\mu}$ and covariance matrix $\mathbf{V}$, then any linear combination of the $X_i$ is a quantity such as

$$\varepsilon = \sum_{i=1}^{r} b_iX_i \tag{5-89}$$

where the $b_i$ are constants, and it has the distribution of a single variate normal distribution with mean $\sum_{i=1}^{r} b_i\mu_i$ and variance $\sum_{i=1}^{r}\sum_{j=1}^{r} b_ib_j v_{ij}$.

### 7.2.4 *Statistical Independence in Multivariate Normal Distribution*

As we have indicated earlier in the bivariate normal distribution case, two variates are statistically independent whenever $\operatorname{cov}(X_i, X_j) = 0$.

In a multivariate normal distribution we call the variates jointly independent whenever the covariance matrix is diagonal or whenever all the $v_{ii} = 0$. In such cases the distribution function can be written as

$$f(X_1, X_2, \ldots, X_r) = f^*(X_1)f^*(X_2) \cdots f^*(X_r) \tag{5-90}$$

with any $f^*(X_i)$ written as

$$f^*(X_i) = [1/(2\pi)^{1/2}v_{ii}] \exp[(-\tfrac{1}{2}v_{ii})(x_i - \mu_i)]^2 \tag{5-91}$$

### 7.2.5 *Conditional Distributions of the Multivariate Normal Distribution*

Multivariate conditional distributions are used whenever we seek to determine something about certain variates in a distribution where the other variates are prescribed or given. For instance, for the estimation of the mean and covariance matrices of some variates of a normal distribution where the rest of the variates are given we have the following.

*Theorem* 5-1. If an $r \times 1$ vector $\mathbf{X}$ is normally distributed with mean $\boldsymbol{\mu}$ and covariance $\mathbf{V}$ and if the vector $\mathbf{X}$ is partitioned such that

$$\mathbf{X} = \begin{bmatrix} \mathbf{X}_1 \\ \mathbf{X}_2 \end{bmatrix} \tag{5-92}$$

and if

$$\mathbf{X}^* = \begin{bmatrix} \mathbf{X}_1 \\ \mathbf{X}_2^* \end{bmatrix}, \qquad \boldsymbol{\mu} = \begin{bmatrix} \mathbf{U}_1 \\ \mathbf{U}_2 \end{bmatrix}, \qquad \mathbf{V} = \begin{bmatrix} \mathbf{V}_{11} & \mathbf{V}_{12} \\ \mathbf{V}_{21} & \mathbf{V}_{22} \end{bmatrix} \tag{5-93}$$

are the partitions of $\mathbf{X}$, $\boldsymbol{\mu}$ and $\mathbf{V}$ respectively, then the conditional distribution of the vector $\mathbf{X}_1$ given $\mathbf{X}_2^*$ is a multivariate distribution with mean $\mathbf{U}_1 + \mathbf{V}_{12}\mathbf{V}_{22}^{-1}(\mathbf{X}_2^* - \mathbf{U}_2)$ and covariance matrix $(\mathbf{V}_{11} - \mathbf{V}_{12}\mathbf{V}_{22}^{-1}\mathbf{V}_{21})$.

### 7.2.6 *Partial and Multiple Correlations Coefficients*

In discussing some properties of the bivariate normal distribution we previously noted the computation of the correlation coefficient and indicated to some extent its meaning. To some extent those ideas can be extended to multivariate distribution. We use the following notation to extend those ideas.

First we define the quantity $v_{ij \cdot st \cdots r}$ as the covariance of $X_i$ and $X_j$ in the conditional distribution $\mathbf{X}_1$ given $\mathbf{X}_2$, which means that $v_{12 \cdot 567}$, for example, is the covariance between $X_1$ and $X_2$ in the joint conditional distribution of $X_1X_2$ and $X_3$ and $X_4$ given $X_5X_6$ and $X_7$.

The *partial correlation coefficient* $\varrho_{ij \cdot st \cdots r}$ is defined as the correlation between $X_i$ and $X_j$ in the conditional distribution of $\mathbf{X}_1$ given $\mathbf{X}_2^*$.

For example $\varrho_{14\cdot 789}$ represents the correlation between $X_1$ and $X_4$ in the conditional distribution of $X_1, X_2, X_3, X_4, X_5$, and $X_6$ given $X_7$, $X_8$, and $X_9$. We can, show that $\varrho_{ij}$, may be written

$$\varrho_{ij} = v_{ij\cdot st\cdots r}[(v_{ii\cdot st\cdots r})(v_{jj\cdot st\cdots r})]^{1/2} \tag{5-94}$$

The multiple correlation coefficient $\varrho_1$ between two random variables $X_1$, $y$, where $y$ may be considered a composition of more than one variable and which may be written as

$$y = \mathbf{V}_{12}\mathbf{V}_{22}^{-1}(\mathbf{X}_2 - \mathbf{U}_2) \tag{5-95}$$

As in the case of two variables we can write the multiple correlation coefficient

$$\varrho_1 = \frac{E\{(X_1 - \mu_1)[y - E(y)]\}}{\{E[(X_1 - \mu_1)^2]E[y - E(y)]^2\}^{1/2}} \tag{5-96}$$

A number of important relationships between the $v_{ij}$, $v_{ij\cdot rst\cdots}$ and $\varrho_1$ exist. Those are given by

$$v_{11\cdot 23\cdots r} \leq v_{11} \tag{5-97}$$

and

$$\varrho_1 = [1 - (v_{11\cdot 23\cdots r}/v_{11})]^{1/2} \tag{5-98}$$

Lastly, we note that the multiple correlation coefficient satisfies the following inequality

$$0 \leq \varrho_1{}^2 \leq 1 \tag{5-99}$$

### 7.2.7 *Partial Regression Coefficients*

Suppose we write the equation

$$X_1 = \mu_1 + \sum_{i=2}^{r} \alpha_i(X_i - \mu_i) + \delta \tag{5-100}$$

which is writing $X_1$ in terms of $X_2, X_3, \ldots, X_r$. All the symbols in the above equations have the same meaning as before and $\delta$ is the error term with zero mean. Equation (5-100) as indicated previously is called the multiple regression of $X_1$ on $X_2, X_3, \ldots, X_r$. Note that this is the conditional distribution of $X_1$ given $\mathbf{X}_2 = \mathbf{X}_2{}^*$. If we consider the expected value of $X_1$ we obtain

$$E(X_1) = \mu_1 - \sum_{i=2}^{r} \alpha_i\mu_i + \sum_{i=2}^{r} \alpha_i X_i \tag{5-101}$$

Differentiating $E(X_1)$ with respect to $X_i$ we find

$$dE(X_1)/dX_i = \alpha_i \tag{5-102}$$

and the $\alpha_i$ are called the partial regression coefficients. The meaning of this partial regression coefficient is depicted by Eq. (5-102) which indicates that the partial regression coefficient $\alpha_i$ is the change in $E(X_1)$ per unit change in $X_i$.

To show the utility of all the quantities mentioned in this and the previous section suppose we measure the growth of a bacterium for six hours several times at several different temperatures, and suppose we compute the variance of all those numbers which we will call $v_{11}$. Next, suppose we compute the variance at a certain temperature (30°). We call the second number $v_{11\cdot 2}$ indicating the variance of the first variable (growth for six hours) while the second variable temperature was kept constant (at 30°). From this example we hope it is clear that $v_{11} \geq v_{11\cdot 2}$, or that the variance of growth for six hours when all the temperatures are taken into consideration is either greater than or equal to the variance with only one temperature.

For the above example we can have two extremes. Either $v_{11} = v_{11\cdot 2}$, in which case the variance has not been reduced by sampling at only one temperature, or $v_{11} - v_{11\cdot 2} = v_{11}$, in which case the variance is reduced by $v_{11}$ by sampling at only one temperature.

From these considerations and Eq. (5-98), we can say that the square of the multiple correlation coefficient gives us a measure of the fraction of reduction in the variance of the random variable when we use related variables. The multiple correlation coefficient is a very useful quantity in helping us decide which is the best regression equation or what variables are or are not pertinent to the model being analyzed.

A number of additional theorems about variances of multivariate normal distribution, partial correlation coefficients, and multiple correlation coefficients are given without proof.

*Theorem* 5-2. If $v_{11} = v_{12} \cdots = v_{1r}$ then the following is true. The partial regression coefficients $\alpha_i$ in Eq. (5-100) are all zero and the following equalities hold $v_{11} = v_{11\cdot 23\cdots r}$ and $\varrho^2 = 0$.

*Theorem* 5-3. If $v_{13} = v_{14} = \cdots = v_{1r} = v_{23} = \cdots = v_{2r} = 0$, then the partial correlation coefficients of $X_1$ and $X_2$ given $X_3, X_4, \ldots, X_r$ equal the simple correlation coefficients, i.e.,

$$\varrho_{12\cdot 23\cdots r} = \varrho_{12}$$

*Theorem* 5-4. If $X_1, X_2, \ldots, X_r$ are jointly normal with covariance matrix $\mathbf{V}$, then the covariance matrix of the conditional distribution of the vector $\mathbf{X}_1$ given the vector $\mathbf{X}_2$ can be given by deleting the rows and columns that correspond to $\mathbf{X}_2$ from $\mathbf{R} = \mathbf{V}^{-1}$ and then taking the inverse of the remaining matrix.

*Theorem* 5-5. The vector that represents the mean of the conditional distribution $\mathbf{X}_1$ given $\mathbf{X}_2$ is the vector that represents the solution of the equations

$$\partial(\mathbf{X} - \boldsymbol{\mu})'\mathbf{R}(\mathbf{X} - \boldsymbol{\mu})/\partial\mathbf{X}_1 = 0$$

where $(\mathbf{X} - \boldsymbol{\mu})'\mathbf{R}(\mathbf{X} - \boldsymbol{\mu})$ represents the quadratic form of the joint distribution $X_1, X_2, \ldots, X_r$.

*Theorem* 5-6. If $\mathbf{X}$ is normally distributed with mean zero and covariance $v^2\mathbf{I}$ and if $\mathbf{Z} = \mathbf{OX}$ where $\mathbf{O}$ is an orthogonal matrix, then $\mathbf{Z}$ is also normally distributed with zero mean and covariance $v^2\mathbf{I}$.

*Example.* The $4 \times 1$ vector $\mathbf{X}$ is distributed as $N(\boldsymbol{\mu}, \mathbf{V})$ and the quadratic form $(\mathbf{X} - \boldsymbol{\mu})'\mathbf{R}(\mathbf{X} - \boldsymbol{\mu})$ is given by

$$3X_1^2 + 2X_2^2 + 2X_3^2 + X_4^2 + 2X_1X_2 + 2X_2X_4 - 6X_1 - 2X_2 - 6X_3 - 2X_4 + 8.$$

a. Find the conditional distribution of $X_1$ given $X_2$, $X_3$, and $X_4$.
b. Find the partial correlation coefficients $\varrho_{12}$ and $\varrho_{13\cdot 2}$.
c. Find the multiple correlation coefficient.

a. By inspection of the quadratic form using second degree terms only we find that

$$\mathbf{R} = \begin{bmatrix} 3 & 1 & 0 & 0 \\ 1 & 2 & 0 & 0 \\ 0 & 0 & 2 & 1 \\ 0 & 0 & 1 & 1 \end{bmatrix}$$

The inverse of $\mathbf{R}$, the matrix $\mathbf{V}$, is given by

$$\mathbf{V} = \begin{bmatrix} \frac{2}{5} & -\frac{1}{5} & 0 & 0 \\ -\frac{1}{5} & -\frac{3}{5} & 0 & 0 \\ 0 & 0 & 1 & -1 \\ 0 & 0 & -1 & 2 \end{bmatrix}$$

The vector of the means $\boldsymbol{\mu}$ is found by solving

$$\partial(\mathbf{X} - \boldsymbol{\mu})'\mathbf{R}(\mathbf{X} - \boldsymbol{\mu})/\partial\mathbf{X} = 0$$

This yields the following four equations

$$6X_1 + 2X_2 - 6 = 0$$
$$6X_1 + 4X_2 - 2 = 0$$
$$4X_3 + 2X_2 - 6 = 0$$
$$2X_3 + 2X_4 - 2 = 0$$

The solution of these four equations is given by

$$\boldsymbol{\mu} = \begin{bmatrix} X_1 \\ X_2 \\ X_3 \\ X_4 \end{bmatrix} = \begin{bmatrix} 1 \\ 0 \\ 2 \\ -1 \end{bmatrix}$$

To obtain $f(X_1 \mid X_2X_3X_4)$ we use Theorem 5-1. The relevant matrices in this case are

$$\mathbf{V}_{11} = [\tfrac{2}{5}], \qquad \mathbf{V}_{12} = [-\tfrac{1}{5}, \ 0, \ 0], \qquad \mathbf{V}_{22} = \begin{bmatrix} \tfrac{3}{5} & 0 & 0 \\ 0 & 1 & -1 \\ 0 & -1 & 2 \end{bmatrix}$$

$$\mathbf{V}_{22}^{-1} = \begin{bmatrix} \tfrac{5}{2} & 0 & 0 \\ 0 & 2 & 1 \\ 0 & 1 & 1 \end{bmatrix}, \qquad \mathbf{U}_1 = [1], \qquad \mathbf{U}_2' = [0, \ 2, \ -1]$$

The mean of this distribution in accordance with Theorem 5-1 is

$$\mathbf{U}_1 + \mathbf{V}_{12}\mathbf{V}_{22}^{-1}[\mathbf{X}_2 - \mathbf{V}_2] = 1 - (\mathbf{X}_2/3)$$

and the variance according to the theorem is

$$\mathbf{V}_{11} - \mathbf{V}_{12}\mathbf{V}_{22}^{-1}\mathbf{V}_{21} = \tfrac{1}{3}$$

This means that $f(X_1 \mid X_2X_3X_4)$ is normally distributed with mean $1 - (\mathbf{X}_2/3)$ and variance $\tfrac{1}{3}$.

b. To find $\varrho_{12}$ we obtain the covariance matrix of $X_1$ and $X_2$

and $\varrho_{12}$ is given by Eq. (5-94) as

$$\varrho_{12} = -\tfrac{1}{5}/(\tfrac{2}{5}\ \tfrac{3}{5})^{1/2} = -1/\sqrt{6}$$

To find $\varrho_{13\cdot 2}$ we must find the covariance of the pertinent variables in the distribution of $X_1, X_3, X_4$ given $X_2$. This is done by striking the second row and column of the matrix $\mathbf{R}$ above.

$$\mathbf{R}_{134} = \begin{bmatrix} 3 & 0 & 0 \\ 0 & 2 & 1 \\ 0 & 1 & 1 \end{bmatrix} \quad \text{and} \quad \mathbf{R}_{134}^{-1} = \begin{bmatrix} \frac{1}{3} & 0 & 0 \\ 0 & 1 & -1 \\ 0 & -1 & 2 \end{bmatrix}$$

so that

$$\varrho_{13\cdot 2} = 0$$

c. To find $\varrho_1$ we use Eq. (5-98)

$$\varrho_1 = [(v_{11} - v_{11\cdot 234})/v_{11}]^{1/2}$$

Here we note that $v_{11} = \frac{2}{6}$ and $v_{11\cdot 234}$ is obtained by striking the second, third, and fourth rows and columns of $\mathbf{R}$. Inverting this yields $v_{11\cdot 234} = \frac{1}{3}$ so that

$$\varrho_1 = [(\tfrac{2}{5} - \tfrac{1}{5})/\tfrac{2}{5}]^{1/2} = \sqrt{\tfrac{1}{6}}$$

## 8. Statistical Inference

### 8.1 Introduction

In general when one conducts an experiment of a series of experiments one knows or is willing to assume that the outcome of the experiments observe a certain functional form say $f(\mathbf{X}, \mathbf{g})$ where $\mathbf{X}$ is a set of variables under the experimenter's control and $\mathbf{g}$ is a set of parameters to be estimated. All the examples given in subsequent chapters can be represented as the above functional form. The objective of the experimenter is to estimate the unknown parameters $\mathbf{g}$. Furter, the investigator may decide to determine whether or not the estimated parameter is equal to some value. In the first instance the investigator is estimating the parameter and in the second testing a hypothesis about the parameter.

As an example of what is intended we refer the reader to Chapter 1 where we spoke of concentration dependent aggregating systems. The objective there was to determine the association constants and to test whether any given constant is significantly different from zero indicating the absence of certain aggregating species. The question of estimating parameters is dealt with in this chapter and the previous one. The question of hypothesis testing is left to Chapter 6.

## 8.2 Point Estimates and Interval Estimates

When one talks of estimating parameters one can have point estimation or an interval estimation. A *point estimate* of a parameter is a single number. An *interval estimate* is given by two numbers between which the parameter in question is assumed to lie. The larger the distance between the two numbers, for a given estimate, the more certain we are that our estimate is going to lie within that interval. This statement gives rise to the following considerations. A *confidence interval* is an interval whose end points $\theta_1$ and $\theta_2$, where $\theta_2 > \theta_1$, are functions of the observed random variable such that the probability that $\theta_1 < g < \theta_2$ is equal to a predetermined number which we shall denote by $1 - \gamma$. The quantity $1 - \gamma$ can be taken to be 0.75, 0.80, 0.95, 0.99, etc. It will be noted that while $1 - \gamma$ is predetermined, the values of $\theta_1$ and $\theta_2$ will change with every series of observations. We can write the above sentences as follows

$$P(\theta_1 < g < \theta_2) = 1 - \gamma$$

The probability interpretation to be put on the above statement is that if $1 - \gamma$ is 0.95, for example, then if we do the experiment 100 times and if we use the same interval $[\theta_1, \theta_2]$ then in 95 times out of those 100 we can expect the interval to cover the parameter. The above statement does not mean that in any given experiment that gives us a value of $g$ we are 95% certain that the interval covers $g$, because in any given experiment either the interval covers $g$ or it does not cover it.

It is quite possible that there will be many functions $\theta_1$ and $\theta_2$ that will give rise to a confidence interval on $g$ with a coefficient $1 - \gamma$. The trick is to find the best pair of functions that gives us the shortest interval since the larger the interval the less precise we are about the parameter. Thus we can always be sure that our interval covers the parameter by making our interval large enough to cover all possible values of the parameter. For example, the statement that a protein contains 0–100% $\alpha$-helix is certainly correct, but how much information does this interval actually give us. The above considerations lead us to attempt to pick the functions that will give the shortest intervals, unfortunately however, we cannot always find the functions that give the shortest confidence intervals for any value of the parameter. In general, an investigator wants an estimated interval to contain the parameter and be short enough.

Returning to the beginning of the chapter it will be recalled that in the determination of parameters it was pointed out that we obtained a probability distribution for the estimator and in Fig. 5-3 we decided that the first experiment was a better experiment than the second. In what follows we define certain desirable properties of point estimators. Thinking in terms of probability distributions for estimators will simplify further reading.

## 8.3 Desirable Properties of Estimator

*Unbiasedness*

Whenever an estimator $\hat{g}$ equals its mathematically expected value or mean, that is to say whenever $E(\hat{g}) = g$ then this estimator is unbiased.

*Consistency*

A consistent estimator $\hat{g}$ is an estimator which in a series of estimates approaches the true value of the parameter $g$.

*Efficiency*

A sequence of estimates of an estimator $g$ is efficient when the quantity $\sqrt{n}(\hat{g}_n - g)$ is asymptotically normally distributed with zero mean and when the variance of that estimator is less than the variance for any other estimator.

*Minimum Variance*

The minimum variance estimator $\hat{g}$ is defined as the estimator satisfying the following inequality

$$E[\hat{g} - E(\hat{g})]^2 \leq E[g^* - E(g^*)]^2$$

where $g^*$ is any other estimator.

## 8.4 Methods for Making Estimates of Parameters

Although we advocate several methods for estimating parameters in Chapter 4, two methods (which are equivalent in one very important case) will be considered here because of their great statistical advantage. These methods give rise to estimators having desirable properties. The first of these methods is the maximum likelihood estimate and the second to be taken up again in greater detail in Chapter 6 is the least squares method.

### 8.4.1 *The Maximum Likelihood Method for Estimation*

Suppose we have a number of observations $X_j$ ($j = 1, 2, \ldots, m$) from a population having a frequency density denoted by $f(\mathbf{X}, g_1, g_2, \ldots, g_n)$ where $g_1, g_2, \ldots, g_n$ are $n$ parameters to be estimated. The product

$$L(g_1, g_2, \ldots, g_n) = \prod_{j=1}^{m} f(\mathbf{X}, g_1, g_2, \ldots, g_n)$$

considered as a function of the parameters is called the *likelihood function.* In this method of estimation we seek the values of the parameters $g_1$, $g_2, \ldots, g_n$ which maximize the likelihood function $L$. To maximize the function $L$ we take the logarithm of $L$, differentiate it with respect to the parameters and equate to zero, as follows

$$\partial \ln L / \partial g_i = \sum_{j=1}^{m} [\partial \ln f(X_j, g_1, g_2, \ldots, g_n) / \partial g_i] = 0 \qquad (i = 1, 2, \ldots, n)$$

Now if the frequency function $f(x, g_1, g_2, \ldots, g_n)$ is known, then, whether this function is linear or nonlinear, we can use the techniques of function optimization to estimate the parameters $g_1, g_2, \ldots, g_n$ by solving the above equations. Further, under most conditions, there exists a solution for these parameters which for large $m$ is normally distributed and where the inverse of the variance–covariance matrix (the information matrix) has the following structure for the $ij$th element.

$$I(\mathbf{g}) = -E(\partial^2 L / \partial g_i \, \partial g_j) = -mE[\partial^2 \ln f(\mathbf{X}, g_1, g_2, \ldots g_n)] / \partial g_i \, \partial g_j]$$

### 8.4.2 *Least Squares or $L_2$-Norm Estimation*

In the method of least squares we assume that our observations $y_j$ can be given by the following equation

$$y = f_j(\mathbf{X}, \mathbf{g}) + \delta_j \qquad (j = 1, 2, \ldots, m)$$

where $f(\mathbf{X}, \mathbf{g})$ is a function with unknown parameters $g_1, g_2, \ldots, g_n$ and $\delta_j$ are random variables (generally the errors in making an observation). To obtain the least squares estimate we form the sum

$$\sum_{j=1}^{m} \delta_j^2 = \sum_{j=1}^{m} [y_j - f_j(\mathbf{X}, \mathbf{g})]^2$$

and the values of $g_1, g_2, \ldots, g_n$ which minimize this sum are taken as the least square estimate.

The maximum likelihood estimate and the least square estimate are equivalent whenever the errors are independent and normally distributed, i.e., when the vector $\boldsymbol{\delta}$ has zero mean and $v^2\mathbf{I}$ covariance. In that case the likelihood function for the sample $X_1, X_2, \ldots, X_m$ is defined by the product

$$\prod_{j=1}^{m} [1/v(2\pi)^{1/2}] \exp(-\delta_j^2/2v^2) = [1/v^m(2\pi)^{m/2}] \exp[(-\boldsymbol{\delta}'\boldsymbol{\delta})/2v^2]$$

and maximizing the likelihood function is equivalent to minimizing $\boldsymbol{\delta}'\boldsymbol{\delta}$ or $\sum_{j=1}^{m} \delta_j^2$. In order to obtain a maximum likelihood estimate we note that one must know the distribution of the errors whereas one does not need to know the distribution of errors in order to make least square estimate of the parameters. Further, despite the fact that the distribution of errors is not known the least square estimate will yield in many cases unbiased and consistent estimators.

To conclude this section on statistical inference we note that the other important aspect of this subject, the testing of hypotheses is taken up in Chapter 6.

## 9. Distributions of Quadratic Forms

During the course of this work we will have occasion to test hypotheses. In order to be able to do so, we will have to know the probability distribution of various quantities and these, by and large, are quadratic forms. The reader may omit this section if prepared to accept the given distributions for various quantities. Certainly omission is recommended on a first reading.

### 9.1 Noncentral Chi-Square

Earlier we looked at the distribution of $\mathbf{X}'\mathbf{X}$ where $\mathbf{X}$ is distributed normally with zero mean and unit variance. When this is the case, $\mathbf{X}'\mathbf{X} = \sum_{j=1}^{m} X_j^2$ is distributed as chi-square with $m$ degrees of freedom which we symbolize by $\chi^2(m)$. If we examine the distribution of $\mathbf{X}^1\mathbf{X}$ when the distribution of $\mathbf{X}$ is normal with a mean of $\boldsymbol{\mu}$ and a covariance of $\mathbf{I}$, we find that this quantity has the *noncentral chi-square distribution*. We now state without proof the following theorem:

*Theorem* 5-7. If $\mathbf{X} = [X_1, X_2, \ldots, X_n]$ are independent normal variables with mean $\boldsymbol{\mu}$ and with covariance matrix $\mathbf{I}$, then $\sum_{i=1}^{n} X_i^2 = \phi$

is distributed as the noncentral chi-square distribution with $n$ degrees of freedom and noncentrality parameter $\lambda = \sum_{i=1}^{n} (\mu_i/2)$ symbolized by $\chi'^2(n, \lambda)$. The distribution function for $\phi$ is given by

$$f(\phi) = \left[e^{-\lambda} \sum_{i=1}^{\infty} \lambda\, \phi^{[(n+2i)/2]-1} e^{-\phi/2}\right] \Big/ \{i!2^{(n+2i)/2}\Gamma[(n+2i)/2]\} \quad (5\text{-}103)$$

for $0 \leq \phi < \infty$.

In Eq. (5-103) $\Gamma$, as before, stands for the gamma function. Several other theorems of importance about noncentral chi-square can be stated.

*Theorem* 5-8. If $\phi_1, \phi_2, \ldots, \phi_m$ are jointly independent and each distributed as the noncentral chi-square so that each $\phi_i$ has $n_i$ degrees of freedom and noncentrality parameter $\lambda_i$, then $\Phi = \sum_{i=1}^{m} \phi_i$ has the noncentral chi-square distribution with degrees of freedom $n = \sum_{i=1}^{m} n_i$ and noncentrality parameter $\lambda = \sum_{i=1}^{m} \lambda_i$.

*Theorem* 5-9. If $\mathbf{X}$ is an $n \times 1$ normally distributed vector with mean $\boldsymbol{\mu}$ and covariance $v^2\mathbf{I}$, then $(\mathbf{X}'\mathbf{X}/v^2)$ has the noncentral chi-square distribution with $n$ degrees of freedom and noncentrality parameter equal to $\lambda = \boldsymbol{\mu}'\boldsymbol{\mu}/2v^2$.

*Theorem* 5-10. If $\mathbf{X}$ is an $n \times 1$ normally distributed vector with mean $\boldsymbol{\mu}$ and a variance $\mathbf{d}$ where $\mathbf{d}$ is a diagonal matrix, then $\mathbf{X}'\mathbf{d}^{-1}\mathbf{X}$ has the noncentral chi-square distribution with $n$ degrees of freedom and noncentrality parameter $\lambda = (\boldsymbol{\mu}'\, \mathbf{d}^{-1}\boldsymbol{\mu})/2$.

### 9.2 Noncentral $F$-Distributions

Consideration of the ratio of quantities having the noncentral chi-square distribution leads us to the investigation of quantities having properties like Snedecor's $F$-distribution. This sort of thing is quite analogous to the consideration of the ratios of quantities having the central chi-square distribution considered earlier. We state the following theorems about those quantities.

*Theorem* 5-11. If two random variables $\phi$ and $z$ are distributed as the noncentral chi-square distributions with $n$ and $m$ degrees of freedom respectively, and noncentrality parameters $\lambda$ and zero, and if $\phi$ and $z$ are independent, then the quantity

$$\varrho = (\phi m/nz)$$

is distributed as the *noncentral F-distribution* with $n$ and $m$ degrees and with noncentrality parameter $\lambda$. The distribution function of $\varrho$ is given by Eq. (104)

$$f(\varrho) = \frac{\sum_{i=1}^{\infty} \Gamma[(2i+m+n)/2](n/m)^{(2i+n)/2}\lambda\, e^{-2}\varrho^{(2i+n-2)/2}}{\Gamma(m/2)\Gamma[(2i+n)/2]i![1+(n\varrho/m)]^{(2i+n+m)/2}} \quad (0 \leq \varrho < \infty) \tag{5-104}$$

## 9.3 Statistical Theorems on Quadratic Forms

The following theorems about quadratic forms are mostly pertinent to the distribution of various quantities arising during the course of computation of linear models.

*Theorem* 5-12. If the vector $\mathbf{X}$ is normally distributed with zero mean and covariance $\mathbf{I}$, a necessary and sufficient condition that the quadratic form $\mathbf{X}'\mathbf{A}\mathbf{X}$ be distributed as $\chi^2(p)$ is that $\mathbf{A}$ is idempotent with rank $p$.

*Theorem* 5-13. If $X$ is normally distributed with mean $\boldsymbol{\mu}$ and covariance $\mathbf{I}$, then $\mathbf{X}'\mathbf{A}\mathbf{X}$ is distributed as the noncentral chi-square with $p$ degrees of freedom and noncentrality parameter $\lambda$ where $\lambda = \boldsymbol{\mu}'\mathbf{A}\boldsymbol{\mu}$ if and only if $\mathbf{A}$ is idempotent with rank $p$.

*Theorem* 5-14. If $\mathbf{X}$ is normally distributed with mean $\boldsymbol{\mu}$ and covariance $v^2\mathbf{I}$ then $\mathbf{X}'\mathbf{A}\mathbf{X}$ is distributed as the noncentral chi-square with $p$ degrees of freedom and noncentrality parameter $\lambda = \boldsymbol{\mu}'\mathbf{A}\boldsymbol{\mu}/2v^2$ if and only if $\mathbf{A}$ is idempotent with rank $p$.

*Theorem* 5-15. If $\mathbf{X}$ is normally distributed with zero mean and covariance $\mathbf{V}$, then $\mathbf{X}'\mathbf{B}\mathbf{X}$ is distributed as the central chi-square with $p$ degrees of freedom if and only if $\mathbf{BV}$ is idempotent with rank $p$.

*Theorem* 5-16. If $\mathbf{X}$ is normally distributed with mean $\boldsymbol{\mu}$ and covariance $\mathbf{V}$ then $\mathbf{XBX}$ has the noncentral chi-square with $p$ degrees of freedom and parameter $\lambda = \boldsymbol{\mu}'\mathbf{B}\boldsymbol{\mu}/2$ if and only if $\mathbf{BV}$ is idempotent and $\mathbf{B}$ is of rank $p$.

## 9.4 Independence of Quadratic Forms and Linear and Quadratic Forms

The independence of quadratic forms and linear and quadratic forms finally reflects on the distribution properties of those forms. The following theorems cover the properties we are most interested in.

*Theorem* 5-17. If $\mathbf{X}$ is normally distributed with mean $\boldsymbol{\mu}$ and covariance $\mathbf{I}$, then the two positive semidefinite quadratic forms $\mathbf{X}'\mathbf{AX}$ and $\mathbf{X}'\mathbf{BX}$ are independent if and only if $\mathbf{AB}$ is equal to zero.

*Theorem* 5-18, If $\mathbf{X}$ is normally distributed with mean $\boldsymbol{\mu}$ and covariance $\mathbf{I}$, then the set of positive semidefinite quadratic forms $\mathbf{X}'\mathbf{A}_1\mathbf{X}$, $\mathbf{X}'\mathbf{A}_2\mathbf{X}, \ldots, \mathbf{X}'\mathbf{A}_r\mathbf{X}$ are jointly independent if and only if $\mathbf{A}_i\mathbf{A}_j = 0$ for all $i \neq j$.

*Theorem* 5-19. If $\mathbf{X}$ is normally distributed with mean $\boldsymbol{\mu}$ and covariance $\mathbf{I}$, then the set of positive semidefinite quadratic forms $\mathbf{X}'\mathbf{A}_1\mathbf{X}$, $\mathbf{X}'\mathbf{A}_2\mathbf{X}, \ldots, \mathbf{XA}_r\mathbf{X}$ are jointly independent and $\mathbf{XA}_i\mathbf{X}$ is distributed as the noncentral chi-square with degree of freedom $n_i$ and noncentrality parameter $\lambda = \mathbf{A}_i\boldsymbol{\mu}/2$ (where $\mathbf{A}_i$ is of rank $n_i$) if any two of the following conditions are satisfied. (a) the $\mathbf{A}_i$ are idempotent, (b) $\sum_{i=1}^r \mathbf{A}_i$ is idempotent, or (c) $\mathbf{A}_i\mathbf{A}_j = 0$ for all $i \neq j$.

*Theorem* 5-20. If $\mathbf{X}$ is normally distributed with mean $\boldsymbol{\mu}$ and covariance $\mathbf{I}$ and if $\mathbf{X}'\mathbf{X} = \sum_{i=1}^r \mathbf{X}'\mathbf{A}_i\mathbf{X}$, then either of the two following conditions is necessary for $\mathbf{XAX}$ to have the noncentral chi-square distribution with $n_i$ degrees of freedom and noncentrality parameter $\lambda_i = \frac{1}{2}\boldsymbol{\mu}'\mathbf{A}_i\boldsymbol{\mu}$ (where $n_i$ is the rank of the matrix $\mathbf{A}_i$) and for the set $\mathbf{X}'\mathbf{A}_1\mathbf{X}$, $\mathbf{X}'\mathbf{A}_2\mathbf{X}, \ldots, \mathbf{X}'\mathbf{A}_r\mathbf{X}$ to be jointly independent. The first condition is that $\mathbf{A}_1, \mathbf{A}_2, \ldots, \mathbf{A}_r$ are each idempotent, and the second is that $\mathbf{A}_i\mathbf{A}_j = 0$ for all $i \neq j$.

*Theorem* 5-21. If $\mathbf{X}$ is normally distributed with mean $\boldsymbol{\mu}$ and covariance $\mathbf{I}$ and if $\mathbf{X}'\mathbf{X} = \sum_{i=1}^r \mathbf{X}'\mathbf{A}_i\mathbf{X}$, then a necessary and sufficient condition that $\mathbf{X}'\mathbf{A}_i\mathbf{X}$ be distributed as the noncentral chi-square with $n_i$ degrees of freedom and noncentrality parameter $\lambda_i = \boldsymbol{\mu}'\mathbf{A}_i\boldsymbol{\mu}/2$ (where $n_i$ is the rank of $\mathbf{A}_i$) and for $\mathbf{X}'\mathbf{A}_1\mathbf{X}$, $\mathbf{X}'\mathbf{A}_2\mathbf{X}, \ldots, \mathbf{X}'\mathbf{A}_r\mathbf{X}$ to be jointly independent, is that the rank of the matrix obtained by summing the $\mathbf{A}_i$ should be equal to the sum of the ranks of the separate $\mathbf{A}_i$.

*Theorem* 5-22. If $\mathbf{X}$ is normally distributed with mean $\boldsymbol{\mu}$ and covariance $v^2\mathbf{I}$, the positive semidefinite quadratic forms $\mathbf{X}'\mathbf{AX}$ and $\mathbf{X}'\mathbf{BX}$ are independent if the covariance of $\mathbf{X}'\mathbf{AX}$ and $\mathbf{X}'\mathbf{BX}$ is equal to zero.

*Theorem* 5-23. If $\mathbf{X}$ is normally distributed with mean $\boldsymbol{\mu}$ and covariance $v^2\mathbf{I}$, and if $\mathbf{X}'\mathbf{X} = \sum_{i=1}^r \mathbf{X}'\mathbf{A}_i\mathbf{X}$ where the rank of any $\mathbf{A}_i$ is $n_i$, and if any one of the following conditions hold: (a) $\mathbf{A}_i$ is idempotent for all $i = 1, 2, \ldots, r$, (b) $\mathbf{A}_i\mathbf{A}_j = 0$ for all $i \neq j$ and (c) $\sum_{i=1}^r n_i = n$, then it is true that $\mathbf{X}'\mathbf{A}_i\mathbf{X}/v^2$ has the noncentral chi-square distribution with $n_i$

degrees of freedom and noncentrality parameter $\lambda_i = \boldsymbol{\mu}'\mathbf{A}_i\boldsymbol{\mu}'/2v^2$ and it is also true that $\mathbf{X}'\mathbf{A}_i\mathbf{X}$ and $\mathbf{X}'\mathbf{A}_i\mathbf{X}$ are independent for all $i \neq j$.

*Theorem* 5-24. If $\mathbf{B}$ is an $m \times n$ matrix, $\mathbf{A}$ is an $n \times n$ matrix, and $\mathbf{X}$ is normally distributed with mean $\boldsymbol{\mu}$ and covariance $v^2\mathbf{I}$, then the linear forms $\mathbf{BX}$ are independent of the quadratic form $\mathbf{X}'\mathbf{AX}$ if $\mathbf{BA} = 0$.

## References

Brownlee, K. A. (1965). "Statistical Theory and Methodology in Science and Engineering," 2nd ed. Wiley, New York.

Colquhoun, D. (1971). "Lectures on Biostatistics." Oxford Univ. Press (Clarendon), London and New York.

Draper, N. R., and Smith, H. (1966). "Applied Regression Analysis." Wiley, New York.

Graybill, F. A. (1961). "An Introduction to Linear Statistical Models," Vol. I. McGraw-Hill, New York.

Mood, A. M., and Graybill, F. A. (1963). "Introduction to the Theory of Statistics," 2nd ed. McGraw-Hill, New York.

Snedecor, G. W., and Cochran, W. G. "Statistical Methods," 6th ed. Iowa State Univ. Press, Ames, Iowa.

*CHAPTER 6*

# STATISTICS II

## 1. Introduction

In this chapter we start by explaining the basic ideas behind hypothesis testing. Next we deal in a fairly comprehensive manner with much of the statistical aspects of the linear model which occurs quite frequently in this kind of work and which can be represented by

$$f(x) = \sum_i \alpha_i f_i(x) + \delta \tag{6-1}$$

where the symbols are identified in previous chapters.

Lastly, we deal with some of the statistical aspects of nonlinear models and some of the problems caused by them in the sense that ideas for linear models cannot be easily extended to nonlinear models.

# 2. Hypothesis Testing

## 2.1 General Introduction

An investigator lends coherence to his work and enables further progress by postulating a certain hypothesis. This hypothesis may be arrived at from theoretical reasoning or be based on empirical considerations. For example, one may make a model showing the expected behavior of a certain system. Or one can make a number of experiments and advance a hypothesis as to how these experimental facts can be explained. For instance, in Chapter 11 we deal with the optical rotatory dispersion of ribonucleic acids. In that connection we find that the authors performing the experimental work advanced the hypothesis that the optical properties of ribonucleic acids can be explained in terms of the optical properties of dimers. Whether this hypothesis is or is not true is something that has not been examined. It is, however, an example of the formation of a hypothesis from experimental consideration and it should be tested.

Once a hypothesis is made the investigator starts to collect data to investigate how closely they conform to the hypothesis as set up. In the event that the hypothesis is set up from experimental considerations further experimentation might be highly desirable in order to see, on the basis of the gleaned observation, whether or not one accepts or rejects the hypothesis.

One important point is that no general hypothesis can be proved, but it can be disproved. For example, if a crap shooter consistently throws sevens and elevens with a pair of dice on the first throw, the hypothesis that this person can throw sevens and elevens on the first throw is not proved no matter how often this person is capable of turning the trick because one single failure will disprove the hypothesis. The fact that we cannot prove general hypotheses does not mean we cannot accumulate a great deal of evidence in favor of one hypothesis or another. Thus, if the crap shooter in question keeps throwing sevens and elevens on the first throw we can say that for all practical purposes this particular person does what is claimed. But the hypothesis is not proved.

## 2.2 Elementary Considerations in the Setting up of a Hypothesis and the Possibility of Committing Errors

The investigator generally sets the rules as to whether or not a hypothesis is to be accepted or rejected. Suppose an investigator wished to test whether or not a certain coin is fair. To do that the investigator decides to throw the coin in the air ten times to test the hypothesis. If the coin is fair the theoretical probability of the coin turning up heads five times is finite and can be computed. Further, if we consider only ten tosses then there is a finite probability that the events six heads, four heads, and three heads can occur, in fact we can calculate the probability of occurrence of each of the events just mentioned using the binomial formula given in Chapter 5.

$$[N!/X!(N - X)!]p^X q^{N-X} \tag{6-2}$$

Given that there is a finite probability for the occurrence of the events mentioned above, there are a number of possible ways the investigator can set up the hypothesis. Suppose the investigator decides that the coin will be accepted as fair only if five heads turn up. What are the chances of this investigator making a mistake and rejecting the coin as being unfair when it is actually fair. The probability that exactly five heads will turn up in ten throws is given by Eq. (6-2) and is only

$$(10!/5!5!)(\tfrac{1}{2})^5(\tfrac{1}{2})^5 = 63/256 \tag{6-3}$$

So it could very well be that the coin is fair but the hypothesis as set makes the chances that the investigator will make a mistake equal to $1 - (63/256)$. In other words the chances of the investigator rejecting his hypothesis when it is actually correct or rejecting the coin as being unfair when it is actually fair is 193/256. This is a fairly high probability; and one way of diminishing it is to decide to accept the coin as being fair if four, five, or six heads turn up. Again we ask what are the chances of this investigator making a mistake. The probability of four, five, or six heads in ten tosses is

$$\begin{aligned}
&[(10!/4!6!)(\tfrac{1}{2})^6(\tfrac{1}{2})^4] + [(10!/5!5!)(\tfrac{1}{2})^5(\tfrac{1}{2})^5] + [(10!/6!4!)(\tfrac{1}{2})^4(\tfrac{1}{2})^6] \\
&\quad = 105/512 + 63/256 + 105/512 \\
&\quad = 336/512
\end{aligned} \tag{6-4}$$

which means the chances of the investigator making a mistake is $1 - (336/512) = (176/512)$. Consequently by deciding to accept the coin as being fair if four, five or six heads turn up the investigator has lessened the chances of making a mistake. If we carry this reasoning to its ultimate conclusion, we find, that since there is a finite probability for a fair coin to produce *any* event from no heads to ten heads in ten tosses, then if we want to make certain that we do not make the mistake that we are rejecting the coin as being unfair when it is actually fair, we simply accept the coin as being fair whatever the outcome. In other words, if we want to make sure we do not make the mistake of rejecting the coin as being unfair when it is actually fair we set up the hypothesis that any event from no head to ten heads in ten tosses of the coin will make us accept the coin as being fair.

Before we go any further with this discussion we would like to define the kind of error we are attempting to minimize by the above reasoning. The error we are attempting to eliminate here is what is known as a *type I error or the error of rejecting a hypothesis when it is actually true.*

At this stage we might be told that an unfair coin or a loaded coin if tossed ten times can have some or even all the possible outcomes of the fair coin (note we cannot calculate the probable outcomes of ten tosses of an unfair coin unless we known how it is unfair). Clearly, therefore we have to consider that possibility of making the mistake of accepting the coin as being fair when it is actually unfair. *Accepting a hypothesis as true* (*in this case that the coin is fair*) *when it is actually untrue* (*the coin is unfair*) *is making an error of type II.* Consequently the happy state of affairs which we embarked upon by deciding that we will eliminate the possibility of making the mistake of rejecting the coin as being unfair when it is actually fair (avoiding a type I error) has confronted us with the possibility of us making an error of type II. Thus we have increased the possibility of accepting the coin as actually fair, when in fact it is unfair.

Suppose we wish to consider the problem of decreasing our chances of making type II errors. We let the reader decide which of the following hypotheses diminishes the chances of making a type II error. (a) Accept the coin as being fair if anywhere from three to seven heads turn up in ten throws of the coin. (b) Accept the coin as being fair if anywhere from two to eight heads turn up in ten throws of the coin. It should be evident that hypothesis (a) diminishes the chances of making a type II error. It should also be evident that hypothesis (a) increases the probability of making a type I error. The main lesson we learn from such considerations is that attempts to decrease one type of error results in the increase in the error of the other type. These considerations must be dealt with in the testing of hypotheses.

In practice it may well be that one type of error is much more serious than the other and when we make a decision about a hypothesis in such instances we would like to limit the more serious of those errors. Increase of the size of the sample can reduce both types of error but it requires several repeat experiments which might be quite expensive.

In general the hypothesis being tested is called the *null hypothesis*. For example, in the coin tossing experiment the hypothesis that the coin is fair is the null hypothesis. This is in contrast to other hypotheses the experimenter is willing to postulate or accept. Such hypotheses are called *alternative hypotheses* or just alternatives.

## 2.3 The Level of Significance

In the testing of hypotheses the maximum probability that we are willing to risk in a type I error is called the level of significance and is generally given by a number $\gamma$. This level of significance is generally set at 0.05 or 0.01 which means that we are 95% confident or 99% confident that we have not rejected a true hypothesis. The level of significance could also be set at 0.1, 0.2, etc. We merely mention 0.05 and 0.01 first because they are the levels most commonly used by statisticians.

In general the hypothesis that is being tested is called the *null hypothesis* represented by $H_0$ as opposed to another hypothesis an investigator might be willing to accept which is called the *alternative hypothesis* or simply the alternative which is usually denoted by $H_1$. Symbolically we can write

$$\gamma = \text{size of type I error} = P(\text{rejecting } H_0 \mid H_0 \text{ true})$$

## 2.4 Elementary Notions on How a Hypothesis is Tested

As we mentioned earlier the hypotheses about the parameters we estimate can be obtained either from theoretical considerations or from experiment. In the linear model

$$f(x) = \sum_i \alpha_i f_i(x) + \delta$$

when the errors are randomly distributed, the coefficients can be shown to have certain distributions. If we have a parameter or statistic whose distribution we think we know, then what are we doing when we test hypotheses about such a parameter? For example, suppose from theoretical consideration that a parameter we are interested in, call it $S$, is normally distributed with known mean $\mu$ and standard deviation $v$. Then the quantity $Z$ given by

$$Z = (S - \mu)/v \tag{6-5}$$

from previous considerations is normally distributed with mean zero and standard deviation 1 (see Chapter 5). In order to test our hypothesis that the parameter or statistic $S$ is normally distributed we run an experiment to estimate $S$ and we choose a level of significance. If we choose our level of significance at 0.036, we are deciding that we want to be 96.4% certain that we do not commit a type I error. We now insert the value of $S$ we estimated from our experiment in (6-5) to obtain a value of $Z$. Examining Fig. 6-1 which represents a curve of normal distribution, we find that all the values of $Z$ lying between $-2.10$ and 2.10 represent about 96.4% of the unshaded area. Consequently any value of $S$ which yields a value of $Z$ lying between $-2.10$ and 2.10 (these numbers are obtained from statistical tables) allow us to accept the hypothesis that $S$ is normally distributed with indicated mean and standard deviation and we are 96.4% confident that this is the case.

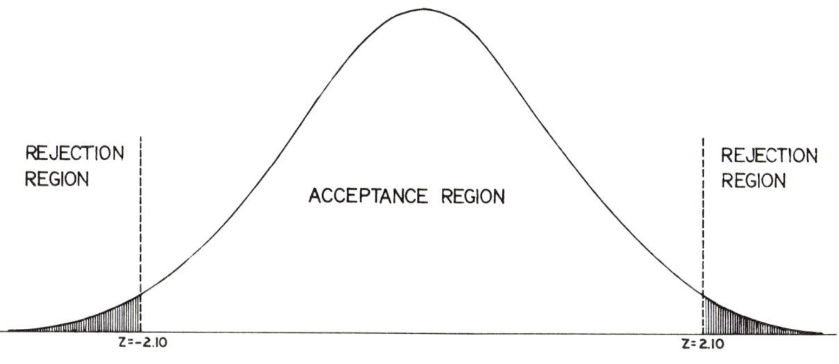


**Fig. 6-1**

A value of $Z$ lying outside the $-2.10$ to 2.10 interval falls in the *critical region* or region of rejection at the 0.036 significance level. Suppose that we wanted to be 99% certain that we will not reject our hypothesis when it is true. Then any values of $Z$ lying between $-2.58$ and 2.58 would have been an acceptable value and in this case the area under the curve lying outside $Z = -2.58$ and $Z = 2.58$ is 1% of the total area.

From the consideration of the acceptance or rejection of the hypothesis at 0.036 and 0.01 level of significance the reader will see that by expanding the range of $Z$ or the *confidence interval* in this case we diminish the probability of making an error of type I but, as in the case of the coin tossing experiment, the greater the size of the confidence interval the greater the chances of committing a type II error. Narrowing the confidence interval will lessen the probability of making a type II error and, if we desire to avoid making a type II error, we can narrow the confidence interval until it vanishes, and that is tantamount to never accepting any hypotheses.

In the testing of hypotheses we have shown how we can limit type I errors and we have shown intuitively that the more we try to limit type I error the greater the probability of making a type II error. Under any given hypothesis it would be extremely helpful if we knew the probability of making a type II error. Graphs showing the probability of making type II errors under various hypotheses are called *operating characteristic curves* and they indicate how any given test will minimize an error of type II or the power of a test.

*Example.* 1. In a sample of 100 tosses of a coin what is the probability of obtaining between 40 and 60 heads inclusive if the coin is fair?

2. Suppose our hypothesis is that the coin is fair if we obtain any outcome between 40 and 60 heads inclusive. What is the probability of making a type I error?

3. What is the level of significance of the decision rule that we have adopted?

4. If the coin is, in fact, not a fair coin and the probability of getting a head is 0.7, what is the probability of making a type II error on the basis of our adopted decision rule?

1. To obtain the probability of getting 40 and 60 heads inclusive we use the binomial distribution and the required probability is,

$$ {}_{100}C_{40}(\tfrac{1}{2})^{40}(\tfrac{1}{2})^{60} + {}_{100}C_{41}(\tfrac{1}{2})^{41}(\tfrac{1}{2})^{59} + \cdots + {}_{100}C_{59}(\tfrac{1}{2})^{59}(\tfrac{1}{2})^{41} + {}_{100}C_{60}(\tfrac{1}{2})^{60}(\tfrac{1}{2})^{40} $$

Recalling that the binomial distribution approximates the normal distribution when both $Np$ and $Nq$ are greater than 5 we can, in this case, use the normal approximation since $Np = Nq = 100(\frac{1}{2}) = 50$.

The mean of our binomial distribution is $100(\frac{1}{2}) = 50$ and the standard deviation $v = [(100)(\frac{1}{2})(\frac{1}{2})]^{1/2} = 5$ on a continuous scale between 40 and 60 heads is the same as between 39.5 and 60.5. Transforming to standard units we find that

$$ (39.5 - 50)/5 = -2.10 $$

and

$$ (60.5 - 50)/5 = 2.10 $$

The required probability of getting anywhere between 40 and 60 heads is the area under the normal curve $Z = +2.10$ and $Z = -2.10$ from the tables we find this to be 0.9642.

2. A type I error is the error of rejecting the hypothesis when it is actually correct. In this case the probability of *not* getting between 40 and 60 heads inclusive. This we compute as follows

$$1 - 0.9642 = .0358$$

To interpret our decision rule graphically we show the normal curve in Fig. 6-1.

3. For any sample of 100 tosses we compute a sample statistic as follows $Z = (S - 50)/5$, where $S$ is the number of heads. If $Z$ is greater than $+2.10$ or less than $-2.10$ we reject the coin as being unfair since the scores fall in the region of rejection. The shaded area in the figure is of course the probability of making a type I error and it is also called the level of significance. Thus we can say, depending on the $Z$ score that we have accepted or rejected, the hypothesis at 3.58% level of significance.

4. If the coin is in fact unfair with $P = 0.7$, what are our chances of accepting it being fair on the basis of our decision rule? The distribution of heads in 100 tosses of a fair coin is shown in Fig. 6-2. The distribution of heads in 100 tosses of a coin where the probability of getting a head is shown by the normal curve to the right of our first normal curve. From Fig. 6-2 we note that the probability of finding the coin to be fair when in fact $P = 0.7$ is given by the area $\beta$. To compute this area we note that $P = 0.7$ and $q = 0.3$, making the mean of the distribution under $P = 0.7$ equal to 70 and the standard deviation $[(100)(0.7)(0.3)]^{1/2} = 4.58$. In standard units

$$(60.5 - 70)/4.58 = -2.07 \qquad \text{and} \qquad (39.5 - 70)/4.58 = -6.66$$

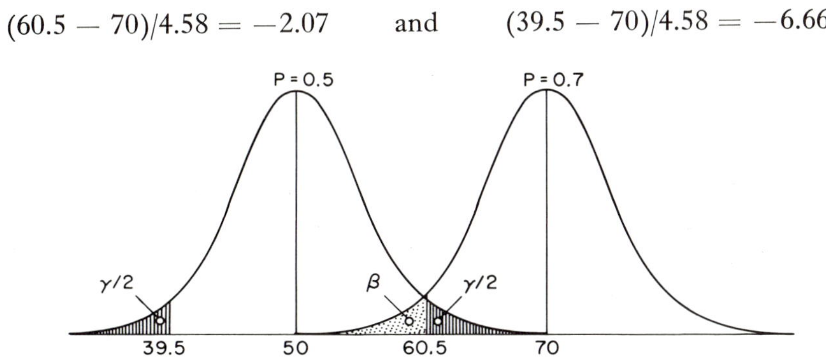


**Fig. 6-2**

This area is therefore the area under the normal curve between $Z = -2.07$ and $Z = -6.66$ which is found in the table to be 0.0192. Accordingly there is very little chance of accepting the coin as being fair under our decision rule if in fact $P = 0.7$. However, when $P = 0.6$ then one computes $\beta$ to be 0.5040 and the chance of making a type II error under our decision rule is greatly increased. If we make a plot of $P$ vs. $\beta$ we obtain what is sometimes called an *operating characteristic curve* and it will be recognized that the sharper the peak of the operating characteristic curve the better our decision rule. Lastly the curve of $1 - \beta$ vs. $P$ is called a *power curve* and the quantity $1 - \beta$ is called the *power function* and it indicates the power of test in rejecting hypotheses which are false. Figures 6-3 and 6-4 show the relevant curve.

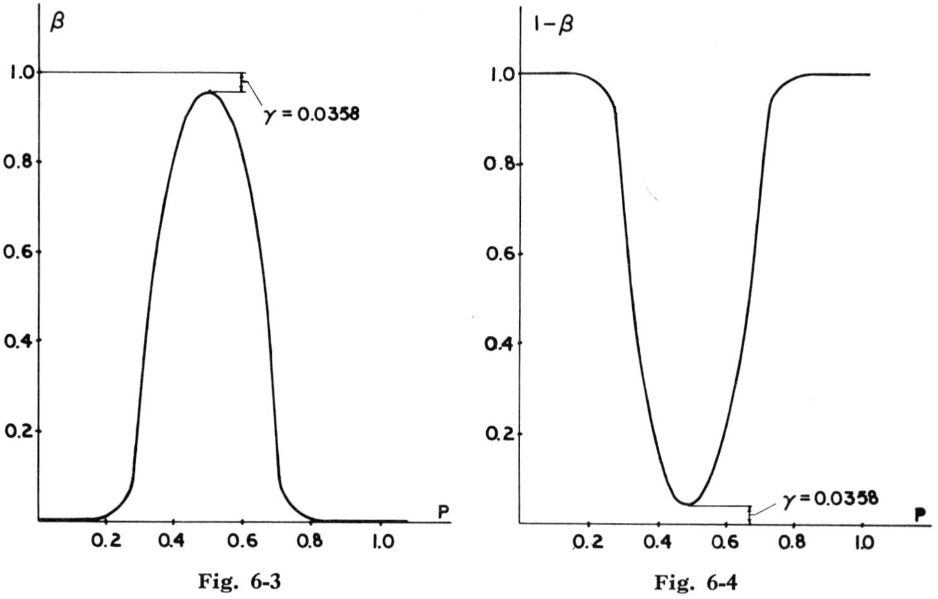


**Fig. 6-3** **Fig. 6-4**

## 2.5 An Example for Testing a Parameter

In the above example hypotheses about the fairness of a coin were tested whereas in most of the work we do we will want to test hypotheses about parameters. Much the same procedure is followed in such cases if we know the distribution of the parameter. To work out an example of this sort let us assume we have a set of pairs of data $(x_i, y_i)$ and we want to fit this data to a model

$$y = \alpha_1 + \alpha_2 x$$

and we want to test a hypothesis about the parameter $\alpha_2$. First we note the distribution of $\alpha_2$. If the hypothesis is that $\alpha_2$ is equal to some value $\alpha_2^0$ is true and we know the theoretical variance $v_\alpha$ of $\alpha_2$, then the random variable

$$(\hat{\alpha}_2 - \alpha_2^0)/v_\alpha$$

is normally distributed. Here $\hat{\alpha}_2$ is the estimated value of $\alpha_2$ from a regression problem. On the other hand, if we don't know the variance and have to estimate $v_\alpha$ by $S_\alpha$, then the random variable

$$(\hat{\alpha}_2 - \alpha_2^0)/S_\alpha$$

has a Student's $t$-distribution with $m - 2$ degrees of freedom where $m$ is the number of data points and where

$$S_\alpha^2 = \frac{[\sum_{i=1}^{m} (y_i - \hat{y}_i)/(m-2)]}{\sum_{i=1}^{m} (x_i - \bar{x})^2}$$

where $\bar{x}$ is the mean of $x$ and $\hat{y}_i$ is the estimated value of $y$. The general procedure of testing hypothesis about $\alpha_2$ from a regression problem can be outlined as follows: Set up the null hypothesis $\alpha_2 = \alpha_2^0$ against an alternative hypothesis $\alpha_2 \neq \alpha_2^0$.

1. Decide on a level of significance $\gamma$.
2. Find $t_{\gamma/2,m-2}$ from Student's $t$ tables.
3. The critical region will be $t \leq -t_{\gamma/2,m-2}$ and $t \geq t_{\gamma/2,m-2}$
4. Estimate $\hat{\alpha}_2$ from the data by finding the regression line and calculate

$$t = (\hat{\alpha}_2 - \alpha_0)/S_\alpha$$

5. Find if $t$ falls in the critical region and accept or reject accordingly.

For the example of a least squares fit shown in Chapter 4, let us test the hypothesis that $\alpha_2 = 0$ against $\alpha_2 \neq 0$ at a level of significance $\gamma = 0.01$. Here

$$t_{\gamma/2,m-2} = t_{0.005,(8)} = 3.355$$

The regression line it will be recalled is

$$\hat{y}_i = 46.5 + 499x_i$$

Here

$$t = (0.4994 - 0)/0.112 = 4.46$$

this makes $t$ in the critical region so that we may reject that hypothesis that $\alpha_2 = 0$. If on the other hand we had estimated $\alpha_2$ to be 0.2 then $t$ would not be in the critical region and one should accept the hypothesis that $\alpha_2 = 0$ or accept that the slope of the straight line is probably zero at the level of significance chosen.

## 3. Formalizing Notions on Hypothesis Testing

### 3.1 Definitions

To formalize some of the notions we have been discussing, we define a statistical hypothesis as an assertion (by the investigator) about the distribution of one or more random variables.

If the statistical hypothesis about the parameters completely specifies the distribution of these parameters it is called a simple *statistical hypothesis*; if it does not it is a *composite statistical hypothesis.*

A *test* of a statistical hypothesis involves a rule which leads to the acceptance or rejection of the hypothesis. The probability of making a type I error (rejecting the hypothesis when it is true) is given by $\gamma$.

In testing any hypothesis on the basis of random sampling the sample space is divided into two regions (subspaces). If the observed sample point falls in one region (subspace) $\omega$ we shall reject the hypothesis, and if the sample point falls in the complimentary region (subspace) $\Omega - \omega$, where $\Omega$ is the whole sample space, the hypothesis is accepted. The region $\omega$ is called the *critical region* and $\gamma$ of the previous paragraph is the size of the critical region. The region $\Omega - \omega$ is called the *region of acceptance.* See Fig. 6-5 for explanation of terms. In this figure the distribution $P(\hat{\theta}, \theta_0)$ is the expected distribution of the estimated parameter denoted by $\hat{\theta}$ under the hypothesis that $\theta = \theta_0$ or the null hypothesis and $\Omega$ is the whole space and may be thought of as the whole of the $\theta$ axis.

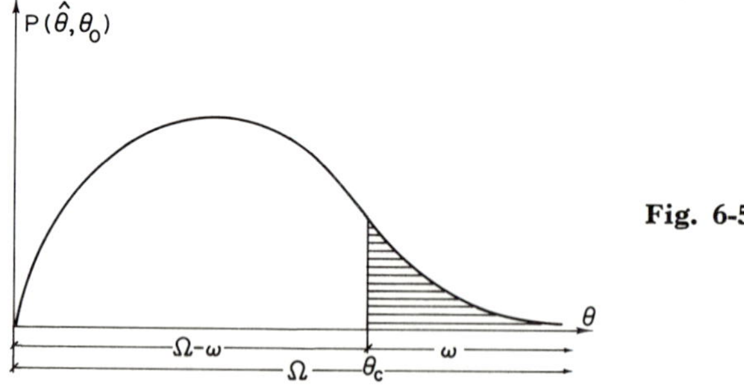


Fig. 6-5

The point $\theta_c$ divides the sample space into a region of acceptance and a critical region (sometimes called the region of rejection).

The probability of a type I error in the previous notation becomes

$$\int_{\theta_c}^{\infty} P(\hat{\theta}, \theta_0)\, d\theta = \gamma \tag{6-6}$$

The probability of a type II error depends on the alternative hypothesis. The distribution of the estimated parameter under the alternative hypothesis, $\theta = \theta_1$, will be given by $P(\hat{\theta}, \theta_1)$. The probability of making a type II error denoted by $\beta$ can then be written as

$$\int_{-\infty}^{\theta_c} P(\hat{\theta}, \theta_1)\, d\theta = \beta \tag{6-7}$$

The power of a test of a hypothesis $\theta = \theta_0$ or the null hypothesis against an alternative hypothesis $\theta = \theta_1$ is given mathematically by

$$\pi(\theta) = 1 - \beta = \int_{\theta_c}^{\infty} P(\hat{\theta}, \theta_1)\, d\theta \tag{6-8}$$

The notation of Eq. (6-8) makes it clear that the power of the test depends on the alternative hypothesis ($\pi(\theta)$ is also called the power function of the test). For example, in the case of 100 tosses of the coin the power of the test is going to be different in the following two cases. If the coin is not fair and the true probability is 60 heads in 100 tosses the test will have a certain power. In another case if the true probability is 70 heads in 100 tosses the power of the test is going different from the first case and what is illustrated here is that the power of the test depends on the probability of the alternative hypothesis.

In the testing of hypotheses we generally choose the probability of type I error to be a small number such as 0.05 or 0.01 and then choose the critical region to minimize type II error and maximize the power $\pi(\theta)$.

In addition to its dependence on an alternative hypothesis the power of a test depends on the choice of the critical region. In fixing the magnitude of the probability of a type I error the reader will observe that *several* critical regions can be constructed to conform to that specification. We note some of those in Fig. 6-6 which can be represented as

$$\int_{\theta_a}^{\theta_b} P(\hat{\theta}, \theta_0)\, d\theta = \int_{\theta_c}^{\theta_d} P(\hat{\theta}, \theta_0)\, d\theta = \gamma$$

Our stipulation, however, is that the critical region must be chosen to maximize the power function. Suppose we have two tests based on two different critical regions of the same size. Inasmuch as the probability of committing a type I error in those two tests is the same, the best way to choose between them is the relative magnitude of the probabilities of committing a type II error. This is stated symbolically as follows; If the power function of a given test $\pi_2(\theta) \geq \pi_1(\theta)$ where $\pi_1(\theta)$ is the power function of any other test, then the test function corresponding to $\pi_2(\theta)$ is uniformly most powerful. This ideal of using a uniformly most powerful test is attained in only a few cases.

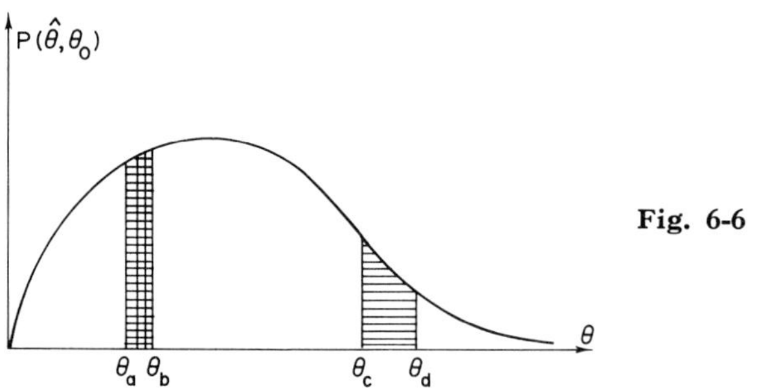


**Fig. 6-6**

## 3.2 The Likelihood Ratio Method

Since we would like our tests to be as powerful as we can get them, we investigate a method based on the ratio of likelihoods. Suppose we have a sample of size $m$ obtained from a distribution $f(y, \theta_1, \theta_2)$ which has two parameters $\theta_1$ and $\theta_2$. We use such a distribution for illustrative purposes but the reasoning we employ will apply to distributions having many more parameters. In the example considered, any possible value of $\theta_1$ and $\theta_2$ can be considered to lie in the space $\Omega$. Thus, if $\theta_1$ and $\theta_2$ can have any positive values including zero, then the space $\Omega$ if the first quadrant of the plane.

Our test of a hypothesis $H_0$ means that we are going to decide on the basis of our experiments whether the parameters are located within a certain region or subspace $\omega$ of the entire parameter space $\Omega$. The subspace $\omega$ is also called *the parameter space under the null hypothesis* $H_0$. When this subspace consists of a single point the hypothesis is considered to be a simple hypothesis and when $\omega$ is more than one point the hypothesis is called composite.

To test the hypothesis we define a test function $\psi(y)$ (which depends on our observations) and which is a rule stating whether the parameters lie in $\omega$ or $\Omega - \omega$. Our procedure it to take the likelihood function (see chapter 5) and to test the hypothesis $H_0$ that $\theta_1$ and $\theta_2$ lie in $\omega$ and the alternative hypothesis $H_1$, is that they do not lie in $\omega$ that is they are in $\Omega - \omega$.

We define $L$ the *likelihood ratio* by

$$L = L(\omega_k)/L(\Omega_k) \tag{6-9}$$

where $L(\Omega_k)$ is the maximum of the likelihood function with respect to the parameters when the parameters occur in $\Omega$ and $L(\omega_k)$ is the maximum of the likelihood function with respect to the parameters when the parameters occur in $\omega$ only. These statements can be understood from consideration of a function of one variable. If we draw the graph of such a function (Fig. 6-7) we see that a function can attain a maximum in a certain region of space $\Omega$ and another maximum in another region of space $\omega$.

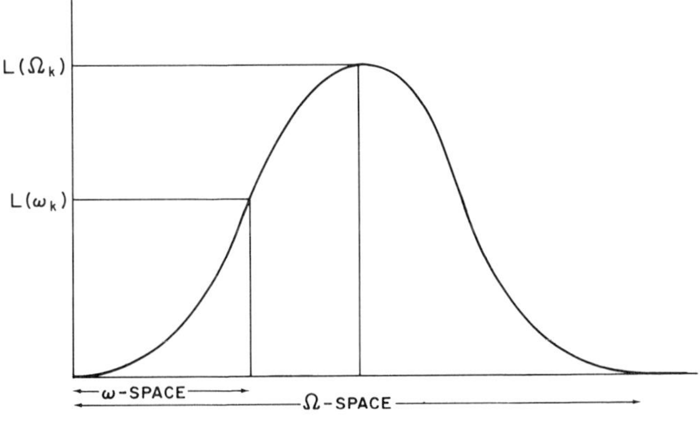


**Fig. 6-7**

The quantity $L$ given by Eq. (6-9) is quite important and essentially the test is: We reject $H_0$ if $L$ is small and accept it if $L$ is large, where $0 \leq L \leq 1$. This is clear from the fact that when $\omega = \Omega$ then $L = 1$ and from the fact that it is always true that $L(\omega_k) \leq L(\Omega_k)$ since $\omega$ is a subspace of $\Omega$.

The likelihood ratio test has a number of important properties. For example, in the event that a uniformly most powerful test exists it is often given by the likelihood ratio test. The likelihood ratio test is very useful for linear models and in this chapter we use it for testing hypotheses on linear models. If our sample size is large the quantity $-2 \ln L$ is approximately distributed as a chi-square variable.

If $L(\omega_k)$ is close to $L(\Omega_k)$ we expect $H_0$ to be true. If on the other hand $L(\omega_k)$ is quite different from $L(\Omega_k)$ then it is reasonable to expect that $H_0$ is false. Therefore we accept $H_0$ if $L$ is large and reject it if it is small. We therefore choose a constant $R$ such that the probability of making a type I error is $\gamma$. The hypothesis is then rejected if $0 < L \leq R$ and accepted if $L > R$. In terms of our previous notation we have

$$\int_0^R f(L, H_0)\, dL = \gamma \tag{6-10}$$

where $f(L, H_0)$ is the frequency distribution under the null hypothesis. The power of the test under an alternative hypothesis $H_1$ is given by

$$\int_0^R f(L, H_1)\, dL \tag{6-11}$$

where $f(L, H_1)$ is the frequency distribution under the alternative hypothesis.

In certain cases it is more convenient to use another variable instead of $L$. In such instances, provided certain conditions are satisfied, this can be done. Suppose $x = g(L)$ where $g(L)$ in a monotonic function of $L$ (see Chapter 3). Then if we change variables such that $L = g(x)$ then

$$\gamma = \int_0^R f(L, H_0)\, dL = \int_{g(0)}^{g(R)} f[g(x)]\, |dg(x)/dx|\, dx \tag{6-12}$$

The above means that as long as we are using monotonic functions of the likelihood ratio we can accept or reject the hypothesis using $x$ as a variable in the same manner as when we are using the likelihood ratio or the variable $L$.

### 3.3 Example of the Use of the Likelihood Ratio

To fix some of the ideas we give the following example. Suppose we select a random sample of size $m$ from a normal population (we use a normal population because it is easy to work with when maximizing likelihood functions). Our normal distribution has a mean of $\mu$ and variance $v^2$ and our sample values are $x_1, x_2, \ldots, x_m$. We desire to test the hypothesis $H_0$ that $\mu = 0$ and the alternative hypothesis is $\mu \neq 0$. We can depict the spaces $\Omega$ and $\omega$ diagramatically in Fig. 6-8.

The parameter space $\Omega$ consists of all points with coordinates $(\mu, v^2)$ such that $-\infty < \mu < \infty$ and $0 \leq v^2 < \infty$ which is to say all points in the plane on and above the line $AB$. On the other hand the space $\omega$ is

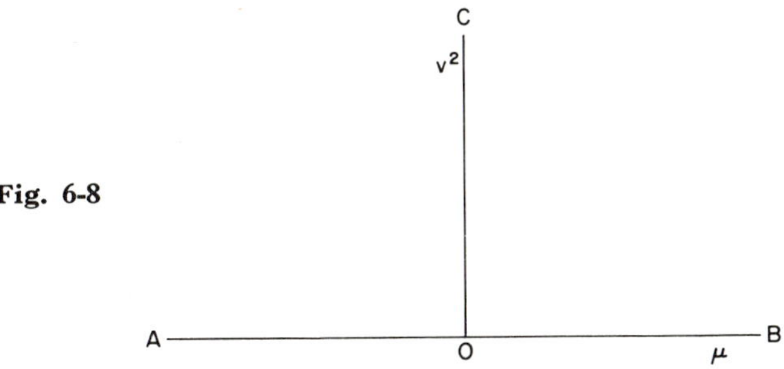


Fig. 6-8

all the points with coordinates $(\mu, v^2)$ such that $\mu = 0$ and $0 \leq v^2 < \infty$. This space consists of the line $OC$ only. From this it is clear that our hypothesis is composite, since the space $\omega$ contains more than one point. Our hypothesis would have been a simple one if, for example, the space $\omega$ was the single point $\mu = 0$ and $v^2 = \lambda$. For the above hypothesis the likelihood function by definition is

$$[1/(v^2 2\pi)^{m/2}] \exp\left[-(1/2v^2) \sum_{j=1}^{m} (x_j - \mu)^2\right] \tag{6-13}$$

Maximizing the likelihood function by differentiating with respect to the parameters $v^2$ and $\mu$ and equating to zero in $\Omega$ space we obtain

$$L(\Omega_k) = \frac{1}{[(2\pi/m) \sum_{j=1}^{m} (x_j - \bar{x})^2]^{m/2}} \exp\left[\frac{-\sum_{j=1}^{m} (x_j - \bar{x})^2}{(2/m) \sum_{j=1}^{m} (x_j - \bar{x})^2}\right] \tag{6-14}$$

$$L(\Omega_k) = e^{-m/2} \Big/ \left[(2\pi/m) \sum_{j=1}^{m} (x_j - \bar{x})^2\right]^{m/2} \tag{6-15}$$

In the smaller space $\omega$, we have $\mu = 0$ and consequently we need only maximize with respect to $v^2$. This yields

$$L(\omega_k) = m^{m/2} e^{-m/2} \Big/ \left(2\pi \sum_{j=1}^{m} x_j^2\right)^{m/2} \tag{6-16}$$

This makes the likelihood ratio

$$\begin{aligned} L &= \left[\sum_{j=1}^{m} (x_j - \bar{x}_j)^2 \Big/ \sum_{j=1}^{m} x_j\right]^{m/2} \\ &= \left\{\sum_{j=1}^{m} (x_j - \bar{x})^2 \Big/ \left[\sum_{j=1}^{m} (x_j - \bar{x})^2 + m\bar{x}^2\right]\right\}^{m/2} \\ &= \left[1 \Big/ \left\{1 + \left[m\bar{x}^2 \Big/ \sum_{j=1}^{m} (x_j - \bar{x})^2\right]\right\}\right]^{m/2} \end{aligned} \tag{6-17}$$

If we put

$$\nu = [m(m-1)\bar{x}^2] \Big/ \Big[ \sum_{j=1}^{m} (x_j - \bar{x})^2 \Big] \qquad (6\text{-}18)$$

then

$$L = \{1/[1 + \nu/(m-1)]\}^{m/2} \qquad (6\text{-}19)$$

and we observe that $L$ is a monotonically decreasing function of $\nu$. Consequently we can set our test with respect to $\nu$ as we would with respect to $L$. Now if the critical region for $L$ is such that $0 \leq L \leq R$ then gives us the critical region for $\nu$ is such that $k \leq \nu < \infty$ where

$$k = [(1/R^{m/2}) - 1](m-1) \qquad (6\text{-}20)$$

The above illustrates some of the ideas about the change in the critical region of the likelihood ratio test upon changing the variable to be tested.

## 4. The General Linear Model of Full Rank

### 4.1 General Considerations: The Normal Theory Case

Consider the linear model

$$f(x_j) = \sum_{i=1}^{n} \alpha_i f_i(x_j) + \delta_j \qquad (j = 1, 2, \ldots, m) \qquad (6\text{-}21)$$

which we will write for simplicity of notation as we wrote it in Chapter 4.

$$\mathbf{Y} = \mathbf{X}\boldsymbol{\alpha} + \boldsymbol{\delta}$$

where $\mathbf{Y}$ is the measured response, $\mathbf{X}$ is an $n \times m$ matrix of known quantities (these are in our previous notation $f_1(x_1), f_1(x_2), \ldots, f_n(x_m)$) and the vector $\boldsymbol{\alpha}$ is the vector of our parameters.

This model is called the general linear hypothesis of full rank if the rank of $\mathbf{X}$ is equal to $n$ where $n \leq m$, i.e., there are more (or the same number of) experiments than unknowns.

The distribution of the vector $\boldsymbol{\delta}$ is of some importance. We consider the case in which the vector $\boldsymbol{\delta}$ has normal distribution with zero mean and variance $v^2$. In the notation of chapter 5 this would be $N(0, v^2\mathbf{I})$ and that means, once again, that all components of our vector $\boldsymbol{\delta}$ have zero mean and variance $v^2$. Here the variance is unknown.

We also consider the case where the vector is such that the expected value of each $\delta_j$ is zero and

$$\text{cov}(\boldsymbol{\delta}) = E(\boldsymbol{\delta}'\boldsymbol{\delta}) = v^2\mathbf{I}$$

The first case in which the vector $\boldsymbol{\delta}$ is normally distributed is of extreme importance. A great deal can be said about this particular case which is called the *normal theory case*. The other case where the distribution of errors is not known has much less statistical theory attached to it but it is still possible to obtain point estimates for the parameters which have desirable properties.

The information which we wish to obtain in the normal theory case is the following: (1) Point estimates of the parameter $\alpha_1, \alpha_2, \ldots, \alpha_n$ and the parameter $v^2$; (2) interval estimates on those parameters, that is to say we want to be confident with a specified degree of probability that the parameter estimated lies in a given interval; (3) we would like point and interval estimates of our response function $f(x)$; (4) we would also like to test the hypotheses that $\alpha_1 = \alpha_1^*$, $\alpha_2 = \alpha_2^*, \ldots, \alpha_n = \alpha_n^*$ where $\alpha_1^*$, $\alpha_2^*, \ldots, \alpha_n^*$ are known constants; (5) in addition to the above hypotheses we test the hypothesis that $\mathbf{d}'\boldsymbol{\alpha}$ is equal to $d_0$, where $\mathbf{d}$ is a known vector and $d_0$ is a known scalar; (6) we may also like to test the hypothesis that $\alpha_1 = \alpha_1^*$, $\alpha_2 = \alpha_2^*, \ldots, \alpha_r = \alpha_r^*$ where $\alpha_1^*, \alpha_2^*, \ldots, \alpha_r^*$ are known constants and the remaining $\alpha_{r+1}, \alpha_{r+2}, \ldots, \alpha_n$ are unspecified; (7) lastly we might wish to test the hypothesis that $\boldsymbol{\lambda}_1\boldsymbol{\alpha} = \boldsymbol{\lambda}_2\boldsymbol{\alpha} = \cdots = \boldsymbol{\lambda}_p\boldsymbol{\alpha} = 0$ where $\boldsymbol{\lambda}_1\boldsymbol{\alpha}, \boldsymbol{\lambda}_2\boldsymbol{\alpha}, \ldots, \boldsymbol{\lambda}_p\boldsymbol{\alpha}$ are linear functions of the $\boldsymbol{\alpha}$ and $\boldsymbol{\lambda}_1, \boldsymbol{\lambda}_2, \ldots, \boldsymbol{\lambda}_p$ are linearly independent vectors.

## 4.2 Point Estimation of Parameters Using Maximum Likelihood

In the normal theory case we shall use the maximum likelihood method to estimate $\alpha_1, \alpha_2, \ldots, \alpha_n$ and $v^2$ (see Chapter 5). The likelihood equations in this case are

$$\begin{aligned} f(\boldsymbol{\delta}, \boldsymbol{\alpha}, v^2) &= \prod_{j=1}^{m} [1/(2\pi v^2)^{m/2}] \exp(-\delta_j^2/2v^2) \\ &= [1/(2\pi v^2)^{m/2}] \exp(-\boldsymbol{\delta}'\boldsymbol{\delta}/2v^2) \\ &= [1/(2\pi v^2)^{m/2}] \exp[-(\mathbf{Y} - \mathbf{X}\boldsymbol{\alpha})'(\mathbf{Y} - \mathbf{X}\boldsymbol{\alpha})/2v^2] \end{aligned} \tag{6-22}$$

Taking the logarithm of $f(\boldsymbol{\delta}, \boldsymbol{\alpha}, v^2)$ we obtain

$$\ln f(\boldsymbol{\delta}, \boldsymbol{\alpha}, v^2) = -(m/2) \ln 2\pi - (m/2) \ln v^2 - (1/2v^2)(\mathbf{Y} - \mathbf{X}\boldsymbol{\alpha})'(\mathbf{Y} - \mathbf{X}\boldsymbol{\alpha}) \tag{6-23}$$

As we outlined in Chapter 5 the maximum likelihood estimates of $\boldsymbol{\alpha}$ and $v^2$ are the solutions to the following equations, in vector notation,

$$\frac{\partial}{\partial \boldsymbol{\alpha}} [\ln f(\boldsymbol{\delta}, \boldsymbol{\alpha}, v^2)] = 0, \qquad \frac{\partial}{\phi v^2} [\ln f(\boldsymbol{\delta}, \boldsymbol{\alpha}, v^2)] = 0 \tag{6-24}$$

Carrying out the differentiation we obtain

$$(1/v^2)(\mathbf{X}'\mathbf{Y} - \mathbf{X}'\mathbf{X}\boldsymbol{\alpha}) = 0$$

and

$$(-m/2v^2) + (\tfrac{1}{2}v^2)[(\mathbf{Y} - \mathbf{X}\boldsymbol{\alpha})'(\mathbf{Y} - \mathbf{X}\boldsymbol{\alpha})] = 0$$

Suppose $\hat{\boldsymbol{\alpha}}$ and $\hat{v}^2$ are solutions to the above equation, this gives

$$\mathbf{X}'\mathbf{X}\hat{\boldsymbol{\alpha}} = \mathbf{X}'\mathbf{Y} \tag{6-26}$$

and

$$\hat{v}^2 = \frac{1}{m} (\mathbf{Y} - \mathbf{X}\hat{\boldsymbol{\alpha}})'(\mathbf{Y} - \mathbf{X}\hat{\boldsymbol{\alpha}}) \tag{6-27}$$

The symbol ^ denotes that we are referring to estimates of parameters. Equation (6-26), we note, is the same equation we obtained for the least squares estimate of the $\boldsymbol{\alpha}$ in Chapter 4 and under those circumstances we note the important fact that when the errors are normally distributed the maximum likelihood estimate of the parameters $\boldsymbol{\alpha}$ is the same as the least squares estimate. Since the matrix $\mathbf{X}$ is of rank $n$ then $\mathbf{X}'\mathbf{X}$ is also of rank $n$. Further, the matrix inverse $(\mathbf{X}'\mathbf{X})^{-1} = \mathbf{C}$ also exists, see Chapter 2. The estimate of the parameters are given by:

$$\hat{\boldsymbol{\alpha}} = \begin{bmatrix} \hat{\alpha}_1 \\ \hat{\alpha}_2 \\ \cdot \\ \cdot \\ \cdot \\ \hat{\alpha}_n \end{bmatrix} = \mathbf{C}\mathbf{X}'\mathbf{Y} \tag{6-28}$$

In chapter 5 we indicated a number of desirable properties of estimators and we defined what we meant by those properties. It will be recalled that among the desirable properties were unbiasedness, consistency, efficiency, and minimum variance. In what follows we make some statements, without proof, about the properties of the estimators we have just obtained.

It can be shown that $\hat{\boldsymbol{\alpha}}$ is an unbiased estimate of $\boldsymbol{\alpha}$ which means that $E(\hat{\boldsymbol{\alpha}})$ the expected value of $\hat{\boldsymbol{\alpha}}$ is actually equal to $\boldsymbol{\alpha}$. However, the estimated quantity $\hat{v}^2$ is not an unbiased estimate, but it can be shown that the quantity $v^{\dagger 2}$

$$v^{\dagger 2} = [m/(m-n)]\hat{v}^2 = [(\mathbf{Y} - \mathbf{X}\hat{\boldsymbol{\alpha}})'(\mathbf{Y} - \mathbf{X}\hat{\boldsymbol{\alpha}})]/(m-n) \quad (6\text{-}29)$$

is an unbiased estimator of $v^2$.

The estimators $\hat{\boldsymbol{\alpha}}$ and $\hat{v}^2$ can also be shown to be consistent, efficient, and minimum variance unbiased. In general an investigator will have more use for some of these properties than for others but it is a good idea to known that they have these properties in the event one wishes to argue about them.

Since $\hat{\boldsymbol{\alpha}}$ is equal to the product of a constant matrix $\mathbf{CX}'$ and a normally distributed vector $\mathbf{Y}$, $\hat{\boldsymbol{\alpha}}$ has an $n$-variate normal distribution.

### 4.3 Point Estimation of Parameter Using Least Squares

Proceeding to the case of the least squares, where the frequency function of the $\boldsymbol{\delta}$ is not known, we want to estimate the vector so that $\sum_{j=1}^{m} \delta_i^2$ is a minimum. This has already been done in Chapter 4, and we can write the least square estimate of $\boldsymbol{\alpha}^{\ddagger}$ as

$$\boldsymbol{\alpha}^{\ddagger} = \mathbf{CX'Y} \quad (6\text{-}30)$$

To examine the properties of the estimator $\boldsymbol{\alpha}^{\ddagger}$ obtained from the least squares estimate we are going to run into difficulties because we do not know the distribution function of the errors and consequently cannot make the same statements which we made about the maximum likelihood estimate. However, a theorem known as the Gauss–Markoff theorem tells us something about the properties of those estimators. When the distribution of the error is unknown and the following conditions are satisfied

$$E(\boldsymbol{\delta}) = \mathbf{0} \quad (6\text{-}31)$$

$$E(\boldsymbol{\delta}'\boldsymbol{\delta}) = v^2\mathbf{I} \quad (6\text{-}32)$$

then the Gauss–Markoff theorem states that if the general linear hypothesis model of full rank $\mathbf{Y} = \mathbf{X}\boldsymbol{\alpha} + \boldsymbol{\delta}$ satisfies the conditions given by Eqs. (6-31) and (6-32), then the best (minimum variance) linear (when the parameters are linear functions of the response) unbiased estimate of $\boldsymbol{\alpha}$ is given by the least square estimate i.e., by $\boldsymbol{\alpha}^{\ddagger}$.

It can also be shown that the best linear unbiased estimate of any linear combination of the $\alpha_i^{\ddagger}$ is the same linear combination of the best linear unbiased estimates of the $\alpha_i$. For example, if the least squares estimator of $\alpha_1^\ddagger$ and $\alpha_2^\ddagger$ are 4 and 5 respectively then the best linear unbiased estimate of $4\alpha_1 + 6\alpha_2 = 4(4) + 6(5) = 46$, but it must be emphasized that this is not true for all functions of the parameter. For instance, if the least squares estimator for $\alpha_1^\ddagger$, $\alpha_2^\ddagger$, and $\alpha_3^\ddagger$ are 1, 2, and 3 respectively and if $\beta = (\alpha_1^\ddagger + \alpha_2^\ddagger)/3\alpha_3$ then the best linear estimate of $\beta$ is *not*

$$\beta = (\alpha_1^\ddagger + \alpha_2^\ddagger)/3\alpha_3^\ddagger = (1 + 2)/3(3) = \tfrac{1}{3} \tag{6-33}$$

In addition to these properties of least squares estimators $\alpha^\ddagger$ we can show that $v^{\dagger 2}$ defined by Eq. (6-29) is an unbiased estimate of $v^2$.

### 4.4 Controlling the Independent Variables, Planning the Experiment

An important consideration in solving the normal equations

$$\mathbf{X}'\mathbf{X}\boldsymbol{\alpha} = \mathbf{X}'\mathbf{Y} \tag{6-34}$$

is the ways of choosing the matrix $\mathbf{X}$, which is also called the *design matrix*. This matrix is a matrix of known quantities and it may or may not be the case that we can control the choice of this matrix. In the event that we have some say in the matter of the choice $\mathbf{X}$ we would like to choose it so as to minimize the variance of as many estimates of the parameters as possible. To probe this matter further let us consider the following simple example. Suppose our model is a simple linear model given by

$$y_j = \alpha_1 + \alpha_2 x_j + \delta_j \qquad (j = 1, 2, \ldots, m) \tag{6-35}$$

where the $\alpha$ are unknown and are to be estimated, and the $y_j$ and $x_j$ are known. Writing the matrices corresponding to the above quantities in full we have

$$\mathbf{Y} = \begin{bmatrix} y_1 \\ y_2 \\ \vdots \\ y_m \end{bmatrix}, \quad \mathbf{X} = \begin{bmatrix} 1 & x_1 \\ 1 & x_2 \\ \vdots & \vdots \\ 1 & x_m \end{bmatrix}, \quad \boldsymbol{\alpha} = \begin{bmatrix} \alpha_1 \\ \alpha_2 \end{bmatrix}, \quad \boldsymbol{\delta} = \begin{bmatrix} \delta_1 \\ \delta_2 \\ \vdots \\ \delta_m \end{bmatrix} \tag{6-36}$$

Matrix multiplication reveals that

$$\mathbf{X}'\mathbf{X} = \begin{bmatrix} m & \sum_{j=1}^m x_j \\ \sum_{j=1}^m x_j & \sum_{j=1}^m x_j^2 \end{bmatrix}$$

and

$$(\mathbf{X}'\mathbf{X})^{-1} = \mathbf{C} = \frac{1}{m \sum_{j=1}^{m} (x_j - \bar{x})^2} \begin{bmatrix} \sum_{j=1}^{m} x_j^2 & -\sum_{j=1}^{m} x_j \\ -\sum_{j=1}^{m} x_j & m \end{bmatrix} \tag{6-37}$$

where $\bar{x}$ is the mean value of $x_1, x_2, \ldots, x_m$.

We also note that

$$\mathbf{X}'\mathbf{Y} = \begin{bmatrix} \sum_{j=1}^{m} y_j \\ \sum_{j=1}^{m} x_j y_j \end{bmatrix} \tag{6-38}$$

According to Eq. (6-28)

$$\hat{\boldsymbol{\alpha}} = \begin{bmatrix} \hat{\alpha}_1 \\ \hat{\alpha}_2 \end{bmatrix} = \mathbf{CX'Y}$$

$$= \frac{1}{m \sum_{j=1}^{m} (x_j - \bar{x})^2} \begin{bmatrix} \sum_{j=1}^{m} x_j^2 \sum_{j=1}^{m} y_j & -\sum_{j=1}^{m} x_j \sum_{j=1}^{m} x_j y_j \\ -\sum_{j=1}^{m} x_j \sum_{j=1}^{m} y_j & m \sum_{j=1}^{m} y_j x_j \end{bmatrix} \tag{6-39}$$

This yields, for the parameters,

$$\hat{\alpha}_1 = \bar{y} - \hat{\alpha}_2 \bar{x} \tag{6-40}$$

and

$$\hat{\alpha}_2 = \left[ \sum_{j=1}^{m} (x_j - \bar{x})(y_j - \bar{y}) \right] \Big/ \sum_{j=1}^{m} (x_j - \bar{x})^2 \tag{6-41}$$

In order to minimize the variance of the estimated parameters it is necessary to obtain equations for the variance of those parameters. To do so requires that we estimate the covariance matrix $\text{cov}(\boldsymbol{\alpha})$ (see Chapter 5). From the definition of the covariance matrix

$$\text{cov}(\boldsymbol{\alpha}) = E[(\hat{\boldsymbol{\alpha}} - \boldsymbol{\alpha})(\hat{\boldsymbol{\alpha}} - \boldsymbol{\alpha})'] = E[(\mathbf{CX'Y} - \boldsymbol{\alpha})(\mathbf{CX'Y} - \boldsymbol{\alpha})'] \tag{6-42}$$

By substituting $\mathbf{X}\boldsymbol{\alpha} + \boldsymbol{\delta}$ for $\mathbf{Y}$ we obtain

$$\begin{aligned} \text{cov}(\boldsymbol{\alpha}) &= E\{[\mathbf{CX}'(\mathbf{X}\boldsymbol{\alpha} + \boldsymbol{\delta}) - \boldsymbol{\alpha}][\mathbf{CX}'(\mathbf{X}\boldsymbol{\alpha} + \boldsymbol{\delta}) - \boldsymbol{\alpha}]'\} \\ &= E[(\mathbf{CX}'\,\boldsymbol{\delta})(\mathbf{CX}\,\boldsymbol{\delta})'] = E(\mathbf{CX}'\,\boldsymbol{\delta}\boldsymbol{\delta}'\mathbf{XC}) \\ &= \mathbf{CX}'[E(\boldsymbol{\delta}\boldsymbol{\delta}')]\mathbf{XC} \\ &= v^2\mathbf{CX'XC} = v^2\mathbf{C} \end{aligned} \tag{6-43}$$

Some of the steps in the above procedure will be clearer if we observe that $(\mathbf{X}'\mathbf{X})^{-1} = \mathbf{C}$ and $E(\boldsymbol{\delta}\boldsymbol{\delta}') = v^2 I$. The diagonal elements of $\text{cov}(\hat{\boldsymbol{\alpha}})$, are the variances which can be written as

$$\operatorname{var}(\hat{\alpha}_1) = \left(v^2 \sum_{j=1}^{m} x_j^2\right) \Big/ m \sum_{j=1}^{m} (x_j - \bar{x}_j)^2 \tag{6-44}$$

and

$$\operatorname{var}(\hat{\alpha}_2) = v^2 \Big/ \sum_{j=1}^{m} (x_j - \bar{x})^2 \tag{6-45}$$

From Eq. (6-45) we note that to minimize $\operatorname{var}(\hat{\alpha}_2)$ we choose the $x_j$ so that $\sum_{j=1}^{m}(x_j - \bar{x})^2$ is as large as possible, which means that we should take an $x_j = -\infty$ and another $x_j = \infty$ if possible. This is tantamount to saying we have to take our $x_j$ over as wide a range as possible. This is always a sensible thing to do. However it was pointed out by Box and Draper (1959) that, while this is true if our model is the exact linear model represented by Eq. (6-1), it is not generally true in the case where Eq. (6-1) is not the true representation of our model. For instance, if there is a second order tendency in the model then the quantity $\sum_{j=1}(x_j - \bar{x})^2$ is not infinity but a finite value.

To minimize $\operatorname{var}(\hat{\alpha}_1)$ our $x_j$ must be chosen so that the quantity $\sum_{j=1}^{m} x_j^2/(x_j - \bar{x})^2$ is minimized. Inasmuch as

$$\sum_{j=1}^{m} (x_j - \bar{x})^2 \leq \sum_{j=1}^{m} x_j^2$$

to minimize the quantity in question the $x_j$ must be chosen so that $\bar{x} = 0$. In the practical situation it is often possible that we can arrange to have it so by a simple linear transformation.

### 4.5 Interval Estimation of Parameters and Functions of Parameters

To determine confidence intervals for parameters one must know something about the probability distribution of the estimates of these parameters. For the linear model, when the errors are normally distributed, the covariance matrix can be written as

$$\operatorname{cov}(\hat{\boldsymbol{\alpha}}) = v^2\mathbf{C} \tag{6-46}$$

as indicated earlier in Eq. (6-43). Here the $\boldsymbol{\alpha}$ are normally distributed with a mean of $\hat{\boldsymbol{\alpha}}$ and covariance of $v^2\mathbf{C}$. In addition to this important fact the distribution of certain other quantities of interest may be taken up. From Eq. (6-29) we note that

$$\begin{aligned}\hat{v}^2 &= (\mathbf{Y} - \mathbf{X}\hat{\boldsymbol{\alpha}})'(\mathbf{Y} - \mathbf{X}\hat{\boldsymbol{\alpha}})/(m - n) \\ &= \mathbf{Y}'(\mathbf{I} - \mathbf{XCX}')\mathbf{Y}/(m - n)\end{aligned} \tag{6-47}$$

In view of the fact that $\mathbf{I} - \mathbf{XCX}'$ is an idempotent matrix of rank $m - n$ we can show from consideration of the distribution of quadratic forms in Chapter 5 that $(m - n)(\hat{v}^2/v^2)$ is distributed as the noncentral chi-square with the degrees $m - n$ and noncentrality parameter $\lambda$ i.e., is distributed as $\chi^2(m - n, \lambda)$. In this case our noncentrality parameter is given by

$$\lambda = (1/2v^2)\boldsymbol{\alpha}'\mathbf{X}'(\mathbf{I} - \mathbf{XCX}')\mathbf{X}\boldsymbol{\alpha} \tag{6-48}$$

which makes $\lambda = 0$ since $(\mathbf{I} - \mathbf{XCX}')\mathbf{X} = 0$, and consequently the quantity $(m - n)(\hat{v}^2/v^2)$ is distributed as the central chi-square with $m - n$ degrees or $\chi^2(m - n)$. Further reference to Chapter 5 is required to show that $\hat{v}^2$ and $\boldsymbol{\alpha}$ are independent. The required theorem showing that independence is Theorem 5-24. In the application of this theorem to the above case we take $\hat{\boldsymbol{\alpha}} = (\mathbf{CX}')\mathbf{Y}$ as the linear form and

$$v^2 = [1/(m - n)][(\mathbf{Y} - \mathbf{X}\hat{\boldsymbol{\alpha}})'(\mathbf{Y} - \mathbf{X}\hat{\boldsymbol{\alpha}})]$$

as the quadratic form.

### 4.5.1 *Interval Estimation of Linear Functions of* $\boldsymbol{\alpha}$

With the facts of the previous section in mind we can now proceed to a confidence interval on the $\alpha_i$. The $\hat{\alpha}_i$ are distributed normally with mean $\alpha_i$ and variance $c_{ii}v^2$ where $c_{ii}$ is the $ii$th element of $\mathbf{C}$ (note that we use only the diagonal elements in our argument). A linear transformation then shows that $(\hat{\alpha}_i - \alpha_i)/v(c_{ii})^{1/2}$ is normally distributed with zero mean and variance $v^2$ and this quantity is independent of $(m - n)(\hat{v}^2/v^2)$ which has the central chi-square distribution. The quantity

$$u = [(\hat{\alpha}_i - \alpha_i)/v(c_{ii})^{1/2}](v^2/\hat{v}^2)^{1/2} = (\hat{\alpha}_i - \alpha_i)/\hat{v}\sqrt{c_{ii}} \tag{6-49}$$

is distributed as Student's $t$-distribution with $m - n$ degrees of freedom. This follows from the definition of Student's distribution. If we choose to set $1 - \gamma$ confidence interval on the $\alpha_i$ we have

$$\int_{-t_{\gamma/2}}^{t_{\gamma/2}} t(u)\, du = P[t_{-\gamma/2} \le (\hat{\alpha}_i - \alpha_i)/(v^2 c_{ii})^{1/2} \le t_{\gamma/2}] = 1 - \gamma \tag{6-50}$$

Equation (6-50) can also be written as

$$P[\hat{\alpha}_i - t_{\gamma/2}(c_{ii}\hat{v}^2)^{1/2} \leq \alpha_i \leq \hat{\alpha}_i + t_{\gamma/2}(c_{ii}\hat{v}^2)^{1/2}] = 1 - \gamma \tag{6-51}$$

It therefore follows that the $1 - \gamma$ confidence interval on any given $\alpha_i$ is

$$\hat{\alpha}_i - t_{\gamma/2}(\hat{v}^2 c_{ii})^{1/2} \leq \alpha_i \leq \hat{\alpha}_i + t_{\gamma/2}(\hat{v}^2 c_{ii})^{1/2} \tag{6-52}$$

In this inequality the quantities $\hat{\alpha}_i c_{ii}$ and $\hat{v}^2$ are determined during the course of our computations of the pertinent matrices and the quantity $t_{\gamma/2}$ is read from the statistical tables. The width of the confidence interval is $2t_{\gamma/2}(c_{ii}\hat{v}^2)^{1/2}$. Equation (6-52) shows how one can compute confidence intervals. Judgments on its adequacy and narrowness depends, of course, on the nature of the experiments and the investigator.

To set a confidence interval on a linear function of the parameter such as

$$d_1\alpha_1 + d_2\alpha_2 + \cdots + d_n\alpha_n \tag{6-53}$$

we let $\mathbf{D}'$ be a vector of known constants $[d_1, d_2, \ldots, d_n]$. A confidence interval $1 - \gamma$ on the vector $\mathbf{D}$ is set by using identical reasoning to the previous paragraphs. The distribution of $\mathbf{D}'\boldsymbol{\alpha}$ is normal with mean $\mathbf{D}'\hat{\boldsymbol{\alpha}}$ and covariance $v^2\,\mathbf{D}'\mathbf{C}\,\mathbf{D}$. A linear transformation shows that $(\mathbf{D}'\hat{\boldsymbol{\alpha}} - \mathbf{D}'\boldsymbol{\alpha})/v\sqrt{\mathbf{D}'\mathbf{C}\,\mathbf{D}}$ is normally distributed with zero mean and variance 1. Thus the $1 - \gamma$ confidence interval on $\mathbf{D}'\alpha$, using reasoning similar to the above, is

$$\mathbf{D}'\hat{\boldsymbol{\alpha}} - t_{\gamma/2}(\hat{v}^2\mathbf{D}'\mathbf{C}\mathbf{D})^{1/2} \leq \mathbf{D}'\boldsymbol{\alpha} \leq \mathbf{D}'\hat{\boldsymbol{\alpha}} + t_{\gamma/2}(\hat{v}^2\mathbf{D}'\mathbf{C}\mathbf{D})^{1/2} \tag{6-54}$$

and as before the width of the confidence interval is

$$2t_{\gamma/2}(\hat{v}^2\mathbf{D}'\mathbf{C}\,\mathbf{D})^{1/2}$$

4.5.2 *Point and Interval Estimation of the Expected Value of the Response.*

A point estimate for the expected value of mean of the response $f(x)$ or $y$ as we are now calling it is simply written

$$E(\hat{y}) = \sum_{i=1}^{n} \hat{\alpha}_i x_i \tag{6-55}$$

and $E(\hat{y})$ is, in such cases, called both the least squares estimator (when the distribution of errors is not known) and the maximum likelihood estimator (when the errors are normally distributed). For an interval estimate of the response at any given point in the space of the known variables we proceed as follows letting

$$h = \sum_{i=1}^{n} (\hat{\alpha}_i - \alpha_i)x_i = \mathbf{x}'(\hat{\boldsymbol{\alpha}} - \boldsymbol{\alpha}) \tag{6-56}$$

where $\mathbf{x}$ is a vector of $n$ elements whose $i$th element is $x_i$. The distribution of the quantity $h$ is normal with zero mean and a covariance $v^2\mathbf{X}'\mathbf{CX}$. It therefore follows that the quantity

$$\sum_{i=1}^{n} (\hat{\alpha}_i - \alpha_i x_i)/[v^2(\mathbf{x}'\mathbf{Cx})^{1/2}] \tag{6-57}$$

has a normal distribution with zero mean and a variance of 1. Reasoning similar to that in the previous sections yields a $1 - \gamma$ confidence interval for the response $E(y) = \sum_{i=1}^{n} \hat{\alpha}_i x_i$ given by

$$\sum_{i=1}^{n} \hat{\alpha}_i x_i - t_{\gamma/2}(\hat{v}^2\mathbf{x}'\mathbf{Cx})^{1/2} \leq \sum_{i=1}^{n} \alpha_i x_i \leq \sum_{i=1}^{n} \hat{\alpha}_i x_i + t_{\gamma/2}(\hat{v}^2\mathbf{x}'\mathbf{Cx})^{1/2} \tag{6-58}$$

In this equation we note once again that all the quantities on the left and right hand side of the inequalities are either known or can be looked up in the statistical tables.

### 4.5.3 *Simultaneous Confidence Intervals for Estimated Parameters*

To consider the question of simultaneous confidence intervals let us examine what we have done up to this point in determining the confidence interval for a parameter. What we have done is to establish for a given probability that any given parameter will lie in the confidence

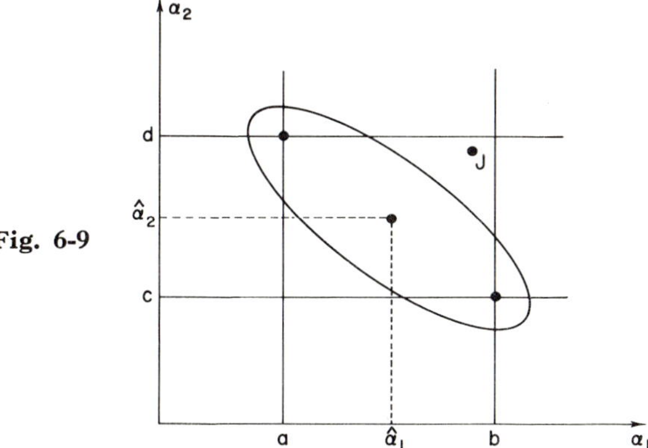


**Fig. 6-9**

interval. To obtain some idea of simultaneous confidence intervals, suppose we calculate 95% confidence intervals for two parameters $\alpha_1$ and $\alpha_2$. In Fig. 6-9 we have depicted the separate confidence intervals, so that $(a, b)$ is the interval for $\alpha_1$ and $(c, d)$ is the interval for $\alpha_2$. The area in the rectangle $ABCD$, however, is not the *joint* 95% confidence interval for $\alpha_1$ and $\alpha_2$. The joint 95% interval for those two parameters is the ellipse shown in the figure and a point such as $J$, while lying inside the rectangle, is not in the ellipse and consequently the values of the coordinates of $J$ do not represent part of the joint 95% confidence interval. Thus while we are 95% confident that the interval $(a, b)$ covers $\alpha_1$ and the interval $(c, d)$ covers $\alpha_2$ we are not 95% confident that points within the rectangle are jointly covered. Whether or not the rectangle provides a reasonable 95% joint confidence interval depends on the size of the correlation between $\hat{\alpha}_1$ and $\hat{\alpha}_2$ since

$$\varrho_{12} = \mathrm{cov}(\hat{\alpha}_1, \hat{\alpha}_2)/[\mathrm{var}(\hat{\alpha}_1)\ \mathrm{var}(\hat{\alpha}_2)]^{1/2} \qquad (6\text{-}59)$$

If $\varrho_{12} = 0$, then the rectangle will approximate the joint confidence region. If not, then the rectangle will not approximate the joint confidence region. Two relevant situations are observed in Fig. (6-8) depending on the sizes of $\varrho_{12}$, $\mathrm{var}(\hat{\alpha}_1)$ and $\mathrm{var}(\hat{\alpha}_2)$. A third situation is observed in the previous figure. The method for obtaining joint confidence intervals on all

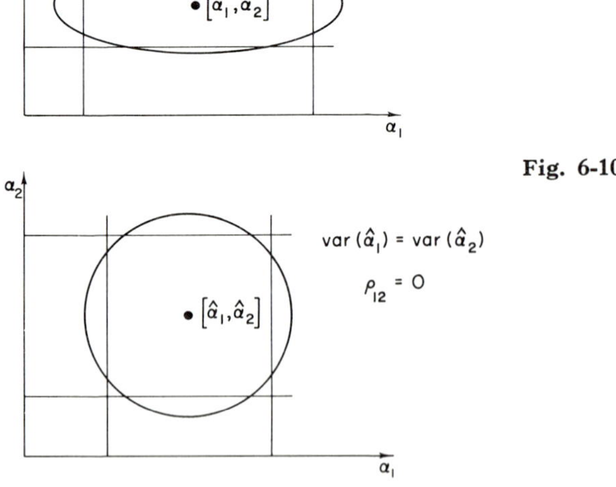


Fig. 6-10

simultaneously is to solve the following quadratic (in all parameters) equation

$$(\boldsymbol{\alpha} - \hat{\boldsymbol{\alpha}})\mathbf{XX}'(\boldsymbol{\alpha} - \hat{\boldsymbol{\alpha}}) \leq ns^2F(n, m - n, 1 - \gamma) \tag{6-60}$$

for the vector $\boldsymbol{\alpha}$. In the inequality (6-60) all quantities are known except the vector $\boldsymbol{\alpha}$. The quantity $F(n, m - n, 1 - \gamma)$ is obtained from statistical tables. The inequality provides an equation for an ellipse when we have two parameters only, for an ellipsoid when we have three parameters, and lastly for an "elliptically shaped" region in $n$-dimension space for $n$ parameters.

Figure 6-10 shows two different situations for the variation of $\varrho_{12}$ with $\mathrm{var}(\hat{\alpha}_1)$ and $\mathrm{var}(\hat{\alpha}_2)$. A third situation illustrated in Fig. 6-9 is the case when $\varrho_{12} \neq 0$.

## 5. Testing General Linear Model Hypotheses

### 5.1 Hypotheses To Be Tested

Earlier it was stated that we would like to test the following hypothesis on our linear model: $\alpha_1 = \alpha_1{}^*, \alpha_2 = \alpha_2{}^*, \ldots, \alpha_n = \alpha_n{}^*$ where $\alpha_1{}^*, \alpha_2{}^*, \ldots, \alpha_n{}^*$ are known constants. Next, we would like to test the hypothesis that $\mathbf{D}'\boldsymbol{\alpha}$ is equal to some scalar $d$. We also want to test the hypothesis that some subset of $\alpha_1, \alpha_2, \ldots, \alpha_n$ is equal to certain constants with the rest of the $\alpha_i$ unspecified, and lastly we might want to test the hypothesis that $\boldsymbol{\lambda}_1{}'\boldsymbol{\alpha} = \boldsymbol{\lambda}_2{}'\boldsymbol{\alpha} = \cdots = \boldsymbol{\lambda}_n{}'\boldsymbol{\alpha} = 0$ where $\boldsymbol{\lambda}_1{}'\boldsymbol{\alpha}, \boldsymbol{\lambda}_2{}'\boldsymbol{\alpha}, \ldots, \boldsymbol{\lambda}_n{}'\boldsymbol{\alpha}$ are linear functions of the vector such that $\lambda_1{}', \lambda_2{}', \ldots, \lambda_n$ are linearly independent. To test these hypotheses we have to find certain quadratic forms that are functions of the data and the parameters and determine how these quadratic forms are distributed.

### 5.2 Testing the Hypothesis that $\boldsymbol{\alpha} = \boldsymbol{\alpha}^*$

To test the first hypothesis of interest, $\boldsymbol{\alpha} = \boldsymbol{\alpha}^*$, we must find a test function which is a function of all the observations $y_i, x_{11}, x_{12}, \ldots, x_{mn}$. Next, we must know the distribution of this function when $H_0$ is the null hypothesis or when $H_0$ is true. If we desire to know the power of the test we must also know what the distribution is when the alternative hypothesis $H_1$ $\boldsymbol{\alpha} \neq \boldsymbol{\alpha}^*$ is true. The best way of handling this problem is to use the likelihood ratio as a test function.

To do this we form the ratio

$$L = L(\omega_k)/L(\Omega_k) \tag{6-61}$$

where $L(\omega_k)$ is the maximum of the likelihood function in the space defined $\alpha_1 = \alpha_1{}^*$, $\alpha_1 = \alpha_2{}^*$, $\ldots$, $\alpha_n = \alpha_n{}^*$ and $0 \leq v^2 < \infty$, and $L(\Omega_k)$ is the maximum likelihood in the space where the $\alpha_i$ can take any value and $v^2$ can take any value such that $0 \leq v^2 < \infty$. The likelihood function, as before is

$$f(\boldsymbol{\alpha}, v^2, \boldsymbol{\delta}) = [1/(2\pi v^2)^{m/2}] \exp(-\boldsymbol{\delta}'\boldsymbol{\delta}/2v^2) \tag{6-62}$$

Substituting for $\boldsymbol{\delta}'\boldsymbol{\delta}$ from Eq. (6-1) we obtain

$$f(\boldsymbol{\alpha}, v^2, \boldsymbol{\delta}) = [1/(2\pi v^2)^{m/2}] \exp[-(\mathbf{Y} - \mathbf{X}\boldsymbol{\alpha})'(\mathbf{Y} - \mathbf{X}\boldsymbol{\alpha})/2v^2] \tag{6-63}$$

To find the maximum of $L(\Omega)$ in the entire $\Omega$ space, we take the logarithm of the likelihood function written as follows

$$\begin{aligned} &\ln f(\mathbf{Y}, \mathbf{X}, \boldsymbol{\alpha}, v^2) \\ &\quad = -(m/2) \ln 2\pi - (m/2) \ln v^2 - [(\mathbf{Y} - \mathbf{X}\boldsymbol{\alpha})'(\mathbf{Y} - \mathbf{X}\alpha)/2v^2] \end{aligned} \tag{6-64}$$

and differentiate with respect to $\boldsymbol{\alpha}$ and $v^2$. Therefore the maximum value of $\ln f(\mathbf{Y}, \mathbf{X}, \boldsymbol{\alpha}, v^2)$ is obtained as the solution of the equations

$$\partial \ln f(\mathbf{Y}, \mathbf{X}, \boldsymbol{\alpha}, v^2)/\partial\boldsymbol{\alpha} = (\mathbf{X}'\mathbf{Y} - \mathbf{X}'\mathbf{X}\hat{\boldsymbol{\alpha}})/\hat{v}^2 = 0 \tag{6-65}$$

and

$$\partial \ln f(\mathbf{Y}, \mathbf{X}, \boldsymbol{\alpha}, v^2)/\partial v^2 = [(\mathbf{Y} - \mathbf{X}\hat{\boldsymbol{\alpha}})'(\mathbf{Y} - \mathbf{X}\boldsymbol{\alpha})/2v^4] - (m/2v^2) = 0 \tag{6-66}$$

where we have put $\hat{\boldsymbol{\alpha}} = \mathbf{C}\mathbf{X}'\mathbf{Y}$.

This leads to

$$L(\Omega_k) = m^{m/2}e^{-m/2}/\{(2\pi)^{m/2}[(\mathbf{Y} - \mathbf{X}\hat{\boldsymbol{\alpha}})'(\mathbf{Y} - \mathbf{X}\hat{\boldsymbol{\alpha}})]^{m/2}\} \tag{6-76}$$

To maximize the likelihood function in $\omega$ space we need only differentiate with respect to $v^2$ and set $\boldsymbol{\alpha} = \boldsymbol{\alpha}^*$. Carrying out these operations

$$d[\ln f(\mathbf{Y}, \mathbf{X}, \alpha^*, v^2)]/dv^2 = \{[(\mathbf{Y} - \mathbf{X}\boldsymbol{\alpha}^*)'(\mathbf{Y} - \mathbf{X}\boldsymbol{\alpha}^*)]/2\hat{v}^4\} - (m/2\hat{v}^2) = 0$$

The solution for this equation is

$$\hat{v}^2 = [(\mathbf{Y} - \mathbf{X}\boldsymbol{\alpha}^*)'(\mathbf{Y} - \mathbf{X}\boldsymbol{\alpha}^*)]/m \tag{6-68}$$

and that makes

$$L(\omega_k) = (m^{m/2}e^{-m/2})/\{(2\pi)^{m/2}[(\mathbf{Y} - \mathbf{X}\boldsymbol{\alpha}^*)'(\mathbf{Y} - \mathbf{X}\boldsymbol{\alpha}^*)]^{m/2}\} \tag{6-69}$$

Our likelihood ratio is formed by dividing Eq. (6-69) by Eq. (6-67). This gives

$$\begin{aligned} L &= L(\omega_k)/L(\Omega_k) \\ &= \{[(\mathbf{Y} - \mathbf{X}\hat{\boldsymbol{\alpha}})'(\mathbf{Y} - \mathbf{X}\hat{\boldsymbol{\alpha}})]/[(\mathbf{Y} - \mathbf{X}\boldsymbol{\alpha}^*)'(\mathbf{Y} - \mathbf{X}\boldsymbol{\alpha}^*)]\}^{m/2} \end{aligned} \tag{6-70}$$

Our objective is to determine the distribution of $L$ under the null hypothesis $H_0$ when $\boldsymbol{\alpha} = \boldsymbol{\alpha}^*$ and to determine an $R$ such that

$$\int_0^R f(L, \alpha^*)\, dL = \gamma \tag{6-71}$$

where $\gamma$ is the probability of making a type I error and the critical region is $0 \leq L \leq R$ (i.e., once $L$ is determined the null hypothesis is rejected if $L \leq R$. To determine the distribution of $L$ under the null hypothesis we look at the various quadratic forms involved in Eq. (6-70).

At this stage our expositions of the various theorems on quadratic forms and their independence in Chapter 5 begin to have their function. To simplify reading we focus on the fact that the matrix **Z**, defined later, has a multivariate normal distribution. This will enable following the implications of the theorems of Chapter 5 and seeing how they are applied in the next paragraphs. If the reader can accept the statement about the distribution of the quantity in Eq. (6-78), then one can proceed to test hypotheses about it by using analysis of variance tables without having to go through Eqs. (6-72)–(6-77). A practical example of hypothesis testing using analysis of variance tables is given in Chapter 12.

From Eq. (6-70) we have the following quantities to look at

$$(\mathbf{Y} - \mathbf{X}\alpha^*)'(\mathbf{Y} - \mathbf{X}\boldsymbol{\alpha}^*) \tag{6-72}$$

and

$$(\mathbf{Y} - \mathbf{X}\hat{\boldsymbol{\alpha}})'(\mathbf{Y} - \mathbf{X}\hat{\boldsymbol{\alpha}}) \tag{6-73}$$

If we let $\mathbf{Z} = \mathbf{Y} - \mathbf{X}\boldsymbol{\alpha}^*$ it can be shown that

$$\mathbf{Z}'\mathbf{Z} = (\mathbf{Y} - \mathbf{X}\boldsymbol{\alpha}^*)'(\mathbf{Y} - \mathbf{X}\boldsymbol{\alpha}^*) = \mathbf{Z}'\mathbf{A}_1\mathbf{Z} + \mathbf{Z}'\mathbf{A}_2\mathbf{Z} \tag{6-74}$$

where

$$\mathbf{A}_1 = \mathbf{I} - \mathbf{X}\mathbf{C}\mathbf{X}' \tag{6-75}$$

and

$$\mathbf{A}_2 = \mathbf{X}\mathbf{C}\mathbf{X}' \tag{6-76}$$

and both $\mathbf{A}_1$ and $\mathbf{A}_2$ are idempotent matrices. Noting that the distribution of $\mathbf{Y}$ is the multivariate normal distribution with a mean $\mathbf{X}\boldsymbol{\alpha}$ and variance $v^2\mathbf{I}$, it follows that the distribution of the matrix $\mathbf{Z}$ is normal with mean $\mathbf{X}\boldsymbol{\alpha} - \mathbf{X}\boldsymbol{\alpha}^*$ and variance $v^2\mathbf{I}$. Inasmuch as $\mathbf{A}_1$ and $\mathbf{A}_2$ are idempotent it follows from our theorems on quadratic forms (see Theorem 5-23) that $(\mathbf{Z}'\mathbf{A}_1\mathbf{Z}/v^2)$ is distributed as $\chi^2(m - n)$ and $(\mathbf{Z}'\mathbf{A}_2\mathbf{Z}/v^2)$ is distributed as the noncentral chi-square with $n$ degrees of freedom and noncentrality parameter $\lambda$, where the scalar $\lambda$ is given by

$$\lambda = [(\boldsymbol{\alpha} - \boldsymbol{\alpha}^*)\mathbf{X}'\mathbf{A}_2\mathbf{X}(\boldsymbol{\alpha} - \boldsymbol{\alpha}^*)/2v^2] \qquad (6\text{-}77)$$

and since $\mathbf{A}_2$ is idempotent and nonsingular then

$$\lambda = (\boldsymbol{\alpha} - \boldsymbol{\alpha}^*)\mathbf{X}'\mathbf{X}(\boldsymbol{\alpha} - \boldsymbol{\alpha}^*)/2v^2$$

Further if we note that $\mathbf{X}'\mathbf{X}$ is positive definite then $(\mathbf{Z}'\mathbf{A}_2\mathbf{Z}/v^2)$ will have a central chi-square distribution if and only if $\boldsymbol{\alpha} = \boldsymbol{\alpha}^*$ or if the null hypothesis is true. Lastly we note that $(\mathbf{Z}'\mathbf{A}_1\mathbf{Z}/v^2)$ and $(\mathbf{Z}'\mathbf{A}_2\mathbf{Z}/v^2)$ are independent. Knowing the distribution properties of $(\mathbf{Z}'\mathbf{A}_1\mathbf{Z}'/v^2)$ and $(\mathbf{Z}'\mathbf{A}_2\mathbf{Z}/v^2)$ we form the function

$$\nu = [(\mathbf{Z}'\mathbf{A}_2\mathbf{Z})/(\mathbf{Z}'\mathbf{A}_1\mathbf{Z})]/[(m - n)/n] \qquad (6\text{-}78)$$

and note that $\nu$ has a noncentral $F$-distribution with $n$ and $m - n$ degrees of freedom, and noncentrality parameter $\lambda$. We note also that $\nu$ will reduce to Snedecor's $F$-distribution or the central $F$ when $\lambda = 0$ or if and only if $\boldsymbol{\alpha} - \boldsymbol{\alpha}^* = 0$ (i.e., when the null hypothesis is true).

For the likelihood ratio we have from (6-70)

$$L = \left[\frac{(\mathbf{Y} - \mathbf{X}\hat{\boldsymbol{\alpha}})'(\mathbf{Y} - \mathbf{X}\hat{\boldsymbol{\alpha}})}{(\mathbf{Y} - \mathbf{X}\boldsymbol{\alpha}^*)'(\mathbf{Y} - \mathbf{X}\boldsymbol{\alpha}^*)}\right]^{m/2} = \left\{\frac{1}{1 + [n/(m - n)]\nu}\right\}^{m/2} \qquad (6\text{-}79)$$

Inspection of (6-79) shows that $L$ is a monotonic function of $\nu$ so we can use $\nu$ to test the hypothesis that $\boldsymbol{\alpha} = \boldsymbol{\alpha}^*$. For the distribution of $\nu$ (which is Snedecor's $F$-distribution) the critical region that corresponds to $0 \le L \le A$ is $F_\gamma \le \nu < \infty$ where $F_\gamma$ is a constant such that

$$\int_{F_\gamma}^{\infty} F(\nu, n, m - n, \lambda)\, d\nu = \gamma \qquad (\lambda = 0)$$

## 5.3 Analysis of Variance Tables for Testing the Hypothesis that $\boldsymbol{\alpha} = \boldsymbol{\alpha}^*$

For the sake of convenience in making the test we arrange the quantities involved in our computation in what is called an analysis of variance table (Table 6-1). In the table we note that

$$SS_1 = (\mathbf{Y} - \mathbf{X}\boldsymbol{\alpha}^*)'(\mathbf{Y} - \mathbf{X}\boldsymbol{\alpha}^*) \tag{6-80}$$

and

$$SS_3 = (\mathbf{Y} - \mathbf{X}\hat{\boldsymbol{\alpha}})'(\mathbf{Y} - \mathbf{X}\hat{\boldsymbol{\alpha}}) \tag{6-81}$$

**Table 6-1**[a]

| | SV | *DF* | *SS* | *MS* | *F* |
|---|---|---|---|---|---|
| 1 | Total | $DF_1 = m$ | $SS_1 = (\mathbf{Y} - \mathbf{X}\boldsymbol{\alpha}^*)' \times (\mathbf{Y} - \mathbf{X}\boldsymbol{\alpha}^*)$ | | |
| 2 | Due to $\boldsymbol{\alpha}$ | $DF_2 = n$ | $SS_2 = SS_1 - SS_3$ | $MS_2 = SS_2/DF_2$ | $MS_2/MS_3$ |
| 3 | Due to error | $DF_3 = m - n$ | $SS_3 = \mathbf{Y}'(\mathbf{I}-\mathbf{XCX}')\mathbf{Y}$ | $MS_3 = SS_3/DF_3$ | |

[a] SV = source of variation, *DF* = degree of freedom, *SS* = sum of squares, *MS* = mean square. Subscripts indicate corresponding sources of variation.

which means that the likelihood ratio given by Eq. (6-70) is

$$L = (SS_3/SS_1)^{m/2} \tag{6-82}$$

and the quantity $SS_2$ can be obtained from $SS_1$ and $SS_3$ by direct subtraction giving

$$SS_2 = SS_1 - SS_3 \tag{6-83}$$

This latter quantity is of importance because in the testing of our hypothesis we actually consider the quantity $\nu = [(m - n)/n](SS_2/SS_3)$ and we have the following theorem which can be deduced from earlier considerations. In the general linear hypothesis, when the error vector $\boldsymbol{\delta}$ is distributed normally with zero mean and covariance $v^2\mathbf{I}$, the quantity $\nu = [(m - n)/n](SS_2/SS_3)$ has the noncentral $F$-distribution with noncentrality parameter

$$\lambda = (\boldsymbol{\alpha} - \boldsymbol{\alpha}^*)'\mathbf{C}^{-1}(\boldsymbol{\alpha} - \boldsymbol{\alpha}^*)/2v^2 \tag{6-84}$$

Consideration of $\nu$ and where it lies in the distribution forms the basis of testing of this hypothesis. In actual practice therefore, computation of $SS_1$ and $SS_3$ is all that is required to proceed with the testing of the hypothesis.

### 5.4 Testing the Hypothesis that $\mathbf{d}'\boldsymbol{\alpha}$ (or a Linear Function of the Parameters) Is Equal to Some Scalar $d_0$.

The test here is that $\sum_{i=1}^{n} d_i \alpha_i = d_0$. Defining the quantity

$$q = [\mathbf{d}'\hat{\boldsymbol{\alpha}} - \mathbf{d}'\boldsymbol{\alpha}]/[(\mathbf{d}'\mathbf{C}\mathbf{d}\hat{v}^2)^{1/2}] \tag{6-85}$$

This quantity $q$ is distributed as Student's $t$-distribution with $m - n$ degrees of freedom. For the purpose of our test replace $\mathbf{d}'\boldsymbol{\alpha}$ by $d_0$ to obtain

$$q = (\mathbf{d}'\hat{\boldsymbol{\alpha}} - d_0)/(\hat{v}^2\mathbf{d}'\mathbf{C}\mathbf{d})^{1/2} \tag{6-86}$$

and compare it with the tabulated values of Student's distribution. Another possible test quantity in connection with this problem arises when we consider that

$$r = (\mathbf{d}'\hat{\boldsymbol{\alpha}} - d_0)/[v^2\mathbf{d}'\mathbf{C}\mathbf{d}]^{1/2} \tag{6-87}$$

is distributed normally with mean $\mu = (\mathbf{d}'\boldsymbol{\alpha} - d_0)/(v^2\mathbf{d}'\mathbf{C}\mathbf{d})^{1/2}$ and variance 1. This makes the distribution of $r^2$ noncentral chi-square with one degree of freedom and noncentrality parameter

$$\lambda = \frac{(\mathbf{d}'\boldsymbol{\alpha} - d_0)^2}{2v^2\mathbf{d}'\mathbf{C}\mathbf{d}}$$

Therefore,

$$z = r^2/(\hat{v}^2/v^2) = (\mathbf{d}'\hat{\boldsymbol{\alpha}} - d_0)^2/(v^2\mathbf{d}'\mathbf{C}\mathbf{d}) \tag{6-88}$$

has the noncentral $F$-distribution with degrees of freedom 1 and $m - n$ and noncentrality parameter $\lambda$. In the event that the null hypothesis is true (i.e., $\mathbf{d}'\boldsymbol{\alpha} - d_0 = 0$) then $z$ has Snedecor's $F$-distribution.

### 5.5 Testing of the Hypothesis that $\boldsymbol{\alpha}_1 = \boldsymbol{\alpha}_1^*$, $\boldsymbol{\alpha}_2 = \boldsymbol{\alpha}_2^*, \ldots, \boldsymbol{\alpha}_p = \boldsymbol{\alpha}_p^*$, with the Remaining $\boldsymbol{\alpha}_i$ Unspecified

The testing of this hypothesis plays a very important role in our future work. The reason for this is that we must often decide how many terms are required to adequately fit any response function. It is clear that the more terms and hence parameters the better our fit, but it could very well be that a limited number of terms could make an adequate fit. There will be many situations where the number of terms have physical significance. Thus in concentration-dependent aggregating systems the number of terms will determine the highest aggregating species of significant concentration and so on for other models.

Our model as before is

$$\mathbf{Y} = \mathbf{X}\boldsymbol{\alpha} + \boldsymbol{\delta}$$

However, the best way to consider this model is to use partitioned matrices as follows

$$\mathbf{X} = [\mathbf{X}_1, \mathbf{X}_2] \qquad \text{and} \quad \boldsymbol{\alpha} = \begin{bmatrix} \boldsymbol{\alpha}_1 \\ \boldsymbol{\alpha}_2 \end{bmatrix}$$

then the model can be written as

$$\mathbf{Y} = [\mathbf{X}_1, \mathbf{X}_2] \begin{bmatrix} \boldsymbol{\alpha}_1 \\ \boldsymbol{\alpha}_2 \end{bmatrix} + \boldsymbol{\delta} \tag{6-89}$$

As set up the hypothesis to be tested is $\boldsymbol{\alpha}_1 = \boldsymbol{\alpha}_1^*$ ($\boldsymbol{\alpha}_1^*$ is known) with $\boldsymbol{\alpha}_2$ unspecified. Here we have

$$\boldsymbol{\alpha}_1 = \begin{bmatrix} \alpha_1 \\ \alpha_2 \\ \cdot \\ \cdot \\ \cdot \\ \alpha_p \end{bmatrix} \qquad \text{and} \quad \boldsymbol{\alpha}_1^* = \begin{bmatrix} \alpha_1^* \\ \alpha_2^* \\ \cdot \\ \cdot \\ \cdot \\ \alpha_p^* \end{bmatrix}$$

For this hypothesis we have the following theorem.

Let the model $\mathbf{Y} = \mathbf{X}\boldsymbol{\alpha} + \boldsymbol{\delta}$ be partitioned in the following manner

$$\mathbf{Y} = \mathbf{X}_1\boldsymbol{\alpha}_1 + \mathbf{X}_2\boldsymbol{\alpha}_2 + \boldsymbol{\delta}$$

where $\boldsymbol{\alpha}_1$ is of dimension $p \times 1$ and let $\boldsymbol{\delta}$ be distributed normally with zero mean and variance $v^2\mathbf{I}$. In order to test the hypothesis $\boldsymbol{\alpha}_1 = \boldsymbol{\alpha}_1^*$, we obtain an estimate of $\boldsymbol{\alpha}$ in the model $\mathbf{Y} = \mathbf{X}\boldsymbol{\alpha} + \boldsymbol{\delta}$ minimizing $\boldsymbol{\delta}'\boldsymbol{\delta}$. The estimate is called $\hat{\boldsymbol{\alpha}}$ as before and the minimum of the quantity $\boldsymbol{\delta}'\,\boldsymbol{\delta}$ is given by $Q_0$. Next minimize $\boldsymbol{\delta}'\,\boldsymbol{\delta}$ with respect to $\boldsymbol{\alpha}_2$ in the model $\mathbf{Y} = \mathbf{X}_1\boldsymbol{\alpha}_1^* + \mathbf{X}_2\boldsymbol{\alpha}_2 + \boldsymbol{\delta}$. Let the value of the estimate be $\hat{\boldsymbol{\alpha}}_2$ and let the minimum of the quantity $\boldsymbol{\delta}'\,\boldsymbol{\delta}$ under those circumstances be $R = Q_0 + Q_1$. The quantity $\nu$ is given by

$$\nu = [(m - n)/p](Q_1/Q_0)$$

and is distributed as the noncentral $F$-distribution with degrees of freedom $p$ and $m - n$ and noncentrality parameter $\lambda$ given by

$$\lambda = (\boldsymbol{\alpha}_1 - \boldsymbol{\alpha}_1^*)'\mathbf{G}(\boldsymbol{\alpha}_1 - \boldsymbol{\alpha}_1^*)/2v^2 \tag{6-90}$$

and where

$$\mathbf{G} = \mathbf{X}_1'\mathbf{X}_1 - \mathbf{X}'\mathbf{X}_2(\mathbf{X}_2'\mathbf{X}_2)^{-1}(\mathbf{X}_2'\mathbf{X}_1) \tag{6-91}$$

Note that the quantity $Q_1$ is obtained by substracting $Q_0$ from $R$ where $R$ is obtained by minimizing subject to the conditions of our hypothesis (i.e., we insert the values of $\boldsymbol{\alpha}_1^*$ and carry out the minimization). Our

**Table 6-2**

Testing the Hypothesis $\boldsymbol{\alpha}_1 = \mathbf{0}$

| SV | *DF* | *SS* | *MS* | *F* |
|---|---|---|---|---|
| Total | $DF_1 = m$ | $SS_1 = \mathbf{Y}'\mathbf{Y}$ | | |
| Due to $\boldsymbol{\alpha}$ | $DF_2 = n$ | $SS_2 = \hat{\boldsymbol{\alpha}}'\mathbf{X}'\mathbf{Y}$ | | |
| Due to $\boldsymbol{\alpha}_2$ (unadjusted) | $DF_3 = n - p$ | $SS_3 = \hat{\boldsymbol{\alpha}}_2\mathbf{X}_2'\mathbf{Y}$ | | |
| Due to $\boldsymbol{\alpha}_1$ (adjusted) | $DF_4 = p$ | $SS_4 = \hat{\boldsymbol{\alpha}}'\mathbf{X}'\mathbf{Y} - \hat{\boldsymbol{\alpha}}_2'\mathbf{X}_2\mathbf{Y} = Q_1$ | $MS_4 = Q_1/p$ | $[(m - n)/p] \times (Q_1/Q_0)$ |
| Error | $DF_5 = m - n$ | $SS_4 = \mathbf{Y}'\mathbf{Y} - \hat{\boldsymbol{\alpha}}'\mathbf{X}'\mathbf{Y} = Q_0$ | $MS_5 = Q_0/(m - n)$ | |

statement about the distribution of $v$ makes the noncentrality parameter $\lambda$ equal to zero if and only if $\boldsymbol{\alpha}_1 - \boldsymbol{\alpha}_1^* = 0$ since the matrix $\mathbf{G}$ is positive definite. To test our hypothesis therefore we compute the required quantity $v$ and compare it with the tables of Snedecor's $F$-distribution. This is done in an analysis of variance scheme shown in Table 6-2 for the subhypothesis $\boldsymbol{\alpha}_1 = 0$. See Chapter 12 for examples.
When $\boldsymbol{\alpha}_1 \neq \mathbf{0}$ then specific values of $\boldsymbol{\alpha}_1$ are inserted in our computation and the quantity $v$ is computed as prescribed earlier.

## 5.6 Testing the Hypothesis That $\boldsymbol{\lambda}_1'\boldsymbol{\alpha} = \boldsymbol{\lambda}_2'\boldsymbol{\alpha} = \cdots = \boldsymbol{\lambda}'_p\boldsymbol{\alpha} = 0$

The general idea involved here is to reduce this model to a mathematical formulation equivalent to the previous situation. In order to do this we make the following definitions of partitioned matrices. Define

$$\mathbf{H} = \begin{bmatrix} \mathbf{H}_1 \\ \mathbf{H}_2 \end{bmatrix}, \quad \mathbf{H}_1 = \begin{bmatrix} \lambda_1' \\ \lambda_2' \\ \vdots \\ \lambda_p' \end{bmatrix}, \quad \mathbf{H}^{-1} = \mathbf{G} = [\mathbf{G}_1, \mathbf{G}_2] \tag{6-92}$$

further let

$$\mathbf{XH}^{-1} = \mathbf{XG} = [\mathbf{XG}_1, \mathbf{XG}_2] = [\mathbf{Z}_1, \mathbf{Z}_2] = \mathbf{Z} \tag{6-93}$$

and

$$\mathbf{H}\boldsymbol{\alpha} = \begin{bmatrix} \mathbf{H}_1\boldsymbol{\alpha} \\ \mathbf{H}_2\boldsymbol{\alpha} \end{bmatrix} = \begin{bmatrix} \boldsymbol{\gamma}_1 \\ \boldsymbol{\gamma}_2 \end{bmatrix} \tag{6-94}$$

Here $\mathbf{H}$ is an $n \times n$ matrix partitioned into two matrices $\mathbf{H}_1$, a $p \times n$ matrix with rank $p$ and $\mathbf{H}_2$. The matrix $\mathbf{H}$ is of rank $n$ and is known, as is its inverse $\mathbf{H}^{-1}$. The rest of the matrices as defined in Eqs. (6-93) and (6-94) are also known. With the above definitions we can rearrange our model $\mathbf{Y} = \mathbf{X}\boldsymbol{\alpha} + \boldsymbol{\delta}$ to

$$\mathbf{Y} = \mathbf{XH}^{-1}\mathbf{H}\boldsymbol{\alpha} + \boldsymbol{\delta} \tag{6-95}$$

or

$$\mathbf{Y} = \mathbf{Z}\boldsymbol{\gamma} + \boldsymbol{\delta} \tag{6-96}$$

Which can be written as

$$\mathbf{Y} = \mathbf{Z}_1\boldsymbol{\gamma}_1 + \mathbf{Z}_2\boldsymbol{\gamma}_2 + \boldsymbol{\delta} \tag{6-97}$$

This is equivalent to the model in the previous section with our hypothesis being $\boldsymbol{\gamma}_1 = \mathbf{0}$ which is equivalent to $\mathbf{H}_1\boldsymbol{\alpha} = \mathbf{0}$.

This completes the procedure for testing hypotheses in the linear model, and if the reader will focus attention on the quantities to be computed, mainly of the previous sections, and the theorems about the distribution of those quantities, then little difficulty will be encountered in application. Proofs of the theorems are given by Graybill (1961).

## 6. Some Notes on Nonlinear Models

### 6.1 General Considerations

In Chapter 4 we have shown how to determine the parameter of a nonlinear model which was written, in our previously defined notation, as

$$\mathbf{Y} = f(\mathbf{X}, \boldsymbol{\alpha}) + \boldsymbol{\delta} \tag{6-98}$$

In this section we elaborate on a number of points pertinent to the application of such models. Primarily we attempt to estimate the confidence interval and confidence contours for the parameters of the model represented by Eq. (6-98). In addition, the number of terms required to fit our response function has, in most models we consider, some physical significance, and an estimate of this quantity is also desirable.

## 6.2 Equivalence of Least Squares and Maximum Likelihood

An important point to note is that when the error vector is normally distributed with zero mean and variance $v^2\mathbf{I}$, as in the linear case, the maximum likelihood estimate in the nonlinear case is also the least squares estimate. If there are $m$ observations we can write

$$\sum_{j=1}^{m} \delta_j^2 = F(\boldsymbol{\alpha}) = \sum_{j=1}^{m} [y_j - f(\mathbf{X}, \boldsymbol{\alpha})]^2 \tag{6-99}$$

The likelihood function of this problem is

$$L(\boldsymbol{\alpha}, v^2) = [1/(2\pi v^2)^{-m/2}] \exp[-F(\boldsymbol{\alpha})/2v^2] \tag{6-100}$$

So that maximizing $L(\boldsymbol{\alpha}, v^2)$ with respect to $\boldsymbol{\alpha}$ is equivalent to minimizing $F(\boldsymbol{\alpha})$ with respect to $\boldsymbol{\alpha}$.

## 6.3 Confidence Contours on Estimated Parameters

The nonlinear methods given in Chapter 4 can estimate the parameters of interest. In this section we discuss the computation of confidence contours on the estimated parameters. If we assume the success of the linearized Taylor method, the method of steepest descent, the Newton–Raphson method, or Marquardt's method in the estimation of our parameters, we will have at some point in those computations estimated the first derivative of our function $F(\mathbf{X}, \boldsymbol{\alpha})$ with respect to the parameter $\boldsymbol{\alpha}$. If we have $m$ $(j = 1, \ldots, m)$ observations and $n$ $(i = 1, \ldots, n)$ parameters then

$$\hat{Q}_{ij} = [\partial F(\mathbf{X}_j, \boldsymbol{\alpha})/\partial \alpha_i]_{\boldsymbol{\alpha}=\hat{\boldsymbol{\alpha}}} \tag{6-101}$$

where $\mathbf{X}_j$ could take the form

$$\mathbf{X}_j = \begin{bmatrix} f_1(x_j) \\ f_2(x_j) \\ \vdots \\ f_k(x_j) \end{bmatrix} \tag{6-102}$$

and $\hat{\boldsymbol{\alpha}}$ is the value of the vector $\boldsymbol{\alpha}$ which we have accepted as our solution vector or $\hat{\boldsymbol{\alpha}}$ is the actual estimator of $\boldsymbol{\alpha}$.

Inspection of Eq. (6-101) shows that elements of the matrix $\mathbf{Q}$ are known when the equation that describes the model is known.

For example, suppose the model is described by the following equation

$$y = \alpha_1 e^{-\alpha_2/x} + \alpha_3 e^{-\alpha_4/x} \tag{6-103}$$

so that we have four parameters $\alpha_1, \alpha_2, \alpha_3, \alpha_4$ and only one independent variable $f_1(x) = x$. The structure of the matrix $\mathbf{Q}$ can be written as follows

$$\hat{\mathbf{Q}} = \begin{bmatrix} e^{-\hat{\alpha}_2/x_1} & -\hat{\alpha}_1 e^{-\hat{\alpha}_2/x_1} & e^{-\hat{\alpha}_4/x_1} & -\hat{\alpha}_3 e^{-\hat{\alpha}_4/x_1} \\ e^{-\hat{\alpha}_2/x_2} & -\hat{\alpha}_1 e^{-\hat{\alpha}_2/x_2} & e^{-\hat{\alpha}_4/x_2} & -\hat{\alpha}_3 e^{-\hat{\alpha}_4/x_2} \\ \vdots & \vdots & \vdots & \vdots \\ e^{-\hat{\alpha}_2/x_m} & -\hat{\alpha}_1 e^{-\hat{\alpha}_2/x_m} & e^{-\hat{\alpha}_4/x_m} & -\hat{\alpha}_3 e^{-\hat{\alpha}_4/x_m} \end{bmatrix} \tag{6-104}$$

where all the matrix elements are known. To obtain confidence contours on the parameters we have the following formula

$$(\boldsymbol{\alpha} - \hat{\boldsymbol{\alpha}})' \hat{\mathbf{Q}}' \hat{\mathbf{Q}} (\boldsymbol{\alpha} - \hat{\boldsymbol{\alpha}}) \leq n R^2 F(n, m - n, 1 - \gamma) \tag{6-105}$$

In the above inequality the left-hand side forms a quadratic form in the elements of the vector $\boldsymbol{\alpha}$ and the right-hand side is made up of known quantities since

$$R^2 = \sum_{j=1}^{m} [\mathbf{Y} - f(\mathbf{X}, \hat{\boldsymbol{\alpha}})]^2/(m - n) \tag{6-106}$$

is made up of known quantities. The quantity $F$ is found in the statistical tables at the required level of significance and degrees of freedom $n$ and $m - n$. Equation (6-105) is similar to Eq. (6-60) in the linear case, and, as before, if $\boldsymbol{\alpha}$ has only two elements we obtain an ellipse as a confidence region, if $\boldsymbol{\alpha}$ has three elements we obtain an ellipsoid, and so on.

Since the model is nonlinear the confidence region will not be a true confidence region since we do not know the probability of confidence. However, if we want an exact confidence contour we can set

$$F(\boldsymbol{\alpha}) = \sum_{j=1}^{m} [\mathbf{Y} - f(\mathbf{X}, \boldsymbol{\alpha})]^2 = \text{constant} \tag{6-107}$$

thus defining an exact confidence contour but not a true confidence region. The following equation

$$F(\boldsymbol{\alpha}) = F(\hat{\boldsymbol{\alpha}})\{1 + [n/(m - n)]F(n, m - n, 1 - \gamma)\} = \lambda_\gamma \qquad (6\text{-}108)$$

defines an exact confidence contour. In Eq. (6-108) we note that $F(\hat{\boldsymbol{\alpha}})$ is determined by replacing the elements of $\boldsymbol{\alpha}$ by estimated quantities $\hat{\boldsymbol{\alpha}}$. The quantity in braces can be determined from the statistical tables so that $\lambda_\gamma$ is known.

On the other hand $F(\boldsymbol{\alpha})$ has as undetermined parameters the entire vector $\boldsymbol{\alpha}$. However, $F(\boldsymbol{\alpha})$ is a quadratic form in the parameters $\boldsymbol{\alpha}$ so we can write Eq. (6-108) as follows

$$\xi_{11}(\alpha_2, \alpha_3, \ldots, \alpha_n)\alpha_1{}^2 + 2\xi_{12}(\alpha_2, \alpha_3, \ldots, \alpha_n)\alpha_1 + \xi_{13}(\alpha_2, \alpha_3, \ldots, \alpha_n) - \lambda_\gamma = 0 \qquad (6\text{-}109)$$

where $\xi_{11}$, $\xi_{12}$, and $\xi_{13}$ are functions of all the parameters but $\alpha_1$. By selecting a value for $\gamma$ (and hence $\lambda_\gamma$) and evaluating

$$\alpha_1 = \{-\xi_{12} \pm [\xi_{12}^2 - \xi_{11}(\xi_{13} - \lambda_\gamma)]^{1/2}\}/\xi_{11} \qquad (6\text{-}110)$$

for a range of the parameters $(\alpha_2, \alpha_3, \ldots, \alpha_n)$ we obtain points on the boundaries of the confidence regions (the range of the parameters $\alpha_2$, $\alpha_3, \ldots, \alpha_n$ should be close to their estimated values).

To illustrate the above further for two parameters, we note that, in this case, $\xi_{11}$, $\xi_{12}$, $\xi_{13}$ are functions of the parameter $\alpha_2$ only. By picking $\gamma = 0.05$ we can determine $\lambda_\gamma$ and hence $\alpha_1$ for a certain range of values of $\alpha_2$ near the solution. Figure 6-9 shows us a confidence region for $\alpha_1$ and $\alpha_2$; for every value of $\alpha_2$ near $\hat{\alpha}_2$ we can obtain two values of $\alpha_1$ using Eq. (6-110).

It is not difficult to extend this concept to higher dimensions for a larger number of parameters. It is important to note in the nonlinear case that while we have defined an *exact* confidence contour we only have an *approximate* probability level.

## 6.4 The Question of How Many Terms to Use in a Model

As indicated earlier we will have occasion to determine how many terms are sufficient to describe the response of a system. In certain cases one would like to know how many exponentials are required to analyze the phosphorescent decay of a protein, the number of relaxation times in an enzyme reaction, the number of Gaussian coefficients to fit a circular dichroism curve, and so forth.

Unlike the case in which the parameters occur linearly we cannot apply $F$-tests to parameters we might consider superfluous. One thing that can be done in such instances is to make visual comparisons between the raw data and the model with computed parameters. This however is subject to the naked eye bias of the investigator.

Although the $F$-test in nonlinear cases is inapplicable it does provide us with a numerical result which can serve as the basis for comparison with similar results obtained by the same investigator or other investigators. How good the $F$-test is in any given instance depends on how much our model has deviated from linearity. Beale (1960) has suggested several measures of nonlinearity. Beale's measures of nonlinearity have been discussed by Guttman and Meeter (1965).

In examining hypothesis testing in nonlinear situations we must distinguish between *linear hypotheses* about estimates coming from nonlinear methods, nonlinear hypotheses about linear models, and the situation which involves both nonlinear estimation and tests of nonlinear hypotheses.

Nonlinear hypotheses are generally of the form

$$H_p(\alpha_1, \alpha_2, \ldots, \alpha_n) = J_p \tag{6-111}$$

where $H_p$ is a function nonlinear in the parameters.
To date there is no satisfactory way of tackling the problem of testing hypotheses in the nonlinear cases, and we have to resort to linearization in order to formulate the problem. Such an approach has its limitations and the investigator must bear in mind some of those limitations.

If we expand $H_p$ in a Taylor series about the point $[\alpha_1^{(j)}, \alpha_2^{(j)}, \ldots, \alpha_n^{(j)}]$ where $\boldsymbol{\alpha}^{(j)}$ is the solution vector we obtain

$$H_p - H_p^{(j)} = \sum_{i=1}^{n} [(\partial H_p/\partial \alpha_i)(\alpha_i - \alpha_1^{(j)}) + \cdots] \tag{6-112}$$

If we neglect higher terms in the expansion we reduce a nonlinear hypothesis into a linear one. To apply statistical tests to parameters represented by Eq. (6-112), our test must be valid over the range of the parameters where it is expected that the linearization process is valid. We also note that if we wish to test nonlinear hypotheses about parameters obtained from the linear least squares we must do the same thing.

A general guideline to the problem of nonlinear hypothesis testing is that a few conclusions can be made provided we restrict ourselves to regions where our linearization processes are valid.

## 6.5 Variance–Covariance Matrix in Nonlinear and Other Cases

In Chapter 5 we noted that the information matrix which we shall call $\mathbf{I}(\boldsymbol{\alpha})$ is the inverse of the variance–covariance matrix. It has been suggested that one could estimate the parameter vector by one or other of the methods given in Chapter 5. Then using the estimated parameters one could construct the likelihood function close to the minimum. This could then be followed by determining the information matrix whose elements are given by

$$I_{ij}(\boldsymbol{\alpha}) = -E(\partial^2 \ln L / \partial\alpha_i \, \partial\alpha_j) \tag{6-113}$$

where $L$ is the likelihood function.

Here again we emphasize that the variance–covariance matrix may not reflected the actual variations the parameters may undergo since, as one varies the parameters over a wide range, the nonlinearities influence the solution and the variations in the fit may be better or worse than those given by the variance–covariance matrix.

To estimate the variance–covariance matrix from the method of Nelder and Mead (1965) those authors extended a method due to Spendley *et al.* (1962). The points of the simplex in $n$ dimension that are accepted as the solution to the minimization problem (the points representing the vertices of the last simplex) are given by $P_0, P_1, \ldots, P_n$. The "halfway points" between those points are given by $P_{ij} = (P_i+P_j)/2$. All the points are taken and a quadratic surface is made out of all $(n+1)(n+2)/2$ points. The final simplex provides a ready-made set of oblique coordinates whose vertices may be taken as the points

$$(0, 0, 0, \ldots, 0),\ (1, 0, 0, \ldots, 0),\ (0, 1, 0, \ldots, 0), \ldots, (0, 0, 0, \ldots, 1) \tag{6-114}$$

The quadratic approximation to the function in the neighborhood of the minimum can be written in vector notation as

$$\mathbf{y} = A_0 + 2\mathbf{A}'\boldsymbol{\alpha} + \boldsymbol{\alpha}'\mathbf{B}\boldsymbol{\alpha} \tag{6-115}$$

where $A_0 = Y_0$ (the response evaluated at $P_0$). The vector $\mathbf{A}$ and the matrix $\mathbf{B}$ have the following elements

$$\begin{aligned} A_i &= 2y_{0i} - (y_i + 3y_0)/2 && (i = 1, 2, \ldots, n) \\ B_{ii} &= 2(y_i + y_0) - 2y_{0i} && (i = 1, 2, \ldots, n) \\ B_{ij} &= 2(y_{ij} + y_0 - y_{0i} - y_{0j}) && (i \neq j) \end{aligned} \tag{6-116}$$

In Eq. (6-116) $y_i$ is the value of the response at the point $P_i$ and $y_{ij}$ is the value of the response at the point $P_{ij}$. The estimated minimum in this case will be given by

$$\boldsymbol{\alpha}_{\min} = -\mathbf{B}^{-1}\mathbf{A} \tag{6-117}$$

where **B** is the information matrix.

## References


Aitken, A. C. (1939). "Statistical Mathematics." Oliver and Boyd, Edinburgh.

Beale, E. M. L. (1960). *J. Roy Stat. Soc. Ser. B.* **22**, 41.

Box, G. E. P., and Draper, N. R. (1959). *J. Amer. Stat. Ass.* **54**, 622.

Brownlee, K. A. (1960). "Statistical Theory and Methodology in Science and Engineering." Wiley, New York.

Draper, N. R., and Smith, H. (1966). "Applied Regression Analysis." Wiley, New York.

Graybill, F. A. (1961). "An Introduction to Linear Statistical Models." McGraw-Hill, New York.

Guttman, I., and Meeter, D. A. (1965). *Technometrics* **7**, 623.

Kendall, M. A., and Stuart, A. (1963). "The Advanced Theory of Statistics," Vol. 1 and 2. Griffin, London.

Lindgern, B. W. (1968). "Statistical Theory," 2nd ed. Macmillan, New York.

Nelder, J. A., and Mead R. (1965). *Comput. J.* **7**, 155.

Spendley, W., Hext, G. R., and Himsworth, A. (1962). *Technometrics* **4**, 441.

Weatherburn, C. E. (1946). "First Course in Mathematical Statistics." Cambridge Univ. Press, London and New York.

*CHAPTER 7*

# ABSORPTION SPECTRA OF MIXTURES

## 1. Introduction

In this chapter we generalize one of the problems mentioned in the introduction. The basic question to answer here is, given the absorption spectrum of a mixture made up of several components with *known* absorption spectra, how much of each component is present. At the outset it is important to emphasize that such analyses will not work if there exists a component in the mixture which is not cataloged or tabulated in the standard library of *known* compounds. For example, if we take the case of a hydrolysate of RNA, the general procedure would be to analyze it in terms of the four basic nucleotides, adenylic acid, guanylic acid, cytidylic acid, and uridylic acid. Our analysis will be sucessful if the hydrolysate contains only the above four components or just three, two, or a single component. If the hydrolysate contains a component which is not in the library we are using to analyze the mixture, then obviously this component will not appear in our final result, and it will spoil the computation.

## 2. Absorption Spectra of Mixtures

Authors who have used such analyses are Reid and Pratt (1960) and Pratt *et al.* (1966) who used the method of the least squares and linear programming techniques to analyze RNA hydrolysate. Lee *et al.* (1965) used the least squares method ($L_2$ norm) to analyze S-RNA in terms of nine components namely adenine (A), uridine (G), cytidine (C), 5-ribosuracil ($\psi U$) 7-methyl guanine (7MG), 6-methylaminopurine riboside (GMA), 6-dimethyl amino purine riboside (DM), and thymine riboside (T).

Both Pratt *et al.* (1964) and Lee *et al.* (1965) used their methods for the analysis of nucleotides, however it is clear that the methods of analysis can be applied to any kind of mixture no matter what the composition is as long as the components have known spectra, are noninteracting, and are linearly independent.

Basically the techniques to find the concentration of each component have been described in Chapter 4. The problem for the analysis of a mixture may be stated as follows: If the absorption spectrum $f(x)$ of a mixture is composed of several compounds whose absorption spectra are $f_1(x), f_2(x), \ldots, f_n(x)$, then it is required to select the "best" value of the coefficients $\alpha_i$ in the following equations.

$$\begin{aligned} f(x_1) &= \alpha_1 f_1(x_1) + \alpha_2 f_2(x_1) + \cdots + \alpha_n f_n(x_1) + \delta_1 \\ f(x_2) &= \alpha_1 f_1(x_2) + \alpha_2 f_2(x_2) + \cdots + \alpha_n f_n(x_2) + \delta_2 \\ &\vdots \\ f(x_m) &= \alpha_1 f_1(x_m) + \alpha_2 f_2(x_m) + \cdots + \alpha_n f_n(x_m) + \delta_m \end{aligned} \tag{7-1}$$

For the sake of clarity we will make the following identification

$$\begin{aligned} &f(x_1) = A_1, \quad f_1(x_1) = \varepsilon_{11}, \quad f_2(x_1) = \varepsilon_{21}, \ldots, f_n(x_1) = \varepsilon_{n1} \\ &f(x_2) = A_2, \quad f_1(x_2) = \varepsilon_{12}, \quad f_2(x_2) = \varepsilon_{22}, \ldots, f_n(x_2) = \varepsilon_{n2} \\ &\quad\vdots \qquad\qquad\quad \vdots \qquad\qquad\qquad \vdots \\ &f(x_m) = A_m, \quad f_1(x_m) = \varepsilon_{1m}, \quad f_2(x_m) = \varepsilon_{2m}, \ldots, f_n(x_m) = \varepsilon_{nm} \\ &\qquad \alpha_1 = C_1, \quad \alpha_2 = C_2, \quad \ldots, \quad \alpha_n = C_n \end{aligned} \tag{7-2}$$

where $A_1, A_2, \ldots, A_n$ are the measured absorptions at the indicated wavelength, $\varepsilon_{ij}$ is extinction coefficient of the $i$th component at the $j$th wavelength, and $C_1, C_2, \ldots, C_n$ are the concentrations of the $n$ components. The meaning of the word best is explained to some extent in Chapter 4, and it depends essentially on what function of the experimental error we minimize. In the case where we are using least squares the word best means that we are determining the $\alpha_i$ such that

$$\sum_{j=1}^{m} [f(x_j) - \sum_{i=1}^{n} \alpha_i f_i(x_j)]^2 = \sum_{j=1}^{m} \delta_j^2 \tag{7-3}$$

is a minimum.

In the linear programming technique used by Pratt *et al.* (1964) we write the equation of absorption for one wave length

$$f(x_j) = \sum_{i=1}^{n} \alpha_i f_i(x_j) + \delta_j \tag{7-4}$$

where $\delta_j$ is the error made in making a measurement at wavelength $x_j$ The technique of linear programming is to obtain the "best" $\alpha_i$, such that all the $\alpha_i$ are *positive* and $\sum_{j=1}^{m} | \delta_j |$ is a minimum. In the linear programming technique $\delta_j$ is represented as

$$\delta_j = s_j' - s'' \tag{7-5}$$

when $s_j'$ and $s_j''$ are called slack variables see Chapter 4 for details. Pratt *et al.* (1964), in carrying out their computation, use both positive and negative slack.

The least squares technique can also be used to minimize the quantity in Eq. (7-3) such that the $\alpha_i$ are restricted to

$$\sum_{i=1}^{n} \alpha_i = k \tag{7-6}$$

where $k$ is some positive constant. This constant will depend on the nature of the problem under consideration. In the case of the absorption spectra of mixtures $k$ will equal the total concentration of all components. It must be noted however that the fact that Eq. (7-6) is constrained to hold does not mean that all the $\alpha_i$ are going to be positive.

In Chapter 4 we have compared the advantages and disadvantages of all methods mentioned above and we refer the reader to this particular section of Chapter 4. In Chapter 6 we have discussed the methods of computation of the standard error for one method and the analysis of variance to determine how many components are present if such an analysis is required.

In the application of the least squares method to RNA hydrolysates Lee *et al.* (1965) have found that a negative coefficient often appears in the results. Lee *et al.* overcame the problem of negative coefficients by simply deleting the component with negative coefficient and starting the analysis anew. In the specific examples analyzed by those authors this procedure worked well, however we have some reservations as to whether or not such a procedure will always work. We believe it might be possible to compute a negative coefficient for a component and find that this component is actually present in the mixture.

Another interesting procedure adopted by Lee *et al.* for deleting some of the components from the library is based on the following reasoning. In view of the fact that greater accuracy is achieved when a restricted library is used a component was considered to be absent when

$$\alpha_j + 1.5\sigma_{\alpha_j} = C_j + 1.5\sigma_{C_j} < C/n \tag{7-7}$$

where $n$ is the number of nucleotides in the oligonucleotide, $C$ is the total nucleoside concentration, and $\sigma_{\alpha_j}$ is the standard deviation of the calculated $j$th concentration. Lee *et al.* reasoned that, inasmuch as the smallest possible concentration of any component is either $C/n$ or zero, then there is 86% probability that the $j$th component is not present if

$$C_j < C/n - 1.5\sigma_{C_j}$$

The probability is 86% since the computed concentrations are normally distributed. This is one of the consequences of a theorem of the general linear hypothesis of full rank stated in Chapter 6. If that is the case and the criterion in Eq. (7-7) is satisfied, then there is an 86% probability that component $j$ is missing because $C/n$ is the *minimum* possible concentration of the $j$th component if it is not absent.

If Lee *et al.* had wanted to set a more stringent criterion for the deletion of a component, say 95% probability that component $j$ is missing, Eq. (7-7) would become approximately

$$\alpha_j + 2.0\sigma_{\alpha_j} = C_j + 2.0\sigma_{C_j} < C/n \tag{7-8}$$

A number of examples from the analysis of Lee *et al.* based on mixtures of known constituents are shown in Table 7-1.

Pratt *et al.* (1964) used the least square method and the linear programming technique. The linear programming technique as explained in Chapter 4 can put any number of constraints on the computed coefficients. Thus, in addition to requiring the coefficients to be positive, we

**Table 7-1**

| Mixture | | A | U | G | C | U | 7MG | MA | DMA | T |
|---|---|---|---|---|---|---|---|---|---|---|
| | Composition (actual) | 0 | 0 | 0 | 0 | 0 | 1.00 | 0 | 0 | 0 |
| 1 | 1st Calculation | 0.01 | −0.08 | 0.02 | 0.00 | 0.01 | 0.99 | −0.01 | −0.00 | 0.02 |
| | 2nd Calculation[a] | −0.01 | | 0.01 | 0.01 | 0.00 | 0.98 | | | 0.01 |
| | Composition (actual) | 0.20 | 0.20 | 0.21 | 0.17 | 0.24 | 0 | 0 | 0 | 0 |
| 2 | 1st Calculation | 0.09 | 0.50 | 0.11 | 0.29 | 0.31 | −0.11 | 0.21 | −0.12 | −0.18 |
| | 2nd Calculation[a] | 0.22 | 0.16 | 0.18 | 0.22 | 0.24 | | −0.00 | | |
| | Composition (actual) | 0.24 | 0.24 | 0.26 | 0.20 | 0 | 0 | 0 | 0 | 0 |
| 3 | 1st Calculation | 0.10 | 0.58 | 0.17 | 0.30 | 0.10 | −0.11 | 0.25 | −0.15 | −0.18 |
| | 2nd Calculation[a] | 0.24 | 0.27 | 0.24 | 0.20 | 0.00 | | 0.00 | | |

[a] All numbers having a negative value in the second calculation have been deleted.

may restrict them to any range of values consistent with a feasible solution. Pratt *et al.* used their program with both positive and negative slack and minimized the absolute sum of the slack. Table 7-2 shows a linear programming solution of a mixture resulting from the hydrolysis of the trinucleotide CpApGp.

**Table 7-2**

| Constraint | Slack[a] | Ap | Cp | Gp | Up | Slack |
|---|---|---|---|---|---|---|
| None | ± | 34.6 | 30.5 | 34.8 | 0 | 0.8 |
| Ap=Cp=Gp | ± | 33.3 | 33.3 | 33.3 | 0 | 1.5 |

[a] Slack = absolute sum of residuals.

In the first attempt at linear programming no constraints were put on the coefficients with the result obtained in the first line. In the next attempt the constraints indicated were put on the coefficient with the result that we have equal quantities of all the nucleotides, however it will be noticed that the slack for such a computation has increased indicating a worse fit in the mathematical sense.

Some other restrictions that may be useful when using linear programming for this type of problem are the following. If the *i*th and *j*th compounds are known to be present in equal quantities we can put

$$\alpha_i - \alpha_j = 0.000 \tag{7-9}$$

We can restrict the coefficients to make their sum equal to the total concentration

$$\sum_{i=1}^{n} \alpha_i = C \tag{7-10}$$

or, if a component gives a very small value for its concentration, it may be expected that this component is actually missing, and therefore we can adopt the constraint that whenever $\alpha_i$ is less than some number it can be put equal to zero. Unlike the least squares there is no statistical criteria for rejection of any component, and it becomes a matter of judgement as to what criteria we can set for rejecting any component when the linear programming technique is used. In Table 7-3 the least squares method and linear programming techniques are compared for a number of mixtures. (The mixtures used by Pratt *et al.* were of known composition. For the accuracy of the data see original papers.)

**Table 7-3**

Calculated Composition (%)[a]

| | | A | C | G | U | Slack |
|---|---|---|---|---|---|---|
| Ap/Cp | LS[b] | 54.8 | 46.1 | 1.2 | −2.0 | |
| | LP[c] | 50.1 | 50.1 | 0.0 | 0.0 | 2.3 |
| Ap/Cp | LS | 54.5 | 46.8 | 0.5 | −1.9 | |
| | LP | 50.2 | 50.1 | 0.0 | 0.0 | 2.0 |
| Ap/Gp | LS | 47.1 | −1.5 | 54.2 | 0.1 | |
| | LP | 50.1 | 0.0 | 49.9 | 0.0 | 1.0 |
| Ap/Ap/Gp | LS | 63.9 | −0.2 | 36.1 | 0.2 | |
| | LP | 66.4 | 0.0 | 33.1 | 0.0 | 0.8 |
| Cp/Ap/Gp | LS | 34.9 | 30.7 | 35.1 | −0.9 | |
| | LP | 33.3 | 33.3 | 33.3 | 0.0 | 1.5 |
| pAp/A | LS | 101.0 | 0.0 | −1.9 | 0.9 | |
| | | 100.1 | 0.0 | 0.0 | 0.0 | 1.2 |

[a] From Pratt *et al.* (1966).
[b] LS = least squares solution
[c] LP = linear programming solution.

To date only least squares and linear programming have been used to compute the concentration of the various components in any given mixture. It is possible that is all that is required. The other methods in Chapter 4 may be used if least squares and linear programming are found to be inadequate for some reason. As stated at the beginning of this chapter, there are a number of areas of application where data from absorption spectroscopy can be analyzed using the methods of Chapter 4. These areas, to be considered next, include the determination of the extent of modification of amino acid residues in proteins, solvent perturbation studies, and binding studies.

## 3. Applications in Protein Modification to Determine Reactivity of Amino Acid Residues

Modification in proteins of the amino acid residues histidine by diazo-1-*H*-tetrazole (Horinishi *et al.*, 1964), tryptophan by *N*-bromosuccinamide (Witkop, 1961), tyrosine by *N*-acetyl imidazole (Riordan *et al.*, 1965) and tetranitromethane (Riordan *et al.*, 1966), lysine ε-amino groups and free amino group by trinitrobenzesulfonate, and others, frequently requires the recording of the absorption spectrum of two or more components in presence of each other.

For instance, the amount of histidine bisazo-1-*H*-tetrazole can in theory be measured at its absorption maximum of 480 mμ, but this has to be done before a recognizable shoulder in the spectrum at 550 mμ appears. This shoulder is due to tyrosine bisazo-1-*H*-tetrazole, and will interfere with the analysis if the absorption is taken at only one wavelength. We maintain that this need not be the case, for, if we can characterize the spectra of both the histidine derivative as well as the tyrosine derivative, then we ought to be able to use the analysis presented above to tell one in presence of the other.

In the case of the reaction of tyrosine with tetranitromethane we have the following reaction due to Riordan *et al.* (1966).

$$R-C_6H_4-OH + C(NO_2)_4 \longrightarrow R-C_6H_3(NO_2)-O^- + C(NO_2)_3^- + 2\,H^+$$

Since both the nitrotyrosine and the nitroform anion absorb in the vicinity of 350 mμ region, we can tell one in presence of the other if we characterize the spectra of both. Riordan *et al.* (1966) has also shown that cysteine is also affected by tetranitromethane. If the absorption spectrum of this derivative of cysteine is known, then we can monitor its modification in presence of other residues.

In view of the fact that we can determine all components in the presence of each other, the case where the reagent used to modify the protein absorbs at a wavelength necessary for the observation of the modification reaction, and the case where the protein itself absorbs at a wavelength necessary for the observation of the modification reaction do not present problems, but can be handled effectively by the above methods.

Previous procedures employed to measure the extent of modification of amino acid residues in proteins generally tend to separate (by dialysis or passage through molecular sieves) all the components in the reaction mixture except the one which is deemed important (generally the modified protein). This was followed by a single observation on the important component at one wavelength. From this data the number of residues modified (generally of one type) is deduced.

We have no quarrel with this procedure except to say that determining the components in presence of each other can give us this information without having to resort to separation procedures and further handling of the reaction mixture. In addition, if the reaction is not instantaneous, then we can follow the kinetics of modification far more easily than if we were forced to separate the components. Lastly, if some of the reagents used are nonspecific and combine with more than one residue, we should regard this as an advantage instead of a disadvantage, because we can tell how much of *each* residue is being modified by rapidly scanning the entire spectrum at various time intervals.

## 4. Application to Solvent Perturbation

Another application of function minimization is in the domain of solvent perturbation. Herskovits and Laskowski (1960) introduced this technique to determine the number of exposed and buried tyrosine residue in a protein (strictly speaking this classification is over simplification because a residue can be partially exposed). The technique of solvent perturbation is to take the difference spectra of a protein under some condition (e.g., 8 *M* urea or plain water at pH 7) and the protein under similar conditions in presence of a perturbant (e.g., sucrose, glycerol, dimethyl sulfoxide). A model mixture containing the same number of residues of tryptophan, tyrosine and phenylalanine under the same conditions is now subjected to the same treatment as the protein. This is illustrated diagramatically as follows.

Difference spectra of protein taken by reading

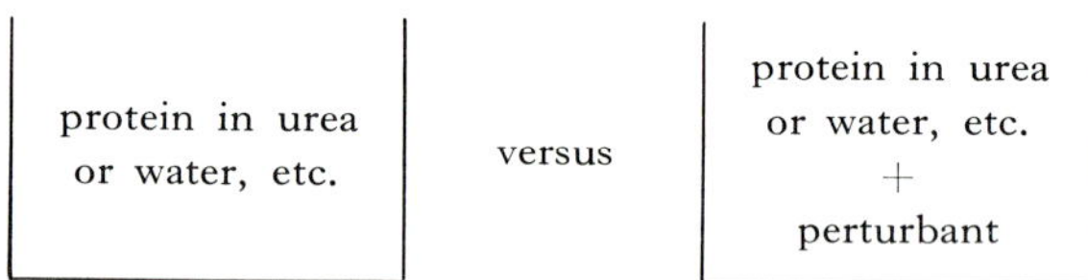


and difference spectra of model mixture

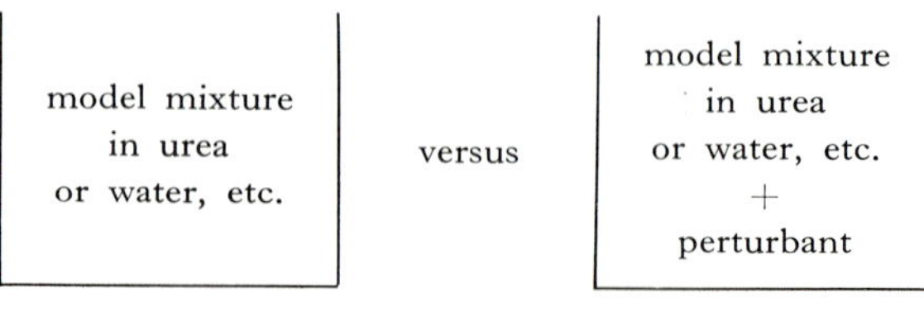

In both cases the perturbant will cause a significant difference spectrum. Call these spectra $\Delta\varepsilon(\lambda)$ (protein) and $\Delta\varepsilon(\lambda)$ (model mixture) for the protein and the model mixture respectively. The percentage exposure of the aromatic amino acids in the protein may be determined by the ratio $R_m$ suggested by Herskovits (1965)

$$R_m = \frac{[\Delta\varepsilon_{\lambda_1}(\text{protein})/\varepsilon_{\lambda_2}(\text{protein})]}{[\Delta\varepsilon_{\lambda_1}(\text{model mixture})/\varepsilon_{\lambda_2}(\text{model mixture})]}$$

where $\lambda_1$ is the wavelength of maximum absorption of difference spectra for both protein and model mixture and $\lambda_2$ is the wavelength of maximum absorption of the *ordinary* spectra of the protein and the model mixture. (In actual practice $\lambda_1$ and $\lambda_2$ are not always constant but are generally confined within a narrow range.)

The exact definition of $\Delta\varepsilon_{\lambda_1}$ is the molar difference absorption coefficient for either the protein or the model mixture. Similarly the exact definition of $\varepsilon_{\lambda_2}$ is the molar absorption coefficient. Model mixtures are generally made with *N*-acetyl-ethyl ester of aromatic amino acids mixed in the same molar ratios as found in the proteins.

A problem arising in this kind of work is the distinction between the amino acids tyrosine and tryptophan since both absorb in the 270–300 m$\mu$, and the difference spectra are taken in that region. Herskovits (1967) suggested that this difficulty may be overcome if we write

$$\begin{aligned} \Delta\varepsilon(\lambda_\alpha)(\text{protein}) &= a\,\Delta\varepsilon(\lambda_\alpha)_{\text{trp}} + b\,\Delta\varepsilon(\lambda_\alpha)_{\text{tyr}} \\ \Delta\varepsilon(\lambda_\beta)(\text{protein}) &= a\,\Delta\varepsilon(\lambda_\beta)_{\text{trp}} + b\,\Delta\varepsilon(\lambda_\beta)_{\text{tyr}} \end{aligned} \qquad (7\text{-}10\text{a})$$

where $\Delta\varepsilon(\lambda_\alpha)_{\text{trp}}$ and $\Delta\varepsilon(\lambda_\alpha)_{\text{tyr}}$ are the difference spectra at wavelength $\lambda_\alpha$ *N*-acetyl tryptophan-ethyl-ester and *N*-acetyl-tyrosine-ethyl ester in the presence and in the absence of perturbant. The parameters *a* and *b* indicate the number of exposed tryptophan and tyrosine residues and the wavelengths in question are $\lambda_\alpha = 292 \pm 1$ m$\mu$ and $\lambda_\beta = 287 \pm 1$ m$\mu$.

In actually applying this method Herskovits and Sorensen (1968a) solve the above equations at two wavelengths for *a* and *b*. Then, using these values for *a* and *b*, they try to reconstruct the whole of $\Delta\varepsilon(\lambda)$ (protein) by adjusting the values of *a* and *b* by trial and error. The authors can attempt to reconstruct the spectrum $\Delta\varepsilon(\lambda)$ for the protein because the entire functions $\Delta\varepsilon(\lambda)_{\text{tyr}}$ and $\Delta\varepsilon(\lambda)_{\text{trp}}$ can be determined for any given perturbant and have, in fact, been tabulated for several perturbants by Herskovits and Sorensen (1968a). At this stage the reader who has read the introduction and this chapter realizes that we are going to recommend that the analysis is best conducted at all wavelengths where the perturbation spectra are significant.

Obviously such a procedure is a far less laborious than trial and error and the parameters $a$ and $b$ can be computed once and for all in a single computation by the techniques outlined in Chapter 4. For this kind of work, the ideal wavelength appears to be 275–305 m$\mu$. This avoids the contribution of phenylalanine. There is, of course, no reason why phenylalanine should not be included in the analysis provided the investigator can characterize its difference spectrum and feels that sufficient accuracy is being achieved to distinguish its small contribution. For the analysis of the entire spectrum for tryptophan and tyrosine we set up the following equations

$$\begin{aligned}
\Delta\varepsilon(\lambda_1) &= a\,\Delta\varepsilon_1(\lambda_1) + b\,\Delta\varepsilon_2(\lambda_1) + \delta_1 \\
\Delta\varepsilon(\lambda_2) &= a\,\Delta\varepsilon_1(\lambda_2) + b\,\Delta\varepsilon_2(\lambda_2) + \delta_2 \\
&\vdots \\
\Delta\varepsilon(\lambda_m) &= a\,\Delta\varepsilon_1(\lambda_m) + b\,\Delta\varepsilon_2(\lambda_m) + \delta_m
\end{aligned} \tag{7-11}$$

where $\Delta\varepsilon(\lambda_i)$ is the molar absorption difference of the protein at $\lambda_i$ in presence and absence of perturbant and $\Delta\varepsilon_1(\lambda_i)$ and $\Delta\varepsilon_2(\lambda_i)$ are the molar absorption difference of $N$-acetyl-$L$-tryptophan and $N$-acetyl-$L$-tyrosine at wavelength $\lambda_i$ in presence and absence of perturbant. The error of each measurement at the indicated wavelength is denoted by the $\delta_i$. Rewriting Eqs. (7-11) in the formalism of Chapter 4 we have

$$\begin{array}{lll}
\Delta\varepsilon(\lambda_1) = f(x_1), & \Delta\varepsilon_1(\lambda_1) = f_1(x_1), & \Delta\varepsilon_2(\lambda_1) = f_2(x_1) \\
\Delta\varepsilon(\lambda_2) = f(x_2), & \Delta\varepsilon_1(\lambda_2) = f_1(x_2), & \Delta\varepsilon_2(\lambda_1) = f_2(x_2) \\
\vdots & \vdots & \vdots \\
\Delta\varepsilon(\lambda_m) = f(x_m), & \Delta\varepsilon_1(\lambda_m) = f_1(x_m), & \Delta\varepsilon_2(\lambda_m) = f_2(x_m)
\end{array} \tag{7-12}$$

$$a = \alpha_1, \qquad b = \alpha_2$$

Whatever has been said about the Eqs. (4-44) in Chapter 4 applies with equal validity to the above. If we want to include phenylalanine in the analysis, all we have to do is add the column vector $\Delta\varepsilon_3(\lambda) = f_3(x)$ to the right-hand side of Eq. (7-11) (this vector being the difference spectra pertaining to phenylalanine). The coefficient $\alpha_3$ is put in front of all the components of $f_3(x)$ and the analysis extended to shorter wavelengths. As stated before, the necessary functions $\Delta\varepsilon_1(\lambda)$ and $\Delta\varepsilon_2(\lambda)$ for nine different perturbants using water as a solvent have been tabulated by Herskovits and Sorensen (1968a) using the above equations to fit the difference spectrum of pepsin.

Figure 7-1 from Herskovits and Sorensen (1968b) has suggested to us another application of linear programming. The choice by those authors of curve VII as a possible fit to the difference spectra of pepsin appears to constrain this particular curve to follow the peaks rather faithfully.

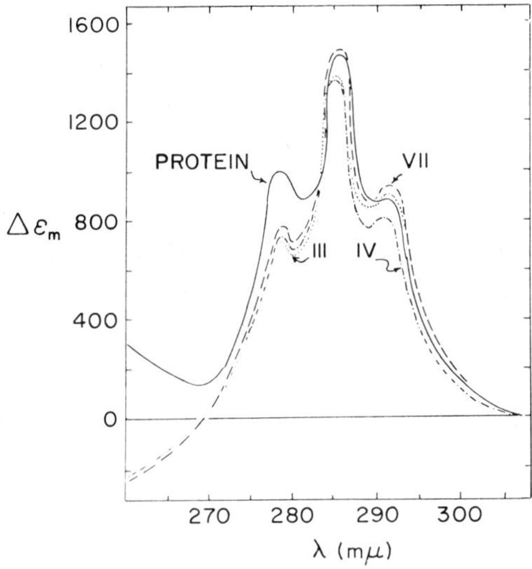


**Fig. 7-1.** A comparison of experimental solvent perturbation difference spectra with calculated curves based on model compound data and $a$ and $b$ parameters obtained from the solutions of Eq. (7-1a). The solid line represents the experimental data of native pepsin, pH 5.6, 0.1 $M$ KCl, obtained with 20% ethylene glycol as perturbant. Curves III, IV, and VII are curves calculated with some trial $a$ and $b$ values and the wavelength-dependent model compound data of Herskovits and Sorensen (1968b). Curve III, a = 2.5, b = 10. Curve IV, a = 2, b = 11. Curve VII, a = 2.5, b = 11. (Reprinted from Herskovits and Sorensen (1968b). *Biochemistry* **7**, 2533. Copyright 1968 by the American Chemical Society. Reprinted by permission of the copyright owner.)

Using linear programming techniques we suggest deleting the slack variable at two or three wavelengths where we have an extremum. How well that will work we do not know, but little will be lost in trying such a procedure if one already has the programs. In Fig. 7-1 we also notice that at wavelengths shorter than 270 mμ there is no agreement between the computed curves and the native protein. This discrepancy may be reduced to some extent by including the phenylalanine spectra, but that will not be adequate because the computed curve and the curve due to native protein actually diverge at those wavelengths.

Several interesting questions arise in connection. If, with the analysis suggested in Chapter 4, the fits obtained are no better than the ones Herskovits and Sorensen obtained, it is pertinent to ask the cause of this discrepancy.

It appears evident from Fig. 7-1 that there is no possibility of agreement at wavelengths below 265 m$\mu$ by using positive coefficients only, since the difference spectrum of the protein increases while the spectra of all the model compounds decrease. This lack of agreement may be due to some kind of interaction between the perturbant and the amide backbone of the protein, as suggested by Herskovits and Laskowski (1962) and Riddiford (1966), in connection with solvent perturbation spectra of ovomucoid and paramyosin respectively. If this is the case one can then ask if the same disagreement is obtained using different perturbants or if this disagreement is obtained when the same perturbant is used on different proteins. Answers to such questions will greatly extend the scope of solvent perturbation and our knowledge about the interaction of proteins.

## 5. Binding Studies

The last application we suggest of the above methods pertains to binding studies. In many dialysis equilibrium studies we can measure the binding of as many substrates and other molecules as we like in presence of each other by simply determining the absorption spectrum of the solution on the outside and the inside of the dialysis bag. Such a procedure will determine the concentration of all components on both sides of the bag. With this information in hand we can determine the effect of other molecules on the binding of the substrate and vice versa. Studies involving binding in the analytical ultracentrifuge can also be undertaken for as many substrates and effectors as one wants. Inasmuch as we can rapidly scan the entire length of the cell at different wavelengths, we determine the concentration of each component at every point in the cell. This will tell us how much binding has occurred of any given substrate or, what is probably more important, any combination of substrate and effectors. For example, if we desire we can determine the association constant of DPNH and a dehydrogense in the presence of DPN, and vice versa in a series of runs where we have *both* the reduced and oxidized coenzymes in the same cell. The overlap of the spectra of these components, as pointed out throughout this chapter, should cause no problem. A more complicated example is the determination of the association constants of all the nucleotides, triphosphates, and DNA, polymers in each others presence in absence of a primer or even its presence.

In concluding this chapter we can state that as long as the spectra of the individual components of our mixture are known and are *linearly independent*, and provided a great deal of experimental accuracy is achieved, the possibilities of determining all components in presence of each other, no matter what the source of the absorption data, appear to be limitless. We believe that with the current improvement in absorption spectroscopy such accuracy can be achieved.

## References


Herskovits, T. T. (1965). *J. Biol. Chem.* **240**, 628.

Herskovits, T. T. (1967). *Methods Enzmol.* **11**, 748.

Herskovits, T. T., and Laskowski, M., Jr. (1960). *J. Biol. Chem.* **235**, PC56.

Herskovits, T. T., and Laskowski, M., Jr. (1962). *J. Biol. Chem.* **239**, 2481.

Herskovits, T. T., and Sorensen, M. (1968a). *Biochemistry* **7**, 2523.

Herskovits, T. T., and Sorensen, M. (1968b). *Biochemittry* **7**, 2533.

Horinishi, H., Hachimori, Y., Kurihara, K., and Shibata, K. (1964). *Biochem. Biophys. Acta.* **86**, 477.

Lee, S., McMullen, D., Brown, G. L., and Slokes, A. R. (1965). *Biochem. J.* **94**, 314.

Pratt, A. W., Toal, J. N., and Rushizky, G. W. (1966). *Ann. N. Y. Acad. Sci.* **128**, 900.

Pratt, A. W., Toal, J. N., Rushizky, G. W., and Sober, H. A. (1964). *Biochemistry* **3**, 1831.

Reid, J. C., and Pratt, A. W. (1960). *Biochem. Biophys. Res. Commun.* **3**, 377.

Riddiford, L. M. (1966). *J. Biol. Chem.* **241**, 2792.

Riordan, J. F., Sokolovsky, M., and Vallee, B. L. (1966). *J. Amer. Chem. Soc.* **88**, 4104.

Riordan, J. F., Wacher, W. E. C., and Vallee, B. L. (1965). *Biochemistry* **4**, 1758.

Witkop, B. (1961). Advan. Prot. Chem. **16**, 221.

*CHAPTER 8*

# ANALYSIS OF NUCLEIC ACID SPECTRA

## 1. Introduction

This chapter deals with some special considerations in the analysis of the ultraviolet spectra of nucleic acids. These studies deal with the determination of the composition of melting regions of nucleic acids. From absorption studies Felsenfeld and Sandeen (1962), Fresco *et al.* (1963), and Mahler *et al.* (1964) have studied the estimation of the composition of the denatured regions and Felsenfeld and Hirschman (1965) have provided us with a crude estimate of the randomness of nucleotide sequences.

## 2. RNA Spectra

The simplest of the formulations mentioned about is the one given by Fresco *et al.* (1963) for the denaturation of the helical structure of RNA.[†]

[†] The authors use the term helical structure to describe the state of RNA before it has melted. Work from the laboratory of Tinoco based on the optical rotatory dispersion

The objective of Fresco *et al.* (1963) is to determine the amount of helical structure in any RNA sample. Basically Fresco *et al.* suggest that when an A–U or a G–C pair is disrupted they each make a separate contribution to the change in absorption spectra. These contributions are different at different wavelengths giving us essentially two functions of wavelength which we call $\Delta E_{AU}(\lambda)$ and $\Delta E_{GC}(\lambda)$ (Fig. 8-1). For any

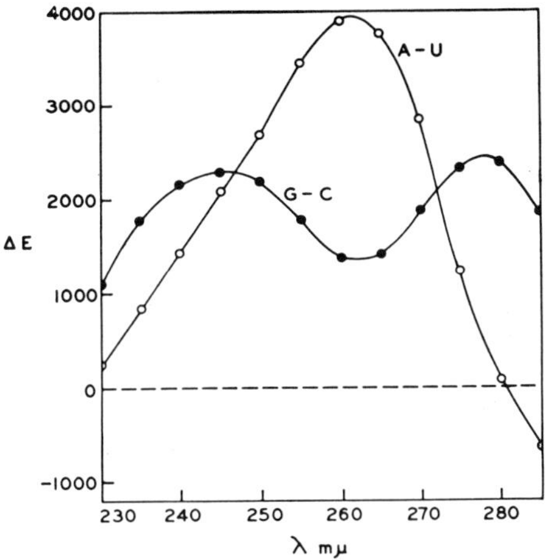


**Fig. 8-1.** Denaturation spectra representing $\Delta E_{AU}$ and $\Delta E_{GC}$. That for the A–U base-pair was obtained using an equimolar mixture of polyriboadenylic acid and polyribouridylic acid in 0.0045 $M$ $Na^{+}$ (phosphate), pH 7.0. At 23° this mixture exists entirely as two-stranded helices containing A–U pairs only. Consequently, $\Delta E_{AU}$ is known on a molar basis. The denaturation spectrum for the G–C base pair was obtained in 0.016 $M$ $Na^{+}$ (phosphate), pH 7.0 using a copolymer containing riboguanylic and ribocytidylic residues in the molar ratio 48 : 52. While it is unlikely that this copolymer contains significant amounts of base pairs other than G–C a small, but unknown proportion of the residues must be unpaired, so that $\Delta E$ in this instance is not known on a molar basis (Fresco *et al.*, 1963).

---

of tobacco mosaic virsus RNA, MS-2-RNA, and various mono-, di-, and trinucleotide models show that the RNA structure may be thought of in terms of three structures. These are a single stranded structure in which we have stacking of the bases, a double stranded structure such as those found in MS-2-RNA, and a "coil structure" whose absorption or rotation is made up of a combination of the absorption or rotation of monomers on mononucleotides. For further information on this subject the reader can consult Chapters 9 and 11. In this chapter we shall continue to use the ideas and notation of Fresco *et al.* (1963) bearing in mind that they could be easily extended to cover the recent ideas of Tinoco *et al.* simply by characterizing the absorption of all the structures suggested by Tinoco *et al.* and attempting to find what linear combination of those structures best describe the RNA sample under consideration.

specific wavelength $\Delta E_{AU}(\lambda_i)$ is the increase in absorption due to the disruption of an A–U base pair at the wavelength $\lambda_i$, and a similar definition must also be made for $\Delta E_{GC}(\lambda_i)$. If we measure the increase in absorption of any RNA sample on denaturation as a function of wavelength, then it should turn out to be a linear combination of the above two functions and should indicate the percentage G–C or A–U base pairs. We must note, however, that the spectra given by Fresco *et al.* as a basis for either analysis are quantitative only in so far as they pertain to the A–U pairs. The spectrum for the denaturation of G–C pairs is not known quantitatively and that raises same complications as we shall see presently.

Mathematically we can set up our problem in the following manner including the experimental error $\delta_i$ at the wavelengths $\lambda_i$.

$$\begin{aligned}
\Delta A(\lambda_1)/c &= F_{AU}\,\Delta E_{AU}(\lambda_1) + F_{GC}\,\Delta E_{GC}(\lambda_1) + \delta_1 \\
\Delta A(\lambda_2)/c &= F_{AU}\,\Delta E_{AU}(\lambda_2) + F_{GC}\,\Delta E_{GC}(\lambda_2) + \delta_2 \\
&\vdots \\
\Delta A(\lambda_m)/c &= F_{AU}\,\Delta E_{AU}(\lambda_m) + F_{GC}\,\Delta E_{GC}(\lambda_m) + \delta_m
\end{aligned} \tag{8-1}$$

In the above formulation $\Delta A(\lambda_i)$ is the change in absorption of the nucleic acid at wavelength $\lambda_i$ and $c$ is the molar concentration of the solution. The quantities $F_{AU}$ and $F_{GC}$ should determine the percentage of the corresponding fractions of the number of nucleotides that are in A–U pairs and G–C pairs respectively. By using the following relations we can put Eq. (8-1) into the notation used throughout the book

$$\Delta A(\lambda_1)/c = f(x_1), \qquad F_{AU} = \alpha_1$$

$$\Delta E_{AU}(\lambda_1) = f_1(x_1), \qquad E_{AU}(\lambda_2) = f_1(x_2), \qquad \ldots, \qquad E_{AU}(\lambda_m) = f_1(x_m)$$

$$\Delta A(\lambda_2)/c = f(x_2), \qquad F_{GC} = \alpha_2$$

$$\Delta E_{GC}(\lambda_2) = f_2(x_2), \qquad E_{GC}(\lambda_2) = f_2(x_2), \qquad \ldots, \qquad E_{GC}(\lambda_m) = f_2(x_m)$$

so that Eq (8-1) becomes

$$\begin{aligned}
f(x_1) &= \alpha_1 f_1(x_1) + \alpha_2 f_2(x_1) + \delta_1 \\
f(x_2) &= \alpha_1 f_1(x_2) + \alpha_2 f_2(x_2) + \delta_2 \\
&\vdots \\
f(x_m) &= \alpha_1 f_1(x_m) + \alpha_2 f_2(x_m) + \delta_m
\end{aligned} \tag{8-2}$$

Solving Eq. (8-2) for $\alpha_1$ and $\alpha_2$ such that

$$\sum_{j=1}^{m} \delta_j^2 \tag{8-3}$$

is a minimum essentially determines the best linear combination of the two functions $\Delta E_{\mathrm{AU}}(\lambda)$ and $\Delta E_{\mathrm{GC}}(\lambda)$ that will fit the data in the $L_2$ norm sense or least squares sense. The techniques for solving these equations are outlined in Chapter 4. Since this case is quite simple the least squares or $L_2$ norm approximation will probably provide the required answer without resort to any of the techniques described in Chapter 4 which constrain $\alpha_1$ and $\alpha_2$ to remain positive. Analysis of the melting profile of RNA may in certain instances require that $\alpha_1 + \alpha_2$ be equal to some number (i.e., we have started with a certain number of bases that were stacked and therefore only that same number of bases may be unstacked). In such instances the $L_2$ norm with the restriction on the sum of the coefficients consistent with the physical situation must be used (see Chapter 4).

To determine the helical content of the RNA it is necessary to calibrate the $\Delta E_{\mathrm{AU}}(\lambda)$ and $\Delta E_{\mathrm{GC}}(\lambda)$ curves on a molar basis. This means we should know the precise quantitative increase in absorption at all wavelengths when so many moles of A–U or G–C are completely disrupted. If those quantities are known then we could determine the helical content. As stated above Fresco *et al.* (1963) could not determine $\Delta E_{\mathrm{GC}}(\lambda)$ on a molar basis so precise estimates of helical content could not be made.

Among the interesting results obtained by Fresco *et al.* is the exhibition by the yeast S-RNA melting profiles of two phases depending on temperature. Each phase can be represented by a different function $\Delta A(\lambda)$. These functions $\Delta A_1(\lambda)$ and $\Delta A_2(\lambda)$ are obtained at the lower and higher temperature respectively. The authors concluded that different fractions of A–U and G–C were melting in the two phases. Table 8-1 exhibits the results for yeast valine S-RNA. In this table we observe several temperature ranges in the same column. To each one of these temperature ranges there corresponds a function $\Delta A(\lambda)$ which is the response of the system in the temperature range indicated. This response function is analysed in the manner indicated in Chapter 4 to determine $F_{\mathrm{GC}}$, and we note that in the lower temperature phase the response functions which have been labeled $\Delta A_{1a}(\lambda)$, $\Delta A_{1b}(\lambda)$, . . . ,[†] give rise to approximately the same $F_{\mathrm{GC}}$. A similar result is true for the six response functions

[†] The second subscript pertains to the temperature range.

**Table 8-1**[a]

| Response function | Temperature range (°C) | $F_{GC}$ | Response function | Temperature range (°C) | $F_{GC}$ |
|---|---|---|---|---|---|
| $A_{1a}(\lambda)$ | 4–17 | 61 | $A_{2a}(\lambda)$ | 58–73 | 79 |
| $A_{1b}(\lambda)$ | 4–28 | 56 | $A_{2b}(\lambda)$ | 58–85 | 73 |
| $A_{1c}(\lambda)$ | 4–40 | 60 | $A_{2c}(\lambda)$ | 58–95 | 74 |
| $A_{1d}(\lambda)$ | 4–45 | 58 | $A_{2d}(\lambda)$ | 58–73 | 86 |
| $A_{1e}(\lambda)$ | 4–50 | 58 | $A_{2e}(\lambda)$ | 68–85 | 75 |
| $A_{1f}(\lambda)$ | 4–54 | 58 | $A_{2f}(\lambda)$ | 68–95 | 79 |
| $A_{1g}(\lambda)$ | 4–54 | 58 | | | |
| | | mean 58.4 | | | mean 77.3 |

[a] Data from Fresco *et al.* (1963).

$\Delta A_{2a}(\lambda)$, $\Delta A_{2b}(\lambda)$, . . . , at the higher temperature range. The close coincidence of the mean values of $F_{GC}$ indicates a uniformity of composition in the melting region of the RNA molecules for the given temperature range.

## 3. DNA Spectra

The method used by Felsenfeld and Hirschman (1965) involves the use of a large body of data on the melting of several DNA molecules at several wavelengths. The objective of this method as before is to determine the composition of the region which melts at any specific temperature. The analysis of Felsenfeld and Hirschman also gives a crude estimate of the deviation from randomness of the base sequences.

The problem dealt with by these authors is more complicated than the previous case and in general terms may be explained in the following manner. When a helical region melts, the contribution to the hyperchromism from the disruption of an A–T pair is going to be different from the contribution of a G–C pair. In addition, if the A–T pair is stacked between two other A–T pairs, its contribution is going to be different from the contribution of an A–T pair stacked between a G–C pair and another G–C pair, and so on. Consequently, if we wish to list the hyperchromism elicited in each instance we will have as many different parameters (dealing with the extent of hyperchromism) as there are distinguishable dinucleotide sequences. Obviously, if this method is to be extended to realize all its potential we will require exact knowledge of the sequence of bases in the model DNA molecules in order to be able to determine the precise contribution to the hyperchromism of any dinucleotide sequence. Since such knowledge is not readily available Felsenfeld and Hirschman had to resort to certain assumptions about the DNA molecules used as models.

These authors assumed that all nucleotides along the chain occur with random frequency. The assumption of random frequency of base pairs adjacent to each other reduces the number of spectral parameters required by this method to only three parameters. These parameters are the contribution of an A–T pair adjacent to an A–T pair, the contribution of a G–C pair adjacent to a G–C pair, and the contribution of an A–T pair adjacent to a G–C pair. Felsenfeld and Hirschman also point out that there is varying relative orientation of base pairs in different sequences and that these varying orientations will undoubtedly contribute differently to the ultimate hyperchromism produced. Inasmuch as the relative orientation of the various base are not known the authors assume a contribution from all base pairs that is an average over all orientations of those pairs.

The theory is set up by considering the disruption of a helical region $j$ with a mean base composition given by the mole fraction set of A–T equal to $\phi_j$ and a total nucleotide concentration $c_j$. When the bases are randomly distributed then the concentration of dinucleotides pairs with an A–T pair adjacent to another A–T pair is $c_j\phi_j^2$. For a G–C pair adjacent to a G–C pair we have a concentration of $c_j(\phi_j - 1)^2$ and for an A–T pair adjacent to a G–C pair the concentration is $2c_j\phi_j(\phi - 1)$. The increase of absorbtion $\Delta A_j(\lambda_i)$ due to the disruption of the helical region $j$ at wavelength $\lambda_i$ is due to three different spectral contributions: $\varepsilon_{\mathrm{AA}}(\lambda_i)$ the average contribution for all the A–T pairs adjacent to A–T pairs averaged over all orientations, $\varepsilon_{\mathrm{GG}}(\lambda_i)$ the average contribution of a G–C pair adjacent to a G–C pair averaged over all orientations, and lastly $\varepsilon_{\mathrm{AG}}(\lambda_i)$ the average contribution due to an A–T pair adjacent to a A–C pair averaged over all orientations. If we define

$$K(\lambda_i) = \varepsilon_{\mathrm{AG}}(\lambda_i)/[\varepsilon_{\mathrm{AA}}(\lambda_i) + \varepsilon_{\mathrm{GG}}(\lambda_i)] \tag{8-4}$$

we can write the following equation for $\Delta A_j(\lambda_i)$

$$\Delta A_j(\lambda_i) = c_j\{\phi_j\varepsilon_{\mathrm{AA}}(\lambda_i) + (1 - \phi_j)\varepsilon_{\mathrm{GG}}(\lambda_i) + (\phi_j^2 - \phi_j)[1 - 2K(\lambda_i)] \times [\varepsilon_{\mathrm{AA}}(\lambda_i) + \varepsilon_{\mathrm{GG}}(\lambda_i)]\} \tag{8-5}$$

Before we proceed further with our analysis of Eq. (8-5) let us derive the equation which includes a parameter denoting the extent of deviation from randomness. A deviation from randomness in such instances means that we have sequences of bases not conforming to a random distribution. This in turn means, to be specific, if we have seventeen bases the probability of having more than four identical bases in a row is low. If we have thirty-three bases the probability of having more than five identical bases in a row would be low, and if we had sixty-five bases the probability of having more than six of them in a row would be low, and so on. The above is very much like the sequence of consecutive heads and tails one would expect from a coin. Felsenfeld and Hirschman (1965) supposed that the fraction of A–T pairs adjacent to A–T pairs is $\phi_j + \varrho_j$. The fraction of A–T pairs which are adjacent to G–C pairs will be proportional to $(1 - \phi_j - \varrho_j)$. Taking these factors into consideration Eq. (8-5) becomes

$$\begin{aligned} \Delta A_j(\lambda_i) = c_j\{\phi_j\varepsilon_{\mathrm{AA}}(\lambda_i) + (1 - \phi_j)\varepsilon_{\mathrm{GG}}(\lambda_i) + \phi_j(\phi_j - 1 + \varrho_i) \\ \times [1 - 2K(\lambda_i)][\varepsilon_{\mathrm{AA}}(\lambda_i + \varepsilon_{\mathrm{GG}}(\lambda_i)]\} \end{aligned} \tag{8-6}$$

The analysis conducted by Felsenfeld and Hirschman proceeded as follows: A large number $m$ of DNA samples of known chemical composition are melted at a given wavelength. Since the chemical composition of these DNA samples is known then $\phi_j$ is known and $c_j$ is also known. Our objective is to determine parameters $\varepsilon_{\mathrm{AA}}(\lambda_i)$, $\varepsilon_{\mathrm{GG}}(\lambda_i)$ and $K(\lambda_i)$ at the wavelength $\lambda_i$ by using Eq. (8-5). Rewriting Eq. (8-5) in our notation at *any given wavelength* with the inclusion of the experimental error $\delta_j$ we have

$$\begin{aligned} f(x_1) &= \alpha_1 f_1(x_1) + \alpha_2 f_2(x_1) + \alpha_3 f_3(x_1) + \delta_1 \\ f(x_2) &= \alpha_1 f_1(x_2) + \alpha_2 f_2(x_2) + \alpha_3 f_3(x_2) + \delta_2 \\ &\vdots \\ f(x_m) &= \alpha_1 f_1(x_m) + \alpha_2 f_2(x_m) + \alpha_3 f_3(x_m) + \delta_m \end{aligned} \tag{8-7}$$

where $f(x_1), f(x_2), f(x_3), \ldots, f(x_m)$ denote the increases in absorption at *the given wavelength* $\lambda_i$ for the first, second, third, ..., $m$th samples of DNA respectively. The rest of the symbols can be identified as follows

$$\begin{aligned} f_1(x) &= c_j\phi_j, & \alpha_1 &= \varepsilon_{\mathrm{AA}}(\lambda_i), \\ f_2(x) &= c_j(1 - \phi_j), & \alpha_2 &= \varepsilon_{\mathrm{GG}}(\lambda_i), \\ f_3(x) &= c_j\phi_j(\phi_j - 1), & \alpha_3 &= [1 - 2K(\lambda_i)][\varepsilon_{\mathrm{AA}}(\lambda_i) + \varepsilon_{\mathrm{GG}}(\lambda_i)] \end{aligned} \tag{8-8}$$

 

We note once again that we conduct the above analysis first for a specific wavelength, then for all wavelengths of interest to determine $\alpha_1$, $\alpha_2$, and $\alpha_3$. Essentially, therefore, we are determining the following functions $\alpha_1(\lambda)$, $\alpha_2(\lambda)$, and $\alpha_3(\lambda)$. These functions will then be used for our stated purpose, namely, to determine the amount, composition, and simple deviations from randomness of any melting region. It is clear that we can use any of the techniques of function minimization to determine $\alpha_1$, $\alpha_2$, and $\alpha_3$ for all wavelengths required such that

$$\sum_{j=1}^{m} \delta_j^2 \qquad \text{or} \qquad \sum_{1=j}^{m} |\delta_j|$$

are a minimum and

$$\alpha_i \geq 0 \qquad \text{for} \quad i = 1, 2, 3.$$

In Chapter 4 we discussed the methods available for such computation. In Chapter 6 we discussed the methods of determining the standard errors of the $\alpha$.

Felsenfeld and Hirschman have evaluated $\varepsilon_{AA}(\lambda)$, $\varepsilon_{GG}(\lambda)$, and $K(\lambda)$ over the wavelength interval 220–290 m$\mu$ so these can be used directly in the following analysis to determine the melting regions in any DNA sample. Experimently the absorption spectrum of the DNA sample under consideration is measured at different temperature increments. Such measurements should yield the parameters $c_j$, $\phi_j$ and $\varrho_j$ for any given temperature on solving Eq. (8-6) which we rewrite below

$$\begin{aligned}
f(x_1) &= \alpha_1 f_1(x_1) + \alpha_2 f_2(x_1) + \alpha_3 f_3(x_1) + \delta_1 \\
f(x_2) &= \alpha_1 f_1(x_2) + \alpha_2 f_2(x_2) + \alpha_3 f_3(x_2) + \delta_2 \\
&\vdots \\
f(x_m) &= \alpha_1 f_1(x_m) + \alpha_2 f_2(x_m) + \alpha_3 f_3(x_m) + \delta_m
\end{aligned} \tag{8-9}$$

where $f(x_1)$, $f(x_2)$, and $f(x_m)$ are the increases in absorption of the DNA sample at wavelengths $\lambda_1, \lambda_2, \ldots, \lambda_m$ for a given temperature increment and where

$$\begin{aligned}
\alpha_1 &= c_j\phi_j, & f_1(x_i) &= \varepsilon_{AA}(\lambda_i), \\
\alpha_2 &= c_j[1 - \phi_j], & f_2(x_i) &= \varepsilon_{GG}(\lambda_i), \\
\alpha_3 &= c_j\phi_j[\phi_j - 1 + \varrho_j], & f_3(x_i) &= [1 - 2K(\lambda_i)][\varepsilon_{AA}(\lambda_i) + \varepsilon_{GG}(\lambda_i)]
\end{aligned} \tag{8-10}$$

Solution of Eq. (8-9) for $\alpha_1$, $\alpha_2$, and $\alpha_3$ yields the parameters $\phi_j$, $c_j$, and $\varrho_j$ from which we can deduce the concentration and composition of the melting region. The parameter $\varrho_j$ in principle should give us the deviation from randomness of the melting regions. The authors point out, however, that the value of $\varrho_j$ is more uncertain than the values of $\phi_j$ and $c_j$. Methods of solution for Eq. (8-9) to determine the $\alpha$ have been described in Chapter 4. Actual application of Eq. (8-5) to calf thymus DNA gave the results shown in Table 8-2.

Felsenfeld and Hirschman (1965) went on to discuss various aspects of this analysis including the changes from helix to nucleotide and coil to nucleotide spectra. For these specific topics we refer the reader to that article. In later work Hirschman and Felsenfeld (1966) used the same techniques as presented here and discussed the determination of the base composition of unknown DNA samples.

**Table 8-2**

| Temperature (°C) | Overall composition of denatured region | | $\varrho_j$ (Overall, for all denatured region) |
|---|---|---|---|
| | A–T mole fraction | Concentration | |
| 59.5 | 0.72 | $0.095 \times 10^{-4}$ | −0.33 |
| 61.8 | 0.74 | 0.2 | −0.38 |
| 63.3 | 0.71 | 0.33 | −0.26 |
| 64.8 | 0.71 | 0.48 | −0.17 |
| 66.0 | 0.70 | 0.61 | −0.19 |
| 67.2 | 0.68 | 0.75 | −0.17 |
| 68.5 | 0.67 | 0.87 | −0.14 |
| 70.7 | 0.64 | 1.05 | −0.13 |
| 73.2 | 0.62 | 1.18 | −0.14 |
| 75.9 | 0.60 | 1.31 | −0.11 |
| 90.5 | 0.587 | 1.44 | −0.08 |

We wish to digress slightly to discuss Fig. 8-2. This figure plots the fraction of A–T pairs melting against the fraction G–C pairs for calf thymus DNA. The slope of the line of this at each point is directly proportional to the A–T : G–C ratio of the region just melting. This graph raises the question of how general is the curve depicted therein.

That is, does the curve depend on the mole fraction of A–T in the DNA samples or do we get different curves when using different DNA molecules with different mole fractions of A–T. The curve in the figure suggests that we can represent $F_{AT}$ as a function of $F_{GC}$ (or vice versa)

$$F_{AT} = f(F_{GC}) \tag{8-11}$$

Suppose we make this function a polynomial and write

$$F_{AT} = a + bF_{GC} + c(F_{GC})^2 + \cdots \tag{8-12}$$

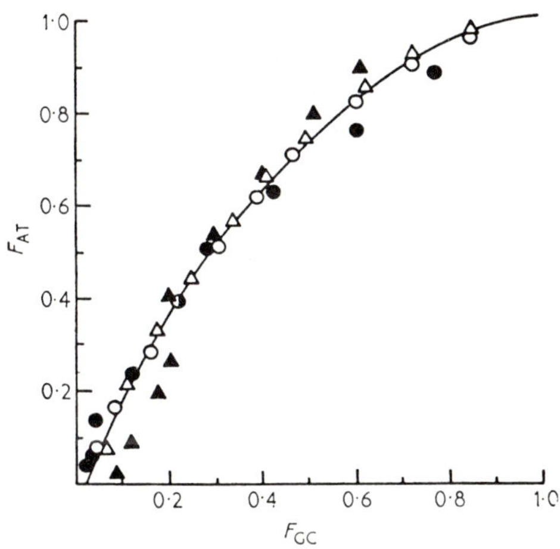


**Fig. 8-2.** Correlation of fraction of all A–T pairs denatured with fraction of all G–C pairs denatured for four samples of calf thymus DNA in 0 · 01 *M* NaCl — 0 · 001 *M* phosphate buffer, (pH 7). Concentrations of DNA: ○, 1 · 43 × $10^{-4}$ *M*; △, 1 · 82 × $10^{-4}$ *M*; ▲, 2 · 06 × $10^{-4}$ *M*; ●, 2 · 05 × $10^{-4}$ *M* (Felsenfeld and Hirschman, 1965).

where $a, b, c, \ldots$ are constants. From the techniques of function minimization described in Chapter 5 we can determine $a, b, c, \ldots$. In such cases, since Eq. (8-12) is a purely phenomenological equation, the $L_2$ norm or least squares method is recommended for the determination $a, b, c, \ldots$ since there are no restrictions on the sign of these coefficients. We asked above whether the coefficients of the polynomial in $F_{GC}$ are the same for all samples (containing different mole fractions of A–T) or different. We would venture to guess that these coefficients would be the same provided the assumption of randomness implicit in this treatment holds for all DNA samples from which the bulk of the analysis is derived. If the coefficients of the polynomial in $F_{GC}$ are the same for most DNA samples, then the DNA samples (if any) which differ drastically might well have a different structure and it would be interesting to discover what that is. For instance, if we consider two DNA samples containing the same mole fraction of A–T with one having a random sequence of bases while the structure of the other is such that there are long sequence of A–T pairs followed by long sequence of G–C pairs, then one would expect that the melting of the first sample could be adequately described by Eq. (8-12) while the melting of the nonrandom sample would have to be described by some other equation. On the assumption however that all the DNA samples exhibit the same coefficients, the whole process of determining the composition of a denaturing region when DNA melts can be represented by a single equation such as (8-12) where the coefficients are fixed once and for all. Another question one might ask is, assuming we get different coefficients in Eq. (8-12) for different mole fractions of A–T, whether there will be any difference in the coefficients between a circular DNA molecule, a supercoiled DNA molecule, and a linear DNA molecule all having the same mole fraction of A–T.

## References

Felsenfeld, G., and Hirschman, S. Z. (1965). *J. Mol. Biol.* **13** 407.
Felsenfeld, G., and Sandeen, G. (1962). *J. Mol. Biol.* **5**, 587.
Fresco, J. R., Klotz, L. C., and Richards, E. G. (1963). *Cold Spring Harbor Symp. Quant. Biol.* **28**, 83.
Hirschman, S. Z., and Felsenfeld, G. (1966). *J. Mol. Biol.* **16**, 347.
Mahler, H. R., Kline, R., and Mehrotra, B. D. (1964), *J. Mol. Biol.* **9**, 801.

*CHAPTER 9*

# MATRIX RANK ANALYSIS

## 1. Introduction

Up on this point and in some of the chapter to follow we are essentially analyzing the response of a system (e.g., its absorption spectra, ORD, weight average molecular weight) in terms of known functions or known components. We have said nothing about the determination of functions or components in terms of which we may analyze the response of a system which we might want to consider.

Matrix rank analysis while not a technique of the minimization of functions of several variables, renders valuable assistance in giving us some idea of the spectra (or to be more general, functions) that we may use to analyze a system when those functions are not known. This method, first proposed by Wallace (1960) and further extended by Wallace and Katz (1964), essentially determines the *number* and perhaps the profile of the *spectra* of the possible components of a mixture from a series of solutions of that mixture under different conditions. The matrix rank analysis method was also proposed by Ainsworth (1961) who used it later on (Ainsworth and Bingham, 1968) to determine the number of different hemoglobin species attached to ligands in partially oxidized haemoglobin solutions. McMullen *et al.* (1967) used this method to analyze the optical rotatory dispersion of tobacco mosaic virus RNA.

In this chapter we describe the method and discuss its use with particlar emphasis on the example given by McMullen *et al.* (1967). In addition the examples given by Wallace and Katz (1964) and Ainsworth and Bingham (1968) will be included. Matrix rank analysis we might remark is not the only available method for determining the number of components.[†]

## 2. The Matrix Rank Analysis Method

In this procedure we represent the spectra (or function) of some system at $n$ wavelengths (or other independent variables) by the following

$$f(\lambda_1), f(\lambda_2), \ldots, f(\lambda_n) \tag{9-1}$$

If we have $m$ solutions of the sample under different conditions (different relative concentration of the component, or conditions that will impose different relative concentrations of the component as for example pH, temperature, and ionic strength), we may represent the spectra by the following matrix

$$\mathbf{F} = \begin{bmatrix} f_1(\lambda_1) & f_1(\lambda_2) & \cdots & f_1(\lambda_n) \\ f_2(\lambda_1) & f_2(\lambda_2) & \cdots & f_2(\lambda_n) \\ f_3(\lambda_1) & f_3(\lambda_2) & \cdots & f_3(\lambda_n) \\ \vdots & \vdots & & \vdots \\ f_m(\lambda_1) & f_m(\lambda_2) & \cdots & f_m(\lambda_n) \end{bmatrix} \tag{9-2}$$

Here $f_1(\lambda_1), f_1(\lambda_2), \ldots, f(\lambda_n)$ are the values of the response function at condition 1 at differing wavelength (or some other independent variable) and $f_m(\lambda_1), f_m(\lambda_2), \ldots, f_m(\lambda_n)$ are the values of the response function at condition $m$ at differing wavelengths (or some other independent variable). The rank of the matrix $\mathbf{F}$ shown above determines the number of linearly independent components (see Chapter 2). Determination of the rank of a matrix can be attained by a process of elementary transformation as we have shown in Chapter 2. The reduction processes we describe

[†] It is the author's current view that component analysis and factor analysis (Lawley and Maxwell 1971, Kendall and Stuart 1968, Harman 1967), two fairly ancient statistical techniques,are superior to matrix rank analysis in determining the number of components. Furthermore, those techniques provide more information than matrix rank analysis (Magar, 1972, *Biopolymers*, in press, and another manuscript, in preparation).

below do essentially the same thing. We observe in the examples in Chapter 2 that a matrix of rank $r$ after elementary transformation has zero elements in all rows after the $r$th. The process of reduction is complete when we arrive at that point. We note, however, that in practice this happens only in an ideal case where there is no experimental error. Due to experimental error encountered during the course of making our measurements we will find that the submatrices formed during the process of reduction or transformation will not have zero components. Inasmuch as this happens to be the case we are faced with following problem. Since we do not know what the rank of our original matrix is and since the elements of the reduced or transformed matrix are not zero beyond a certain row (because of experimental error), at what stage of the reduction process can we consider the remaining elements insignificant? Clearly there must be some criterion to tell us when to stop or to tell that for all practical purposes the matrix $\mathbf{F}$ is of a certain rank, say 3, and all the elements beyond the third row although not zero are insignificant and arise because of experimental error.

Wallace (1960) decided to use an elementary statistical criterion to terminate the reduction process. The argument for this runs as follows: The probable error of a function $\Psi$ of several variables $x, y, z, \ldots$ written as $\Psi(x, y, z, \ldots)$ is given by

$$[P_x{}^2(\partial\Psi/\partial x)^2 + P_y{}^2(\partial\Psi/\partial y)^2 + P_z{}^2(\partial\Psi/\partial z)^2 + \cdots \qquad (9\text{-}3)$$

where $P_x, P_y, P_z, \ldots$ are the probable errors of $x$, $y$, and $z$, respectively (Margenau and Murphy, 1955). Now, a determinant of a square matrix $\mathbf{A}$ is a function of all its elements (see Chapter 2). The standard deviation of a determinant therefore is

$$\left[\sum_i \sum_j (\text{probable error of } a_{ij})^2 \times (\partial A/\partial a_{ij})^2\right]^{1/2} \qquad (9\text{-}4)$$

Since the derivative of a matrix with respect to any one of its elements is its cofactor (see Chapter 2 and possibly verify it for yourself on a $2 \times 2$ determinant), then the standard deviation of a determinant $\sigma_{|\mathbf{A}|}$ is given by

$$\sigma_{|\mathbf{A}|} = \left[\sum_i \sum_j \sigma_{a_{ij}}^2 (a^{ij})^2\right]^{1/2} \qquad (9\text{-}5)$$

where $\sigma_{a_{ij}}$ is the standard deviation of the element $a_{ij}$ and $a^{ij}$ is the cofactor of $a_{ij}$. Now, if the errors of the individual measurements are known then the probability that the matrix is singular can be determined from the ratio $|\mathbf{A}|/\sigma_{|\mathbf{A}|}$ and by consulting any statistical table. Wallace and Katz (1964) though the above criterion for terminating the reduction laborious and decided to use a method which they felt was more suitable for machine computation. The new method used by Wallace and Katz (1964) was used by others subsequently.

The criterion used by Wallace and Katz and later by Ainsworth and Bingham (1968) is to use a companion matrix called the error propagation matrix whose elements are the standard errors of the matrix **F**. McMullen *et al.* (1967) did not use the error propagation matrix but used another criterion to terminate their reduction process.

In what follows we first describe the method using the error propagation matrix, then discuss the criteria used by McMullen *et al.* to terminate the process of reduction. Lastly we suggest another statistical method to determine the rank of the matrix.

For the method using the error propagation matrix, the matrix **F** is pivoted by interchanging rows and columns to place the elements whose absolute value is the greatest in the first row and column. Next by performing the operation

$$f_i^*(\lambda_j) = f_i(\lambda_j) - [f_i(\lambda_1)/f_1(\lambda_1)]f_1(\lambda_i) \tag{9-6}$$

on all elements except those of the first row we obtain the matrix $\mathbf{F}^*$ with the property that all the elements in the first column except the first are zero. Having done that, the operation is repeated for the next row and so on until we obtain the final matrix $\mathbf{F}^\dagger$ given by Eq. (9-7)

$$\mathbf{F}^\dagger = \begin{bmatrix} f_1(\lambda_1) & f_1(\lambda_2) & \cdots & & f_1(\lambda_n) \\ 0 & f_2^*(\lambda_2) & \cdots & & f_2^*(\lambda_n) \\ 0 & 0 & f_3^{**}(\lambda_3) & \cdots & f_3^{**}(\lambda_n) \\ \vdots & \vdots & \vdots & & \vdots \\ 0 & 0 & 0 & \cdots & f_m^{(m-1)*}(\lambda_n) \end{bmatrix} \tag{9-7}$$

The rank of this matrix is the number of significant spectra left after the reduction. The use of the error propagation matrix helps in identifying the number of significant spectra left. To obtain the error propagation matrix or the reduced error propagation matrix after a process of reduction we start with the **S** matrix whose elements are the standard errors of the elements of the **F** matrix.

Thus a typical element of the **S** matrix denoted by $S_{ij}$ is considered to be the standard error of $f_i(\lambda_j)$. The *reduced* error propagation matrix is a matrix whose elements are computed in the following fashion

$$\mathbf{S}_{ij}^* = \left\{S_{ij}^2 + S_{1j}^2\left[\frac{f_i(\lambda_1)}{f_1(\lambda_1)}\right]^2 + S_{i1}^2\left[\frac{f_1(\lambda_j)}{f_1(\lambda_1)}\right]^2 + S_{11}^2\left[\frac{f_i(\lambda_1)f_1(\lambda_j)}{f_1(\lambda_1)^2}\right]^2\right\}^{1/2} \tag{9-8}$$

where the elements $S_{ij}^*$ are the estimated errors of $f_i^*(\lambda_j)$. The process is then continued to obtain other rows until we arrive at the entire reduced error propagation matrix. It must be noted that, similar to the matrix **F**, the elements $S_{ij}^*$ are the elements of the second row of the reduced error propagation matrix **S***. The reduced error propagation matrix has the following structure

$$\mathbf{S}^* = \begin{bmatrix} S_{11} & S_{12} & S_{13} & \cdots & S_{1n} \\ 0 & S_{22}^* & S_{23}^* & \cdots & S_{2n}^* \\ 0 & 0 & S_{33}^{**} & \cdots & S_{3n}^{**} \\ 0 & 0 & 0 & \cdots & S_{mn}^{(m-1)*} \end{bmatrix} \tag{9-9}$$

When the reduced error propagation matrix is found one need only compare the principal diagonal of the reduced matrix **F*** and the reduced error propagation matrix **S*** to determine the number of significant spectra. An example of how this technique works is illustrated in Table 9-1.

From this table Wallace and Katz (1964) concluded that the spectra they were analyzing contained at least three components and possibly four because the fourth element on the diagonal of the reduced **F** matrix, $-0.016$, is a little over three times the fourth element on the diagonal of the reduced **S*** matrix 0.005.

McMullen *et al.* (1967) who used matrix rank analysis to analyze the optical rotatory dispersion of tobacco mosaic virus RNA employed another criterion to determine the number of significant spectra. This criterion, although quite arbitrary in terms of probability, seems to have served them well. Essentially these authors have defined the following quantity

$$(\mathrm{RMS})_i = (1/m)\left\{\sum_{j=1}^{m} [f_i^{(i-1)*}(\lambda_j)]^2\right\}^{1/2} \tag{9-10}$$

where $(\mathrm{RMS})_i$ is the root mean square value of the $i$th reduced dispersion spectra. They decided that the number of significant spectra present after matrix reduction is determined when $(\mathrm{RMS})_i$ reaches a certain arbitrary low value and does not change appreciably as more reduced dispersion spectra are computed. Stated in another manner if there are two significant components then the quantities $(\mathrm{RMS})_3$, $(\mathrm{RMS})_4$, $(\mathrm{RMS})_5$, etc. should not differ greatly in value and should be below a certain arbitrary value. Mathematically this means that the submatrices of the reduced matrix are approaching zero.

Table 9-1[a]

*Original* **F** *matrix*

| | | | | | | | |
|---|---|---|---|---|---|---|---|
| 0.017 | 0.045 | 0.062 | 0.110 | 0.197 | 0.278 | 0.332 | 0.377 |
| 0.058 | 0.080 | 0.100 | 0.152 | 0.238 | 0.320 | 0.372 | 0.417 |
| 0.167 | 0.180 | 0.187 | 0.222 | 0.282 | 0.330 | 0.368 | 0.400 |
| 0.420 | 0.402 | 0.395 | 0.365 | 0.342 | 0.310 | 0.300 | 0.292 |
| 0.770 | 0.735 | 0.690 | 0.565 | 0.437 | 0.278 | 0.192 | 0.131 |
| 1.015 | 0.990 | 0.992 | 0.742 | 0.528 | 0.282 | 0.147 | 0.051 |
| 0.935 | 0.940 | 0.875 | 0.702 | 0.488 | 0.255 | 0.118 | 0.032 |
| 0.443 | 0.480 | 0.462 | 0.342 | 0.235 | 0.108 | 0.049 | 0.015 |

*Reduced* **F** *matrix*

| | | | | | | | |
|---|---|---|---|---|---|---|---|
| 1.015 | 0.051 | 0.922 | 0.742 | 0.528 | 0.282 | 0.990 | 0.147 |
| 0 | 0.414 | 0.047 | 0.110 | 0.208 | 0.304 | 0.023 | 0.364 |
| 0 | 0 | 0.060 | 0.020 | 0.008 | −0.010 | 0.048 | 0.009 |
| 0 | 0 | 0 | −0.016 | −0.007 | −0.007 | −0.005 | −0.003 |
| 0 | 0 | 0 | 0 | −0.006 | −0.005 | −0.006 | 0.001 |
| 0 | 0 | 0 | 0 | 0 | 0.005 | 0.003 | 0.003 |
| 0 | 0 | 0 | 0 | 0 | 0 | 0.003 | 0.004 |
| 0 | 0 | 0 | 0 | 0 | 0 | 0 | 0.004 |

*Reduced* **S** *matrix*

| | | | | | | | |
|---|---|---|---|---|---|---|---|
| 0.003 | 0.003 | 0.003 | 0.003 | 0.003 | 0.003 | 0.003 | 0.003 |
| 0 | 0.003 | 0.004 | 0.004 | 0.003 | 0.003 | 0.004 | 0.003 |
| 0 | 0 | 0.004 | 0.004 | 0.004 | 0.004 | 0.005 | 0.004 |
| 0 | 0 | 0 | 0.005 | 0.005 | 0.005 | 0.007 | 0.005 |
| 0 | 0 | 0 | 0 | 0.006 | 0.006 | 0.008 | 0.006 |
| 0 | 0 | 0 | 0 | 0 | 0.010 | 0.013 | 0.008 |
| 0 | 0 | 0 | 0 | 0 | 0 | 0.017 | 0.012 |
| 0 | 0 | 0 | 0 | 0 | 0 | 0 | 0.021 |

[a] From Wallace and Katz, 1964.

Instead of employing arbitrary criteria to terminate the reduction process we suggest the following statistical test to determine the number of significant components. After determining each reduced spectrum we can use these spectra to fit our original data and use an analysis of variance to compute an $F$ value (see Chapter 6). When this is done the investigator is in a position to state with what probability his reduced spectra fits his original data. To demonstrate this process suppose our original data consisted of the spectra of a solution under six different conditions. Call them

$$\begin{array}{cccc} f_1(\lambda_1), & f_1(\lambda_2), & \ldots, & f_1(\lambda_n) \\ f_2(\lambda_1), & f_2(\lambda_2), & \ldots, & f_2(\lambda_n) \\ \vdots & \vdots & & \vdots \\ f_6(\lambda_1), & f_6(\lambda_2), & \ldots, & f_6(\lambda_n) \end{array} \tag{9-11}$$

and suppose our first two computed reduced spectra were

$$\begin{array}{cccc} g_1(\lambda_1), & g_1(\lambda_2), & \ldots, & g_1(\lambda_n) \\ g_2(\lambda_1), & g_2(\lambda_2), & \ldots, & g_2(\lambda_n) \end{array} \tag{9-12}$$

We can now use the function $g_1(\lambda)$ and $g_2(\lambda)$ to obtain an $L_2$ norm approximation or least square fit to all of the six functions $f_1(\lambda), f_2(\lambda), \ldots, f_6(\lambda)$. Stating it in another way we seek to determine the best coefficients $\alpha_{11}, \alpha_{12}, \ldots, \alpha_{26}$ in the least square sense to use in the following equations

$$\begin{aligned} f_1(\lambda) &\cong \alpha_{11} g_1(\lambda) + \alpha_{21} g_2(\lambda) \\ f_2(\lambda) &\cong \alpha_{12} g_1(\lambda) + \alpha_{22} g_2(\lambda) \\ &\vdots \\ f_6(\lambda) &\cong \alpha_{16} g_1(\lambda) + \alpha_{26} g_2(\lambda) \end{aligned} \tag{9-13}$$

Once these coefficients are determined all six $\alpha_{2i}$ are tested to determine if they are significantly different from zero. The procedure for doing this is outlined in Chapter 6 in the discussion on the analysis of variance. The investigator will have to choose a certain level of significance in order to test for these coefficients. Three possibilities arise in testing for all the $\alpha_{2i}$ or at any given **F** or level of significance. First, all six $\alpha_{2i}$ are significantly different from zero in which case we will know we have *at least two components* and proceed to determine the third reduced spectrum. Next, all six $\alpha_{2i}$ are not significantly different from zero in which case we can conclude that we have *only one component*. Lastly, some of the $\alpha_{2i}$ are significantly different from zero while the rest are not at the chosen level of significance. At this stage the investigator can either change the level of significance so all the $\alpha_{2i}$ are significant or he can retain the level of significance and accept the existence of at least two components and continue the procedure to compute the six $\alpha_{3i}$ and test for their significance. The advantage of this method is that it can tell us with what probability we can expect our computed reduced spectra to fit our data. This represents a more definite piece of information than either the Wallace and Katz (1964) criteria or that of McMullen *et al.* (1967).

## 3. Application of Matrix Rank Analysis to the Optical Rotatory Dispersion Spectra of TMV RNA Taken at a Variety of Ionic Strengths and Temperatures

The objective of this analysis is to determine the number of components contributing to these spectra. McMullen *et al.* (1967) start by considering a series of spectra at a single ionic strength and different temperatures. They do this for four ionic strengths. For each of the four ionic strengths considered matrix rank analysis yields two components. However this preliminary analysis only yielded a first approximation to the shape of the spectra and not readily identifiable physical components. The importance of determining physically identifiable components has been emphasized in the introduction of this book and it was noted there that unless physical identification was effected, then the matrix rank analysis serves a limited purpose. In fact McMullen *et al.*, at the end of the preliminary analysis, indicated that the first approximation to the shape of the spectra was at best a gross approximation to the physical situation.

McMullen *et al.* reasoned with some ingenuity that at some ionic strength and at some temperature they might expect to find a component with a reasonably constant shape that reflected a stable physical structure. On the basis of other empirical evidence and their preliminary matrix rank analysis they concluded that a certain ORD profile which they called spectrum $S$ could be considered to form one component. By the simple expedient of adding this spectrum $S$ to each ionic strength set of spectra and repeating the analysis they determined the presence of a second component which they called spectrum $D$. The reasoning behind the addition of spectrum $S$ to each ionic strength component was to ensure its emergence as a possible component. This procedure is quite valid because in essence we do not add any more linearly independent components. When an investigator feels that a certain component should emerge in his analysis there is no reason why one should not add it to the data and proceed with the analysis. This is perfectly legitimate since, in the final analysis, the ultimate justification for subsequent application is the identification of the components obtained with known physical entities.

The identification of spectrum $S$ with the single-stranded form of RNA was made on the basis of computations made by Cantor *et al.* (1966) for such molecules. By single-stranded RNA one means that the bases or the monomers in this RNA are interacting. This is in contrast to the optical rotatory dispersion of a mixture of single monomers expected of an RNA molecule if the bases are not, in anyway, interacting. Identification of spectrum $D$ with the double-stranded RNA was effected by measuring the optical rotatory dispersion of an RNA molecule known to be in the double-stranded configuration, MS-2 replicative form (Ochoa *et al.*, 1964). Spectrum $D$ and the double-stranded MS-2-RNA are compared in Fig. 9-1. In both cases of identification the authors stated that the identification could not be considered unambiguous. However from a close look at Figs. 9-1 and 9-2 it seems that the authors interpretation of their spectra is quite valid. Certainly it can be considered a reasonable working hypothesis. During the course of the various matrix rank analyses that McMullen *et al.* (1967) conducted, they found that a matrix rank

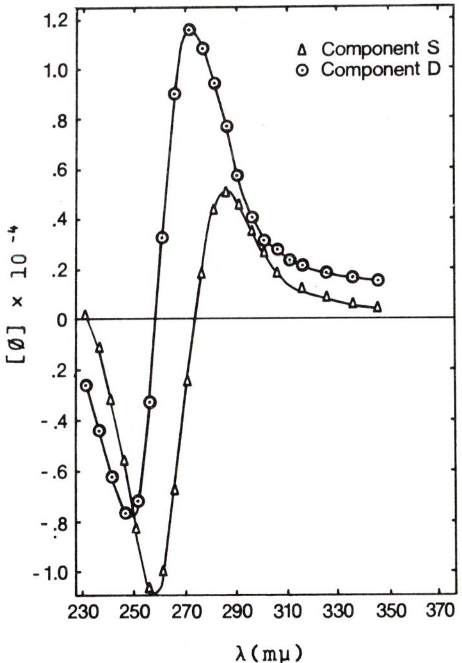


**Fig. 9-1.** Shapes of the basic spectra for TMV RNA, component $S$ and $D$. The crossover wavelength for component $S$ is 275 m$\mu$ (277.5 m$\mu$ for 1.0 $M$ $Na^+$) and for component $D$, 260 m$\mu$. Points are shown every 5 m$\mu$ for reference (McMullen *et al.*, 1967).

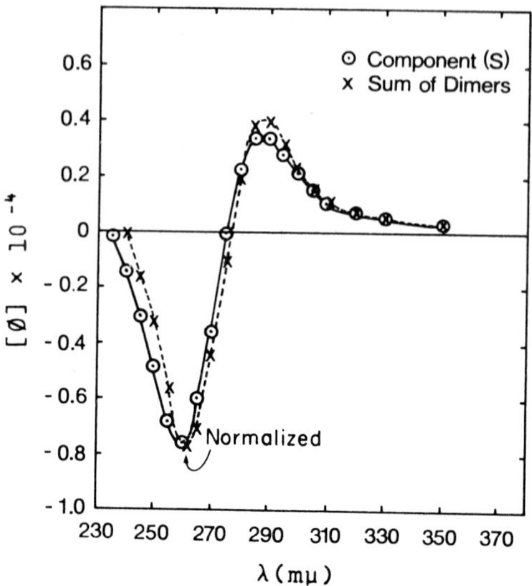


**Fig. 9-2.** Comparison between component $S$ and the calculated spectrum for single-stranded TMV RNA. Points are shown every 5 m$\mu$ for reference (McMullen *et al.*, 1967).

analysis on the entire set of experimental data (at all ionic strengths and temperatures) yielded the two components $S$ and $D$.

Having established that two components are required to fit the data if we want to know how much of each is present at any temperature or ionic strength then we use the techniques of function minimization as follows: If $f(\lambda), f_1(\lambda), f_2(\lambda)$ are the optical rotatory dispersion of, respectively, the RNA sample in question, the single stranded structure, and the double stranded structure, then the equations which we have to analyze are

$$\begin{aligned} f(\lambda_1) &= \alpha_1 f_1(\lambda_1) + \alpha_2 f_2(\lambda_1) + \delta_1 \\ f(\lambda_2) &= \alpha_1 f_1(\lambda_2) + \alpha_2 f_2(\lambda_2) + \delta_2 \\ &\vdots \\ f(\lambda_m) &= \alpha_1 f_1(\lambda_m) + \alpha_2 f_2(\lambda_m) + \delta_m \end{aligned} \tag{9-14}$$

Once again we seek the values of $\alpha_1$ and $\alpha_2$ such that $\sum_{j=1}^{m} \delta_j^2$ is a minimum or $\sum_{j=1}^{m} |\delta_j|$ is a minimum. The techniques described in Chapter 4 will adequately handle this problem. To estimate the variance of the coefficients we refer the reader to Chapter 6. In analyzing any RNA sample in terms of double-stranded and single-stranded structures (i.e., in determining $\alpha_1$ and $\alpha_2$), it would not only be quite in order but almost compulsory to conduct an analysis of variance to determine with what probability one can expect the spectrum of the RNA molecule which one is analyzing to be fitted with the two components $S$ and $D$. It may well be that an unknown RNA sample may require more than two components to obtain a good fit, in which case it would be worthwhile to attempt to discover the third component, once again by matrix rank analysis.

McMullen *et al.* have suggested that we use the following equation to estimate the amount of single- and double-stranded form.

$$\alpha_1 = f(\lambda_{260})/f_1(\lambda_{260}), \qquad \alpha_2 = f(\lambda_{275})/f_1(\lambda_{275}) \tag{9-15}$$

The reasoning behind this procedure is clear from the figures since the crossover point (point of zero rotation) for the single-stranded RNA is 275 m$\mu$ and that of the double-stranded RNA is 260 m$\mu$. These authors find that use of the above two equations provide good estimates for $\alpha_1$ and $\alpha_2$. While this is a simpler procedure we advise using one of the techniques of function minimization which allow us to conduct an analysis of variance when dealing with a sample of unknown RNA.

## 4. Application to Reaction Rates of Partially Oxidized Hemoglobin

Matrix rank analysis was used by Ainsworth and Bingham (1968) to analyze the reaction rates of partially oxidized hemoglobin with carbon monoxide. In their model these authors suggest that in order to account for the contribution of each intermediate to the initial reaction rate of a solution of partially bound protein with free ligand we may write an equation of the type

$$\frac{d}{dt}\,\frac{[\mathrm{PX}]}{[\mathrm{P}][\mathrm{X}]} = k_1a_1 + k_2a_2 + \cdots + k_ia_i + \cdots + k_na_n \tag{9-16}$$

where [X] is the concentration of the ligand, [P] the concentration of the protein, [PX] is the corresponding concentration of bound protein, $a_i$ is the fractional concentration of the intermediate, and $k_i$ is a rate constant. From Eq. (9-16) we note that the initial rate is the sum of products of two parameters $a_i$ and $k_i$. The quantity $d\{[\mathrm{PX}]/[\mathrm{P}][\mathrm{X}]\}/dt$ can be measured under different sets of $a_i$ or $k_j$. The former condition is satisfied by using different concentrations of proteins and ligands. A different set of $k_j$ are obtained by running the reaction at differing pH or temperature. By submitting the system to these variations we can obtain the following equation

$$\left\{\frac{d}{dt}\frac{[PX]}{[P][X]}\right\}_{ij} = \sum_{q=1}^{n} a_{iq}k_{jq} \tag{9-17}$$

If we put

$$\left\{\frac{d}{dt}\frac{[PX]}{[P][X]}\right\}_{ij} = r_{ij}$$

which is the initial rate under the $i$th concentration of protein and ligand and the $j$th condition of temperature pH, or whatever condition that changes the equilibria then Eq. (9-17) becomes in matrix notation

$$\mathbf{R} = \mathbf{AK}$$

By a theorem in Chapter 2 the rank of the matrix **R** can not exceed the rank of **A** or **K**. Consequently a determination of the rank of **R** gives the number of linearly independent variables. Establishing the rank of matrix **R** tells us, in effect, how many terms of Eq. (9-16) are required to describe the observables $r_{ij}$. This gives in turn, the number of components or intermediates in the reaction system.

In actual application Ainsworth and Bingham (1968) used the reactivity of partially oxygenated hemoglobin with carbon monoxide at different percentages of hemoglobin in the reactant solution and various pH. From these conditions they determined the matrix elements $r_{ij}$. In determining the rank of the matrix **R**, Ainsworth and Bingham used the method of Wallace and Katz (1964) described above. The final conclusion is that over a range of pH from 5.72–8.10 and percentage hemoglobin in solution of 26–70% two components or intermediates from a possible thirteen are required to describe the data.

The matrices set up by Ainsworth and Bingham are given in Tables 9-2–9-5. The criterion suggested by Wallace and Katz (1964) is used to complete the analysis, and comparison of the diagonals of the reduced matrix and the error propagation matrix shows

$$r'_{ij} = 2.22 \quad 0.3968 \quad -0.2150 \quad 0.0808$$

$$s'_{ij} = 0.120 \quad 0.0900 \quad 0.1570 \quad 0.1682$$

indicating two components or at most three.

**Table 9-2**

SPECIFIC INITIAL VELOCITIES OF REACTIONS BETWEEN SOLUTIONS OF PARTIALLY OXIDISED HEMOGLOBIN AND CO[a]

| % Hb in reactant solution | $r_{ij} \times 10^{-5}$ (liter mole$^{-1}$ sec$^{-1}$) | | | | | | |
|---|---|---|---|---|---|---|---|
| | pH 8.10 | pH 7.55 | pH 7.12 | pH 6.60 | pH 6.20 | pH 5.72 | Mean of row |
| 26 | 2.08 | 1.18 | 0.78 | 0.84 | 0.78 | 1.12 | |
| 27 | 2.17 2.22 | 1.15 1.11 | 0.88 0.82 | 0.73 0.81 | 0.76 0.80 | 0.94 1.02 | 1.13 |
| 28 | 2.39 | 1.01 | 0.80 | 0.86 | 0.86 | 1.11 | |
| 33 | 1.28 1.40 | 1.01 0.95 | 0.79 0.75 | 0.73 0.75 | 0.76 0.76 | 1.01 1.04 | 0.94 |
| 34 | 1.52 | 0.88 | 0.70 | 0.76 | 0.76 | 1.06 | |
| 43 | 1.67 1.67 | 0.97 0.92 | 0.67 0.67 | 0.67 0.67 | 0.76 0.75 | 0.99 0.99 | 0.94 |
| 45 | 1.68 | 0.87 | 0.67 | 0.67 | 0.73 | 0.98 | |
| 55 | 1.86 1.87 | 0.89 0.87 | 0.80 0.75 | 0.80 0.76 | 0.83 0.78 | 1.03 1.02 | 1.00 |
| 56 | 1.87 | 0.85 | 0.70 | 0.72 | 0.73 | 0.99 | |
| 67 | 1.90 2.02 | 1.02 0.96 | 0.76 0.75 | 0.76 0.80 | 0.96 0.88 | 1.20 1.19 | 1.10 |
| 70 | 2.13 | 0.90 | 0.74 | 0.85 | 0.79 | 1.17 | |
| Mean of column | 1.84 | 0.96 | 0.75 | 0.76 | 0.79 | 1.05 | 1.02 |

[a] From Ainsworth and Bingham (1968). 20°C, 0.225 M phosphate buffer.

**Table 9-3**

AVERAGE NUMERICAL DEVIATIONS FROM $r_{ij}$[a]

| Mean % Hb in reactant solution | $s_{ij} \times 10^{-3}$ (liter mole$^{-1}$ sec$^{-1}$) | | | | | |
|---|---|---|---|---|---|---|
| | pH 8.10 | pH 7.55 | pH 7.12 | pH 6.60 | pH 6.20 | pH 5.72 |
| 27 | 0.120 | 0.070 | 0.040 | 0.053 | 0.040 | 0.090 |
| 34 | 0.120 | 0.065 | 0.045 | 0.015 | — | 0.025 |
| 44 | 0.015 | 0.050 | — | — | 0.015 | 0.005 |
| 56 | 0.015 | 0.020 | 0.050 | 0.040 | 0.050 | 0.020 |
| 69 | 0.155 | 0.060 | 0.010 | 0.045 | 0.085 | 0.015 |

[a] From Ainsworth and Bingham (1968).

**Table 9-4**

THE TRANSFORMED MATRIX DERIVED FROM TABLE 9-2[a]

| | | | | | |
|---|---|---|---|---|---|
| 2.22 | 1.11 | 1.02 | 0.82 | 0.81 | 0.80 |
| 0 | 0.3968 | 0.2556 | 0.2501 | 0.2393 | 0.2330 |
| 0 | 0 | −0.2150 | −0.1499 | −0.0949 | −0.0166 |
| 0 | 0 | 0 | 0.0808 | 0.0541 | 0.0154 |
| 0 | 0 | 0 | 0 | −0.0230 | 0.0165 |

[a] From Ainsworth and Bingham (1968).

**Table 9-5**

PART OF THE TRANSFORMED ERROR MATRIX DERIVED FROM TABLE 9-3[a]

| | | | | | |
|---|---|---|---|---|---|
| 0.120 | 0.070 | 0.090 | 0.040 | 0.053 | 0.040 |
| 0 | 0.0900 | — | 0.1058 | — | 0.0735 |
| 0 | 0 | 0.1570 | 0.1120 | — | — |
| 0 | 0 | 0 | 0.1682 | — | — |
| 0 | 0 | 0 | 0 | — | — |

[a] From Ainsworth and Bingham (1968).

## 5. Alternative Methods for Determining the Shape of the Spectra

An alternative method to matrix rank analysis depends on making certain assumptions about the system under consideration. These assumptions can be based on other experimental data obtained by independent methods, on the intuition and the imagination of the investigator, or they can be based on both of these. This method is best illustrated by a specific example. Suppose we desire to determine the spectra of exposed and buried tryptophan and tyrosine in proteins. In this model we assume that a buried tyrosine residue, a surface tyrosine residue, a buried tryptophan residue, and a surface tryptophan residue will give rise to different spectra. Next we suppose from independent evidence (such as solvent perturbation, reactivity with amino acid reagents, etc.) that we know how many residues of tryptophan and tyrosine are buried in a number $m$ of proteins. Lastly we assume that the absorbance of proteins at any wavelength from 265–305 m$\mu$ can be represented as a linear combination written as follows

$$A_{1\lambda_i} = \varepsilon_{1\lambda_i}\alpha_{11} + \varepsilon_{2\lambda_i}\alpha_{12} + \varepsilon_{3\lambda_i}\alpha_{13} + \varepsilon_{4\lambda_i}\alpha_{14} \tag{9-18}$$

where $A_{1\lambda_i}$ is the absorption of protein 1, at wavelength $\lambda_i$; $\varepsilon_{1\lambda_i}$, $\varepsilon_{2\lambda_i}$, $\varepsilon_{3\lambda_i}$, and $\varepsilon_{4\lambda_i}$ are the contributions of a buried tyrosine, surface tyrosine, buried tryptophan, and surface tryptophan residues respectively in protein 1. For $m$ proteins we have $m$ equations at any given wavelength $\lambda_i$, and the $m$ equations for absorption at this wavelength are

$$\begin{aligned}
A_{1\lambda_i} &= \varepsilon_{1\lambda_i}\alpha_{11} + \varepsilon_{2\lambda_i}\alpha_{12} + \varepsilon_{3\lambda_i}\alpha_{13} + \varepsilon_{4\lambda_i}\alpha_{14} + \delta_1 \\
A_{2\lambda_i} &= \varepsilon_{1\lambda_i}\alpha_{21} + \varepsilon_{2\lambda_i}\alpha_{22} + \varepsilon_{3\lambda_i}\alpha_{23} + \varepsilon_{4\lambda_t}\alpha_{24} + \delta_2 \\
&\vdots \\
A_{m\lambda_i} &= \varepsilon_{1\lambda_i}\alpha_{m1} + \varepsilon_{2\lambda_i}\alpha_{m2} + \varepsilon_{3\lambda_i}\alpha_{m3} + \varepsilon_{4\lambda_i}\alpha_{m4} + \delta_m
\end{aligned} \tag{9-19}$$

Set in the above manner the equations can be solved for $\varepsilon_{1\lambda_i}$, $\varepsilon_{2\lambda_i}$, $\varepsilon_{3\lambda_i}$, and $\varepsilon_{4\lambda_i}$ using any one of the minimization techniques outlined in Chapter 4 with the provision that $\varepsilon_{1\lambda_i}$, $\varepsilon_{2\lambda_i}$, $\varepsilon_{3\lambda_i}$, and $\varepsilon_{4\lambda_i}$ are all positive. Solving the equation for these coefficients gives us the contribution of the residues at wavelength $\lambda_i$. To obtain the entire spectrum of all the residues the above calculation must be done for all pertinent wavelengths. An important feature that can be incorporated in this method is realized when the investigator has some notion of how the function he is trying to elucidate behaves. Thus if the investigator feels that the absorption of a surface tyrosine group at 305 m$\mu$ is zero then $\varepsilon_{2,305}$ is set equal to zero when the analysis is done at 305 m$\mu$. A more sophisticated possibility occurs when the investigator decides that the absorption for a surface tryptophan residue is greater at 295 m$\mu$ than at 300 m$\mu$, in which case the analysis at 295 m$\mu$ is conducted subject to the constraint that $\varepsilon_{4,300} < \varepsilon_{4,295}$. Similarly we might feel that the absorption of a surface tryptophan is greater at 280 m$\mu$ than at 275 m$\mu$ so we set $\varepsilon_{4,275} < \varepsilon_{4,280}$. In essence, every piece of information is utilized in constructing the unknown functions. In the above example, because of all the constraints put on the coefficients, linear programming appears to us to be an excellent technique for minimization. As with the matrix rank analysis method as applied to TMV RNA, final identification of the computed functions with known physical entities constructed from model compounds will provide the confidence required for further application of these functions.

Finally we note the following applications suggested by others for matrix rank analysis (and factor and component analysis): Applications to gel filtration data (Ackers, 1968, see also Chapter 12), to enumeration of components from fluorescent spectra (Weber, 1961), and to the enumeration of components from experiments in the ultracentrifuge (Godschalk, 1971).

## References

Ackers, G. K. (1968). *J. Biol. Chem.* **243**, 10.

Ainsworth, S. (1961). *J. Phys. Chem.* **65**, 1968.

Ainsworth, S., and Bingham, W. S. (1968). *Biochem. Biophys. Acta* **160**, 10.

Cantor, C. R., Jaskunas, S. R., and Tinoco, I., Jr. (1966). *J. Mol. Biol.* **20**, 39.

Godschalk, W. (1971). *Biochemistry* **10**, 2056.

Harman, H. H. (1967). "Modern Factor Analysis," 2nd ed. Chicago Univ. Press., Chicago, Illinois.

Kendall, M. and Stuart A. (1968). "The Advanced Theory of Statistics," Vol. 3. Griffin, London.

Lawley, D. N. and Maxwell, A. E. (1971). "Factor Analysis as a Statistical Method." American Elsevier, New York.

Magar, M. E. (1972). *Biopolymers*, in press.

McMullen, D. W., Jaskunas, S. R., and Tinoco, I., Jr. (1967). *Biopolymers* **5**, 589.

Margenau, H., and Murphy, G. M. (1955). "Mathematics of Physics and Chemistry," Vol. 1. Van Nostrand-Reinhold, Princeton, New Jersey.

Ochoa, S., Weissman, C., Borst, P., Burdon, R. H., and Billeter, M. A. (1964). *Fed. Proc. Fed. Amer. Soc. Exp. Biol.* **23**, 1285.

Wallace, R. (1960). *J. Phys. Chem.* **64**, 899.

Wallace, R., and Katz, S. M. (1964). *J. Phys. Chem.* **68**, 3890.

Weber, G. (1961). Nature *190*, 27

*CHAPTER 10*

# OPTICAL ROTATORY DISPERSION AND CIRCULAR DICHROISM OF PROTEINS

## 1. Introduction

Optical rotatory dispersion or the rotation of the plane of polarized light by optically active substances as a function of wavelength is one of the main tools for obtaining information on the conformation of molecules. In this chapter we emphasize the empirical and phenomenological points of view of the optical rotatory dispersion of proteins. We trace the development of this subject from the time Moffitt (1956) introduced his theoretical treatment of the $\alpha$-helix, considering it as a molecular crystal. This treatment ultimately led from the semiempirical equation of Moffitt and Yang (1956) to current attempts at analyzing the ultraviolet rotatory dispersion of proteins in terms of standard reference conformations. We also deal with the problem of decomposing ORD spectra in terms of their contributing circular dichroism bands (Moscowitz, 1960; Carver *et al.*, 1966b), the decomposition of circular dichrotic bands into their possible components Gaussian components and the analysis of Drude equations.

## 2. Possible Structures of Proteins and Polypeptides (the Ramachandran Diagrams)

Before embarking on the subjects mentioned above, let us examine the possible structures a protein and a polypeptide can take. This will enable us to define more clearly what we are looking for in using optical rotatory dispersion and circular dichroism as a tool for determining the secondary structure of proteins. Primarily due to Edsall *et al.* (1966) a variety of similar (but not identical) rules have been formulated for the construction of protein and polypeptide chains.

Figure 10-1 shows a diagram of part of a polypeptide chain. On account of the following resonating forms

$$\begin{array}{c} \text{O} \\ \| \\ \text{—C—N—} \\ \quad | \\ \quad \text{H} \end{array} \rightleftharpoons \begin{array}{c} \text{OH} \\ | \\ \text{—C=N—} \end{array}$$

the amide group is considered planar. This implies that the rotation is completely restricted about the C–N bond in the amide group. However the N–$C_\alpha$ and C–$C_\alpha$ in the notation of Edsall and his coworkers have effectively unrestricted rotation, and consequently the values taken by the the two angles ($\phi$, $\psi$) illustrated in Fig. 10-1 describe the conformation of the polypeptide chain. In Fig. 10-2 we see a plot of the angle $\phi$ versus the angle $\psi$. This figure [based on Ramachandran *et al.* (1963)] was prepared by computing the interatomic distances between the various atoms in the peptide unit, including the $\beta$-carbon, over all possible values of the angles ($\phi$, $\psi$) and rejecting all values of ($\phi$, $\psi$) which permitted distances between atoms which were less than the Van der Waal contact distances of those atoms. Or stating it crudely, if for any value of ($\phi$, $\psi$) two atoms got in each other's way this value of ($\phi$, $\psi$) is thrown out. The permissable region therefore is the values of ($\phi$, $\psi$) which do not permit overlap of the atoms.

From the physiochemical point of view the approach taken by Ramachandran and coworkers treats the potential of interaction between atoms

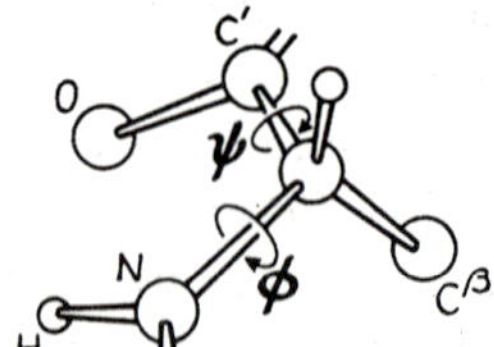


**Fig. 10-1.** Definition of the dihedral angles $\phi$ and $\psi$ (Stryer, 1968).

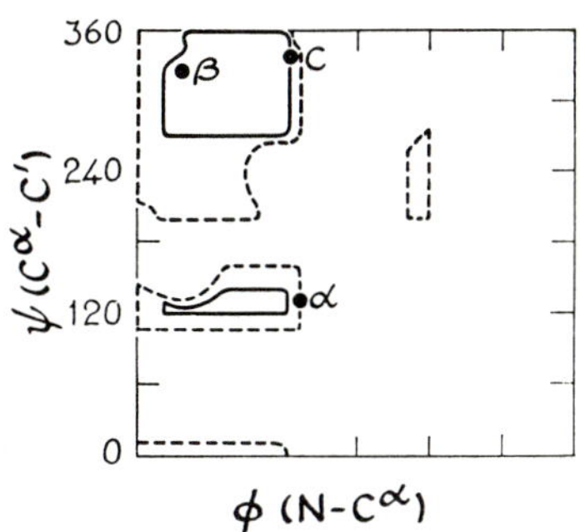


**Fig. 10-2.** Conformational map plotted according to Ramachandran. Here $\psi$ is the ordinate and $\phi$ is the abscissa. Conformations predicted to be fully allowed are enclosed by a solid line, while those predicted to be marginally allowed are enclosed by a dotted line. On this conformational map $\alpha$, $\beta$, and $C$ denote the positions of the right hand $\alpha$-helix, antiparallel pleated sheet, and collagen helix, respectively (Stryer, 1968).

as a hard sphere potential. True potentials of interaction however must treat atoms not as hard spheres but as compressible spheres. For a discussion of this prior to 1963 see Schellman and Schellman (1964). Currently various assumptions are being made as to the nature of the potential of interaction between the various atoms of the polypeptide chain (De Santis *et al.*, 1965; Brant and Flory, 1965; Ramachandran *et al.*, 1966; Nemethy and Scheraga, 1965; Scott and Scheraga, 1966; and others). These authors take into consideration a number of factors that may contribute to the potentials of interaction between various atoms. These factors include dipole–dipole interaction, hydrogen bonding, rotational restriction about the $N–C_\alpha$ and $C–C_\alpha$ bonds, as well as steric hinderance. From the results to date it seems evident that the investigators are on the right track with respect to the ultimate results of their calculations. There is, however, no universal agreement as to the precise nature of the potential functions and it seems likely that we will have to wait some time for some more definite results. To quote Ramachandran (1968) "It appears that reliable conclusions regarding the actual energies of different conformations can only be drawn after more detailed studies are made on the potential functions." Nevertheless, as far as predicting the accepted regions on the Ramachandran diagrams for proteins, the methods of the authors mentioned above appear to have succeeded quite well. This is observed when we look at the Ramachandran plot for the two proteins myoglobin and lysozyme whose crystal structure is known (Fig. 10-3).

From the above considerations it is clear that optical rotatory dispersion (ORD), if it is to tell us anything about the conformation of proteins, must interpret its results in terms of the available conformations depicted by the $(\phi, \psi)$ diagrams. It is, of course, too much to expect optical rotatory dispersion and circular dischroism to tell us precisely where all the dihedral angles between successive peptide groups of any given protein fall on the Ramachandran map. It is not, however, unreasonable to rely on optical rotatory dispersion and circular dichroism to give us some knowledge as to where a fair percentage of the dihedral angles of any given protein lie on the $(\phi, \psi)$ diagram, and we now examine to what extent optical rotatory dispersion has done that.

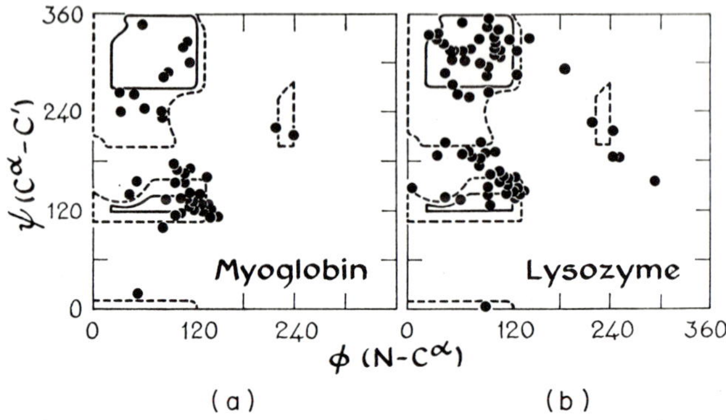


**Fig. 10-3.** (a) Plot of $\phi$ and $\psi$ of the main chain of sperm whale myoglobin. (b) Plot of $\phi$ and $\psi$ of the main chain of hen egg-white lysozyme. Residues that are part of a regular $\alpha$-helical segment are not shown in (a) and (b) (Stryer, 1968).

## 3. Methods for Determining the Secondary Structure of Proteins

### 3.1 Moffitt–Yang Treatment

A number of authors have proposed methods using data from the visible and near ultraviolet to determine the $\alpha$-helical content of polypeptides and proteins as well as deviations from this structure. On the basis of the theoretical treatment of Moffitt (1956), Moffitt and Yang (1956) first proposed the equation

$$[m'] = \frac{a_0 \lambda_0^2}{\lambda^2 - \lambda_0^2} + \frac{b_0 \lambda_0^4}{(\lambda^2 - \lambda_0^2)^2} \tag{10-1}$$

where $[m']$ is the reduced mean residue rotation and $a_0$, $b_0$, and $\lambda_0$ are constants. By setting $\lambda_0$ equal to some fixed value and plotting $[m'](\lambda^2 - \lambda_0^2)$ vs. $1/(\lambda^2 - \lambda_0^2)$, one may obtain a straight line. The slope of this straight line is proportional to $b_0$, which in turn was thought to be proportional to the helical content. In the instances when $\lambda_0 = 212$ m$\mu$ then, if $b_0 = 0$ the protein was supposed to be in completely random coil conformation, and if $b_0 = -630$ protein was supposed to be in a completely $\alpha$-helical configuration. The Moffitt–Yang treatment assumed that only $\alpha$-helical configurations and coil configurations existed in a protein.

### 3.2 Two-Term Drude Equations

A number of two-term Drude equations were used to analyze the secondary structure. The general form of a two-term Drude equation is:

$$[m'] = \frac{\alpha_1 \lambda_1^2}{\lambda^2 - \lambda_1^2} + \frac{\alpha_2 \lambda_2^2}{\lambda^2 - \lambda_2^2} \tag{10-2}$$

where $\alpha_1, \alpha_2, \lambda_1$, and $\lambda_2$ are constants. Yamaoka (1964) first used an equation of this kind to describe the ORD of polypeptides in the visual region. Imahori (1963) also used Eq. (10-2) to describe the ORD of poly-$\gamma$-benzyl-$L$-glutamate. Shechter and Blout (1964a, 1964b) and Shechter *et al.* (1964) proposed the following two-term Drude equation to describe the ORD of proteins and polypeptides,

$$[m'] = \frac{A(\alpha, \varrho)_{193} \lambda_{193}^2}{\lambda^2 - \lambda_{193}^2} + \frac{A(\alpha, \varrho)_{225} \lambda_{225}^2}{\lambda^2 - \lambda_{225}^2} \tag{10-3}$$

and used it to determine helical content of these polymers as well as deviations from structures consisting of a mixture of only $\alpha$-helices and random coils. When $[m'](\lambda^2 - \lambda_{193}^2)/\lambda_{193}^2$ is plotted against $\lambda_{225}^2/(\lambda^2 - \lambda_{225}^2)$, a straight line is obtained whose slope is $A(\alpha\varrho)_{225}(\lambda_{225}^2 - \lambda_{193}^2)/\lambda_{193}^2$ and whose intercept is $A(\alpha\varrho)_{193} + A(\alpha\varrho)(\lambda_{225}^2/\lambda_{193}^2)$. Once $A(\alpha\varrho)_{193}$ and $A(\alpha\varrho)_{225}$ were obtained, one could estimate the helical content from the equation

$$H_{193} = (A(\alpha\varrho)_{193} + 760)/36.5 \tag{10-4a}$$

$$H_{225} = (A(\alpha\varrho)_{225} + 60)/19.9 \tag{10-4b}$$

In addition, these authors proposed that if $H_{193} = H_{225}$, then no structure other than the $\alpha$-helix and random coils was present. The use of the Moffitt–Yang equation and the two-term Drude equations to estimate the helical content has been discussed by Yang (1965) in an article on the phenomenological treatment of optical rotatory dispersion of polypeptides and proteins. A comparison of these two methods and others were discussed by Carver *et al.* (1966a, 1966b) and certain relationships between $a_0$ and $b_0$ of the Moffitt–Yang treatment and $A(\alpha\varrho)_{193}$ and $A(\alpha\varrho)_{225}$ were established. Harrington *et al.* (1966), examining current literature, have been unable to decide whether the two-term Drude equation or the Moffitt–Yang equation yield more accurate results and have recommended the use of both.

The Moffitt–Yang equation and the two-term Drude equation may not adequately represent the true $\alpha$-helical content of proteins, as compared with what is found by the crystallographer. That, however, is a general criticism of the subject we shall take up later in the chapter. In the meantime we might note that we can use function minimization to estimate the parameters $a_0$, $b_0$, $H_{193}$, and $H_{225}$, and we can use statistical tests to test hypotheses about the values of these parameters or to test, for example, whether or not, $H_{193} = H_{225}$ within the limits of statistical significance, and so on.

The above method uses the visible and near ultraviolet regions for estimating the secondary structure. The characteristic features of the ORD of the peptide backbones in the 190–250 mμ region has generated methods to deal specifically with this region of the spectrum.

### 3.3 Method Using Data from Cotton Effects

Simmons *et al.* (1961) discovered that the trough of a Cotton effect at 233 mμ appeared to be a characteristic of the α-helix. A year later Blout *et al.* (1962) showed that the α-helical poly-*L*-glumatic acid possesses a peak at 198 mμ, in addition to the trough at 233 mμ. Subsequently, numerous authors confirmed these two ORD characteristics of the α-helix.

Simmons *et al.* (1961) proposed estimating the helical content of protein by means of the formula

$$H_{233} = \frac{[m']_{233}\ \text{(native protein)} - [m']_{233}\ \text{(unfolded protein)}}{\text{difference in rotation between completely } \alpha\text{-helical protein and completely unfolded protein}} \tag{10-5a}$$

A similar method has been proposed to estimate helical content from the peak at 198 mμ.

$$H_{198} = \frac{[m']_{198}\ \text{(native protein)} - [m']_{198}\ \text{(unfolded protein)}}{\text{difference in rotation between completely } \alpha\text{-helical protein and completely unfolded protein}} \tag{10-5b}$$

It will be noted that the above two equations, like the Moffitt–Yang and the two-term Drude equations, analyze the proteins in terms of α-helices and random coils only, and no other secondary structures are taken into consideration. With the characterization of the β-form and the possibility that it may contribute to the optical rotatory dispersion of proteins, the following simultaneous equations were proposed for the analysis of the secondary structure of proteins in terms of the helix, $\beta$-form and coil:[†]

$$
\begin{aligned}
[m']_{\lambda_1} &= a[m_\alpha']_{\lambda_1} + b[m_\beta']_{\lambda_1} + c[m_c']_{\lambda_1} \\
[m']_{\lambda_2} &= a[m_\alpha']_{\lambda_2} + b[m_\beta']_{\lambda_2} + c[m_c']_{\lambda_2} \\
[m']_{\lambda_3} &= a[m_\alpha']_{\lambda_3} + b[m_\beta']_{\lambda_3} + c[m_c']_{\lambda_3}
\end{aligned}
\tag{10-6}
$$

where $[m']_\lambda$, $[m_\alpha']_\lambda$, $[m_\beta']_\lambda$, and $[m_c']_\lambda$ are the mean residue rotation at the indicated wavelengths of the protein under consideration, the $\alpha$-helix, the $\beta$-form and the coil respectively and $a$, $b$, and $c$ are the fraction helix, $\beta$-form, and coil content respectively. The form of Eq. (10-6) is now easily recognized as the standard form we are analyzing without the inclusion of the experimental error. If the following substitutions are made

$$
\begin{array}{llll}
[m']_{\lambda_1} = f(x_1), & [m_\alpha']_{\lambda_1} = f_1(x_1), & [m_\beta']_{\lambda_1} = f_2(x_1), & [m_c']_{\lambda_1} = f_3(x_1) \\
[m']_{\lambda_2} = f(x_2), & [m_\alpha']_{\lambda_2} = f_1(x_2), & [m_\beta']_{\lambda_2} = f_2(x_2), & [m_c']_{\lambda_2} = f_3(x_2) \\
\vdots & \vdots & \vdots & \vdots \\
[m']_{\lambda_m} = f(x_m), & [m_\alpha']_{\lambda_m} = f_1(x_m), & [m_\beta']_{\lambda_m} = f_2(x_m), & [m_c']_{\lambda_m} = f_3(x_m)
\end{array}
\tag{10-7}
$$

$$a = \alpha_1, \qquad b = \alpha_2, \qquad c = \alpha_3$$

and the experimental errors are included we obtain our familiar equation for all appropriate wavelengths.

$$
\begin{aligned}
f(x_1) &= \alpha_1 f_1(x_1) + \alpha_2 f_2(x_1) + \alpha_3 f_3(x_1) + \delta_1 \\
f(x_2) &= \alpha_1 f_1(x_2) + \alpha_2 f_2(x_2) + \alpha_3 f_3(x_2) + \delta_2 \\
&\vdots \\
f(x_m) &= \alpha_1 f_1(x_m) + \alpha_2 f_2(x_m) + \alpha_3 f_3(x_m) + \delta_m
\end{aligned}
\tag{10-8}
$$

While the above equations pertain to the analysis of the protein in terms of only three reference conformations it is clear that we are not so limited. If more conformations are discovered and are thought to contribute significantly then they can be included without much difficulty.

Solution of these equations for the $\alpha$ using the least squares method or an approximation in the $L_2$ norm was attempted by Magar (1968) using the values of Carver *et al.* (1966b) for the $\alpha$-helix and coil and the values of Sarkar and Doty (1966) for the $\beta$-form. Analysis of the optical rotatory dispersion of myoglobin, from the data of Harrison and Blout (1965),

[†] This procedure was first suggested by Yang.

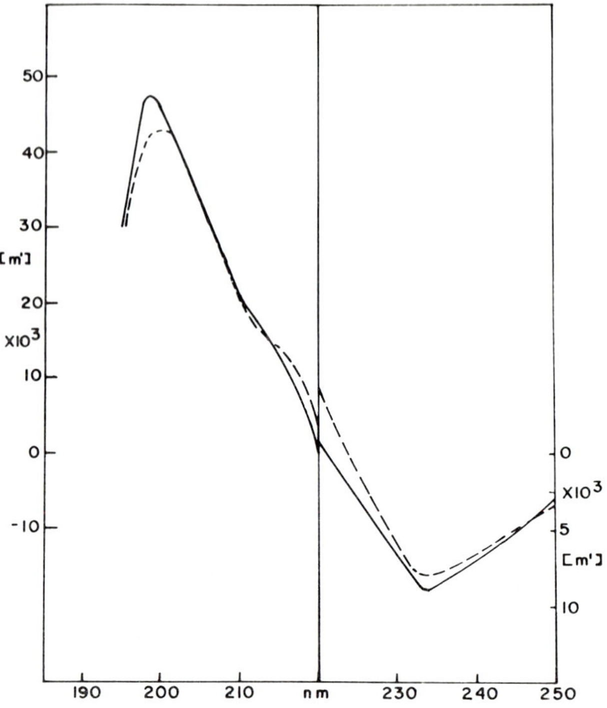


**Fig. 10-4.** Linear programming fit of myoglobin.

and lysozyme, from the data of Tomimatsu and Gaffield (1965), gave the results: myoglobin 55% $\alpha$-helix, 35% $\beta$-form, and 10% coil; and lysozyme 21% $\alpha$-helix, 33% $\beta$-form, and 40% coil. These values were close to the ones obtained by Greenfield *et al.* (1967). The fit shown in Fig. 10-4 is a linear programming fit for myoglobin which we shall discuss presently.

Inasmuch as the possibility that negative coefficients may arise, what we recommend in Chapter 4 may be reiterated here. The weighted least squares may be applied to this analysis and will probably yield better results. Since we wish to have $\sum_{i=1}^{3} \alpha_i = 1$, then the $L_2$ norm, subject to the restriction that $\sum_{i=1}^{3} \alpha_i = 1$, may be attempted. It must be noted, however, that the three techniques, least squares, weighted least squares, and least squares restricted such that $\sum_{i=1}^{3} \alpha_i = 1$, do not guarantee that the coefficients will all be positive. If investigator is faced with a situation where negative coefficients are found and he seeks a solution that will minimize $\sum_{j=1}^{m} \delta_j^2$, then the method of steepest descent, the Nelder and Mead method or the modified Marquardt's procedure may be used.

In the event that we do not wish to minimize $\sum_{j=1}^{m} \delta_j^2$ but $\sum_{j=1}^{m} | \delta_j |$ then the linear programming technique is used. We have attempted a linear programming fit with the following results: 48%, 45%, and 3% for the $\alpha$-helix, random coil, and $\beta$-form respectively. In Fig. (10-4)

**Table 10-1**

| Wavelength (nm) | Myoglobin | $\alpha$-helix | $\beta$-form | Coil |
|---|---|---|---|---|
| 195.0 | 30000 | 57125 | 4832 | 2997 |
| 197.5 | 46250 | 70256 | 14480 | −7675 |
| 198.6 | 47000 | 70856 | 18600 | −11000 |
| 200.0 | 45500 | 69122 | 22978 | −15801 |
| 202.5 | 41600 | 62125 | 26946 | −20838 |
| 204.0 | 38000 | 55000 | 28800 | −21627 |
| 205.0 | 34700 | 49800 | 29103 | −21929 |
| 210.0 | 20600 | 26593 | 21456 | −17295 |
| 212.0 | 17800 | 22277 | 17200 | −14400 |
| 215.0 | 14400 | 18236 | 11157 | −10361 |
| 220.0 | 1180 | 8526 | 862 | −5462 |
| 230.0 | −7200 | −13661 | −6254 | −2261 |
| 233.0 | −9100 | −14723 | −6050 | −2424 |
| 235.0 | −8600 | −13346 | −5739 | −2432 |
| 240.0 | −7000 | −9638 | −4443 | −2346 |
| 245.0 | −4800 | −6365 | −3465 | −2212 |
| 250.0 | −3200 | −4256 | −2688 | −2036 |

we compare our linear result with that of Greenfield *et al.* (1967). Table 10-1 shows the values we used for $f(x)$, $f_1(x)$, $f_2(x)$ and $f_3(x)$ in the linear programming technique.

The estimations for the confidence intervals for the coefficients using the least squares, weighted least squares, restricted least squares, steepest descent, Nelder and Mead, and other methods are all shown in Chapter 6.

## 4. Problems Involved in Optical Rotatory Dispersion Analysis of Proteins†

There are three main problems encountered in determining secondary structure by analysis of optical rotatory dispersion of proteins. The first

† More on this subject can be found in Magar (1971).

problem question whether the contributions of the standard secondary structure to the optical rotatory dispersion of the protein are additive, or whether they somehow interact. Mathematically we are asking whether an equation of the form

$$[m']_\lambda = a[m_\alpha']_\lambda + b[m_\beta']_\lambda + c[m_c']_\lambda \quad (10\text{-}9)$$

is sufficient to describe the optical rotatory dispersion or do we need to add some interaction term making the equation to be analyzed of the form

$$[m']_\lambda = a[m_\alpha']_\lambda + b[m_\beta']_\lambda + c[m_c']_\lambda + \phi_\lambda \quad (10\text{-}10)$$

where $\phi_\lambda$ represents some kind of interaction between the secondary structure. Little is known about whether Eq. (10-9) is correct or something like Eq. (10-10) is required.

The second problem pertains to some of the discrepancies in the literature about the precise magnitude of the rotation of the various standard conformations in terms of which we seek to analyze our proteins. To quote Yang (1967) commenting on a table depicting the magnitude and the position of the extrema of the standard conformations, "the magnitude of the extrema listed here is still tentative." If the extrema are still tentative then it stands to reason that values obtained at wavelengths other than the extrema are also still tentative. It is clear therefore that some kind of agreement must be obtained about the precise magnitude of the rotation of the reference conformation if we are to proceed further. This problem is to some extent complicated by the fact that poly-*L*-glumatic acid, a favorite model compound (at pH 4.5 this polypeptide is taken by many investigators as representing 100% helix), aggregates in concentrated solution as the pH is lowered. This suggests that the quaternary structure of proteins might influence the optical rotatory dispersion and alter the magnitude of the rotation of the standard conformations and also that of the proteins. It is to be noted, however, that the optical rotatory dispersion and circular dichroism of poly-*L*-glutamic acid is essentially constant for pH 4.5–4.8 in dilute solutions (Tomimatsu *et al.*, 1966; Yang, 1967). This fact has been taken to mean that the polymer is unaggregated in this pH range and is at the plateau of the helix–coil transition and consequently is probably 100% helical. Thus, despite the propensity for aggregation we could take the rotation of poly-*L*-glutamic acid at pH 4.5–4.8 as the standard for a 100% helix. The problem of aggregation is not peculiar to $\alpha$-helical conformations since it has been pointed out by Yang (1967) that $\beta$-forms also aggregate.

The possibility that the optical rotatory dispersion of natural proteins may be used as references cannot be discounted. Riddiford (1966) suggested that we take the optical rotatory dispersion of paramyosin as the standard of the 100% $\alpha$-helix and Jirgensons (1966) suggested that the protein phosphoavidin is in 100% coil conformation. Whatever the final values of the optical rotatory dispersion of the specific reference conformation it is clear that they must be agreed upon before we can use them effectively. In this respect, however, it is probable that agreement will be reached without a great deal of difficulty because the values obtained for the reference conformation do not differ greatly.

The third problem, and the most important encountered in the analysis of the optical rotatory dispersion, is the contribution of the side chains. The results obtained by using the methods described above to obtain the amount of secondary structure of myoglobin and lysozyme do not agree with the results obtained from X-ray diffraction data. It is generally thought (Fasman *et al.*, 1964, 1965; Beychok, 1965, 1968; Coleman and Blout, 1967; Yang, 1967; and others) that by and large the major source of disagreement is the aromatic and disulfide contributions.

## 5. Possible Methods for Determination of Side Chain Contributions

The author in a previous publication (Magar, 1968) has labeled all side chain contributions by $\psi(\lambda)$ and suggested that this function may be written as the sum of the following contributions

$$\psi(\lambda) = \psi_{\text{disul}}(\lambda) + \psi_{\text{tyr}}(\lambda) + \psi_{\text{trp}}(\lambda) \tag{10-11}$$

where $\psi_{\text{disul}}(\lambda)$, $\psi_{\text{tyr}}(\lambda)$, $\psi_{\text{trp}}(\lambda)$ are the contribution of the disulfide bonds, tyrosine residues, and tryptophan residues respectively. If $\psi(\lambda)$, which we shall call $\psi(x)$ to be consistent with our notation, can be precisely evaluated then it can be incorporated in the analysis in the following manner: On the left-hand side of Eq. (10-8) instead of using $f(x_1)$, $f(x_2), \ldots, f(x_m)$ we use the vector $f(x_1) - \psi(x_1), f(x_2) - \psi(x_2), \ldots,$ $f(x_m) - \psi(x_m)$ and proceed to solve the equation for $\alpha_1, \alpha_2, \alpha_3$. These equations do not present any problems provided $\psi(x)$ is known, but since it is not, we must direct our attention to the problem of how it might be evaluated.

The determining of $\psi(x)$ in general will be a complicated process. What we do here is give a few suggestions as to how it might be done. These suggestions are based on the acceptance of the fact that the crystalline structure remains largely or wholly intact on going into solution (whether this is or is not the case we shall not go into here, we merely emphasize that this assumption must hold if we are going to carry the analysis further). We are not particularly interested in the $\psi(x)$ per se as they will vary form protein to protein and can be determined by subtracting the contribution of the secondary structure, determined from X-ray crystallography, from the rotation of the given protein. What we are most interested in are the component functions of $\psi(x)$ and we suggest the following way for their determination: Using X-ray crystallography data we determine the contribution of the various secondary structures to the optical rotatory dispersion of the protein. By subtracting those contributions from the actual ORD of the protein we can obtain $\psi(x)$ for any given protein. This process is done for all the proteins where we have a good notion of the contribution of the secondary structure. When that is done for several proteins we will have a series of functions which we shall call $\psi_{\text{myoglobin}}(x)$, $\psi_{\text{lysozyme}}(x)$, $\psi_{\text{carbonic anhydrase}}(x)$, . . . , etc. These functions can then be subjected to matrix rank analysis in order to obtain the number of linearly independent components which make up those spectra. Once the shape of the linearly independent components are determined it may well be possible to determine to what each one is due. In this manner the spectra of tyrosine, tryptophan, and disulfide residues might ultimately be resolved (see Chapter 9 on matrix rank analysis). Alternatively we could use the method we suggested in Chapter 9 to determine those spectra.† Analysis of $\psi_{\text{myoglobin}}(x)$, $\psi_{\text{lysozyme}}(x)$, . . . , etc., to determine the contribution of the aromatic residues then follows in complete analogy to our attempts in that chapter to determine the spectra of exposed and buried trypt ophan and tyrosine from the ultraviolet spectra of a number of proteins. Thus from independent evidence such as reactivity of amino acid residues, solvent perturbation or from X-ray crystallographic evidence we can proceed to represent $\psi_{\text{myoglobin}}(x)$, $\psi_{\text{lysozyme}}(x)$, or any $\psi(x)$ for any other protein as shown below for any wavelength $x_1$.

$$\begin{aligned} \psi_{\text{myoglobin}}(x_1) &= \alpha_{11}\psi_{\text{tyr}}(x_1) + \alpha_{12}\psi_{\text{trp}}(x_1) + \alpha_{13}\psi_{\text{disul}}(x_1) \\ \psi_{\text{lysozyme}}(x_1) &= \alpha_{21}\psi_{\text{tyr}}(x_1) + \alpha_{22}\psi_{\text{trp}}(x_1) + \alpha_{23}\psi_{\text{disul}}(x_1) \end{aligned} \tag{10-12}$$

Here $\alpha_{11}$, $\alpha_{12}$, and $\alpha_{13}$ will be the number of tyrosine residues, the number of tryptophan residues, and the number of disulfide bonds in myoglobin

† See Magar (1971) for details.

and $\alpha_{21}$, $\alpha_{22}$, and $\alpha_{23}$ will be the number of tyrosine residues, the number of tryptophan residues, and the number of disulfide bonds in lysozymes. Similar equations are written for the rest of the proteins. Since the only unknowns in Eq. (10-12) are the components functions of $\psi(x)$ at the wavelength $x_1$, we can solve for them and thus determine their value at that specific wavelength. To obtain the entire component functions one must solve for all pertinent wave lengths. These functions *must be* finally identified with the physical entity they are supposed to represent before they can be used. The final identification of a spectrum or a function with the physical entity producing it is a concept of some import and has been discussed to some extent in Chapter 9.

To digress slightly it should be pointed out that we are by no means limited to attempting to analyse $\psi_{\text{myoglobin}}(x)$ and the rest of the $\psi(x)$ for other proteins in terms of aromatic residues and disulfide bonds. By examining the X-ray crystallographic data an investigator might decide that a positive charge or a negative charge in the vicinity of the tyrosine or tryptophan residues might be causing these residues to contribute more or less to the optical rotatory dispersion of any protein. One can then proceed to analyze the functions $\psi_{\text{protein1}}(x)$, $\psi_{\text{protein2}}(x)$, . . . , etc. in terms of this hypothesis provided enough of these proteins exhibit this physical entity. Proceeding in the same manner as suggested above the investigator may be able to elucidate functions or spectra which will take the effects of those charges into consideration. Having determined these functions one proceeds to test the hypothesis by applying them to other proteins. As stated, this example is hypothetical. It may well be that no tyrosine or tryptophan residues are so situated in a sufficient number of proteins to permit us to carry an analysis of this sort. This, however, is just an example to show that if it is thought that a prominent physical entity in a number of proteins is producing a characteristic optical rotatory dispersion spectrum, we can set out to find its contribution by using matrix rank analysis or the alternative method we suggest. Thus in most instances an investigator can take any reasonable (and that may vary considerably from one investigator to another) hypothesis and put it in mathematical form as shown above. Whether the hypothesis is true or not is determined by further experiments.

Beychok (1968), writing in a recent review on the comparison of estimates of helicity using optical rotatory dispersion and circular dichroism with the results obtained from crystalline structure, states that while there is reasonable agreement between the observed helical content in crystals and optical rotatory dispersion measurements for myoglobin, hemoglobin, lysozyme, ribonuclease, and chymotrypsin, the case of ferricytochrome c and carbonic anhydrase c presents some problems. The $\alpha$-helical content of ferricytochrome c is computed to be about 27% by Urry and Doty (1965) whereas X-ray diffraction at 4 Å resolution shows this protein to have little $\alpha$-helix. On the other hand, carbonic anhydrase c shows little helical content on the basis of its optical rotatory dispersion and circular dichroism while X-ray work at 5.5 Å resolution reveals its $\alpha$-helical content to be slightly less than 30%.

The situation concerning carbonic anhydrase has been explained on the basis that the aromatic contributions for this protein are quite strong in the 200–240 m$\mu$ region and effectively diminish the magnitude of rotation at the trough at 233 m$\mu$. To nail this point down more conclusively we must determine the contribution of the aromatic amino acids and show that to be the case. The methods suggested above (matrix rank analysis or the alternative method) will help. The situation in ferricytrochrome c, however, is different because it appears that non-$\alpha$-helical structures are giving rise to spectra quite like an $\alpha$-helix structure (this might not be the case when all the X-ray data is in hand). It appears that for the time being at least we must accept the conclusion that ferricytochrome c has a peculiar structure which gives rise to this kind of rotation. It is, of course, impossible to attempt to elucidate the optical rotatory dispersion of this peculiar structure by using matrix rank analysis or our procedure if we have only ferricytochrome c to deal with. However, if another protein or proteins were to show the same kind of behavior and if this structure or some reasonable facsimile thereof were present in these proteins also, the methods we have described in Chapter 9 would probably elucidate the optical rotatory dispersion of this structure.

From the above discussion it therefore follows that if we are to prove the statements about carbonic anhydrase c, we have to resort to our methods to determine the spectra of the aromatic amino acids. On the other hand, if we wish to determine whether or not we are correct as far as ferricytochrome c is concerned we still have to use the above methods, provided of course, we know some other proteins which exhibits the same behavior. It is interesting to note that if no other proteins are found which exhibit this kind of behavior and if ferricytochrome c does in fact have some "peculiar structure" it will be difficult to refute an investigator who, after subtracting all the likely contribution from the optical rotatory dispersion of ferricytochrome c, attributes the residual rotation to this "peculiar structure." The only way to refute the contention of this investigator is to show the occurrence of this "peculiar structure" in another protein which does not show the same optical rotatory dispersion characteristic as ferricytochrome after all the likely contributions have been subtracted.

## 6. Circular Dischroism of Proteins

In view of the fact that all circular dichrotic spectra of proteins and polypeptides are functions of wavelength, everything that has been said about the optical rotatory dispersion of proteins applies, with necessary changes having been made, to circular dichroism. In addition, there are some considerations, in circular dichroism which do not apply to optical rotatory dispersion. These considerations are suggested by the work of Hashizume *et al.* (1967) on ribonuclease. Examination of the circular dichrotic spectra of ribonuclease at 280 mμ revealed an ellipticity per mole of protein amounting to −31,000.

These authors attributed this ellipticity to the tyrosine groups in the protein stating that the three buried tyrosine groups gave rise to an ellipticity of −9,500 per residue and the exposed tyrosine groups gave rise to an ellipticity of −800 per residue. The evidence for their claim is based on the fact that poly-*L*-tyrosine, whether in the $\alpha$-helical conformation or in the disordered conformation gives rise to an ellipticity of −800 per residue at 280 mμ. The large contribution of ellipticity by the buried tyrosine groups is attributed to the hinderance to free rotation and not to their being buried or in an environment of low dielectric constant, or in an $\alpha$-helical configuration. There is ample evidence for this, particularly from the work of Gill (1967) which shows that the optical rotatory dispersion of tyrosine peptides in solvents of low dielectric constants does not alter appreciably.

One should ask here (to introduce a particular application of function minimization to circular dichroism), whether all three buried tyrosine groups are contributing equally to the ellipticity or whether it is possible that only one or two tyrosine residues are contributing to the major part of the ellipticity recorded at 280 mμ. We feel that the best way of determining whether or not that is true is to use either matrix rank analysis or our alternative procedure [see Chapter 9 and explanation to Eq. (10-12)]. In order to carry out the analysis, the circular dichroism of several proteins in the pertinent region (i.e., 260–310 mμ) must be determined and the analyses previously advocated applied. It is clear that optical rotatory dispersion cannot be used for such analyses because of the contribution of the rotation of the amide backbone. Once the circular dichroism spectra is available we can attempt to determine the spectra of the buried and exposed aromatic amino acid by using matrix rank analysis or the alternative method outlined in Chapter 9 in an identical manner as the applications in that chapter for the elucidation of the contribution of the buried and exposed aromatic residues from absorption spectroscopy. In this manner the separate contribution of the aromatic amino acids, whatever their situation in the proteins, can be determined. If the disulfide bonds are thought to contribute there will be no difficulty in including them in the analysis. It will be noted that we are now essentially suggesting the same thing we advocated in connection with optical rotatory dispersion except that we are confining ourselves to a region where the peptide backbone of the proteins does not contribute.

The choice of what region to use in conducting an analysis of this kind can be used to the advantage of the investigator. Thus, we can choose our region such that any contribution which will interfere with the analysis can be excluded. In such instances, it is clear that we can only expect to determine the contribution of the chromophore in question in the region we analyze. This, however, is quite a help because having determined the major contribution of the chromophore in a specific region, it may be a relatively easy matter to determine the "tail" of the band in the region where other interfering chromophores start to contribute. For example, if we suspect that the disulphide bonds start to contribute at 270 mμ, then we can analyze our spectra by matrix rank or otherwise for the tyrosines and tryptophans only down to 270 mμ. This will elicit the major contribution of those amino acids down to 270 mμ and the tails of those bands below 270 mμ can be constructed on the assumption that they are Gaussians or Lorentzians.

The techniques of function minimization have several other applications in optical rotatory dispersion and circular dichroism. While not confined to proteins these applications have been used on proteins to a considerable extent. These topics are fitting an optical rotatory dispersion curve with a number of Drude terms, fitting circular dichroism curves in terms of a number of Gaussians, and resolving an optical rotatory dispersion into its contributing Gaussian bands using the Kronig–Kramer transform.

## 7. Fitting an Optical Rotatory Dispersion Curve with Drude Terms

Generally the optical rotatory dispersion far away from a transition may be represented by an equation

$$[m'] = \sum_i K_i/(\lambda^2 - \lambda_i^2)$$

where $K_i$ and $\lambda_i$ are constants. For the simple case of a one term Drude equation

$$[m']_\lambda = K_\alpha/(\lambda^2 - \lambda_\alpha^2) \tag{10-13}$$

which we rearrange to

$$[m']_\lambda \lambda^2 = K_\alpha + [m']_\lambda \lambda_\alpha^2 \tag{10-14}$$

By plotting $[m']_\lambda \lambda^2$ against $[m']_\lambda$ we obtain a straight line whose slope is $\lambda_\alpha^2$ and whose intercept is $K_\alpha$, the two parameters we require. The best least squares straight line is simply one that minimizes $\sum_{j=1}^{m} \delta_j^2$ in the equations

$$\begin{aligned} [m']_{\lambda_1} \lambda_1^2 &= [m']_{\lambda_1} \lambda_\alpha^2 + K_\alpha + \delta_1 \\ [m']_{\lambda_2} \lambda_2^2 &= [m']_{\lambda_2} \lambda_\alpha^2 + K_\alpha + \delta_2 \\ &\vdots \\ [m']_{\lambda_m} \lambda_m^2 &= [m']_{\lambda_m} \lambda_\alpha^2 + K_\alpha + \delta_m \end{aligned} \tag{10-15}$$

For an equation with more than one Drude term we shall use a two-term Drude equation to illustrate the method.

The two-term Drude equation may be written

$$[m'] = [K_\alpha/(\lambda^2 - \lambda_\alpha^2)] + [K_\beta/(\lambda^2 - \lambda_\beta^2)] \tag{10-16}$$

which may be rearranged to

$$\begin{aligned} [m']\lambda^4 = (K_\alpha + K_\beta)\lambda^2 - [m']\lambda^2(\lambda_\beta^2 + \lambda_\alpha^2) - [m']\lambda_\alpha^2\lambda_\beta^2 - K_\alpha\lambda_\beta^2 \\ - K_\alpha\lambda_\beta^2 - K_\beta\lambda_\alpha^2 \end{aligned} \tag{10-17}$$

which can be written in our notation

$$f(x_j) = \sum_{i=1}^{4} \alpha_i f_i(x_j) + \delta_j \qquad (j = 1, 2, \ldots, m) \tag{10-18}$$

where we make the following identifications

$$\begin{aligned} f(x_j) &= [m']_{\lambda_j} \lambda_j^4, \quad f_1(x_j) = \lambda_j^2, \quad f_2(x_j) = [m']_{\lambda_j} \lambda_j^2 \\ f_3(x_j) &= [m']\lambda_j, \quad f_4(x_j) = 1 \\ \alpha_1 &= (K_\alpha + K_\beta), \quad \alpha_2 = -(\lambda^2\beta + \lambda_\alpha^2) \\ \alpha_3 &= -\lambda_\alpha^2 \lambda_\beta^2, \quad \alpha_4 = -K_\alpha \lambda_\beta^2 - K_\beta \lambda_\alpha^2 \end{aligned} \tag{10-19}$$

If the $f_1(x)$, $f_2(x)$, $f_3(x)$, $f_4(x)$ are linearly independent then we can solve for all the $\alpha$'s and consequently determine the constants $K_\alpha$, $K_\beta$, $\lambda_\beta$, and $\lambda_\alpha$. Once again we refer the reader to Chapter 4 for the determination of the $\alpha_j$.

Methods of estimation of the standard error of the parameters are shown in Chapter 6. To determine how many Drude terms one needs we have to resort to an analysis of variance as described in Chapter 6. If more than two Drude terms are required then two additional functions for every Drude term are added to the right-hand side of an equation similar to Eq. (10-17).

For a three-term Drude equation, we analyze $[m']\lambda^6$ in terms of the functions $[m']\lambda^4$, $\lambda^4$, $[m']\lambda^2$, $\lambda^2$, $[m']$, and 1 to obtain $K_\alpha$, $K_\beta$, $K_\gamma$, $\lambda_\alpha$, $\lambda_\beta$ and $\lambda_\gamma$. It will always be possible, in principle, to make an analysis of the optical rotatory dispersion in terms of as many Drude terms as one wishes. In practice, however, since the optical rotatory dispersion outside the region of absorption is a featureless curve we will seldom require more than a two-term Drude equation.

With respect to the relationship of the Drude parameter to physically meaningful parameters it has been noted by Djerassi (1960) and Heller (1958) that the Drude-fit parameters can not be related reliably to physically meaningful parameters of absorption extrema and rotational strength. Buber and Katzin (1966) have made a study of the relationship between the Drude parameters and the circular dichroic data from which they are derived. This study determined the effect of circular dichroism bandwidth, interband separation, experimental error, and the approximations generally made with Drude equations upon the correspondence between circular dichroism parameters and Drude equation parameters. For their conclusions regarding this matter we refer the reader to their article. We also note that in making their fits Buber and Katzin essentially used the linearized Taylor series method (see Chapter 4) thus treating a two-term Drude equation as a nonlinear problem. Other nonlinear methods described in Chapter 4 can also be used to estimate the parameter of (10-16).

## 8. Analysis of Circular Dichroism Spectra

Resolving circular dichroism spectra into a series of Gaussians using a Dupont 310 curve analyzer has been mentioned recently by Urry (1967), Myer (1968), Coleman (1968), and others. There is no doubt that the curve analyzer is a great help in resolving spectra in terms of Gaussians or Lorentzians, however it depends on a naked eye comparison between the raw data and the generated curves. For all practical purposes this could be all right except that curve fitting gives some numerical results as to how far we are off in our fits and therefore serves as an objective criterion to choose between fits considered physically meaningful. In addition, within the limits of the discussion in Chapter 6 we may obtain some statistics on our parameters.

With respect to the actual curve fitting, suppose we have measurements of ellipticity at $m$ wavelengths written as follows

$$[\theta]_{\lambda_j} = \sum_{i=1}^{n} [\theta_i] \exp\{-[(\lambda_j - \lambda_i)^2/\Delta_i^2]\} + \delta_j \quad (j = 1, 2, \ldots, m) \qquad (10\text{-}20)$$

Equations (10-20) represent a set of nonlinear equations and the techniques mentioned in Chapter 4 can be used to estimate the parameters $[\theta_i]$, $\lambda_i$ and $\Delta_i$. As with most of the nonlinear methods mentioned in this book we can not say which will work best in this specific instance without trying. They were, however, discussed in general in Chapter 4. The importance of good starting values for all these iterative techniques cannot be overemphasized and the curve analyzers can provide some of these guesses. In Chapter 6 we show how the confidence regions for the estimated parameters may be obtained and we must bear in mind that while the confidence regions are exact the confidence limits are approximate.

With respect to how many Gaussians we need, the usual $F$-tests while nonvalid in general, do provide us with a better comparison of the data in numerical terms than the naked eye, as has been discussed in Chapter 6.

## 9. Decomposition of Optical Rotatory Dispersion Curves into Their Contributing Circular Dichroic Bands

This is the last topic we deal with in this chapter. Moscowitz (1960) has shown that the Kronig–Kramer transformation of a Gaussian band, written in the same notation as before,

$$[\theta_i] \exp[-(\lambda - \lambda_i^2)/\Delta_i^2]$$

can be approximated by

$$[m']_\lambda \cong (2[\theta_i]/\sqrt{\pi}) \left\{ \exp[-(\lambda - \lambda_i)/\Delta_i^2] \int_0^{[(\lambda-\lambda_i)/\Delta_i]} \exp(y^2)\, dy - [\Delta_i/2(\lambda + \lambda_i)] \right\} \qquad (10\text{-}21)$$

From the above equation it is clear that if we are given the optical rotatory dispersion $[m']_\lambda$ we can obtain the parameters of the contributing circular dichroic band. The only thing that may puzzle the reader is the integral on the right side of Eq. (10-21). This definite integral will become a number after evaluation and does not present any problems. (Since this is a nonlinear problem it will be solved by iteration so the values of $\lambda_i$ and $\Delta_i$ are given to the computer at each iteration and there is no difficulty in evaluating the integral.) Equation (10-21) is best put in terms of rotational strength. The rotational strength $R_i$ of the $i$th band is defined as

$$R_i = (hc/48\pi^2 N) \int_{\text{over band}} ([\theta]_\lambda/\lambda) d\lambda \text{ erg cm}^3 \qquad (10\text{-}22)$$

where $h$ is Planck's constant and $c$ the velocity of light. On evaluating the integral over the whole band we obtain

$$R_i = 1.09 \times 10^{-42} (2[\theta_i]\Delta_i/\pi^{1/2}\lambda_i) \text{ erg cm}^3 \qquad (10\text{-}23)$$

To use a quantity more amenable to computation we define

$$A_i = [R_i/(1.09 \times 10^{-42})] \text{ erg cm}^3 \text{ deg} \qquad (10\text{-}24)$$

With the above definitions we may write the optical rotation in the following manner for a single transition

$$[m']_\lambda = (A_0\lambda_0/\Delta_0)\Big\{\exp[-(\lambda - \lambda_0)^2]/\Delta_0 \int_0^{[(\lambda-\lambda_0)/\Delta_0]} \exp(y^2)\, dy - [\Delta_0/2(\lambda + \lambda_0)]\Big\} \text{ deg cm}^2 \text{ mole} \qquad (10\text{-}25)$$

Moscowitz was the first person to analyze optical rotatory dispersion curves in order to obtain the circular dichroic parameters of the single band giving rise to the optical rotatory dispersion. The equation Moscowitz used has the following form

$$[m']_\lambda = (A_0\lambda_0/\Delta_0)\Big\{\exp[-(\lambda - \lambda_0)/\Delta_0]^2 \int_0^{[(\lambda-\lambda_0)/\Delta_0]} \exp(y^2)\, dy + [B/(\lambda^2 - Q^2)]\Big\} + (C/\lambda^2) \qquad (10\text{-}26)$$

In the above equation we note the introduction of two new terms $B/(\lambda^2 - Q^2)$ and $C/\lambda^2$ and the deletion of $\Delta_0/[2(\lambda + \lambda_0)]$ from Eq. (10-25).

The terms containing constants $B$, $C$, and $Q$ were included to account for the Cotton effects coming from a region other than the one being analysed and which may be called the background rotation. Moscowitz suggested that the deleted term could be incorporated in the background. This procedure was to obtain the ORD data and attempt to obtain the best fit with the parameters $A_0$, $\Delta_0$, $\lambda_0$, $B$, $C$, and $Q$. The best fit in the least squares sense would be the one that minimized the quantity

$$\Big\{[m']_\lambda - (A_0\lambda_0/\Delta_0)\Big\{\exp[-((\lambda - \lambda_0)/\Delta_0)^2] \int_0^{[(\lambda-\lambda_0)/\Delta_0]} \exp(y^2)\, dy\Big\}$$

$$[B/(\lambda - Q^2)] + (C/\lambda^2)\Big\}^2 \qquad (10\text{-}27)$$

summed over all wavelengths.

When the parameters $\lambda_0$, $\Delta_0$, and $A_0$ are obtained as a result of minimizing the quantity in Eq. (10-27) the circular dichroic band is completely characterized. Note that we also determine $B$, $C$, and $Q$.

The application of the Moscowitz method to model polypeptide compounds is due to Carver *et al.* (1966b) who analyzed optical rotatory dispersion of polypeptides in helical and coil conformation using the following equation which incorporate more than one transition.

$$[m']_\lambda = \sum_i (A_i\lambda_i/\Delta_i)\Big\{\exp[-((\lambda - \lambda_i)/\Delta_i)^2] \int_0^{[(\lambda-\lambda_i)/\Delta_i]} \exp(y^2)\, dy$$

$$\times\ [\Delta_i/2(\lambda - \lambda_i)]\Big\} + [BQ/(\lambda^2 - Q)] + (C^2/\lambda^2) \qquad (10\text{-}28)$$

Carver *et al.* determined the parameters characterizing the CD bands of these polypeptides (i.e., $A_i$, $\lambda_i$, $\Delta_i$) for a number of transitions and the parameters characterizing the background rotation $B$, $Q$, and $C$. They weighted (see Chapter 4) all their observations and minimized the quantity

$$\sum_\lambda [W_\lambda{}^2([m']_\lambda^{\text{obs}} - [m']_\lambda^{\text{pred}})^2/(n - n_p)]$$

where the summation is taken over all wavelengths, $W_\lambda$ is the weight at the indicated wavelength, $n$ the number of data points, and $n_p$ the number of parameters involved in computing the predicted rotation from Eq. (10-28). Precisely how the authors managed to minimize the quantity is not clear from the manuscript. The authors merely state they used a nonlinear least squares method.

With respect to the actual analysis of a single Cotton effect using the Eq. (10-26) suggested by Moscowitz or more than one cotton effect as suggested by Carver *et al.* using Eq. (10-28), we generally have measurements at $m$-wavelengths which result in us writing Eq. (10-26) or (10-28) $m$ times yielding $m$ nonlinear equations. The solution of those $m$ nonlinear equations is achieved by using the methods in Chapter 4. In fact what has been said about solving the nonlinear equations arising from circular dichroism in the previous section applies, with necessary changes having been made, to both the above cases so that we shall not repeat it here again.

Some of the results obtained by Carver *et al.* are shown in Tables 10-2–10-5. In Table 10-2 we see the results of a computation for the polypeptide poly-$\gamma$-morpholinethyl-$\alpha$-$L$-glutamic acid in methanol–water (9 : 1), where it is presumed to be $\alpha$-helical using the two Cotton effect solution. In the second case, in Table 10-3 we show a computation for the same polypeptide under identical conditions using a three Cotton effect solution. In Table 10-2 we note that when one fits this polypeptide optical rotation with the two Cotton effect, solution the wavelengths of the cir-

**Table 10-2**

TWO COTTON EFFECT SOLUTIONS FOR PAEMG IN METHANOL–WATER (9:1)[a]

| | 1 | 2 | 3 |
|---|---|---|---|
| RESMS[b] $\times 10^{-8}$ | 1.18 | 1.21 | 1.21 |
| $R \times 10^{40}$ (erg cm$^3$) | $34.7 \pm 0.9$ | $35 \pm 2$ | $34 \pm 3$ |
| $A \times 10^{-4}$ (°C erg cm$^3$) | $0.318 \pm 0.008$ | $0.32 \pm 0.02$ | $0.31 \pm 0.02$ |
| $\lambda$ (m$\mu$) | $192.7 \pm 0.4$ | $192.7 \pm 0.6$ | $192.8 \pm 0.5$ |
| $\Delta$ (m$\mu$) | $6.4 \pm 0.4$ | $6.4 \pm 0.7$ | $6.3 \pm 0.5$ |
| $R = 10^{40}$ (erg cm$^3$) | $-27.4 \pm 0.6$ | $-27.2 \pm 0.9$ | $-27.5 \pm 0.7$ |
| $A \times 10^{-4}$ (°C erg cm$^3$) | $-0.251 \pm 0.005$ | $-0.249 \pm 0.008$ | $-0.252 \pm 0.007$ |
| $\lambda$ (m$\mu$) | $220.0 \pm 0.3$ | $220.0 \pm 0.3$ | $220.0 \pm 0.3$ |
| $\Delta$ (m$\mu$) | $13.9 \pm 0.3$ | $13.8 \pm 0.3$ | $14.0 \pm 1.0$ |
| $B \times 10^{-4}$ (°C) | 0.0[c] | $-0.02 \pm 0.13$ | $0.3 \pm 1.0$ |
| $C \times 10^{-4}$ (°C m$\mu^{-2}$) | 0.0[c] | $500 \pm 2200$ | $-3100 \pm 9200$ |
| $Q^{1/2}$ (m$\mu$) | 0.0[c] | 173.2[c] | 100.0[c] |

[a] Tables 10-2–10-5 are reprinted from Carver *et al.* (1966a). *J. Amer. Chem. Soc.* **88**, 2550. Copyright 1966 by the American Chemical Society. Reprinted by permission of the copyright owner.

[b] Experimental RESMS $= 0.01 \times 10^8$.

[c] Indicates those parameters which were held constant.

**Table 10-3**

THREE COTTON EFFECT SOLUTIONS FOR PAEMG IN METHANOL–WATER (9:1)[a]

| | Type I | | | Type II | | |
|---|---|---|---|---|---|---|
| | 1 | 2 | 3 | 4 | 5 | 6 |
| RESMS[b] $\times 10^{-8}$ | 0.018 | 0.016 | 0.016 | 0.015 | 0.015 | 0.014 |
| $R \times 10^{40}$ | $40.0 \pm 0.3$ | $41.0 \pm 0.70$ | $40.5 \pm 1.0$ | $38.3 \pm 0.4$ | $38.9 \pm 0.9$ | $38 \pm 1$ |
| $A \times 10^{-4}$ | $0.367 \pm 0.003$ | $0.376 \pm 0.006$ | $0.371 \pm 0.009$ | $0.352 \pm 0.004$ | $0.357 \pm 0.008$ | $0.35 \pm 0.01$ |
| $\lambda$ | $192.4 \pm 0.1$ | $192.3 \pm 0.1$ | $192.3 \pm 0.1$ | $192.2 \pm 0.1$ | $192.2 \pm 0.1$ | $192.2 \pm 0.1$ |
| $\Delta$ | $8.2 \pm 0.1$ | $8.5 \pm 0.2$ | $8.4 \pm 0.2$ | $7.9 \pm 0.1$ | $8.1 \pm 0.2$ | $8.0 \pm 0.2$ |
| $R \times 10^{40}$ | $150 \pm 10$ | $100 \pm 20$ | $100 \pm 10$ | $-12.1 \pm 0.6$ | $-12.5 \pm 0.8$ | $-12.4 \pm 0.7$ |
| $A \times 10^{-4}$ | $1.4 \pm 0.1$ | $0.9 \pm 0.2$ | $0.9 \pm 0.1$ | $-0.111 \pm 0.005$ | $-0.115 \pm 0.007$ | $-0.114 \pm 0.007$ |
| $\lambda$ | $216.9 \pm 0.1$ | $216.7 \pm 0.2$ | $216.7 \pm 0.2$ | $208.8 \pm 0.2$ | $208.7 \pm 0.2$ | $207.8 \pm 0.2$ |
| $\Delta$ | $11.1 \pm 0.1$ | $11.0 \pm 0.1$ | $11.0 \pm 0.1$ | $8.7 \pm 0.3$ | $8.9 \pm 0.4$ | $8.8 \pm 0.4$ |
| $R \times 10^{40}$ | $-180 \pm 10$ | $-130 \pm 20$ | $-130 \pm 10$ | $-18.4 \pm 0.3$ | $-18.4 \pm 0.3$ | $-18.5 \pm 0.3$ |
| $A \times 10^{-4}$ | $-1.7 \pm 0.1$ | $-1.2 \pm 0.2$ | $-1.2 \pm 0.1$ | $-0.169 \pm 0.003$ | $-0.169 \pm 0.003$ | $-0.169 \pm 0.003$ |
| $\lambda$ | $216.9 \pm 0.1$ | $216.8 \pm 0.2$ | $216.8 \pm 0.2$ | $224.0 \pm 0.1$ | $224.0 \pm 0.1$ | $224.0 \pm 0.1$ |
| $\Delta$ | $12.0 \pm 0.1$ | $12.3 \pm 0.2$ | $12.3 \pm 0.1$ | $10.9 \pm 0.1$ | $10.9 \pm 0.1$ | $10.9 \pm 0.1$ |
| $B \times 10^{-4}$ | 0.0[b] | $0.08 \pm 0.12$ | $0.01 \pm 0.01$ | 0.0[c] | $0.1 \pm 0.1$ | $0.02 \pm 0.02$ |
| $C \times 10^{-4}$ | 0.0[c] | $900 \pm 1100$ | $500 \pm 200$ | 0.0[c] | $1000 \pm 1000$ | $500 \pm 300$ |
| $Q^{1/2}$ | 0.0[c] | 100.0 | 173.2[c] | 0.0[c] | 100.0[c] | 170.0[c] |

[a] From Carver *et al.* (1966a).

[b] Experimental RESMS = $0.01 \times 10^{8}$. For units see Table 10-2.

[c] Indicates those parameters which were held constant.

**Table 10-4**

TWO COTTON EFFECT SOLUTIONS FOR PGA (pH 7.0) IN WATER[a]

| | 1 | 2 | 3 | 4 | 5 | 6 |
|---|---|---|---|---|---|---|
| RESMS[b] $\times 10^{-8}$ | 0.0125 | 0.00435 | 0.00435 | 0.00430 | 0.00430 | 0.00618 |
| $R \times 10^{40}$ | $-12.5 \pm 0.2$ | $-14.0 \pm 0.2$ | $-14.0 \pm 0.2$ | $-14.0 \pm 0.2$ | $-14.0 \pm 0.2$ | $-14.3 \pm 0.3$ |
| $A \times 10^{-4}$ | $-0.114 \pm 0.002$ | $-0.128 \pm 0.002$ | $-0.128 \pm 0.002$ | $-0.128 \pm 0.002$ | $-0.128 \pm 0.002$ | $-0.131 \pm 0.003$ |
| $\lambda$ | $197.7 \pm 0.1$ | $197.5 \pm 0.1$ | $197.5 \pm 0.1$ | $197.5 \pm 0.1$ | $197.5 \pm 0.1$ | $197.4 \pm 0.1$ |
| $\Delta$ | $7.8 \pm 0.2$ | $8.3 \pm 0.1$ | $8.3 \pm 0.1$ | $8.3 \pm 0.1$ | $8.3 \pm 0.1$ | $8.6 \pm 0.2$ |
| $R \times 10^{40}$ | $1.5 \pm 0.1$ | $1.74 \pm 0.09$ | $1.74 \pm 0.09$ | $1.74 \pm 0.09$ | $1.74 \pm 0.09$ | $1.7 \pm 0.1$ |
| $A \times 10^{-4}$ | $0.014 \pm 0.001$ | $0.0159 \pm 0.0008$ | $0.0159 \pm 0.0008$ | $0.0159 \pm 0.0008$ | $0.0159 \pm 0.0008$ | $0.016 \pm 0.001$ |
| $\lambda$ | $217.6 \pm 0.7$ | $217.7 \pm 0.4$ | $217.7 \pm 0.4$ | $217.7 \pm 0.4$ | $217.7 \pm 0.4$ | $217.4 \pm 0.5$ |
| $\Delta$ | $8.6 \pm 0.6$ | $8.8 \pm 0.3$ | $8.8 \pm 0.3$ | $8.8 \pm 0.3$ | $8.8 \pm 0.3$ | $8.8 \pm 0.4$ |
| $B \times 10^{-4}$ | 0.0[c] | $-300 \pm 2000$ | $-0.0005 \pm 0.0042$ | 0.0[c] | $64 \pm 5$ | $0.022 \pm 0.003$ |
| $C \times 10^{-4}$ | 0.0[c] | $3000 \pm 2000$ | $700 \pm 100$ | $640 \pm 50$ | 0.0[c] | 0.0[c] |
| $Q^{1/2}$ | 0.0[c] | 3.2[c] | 173.2[c] | 0.0[c] | 3.2[c] | 173.2[c] |

[a] From Carver *et al.* (1966a).

[b] Experimental RESMS = $0.01 \times 10^{8}$. For units see Table 10-2.

[c] Indicates those parameters which were held constant.

**Table 10-5**

THREE COTTON EFFECT SOLUTIONS FOR PGA (pH 7.0) IN WATER[a]

| | 1 | 2 | 3 |
|---|---|---|---|
| RESMS[b] $\times 10^{-8}$ | 0.00150 | 0.00144 | 0.00143 |
| $R \times 10^{40}$ | $-14.4 \pm 0.2$ | $14.4 \pm 0.2$ | $-14.2 \pm 0.2$ |
| $A \times 10^{-4}$ | $-0.129 \pm 0.001$ | $-0.132 \pm 0.002$ | $-0.130 \pm 0.002$ |
| $\lambda$ | $197.6 \pm 0.1$ | $197.5 \pm 0.1$ | $197.6 \pm 0.1$ |
| $\Delta$ | $8.4 \pm 0.1$ | $8.5 \pm 0.1$ | $8.5 \pm 0.1$ |
| $R \pm 10^{40}$ | $1.9 \pm 0.1$ | $2.0 \pm 0.5$ | $1.9 \pm 0.2$ |
| $A \times 10^{-4}$ | $0.017 \pm 0.001$ | $0.019 \pm 0.004$ | $0.018 \pm 0.002$ |
| $\lambda$ | $216.8 \pm 0.4$ | $216.7 \pm 0.7$ | $216.6 \pm 0.4$ |
| $\Delta$ | $10.1 \pm 0.8$ | $11 \pm 1$ | $10.3 \pm 0.8$ |
| $R \times 10^{40}$ | $-0.13 \pm 0.05$ | $-0.2 \pm 0.4$ | $-0.13 \pm 0.05$ |
| $A \times 10^{-4}$ | $-0.0012 \pm 0.0005$ | $-0.002 \pm 0.003$ | $-0.0012 \pm 0.0005$ |
| $\lambda$ | $235 \pm 3$ | $230 \pm 10$ | $235 \pm 3$ |
| $\Delta$ | $10 \pm 2$ | $13 \pm 6$ | $10 \pm 2$ |
| $B \times 10^{-4}$ | $500 \pm 200$ | $0.15 \pm 0.07$ | $0.002 \pm 0.001$ |
| $C \times 10^{-4}$ | $50{,}000 \pm 20{,}000$ | $800 \pm 600$ | $660 \pm 40$ |
| $Q^{1/2}$ | 10.0[c] | 100.0[c] | 184.4[c] |

[a] From Carver *et al.* (1966a).
[b] Experimental RESMS $= 0.01 \times 10^8$. For units see Table 10-2.
[c] Indicates those parameters which were held constant.

cular dichroic extrema are at approximately 192 mμ and 220 mμ and the best fit using the predicted rotation from this two cotton effect solution is shown in Fig. 10-5. In Table 10-3 we show the parameters for fitting these polypeptides using the three Cotton effect solution. From this table there appears to be two possibilities discussed below. We note the closeness of the fit with the actual data in Fig. 10-5. The computed points appear virtually indistinguishable from the measured optical rotatory dispersion on the scale shown in Fig. 10-6. A two Cotton effect solution for poly-*L*-glumatic acid at pH 7 (in the coiled conformation) and a three Cotton effect solution for the same polypeptide under the same conditions are shown in Fig. 10-6. We note that both the two Cotton effect and the three Cotton effect solutions coincide quite closely with the measured optical rotatory dispersion. The values of the parameters used in the fits shown in Fig. 10-6 are taken from Tables 10-4 and 10-5.

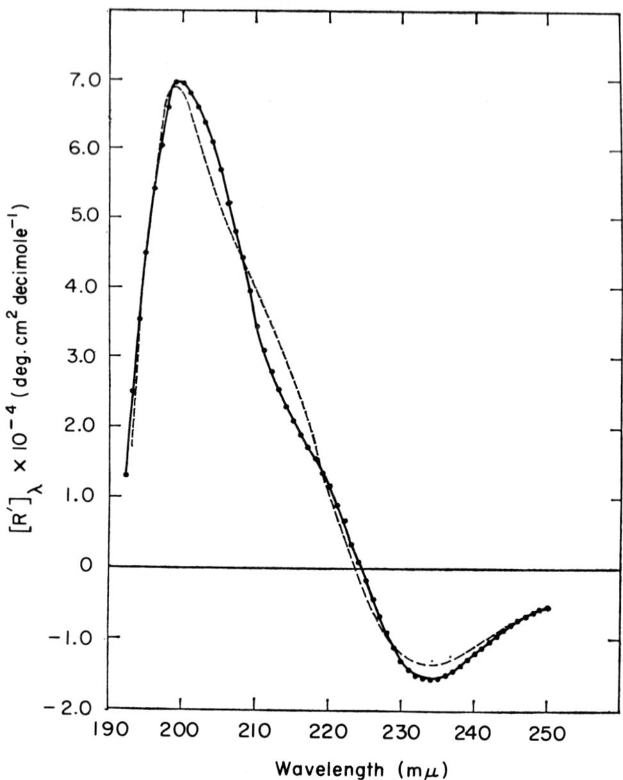


**Fig. 10-5.** Solutions for the optical rotatory dispersion of polymorpholinylethyl-*α-L*-glutamamide in methanol–water (9:1): ----, two Cotton effect solution; ——, three Cotton effect solution; • • •, observed values. (Figures 10-5 and 10-6 are reprinted from Carver *et al.* (1966a). *J. Amer. Chem. Soc.* **88**, 2550. Copyright 1966 by the American Chemical Society. Reprinted by permission of the copyright owner.)

A number of things deserve mention in respect to Tables 10-2–10-5 and Figs. 10-5 and 10-6. The reader would do well to read these comments as they are applicable to just about every other problem we encounter. It is clear from Fig. 10-5 that a two Cotton effect solution missed the actual data points in several places but a three Cotton effect solution serves very well. From this one might argue that a visual comparison is adequate and there is no need to resort to an F-test to determine the number cotton effect solution required. We have no quarrel with this point of view when there is such close coincidence between the computed curves and the actual data. Certainly, one can not do better than actual coincidence provided the plots are not scaled down to the point where coincidence is unavoidable. The $F$-test, which is not valid in general in the nonlinear cases (and we are dealing with nonlinear cases here), does provide us, as stated and explained in Chapter 6, with a less arbitrary measure than the naked eye. This test can be relied on, when the necessary precautions are taken, to help in our decision as to how many Cotton effects we need when the case is not quite clear.

Of far greater significance than any $F$-test is the question of whether or not the solution obtained bears any relationship to reality. This is a particularly appropriate question when, as shown in Table 10-3, there exist two possible computations which yield different results, but fit quite closely. In the first instance the solution, labelled type I by Carver *et al.* (1966b), yielded the following approximate $\lambda_i$: 192, 217, and 217 mμ. The solution labeled type II yielded the following approximate $\lambda_i$: 192, 208, and 224 mμ. The decision between the two or three cotton effect solutions that is to say between type I and type II, must rest in the last analysis, on other experimental data. This fact can not be over-

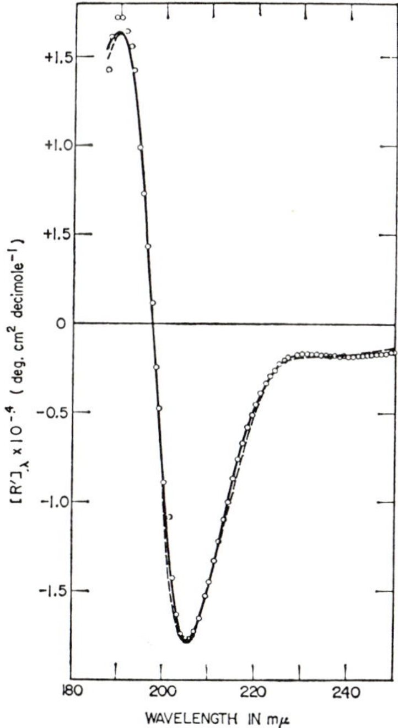


**Fig. 10-6.** Solutions for the optical rotatory dispersion of poly-α-$L$-glutamic acid in water at pH 7: ----, two Cotton effect solution; ——, three Cotton effect solution; • • •, observed values (Carver *et al.*, 1966a).

emphasized. Every single experimental fact that will aid in the solution of the problem must be built into the solution.

We are not attempting to predict the parameters of the circular dichroic bands when using these methods as Carver *et al.* suggest (1966a; 1966b). We merely present possible solutions, one of which because of the closeness of the fit will probably correspond to reality. Thus the fact that Holzwarth (1964) showed that $\alpha$-helical polypeptides have circular dichroic bands with extrema at 190, 206, and 222 m$\mu$, immediately eliminated a two Cotton effect solution and three Cotton effect solution of type I for a helical polymer. Similarly, in the case of the coiled polypeptide conformation the choice between the two or three cotton effect solutions is settled by the fact that there are in fact three circular dichroic bands, the third one being the one measured by Carver *et al.* (1966b) and which appears to be centered around 235 m$\mu$. Better data show it to be situated at 238 m$\mu$ (Yang, 1967).

In the event that we produce a solution showing the existence of a transition in a region inaccessible to our present instrumentation all we can say is that the data provides evidence consistent with the presence of such a transition. It does not mean that such a transition exists. We will know of the existence of this transition when and only when we demonstrate its presence experimentally. The only value these methods have is the suggestion they give as to where to look for such transitions.

### References

Beychok, S. (1965). *Proc. Nat. Acad. Sci. U.S.* **53**, 999.

Beychok, S. (1968). *Annu. Rev. Biochem.* **37**, 435.

Blout, E. R., Schmier, I., and Simmons, N. S. (1962). *J. Amer. Chem. Soc.* **84**, 3139.

Brant, D. A., and Flory, P. J. (1965). *J. Amer. Chem. Soc.* **85**, 2791.

Buber, T., and Katzin, L. I. (1966). *J. Phys. Chem.* **70**, 2662.

Carver, J. P., Shechter, E., and Blout, E. R. (1966a). *J. Amer. Chem. Soc.* **88**, 2550.

Carver, J. P., Shechter, E., and Blout, E. R. (1966b). *J. Am. Chem. Soc.* **88**, 2562.

Coleman, J. E. (1968). *J. Biol. Chem.* **243**, 4575.

Coleman, J. E., and Blout, E. R. (1967). *In* "Conformation of Biopolymers" (G. N. Ramachandran, ed.), Academic Press, New York.

DeSantis, P., Giglio, E., Liquori, A. M., and Ripamonti, A. (1965). *Nature* **206**, 456.

Djerassi, C. (1960). "Optical Rotatory Dispersion." McGraw-Hill, New York.

Edsall, J. T., Flory, P. J., Kendrew, J. C., Liquori, A. M., Nemethy, G., Ramachandran, G. N., and Scheraga, H. A. (1966). *Biopolymers* **4**, 121.

Fasman, G. D., Bodenheimer, E., and Lindblow, C. (1964). *Biochemistry* **3**, 1665.

Fasman, G. D., Landsberg, M., and Buchwald, M. (1965). *Can. J. Chem.* **43**, 1388.

Gill, E. W. (1965). *Biochim. Biophys. Acta.* **102**, 302.

Gill, E. W. (1967). *Biochim. Biophys. Acta.* **133**, 381.

Greenfield, N., Davidson, B., and Fasman, G. (1967). *Biochemistry* **6**, 1630.
Harrington, W. F., Josephs, R., and Segal, D. M. (1966). *Annu. Rev. Biochem.* **35**, 599.
Harrison, S. C., and Blout, E. R. (1965). *J. Biol. Chem.* **240**, 299.
Hashizume, H., Shiraki, M., and Imahori, K. (1967). *J. Biochem.* (*Tokyo*) **62**, 543.
Heller, W. (1958). *J. Phys. Chem.* **62**, 1569.
Holzwarth, G. (1964). Ph. D. Thesis. Harvard Univ., Cambridge, Massachusetts.
Imahori, K. (1963). *Kobunshi Suppl.* **12**, 34.
Jirgensons, B. (1966). *J. Biol. Chem.* **241**, 147.
Magar, M. E., (1968). *Biochemistry* **7**, 617.
Magar, M. E. (1971). *J. Theor. Biol.* **33**, 105.
Moffitt, W. (1956). *J. Chem. Phys.* **25**, 467.
Moffitt, W. and Yang, J. T. (1956). *Proc. Nat. Acad. Sci. U.S.* **42**, 695.
Moscowitz, A. (1960). *In* "Optical Rotatory Dispersion" (C. Djerrasi, ed.). McGraw-Hill, New York.
Myer, Y. P. (1968). *J. Biol. Chem.* **242**, 2115.
Nemethy, G. and Scheraga, H. A. (1965). *Biopdymers* **3**, 155.
Ramachandran, G. N. (1968). *In* "Structural Chemistry and Molecular Biology," (A. Rich and N. Davidson, eds.). Freeman, San Francisco, California.
Ramachandran, G. N., Ramakrishnan, C., and Sasisekharan, V. (1963). *J. Mol. Biol.* **7**, 95.
Ramachandran, G. N., Venkatachalam, C. M., and Krimm, S. (1966). *Biophys. J.* **6**, 849.
Riddiford, L. (1966). *J. Biol. Chem.* **241**, 2792.
Sarkar, P. K., and Doty, P. (1966). *Proc. Natl. Acad. Sci. U.S.* **55**, 981.
Schellman, J., and Schellman, C. (1964). *In* "The Proteins" (H. Neurath, ed.), Vol. 2. Academic Press, New York.
Scott, R. A., and Scheraga, H. A. (1966). *J. Chem. Phys.* **45**, 2091.
Shechter, E., and Blout, E. R. (1964a). *Proc. Natl. Acad. Sci. U.S.* **51**, 695.
Shechter, E., and Blout, E. R. (1964b). *Proc. Natl. Acad. Sci. U.S.* **51**, 794.
Shechter, E., Carver, J. P., and Blout, E. R. (1964). *Proc. Natl. Acad. Sci. U.S.* **51**, 1029.
Simmons, N. S., Cohen, C., Szent-Gyorgi, A. G., Wetlaufer, A. G., and Blout, E. R. (1961). *J. Am. Chem. Soc.* **83**, 4766.
Stryer, L. (1968). *Annu. Rev. Biochem.* **37**, 25.
Tomimatsu, Y., and Gaffield, W. (1965). *Biopolymers* **3**, 509.
Tomimatsu, Y., Vitello, L., and Gaffield, W. (1966). *Biopolymers* **4**, 653.
Urry, D. W. (1967). *J. Amer. Chem. Soc.* **89**, 4190.
Urry, D. W., and Doty, P. (1965). *J. Amer. Chem. Soc.* **87**, 2756.
Yamaoka, K. K. (1964). *Biopolymers* **2**, 219.
Yang, J. T. (1967). *In* "Conformation of Biopolymers" (G. N. Ramachandran, ed.). Academic Press, New York.
Yang, J. T. (1965). *Proc. Natl. Acad. Sci. U.S.* **53**, 438.

*CHAPTER 11*

# OPTICAL ROTATORY DISPERSION AND CIRCULAR DICHROISM OF NUCLEIC ACIDS AND THEIR COMPONENTS

## 1. Introduction

In this chapter we describe some of the recent experimental results in this field and show some of the applications of function minimization. Using function minimization we can elucidate the mole fraction of dinucleoside phosphates in ribonucleic acid (RNA) specimens provided certain assumptions hold. In addition, we can determine the mole fraction of nucleoside in some RNA molecules, also provided certain assumptions hold and the bases are randomly distributed. Most of the work that will be discussed in this chapter has originated in the laboratory of Tinoco, e.g., Warshaw *et al.* (1965), Cantor and Tinoco (1965, 1967), Cantor *et al.* (1966), and Warshaw and Tinoco (1965, 1966).

The reader who is acquainted with Chapter 8, where we discussed the variation of the absorption of desoxyribonucleic acid (DNA) with temperature and the information we can deduce from that kind of study, will readily note that whatever has been said of such analyses can be said, after the necessary changes have been made, for similar analyses on the ORD and CD of DNA. Thus measurement of the ORD or CD of DNA at varying temperatures will yield the same kind of information as the absorption measurements. Since ORD and CD have more features than plain absorption spectra the possibility of running into linearly dependent functions lessens and this is a much better state of affairs from the mathematical point of view. On the other hand, the precision achieved by absorption spectroscopy over ORD might well offset that advantage.

## 2. ORD of RNA Components

With respect to the work of Tinoco and coworkers it was observed that the optical rotatory dispersion of adenylic acid, when compared to the dinucleoside phosphate containing two adenine bases and written as ApA, revealed a marked difference as depicted by Fig. 11-1. This observation prompted Cantor and Tinoco (1965) to the following considerations. For the ORD of a dinucleoside phosphate *ipj* denoted by $\phi_{ij}(\lambda)$ the following expression was written

$$\phi_{ij}(\lambda) = \phi_i(\lambda) + \phi_j(\lambda) + \eta_{ij}(\lambda) \tag{11-1}$$

where $\phi_i(\lambda)$ is the ORD of the base $i$, $\phi_j(\lambda)$ the ORD of base $j$, and

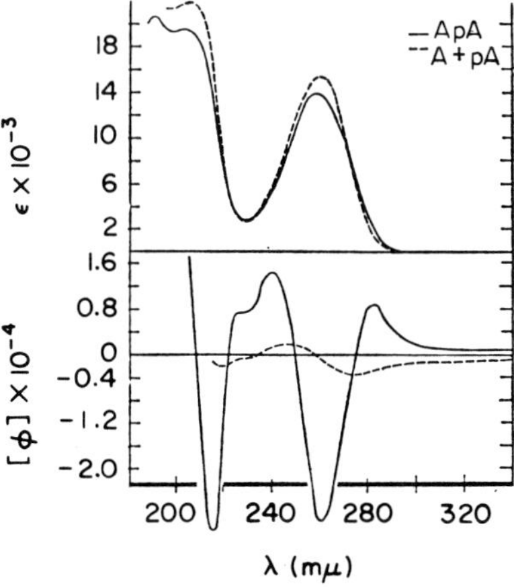


**Fig. 11-1.** Absorption and optical rotation of adenylic acid and of ApA at pH 6.9 and ionic strength 0.1 (Warshaw and Tinoco, 1965).

$\eta_{ij}(\lambda)$ accounts for the interaction between base $i$ and base $j$. Using the same reasoning the investigators went on to write the following equation for the ORD of a trinucleoside diphosphate (*ipjpk*)

$$\phi_{ijk}(\lambda) = \phi_i(\lambda) + \phi_j(\lambda) + \phi_k(\lambda) + m_{ij}(\lambda) + m_{jk}(\lambda) + m_{ik}(\lambda) \quad (11\text{-}2)$$

where the quantity $m_{ij}(\lambda)$ represents the interaction between the *i*th base and *j*th base in the trinucleoside diphosphate and the rest of the quantities have the same meanings as before. A number of modifications were made in Eq. (11-2). Ignoring the interaction of the next nearest neighbor $m_{ik}(\lambda)$ was set equal to zero and the interaction between the bases in the trinucleoside diphosphate was assumed to be the same as in the dinucleosides, which is to say that

$$m_{ij}(\lambda) = \eta_{ij}(\lambda), \qquad m_{ik}(\lambda) = \eta_{ik}(\lambda) \quad (11\text{-}3)$$

and so on. These modifications made it possible to write the ORD of a trinucleoside diphosphate in units of molar rotation per residue as

$$\phi_{ijk}(\lambda) = [2\phi_{ij}(\lambda) + 2\phi_{ik}(\lambda) - \phi_j(\lambda)]/3 \quad (11\text{-}4)$$

With these results in hand Cantor and Tinoco proposed the characterization of the optical properties of oligonucleotides and entire RNA molecules in terms of the optical properties of dinucleosides only. The following equation was written for the ORD of RNA

$$\phi_{\mathrm{RNA}}(\lambda) = \sum_{i=1}^{4} \sum_{j=1}^{4} \chi_{ij}[\phi_{ij}(\lambda)] - \sum_{i=1}^{4} \chi_i[\phi_i(\lambda)] \quad (11\text{-}5)$$

In this equation $\chi_{ij}$ is the mole fraction of the dinucleoside phosphate *ipj* and $\chi_i$ is the mole fraction of the nucleoside $i$. Equation (11-5) ignores the contribution from the ends of the RNA chain. For the use of Eq. (11-5) in the computation of the ORD of RNA one must know the quantities $\chi_{ij}$. In order to do that one must determine the various dinucleoside phosphate sequences in the RNA, a task that is becoming more feasible. If one does not know that and if one assumes that the bases are randomly distributed in the RNA sample, then we can write the following equation for the ORD of RNA

$$\phi_{\mathrm{RNA}}(\lambda) = \sum_{i=1}^{4} \chi_i \left\{2 \sum_{j=1}^{4} \chi_i[\phi_{ij}(\lambda)] - [\phi_i(\lambda)]\right\} \quad (11\text{-}6)$$

Equations similar to (11-1)–(11-6) can be written for other optical properties of RNA, notably the absorption. Although not mentioned by Cantor and Tinoco circular dichroism can be included among the optical properties that can be utilized for the same type of analysis. If the $\phi_{ij}(\lambda)$ are sufficiently different (i.e., if the $\phi_{ij}(\lambda)$ are linearly independent) one can determine the mole fraction of the dinucleoside sequences in an RNA sample. In this particular paper Cantor and Tinoco (1965) do not say how, but we presume that if the ORD of the RNA sample $\phi_{\mathrm{RNA}}(\lambda)$ is considered as the response function $f(x)$ in our notation, if the rest of the functions can be written as

$$\begin{array}{llll} \phi_{11}(\lambda) = f_1(\chi), & \phi_{12}(\lambda) = f_2(\chi), & \ldots, & \phi_{44}(\lambda) = f_{16}(\chi) \\ \phi_1(\lambda) = f_{17}(\chi), & \phi_2(\lambda) = f_{18}(\chi), & \ldots, & \phi_4(\lambda) = f_{20}(\chi) \end{array} \qquad (11\text{-}7)$$

and if the constants $\chi_{ij}$ and $\chi_j$ can be equated to $\alpha_i$ as follows

$$\begin{array}{llll} \chi_{11} = \alpha_1, & \chi_{12} = \alpha_2, & \ldots, & \chi_{44} = \alpha_{16} \\ \chi_1 = \alpha_{17}, & \chi_2 = \alpha_{18}, & \ldots, & \chi_4 = \alpha_{20} \end{array} \qquad (11\text{-}8)$$

then after these substitutions are made the form of Eq. (11-5) is readily recognized as a linear equation in the $\alpha_1, \alpha_2, \ldots, \alpha_{20}$ and can be solved by the linear techniques of Chapter 4 to determine the $\alpha_i$. These quantities represent the mole fraction of the dinucleoside phosphates and the mole fraction of the separate nucleotides. A similar argument can be presented for determining these quantities from absorption or circular dichroism measurements provided the $f_1(\chi), f_2(\chi), \ldots, f_{20}(\chi)$ obtained from these measurements are all linearly independent. The chances that the $f_i(\chi)$ are linearly independent appear to be much greater for the ORD analysis than for the analysis of absorption and circular dichroism spectra since ORD curves have more features than CD or absorption spectra. Inasmuch as all the $\alpha_i \geq 0$ then this problem is much more suited for linear programming than least squares. On the whole however we feel that the large number of functions makes it unlikely that we shall obtain satisfactory resolutions.

## 3. Possible Sequence Determination and Distinguishing between Isomers

Since the above method if successful can determine dinucleotide sequence of any given *trinucleotide* can be determined, but that is as far as it can go in determining RNA sequence.

To test the ability of this method to reproduce the ORD of oligomers Cantor and Tinoco (1965, 1967) have computed the ORD of all 64 trinucleoside phosphates and found excellent agreement in several instances. For instance, one can achieve a precise fit for ApApU and GpApU in the 220–330 mμ region, but this is not true of all trinucleotides. This state of affairs limits the use of the Eqs. (11-5) and (11-6) in sequence determination. The inability of these procedures to accurately reproduce the ORD of all trinucleotides has been attributed to the neglect of higher order interactions such as the interaction between a base and its nearest neighbour.

The investigators have felt that their computations on the ORD of 64 trinucleosides gave a great deal of utility in the resolution of the sequence of various *isomers* of any given trinucleoside phosphate. Thus if one knew the base composition of a trimer without knowing its sequences the question becomes how easily (on the basis of the computation of Cantor and Tinoco) can one distinguish this isomer from a mixture of its isomer having the same base composition but different sequences. This question will in turn depend on whether or not the computed ORD of the different isomers are linearly independent. If they are, then one can determine how much of each isomer is present.† A criterion for the extent of linear independence was set up and it was found that whenever one has the isomers $AG_2$, $CG_2$, $UG_2$, $U_2G$, $A_2G$, $C_2G$, ACG, $A_2G$, $AU_2$, AUG, and $AC_2$ then the amount of each isomer in a mixture could be determined using least squares. On the other hand the isomers UCG, $A_2U$, AUC, $UC_2$, and $U_2G$ could not, according to the criteria of linear independence set up by Cantor and Tinoco, be resolved from each other. One thing to be noted about this work is that all the calculations are based on the formalism in this chapter and that lack of agreement between the actual value of the ORD of the trinucleotide and that based on the formalism of this chapter will correspondingly magnify the *ability or inability* of these authors to resolve the isomers.

## 4. Calculation of ORD of Homopolymers and RNA

With respect to the actual applications to RNA samples, Cantor *et al.* (1966) have calculated the ORD of some RNA samples as well as some

† As indicated earlier, there are 64 possible trinucleosides, four of the form *iii*, 24 are various permutations of the four different bases taken three at a time with each set of three bases having six isomers. Finally 36 trinucleosides are of the form $j_2i$, $ij_2$, or $jij$.

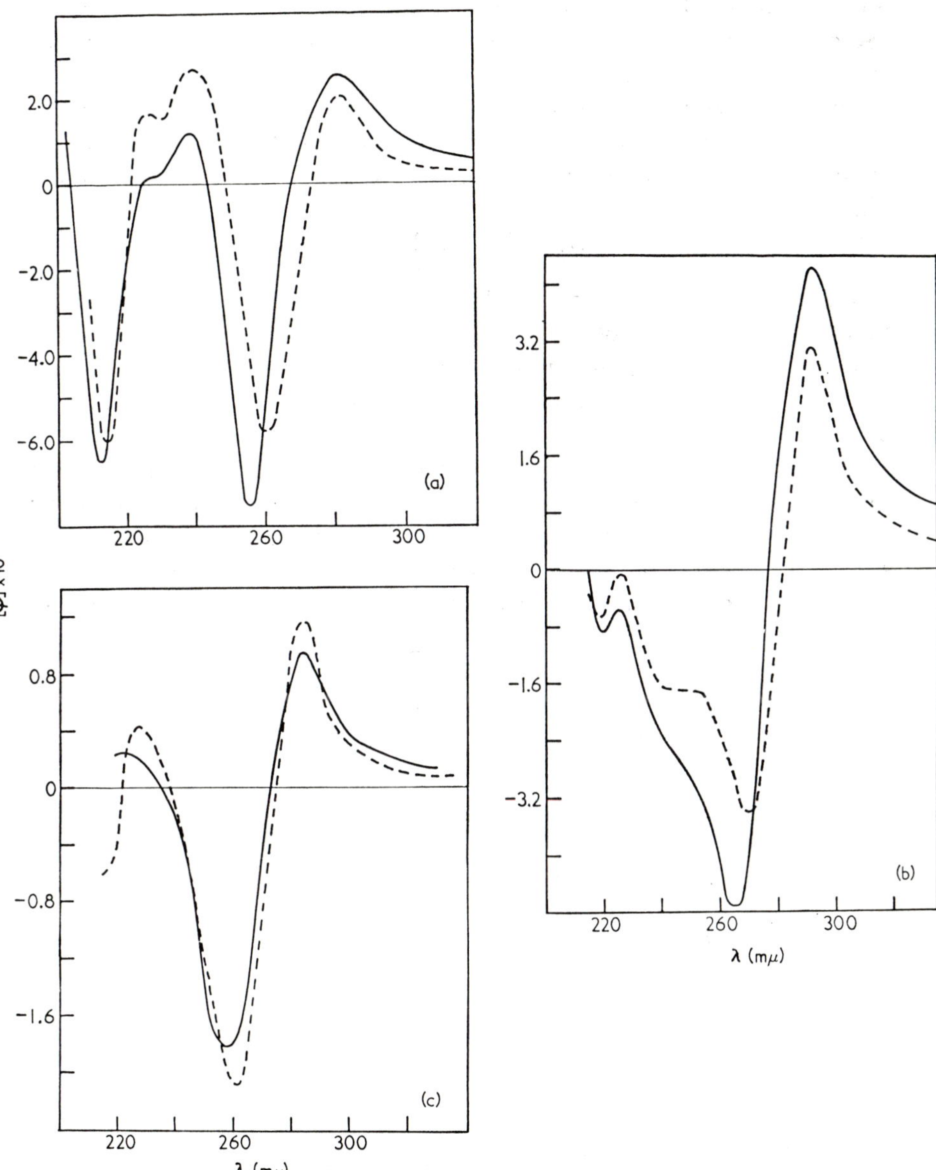


**Fig. 11-2.** (a) Optical rotatory dispersion of poly A at neutral pH. ——, Experimental (Holcomb and Tinoco, 1965); ----, nearest-neighbor calculation. (b) Optical rotatory dispersion of poly C at neutral pH. ——, Experimental; ----, nearest-neighbor calculation. (c) Optical rotatory dispersion of poly U at neutral pH. ——, Experimental (Sarkar and Yang, 1965a); ----, nearest-neighbor calculation (Cantor *et al.*, 1966).

homopolymers. In Fig. 11-2 we observe some of the results on the agreement between the actual and predicted rotation of polyadenylic, polycytidylic, and polyuridylic acid at neutral pH. Fair agreement is observed between the actual and computed ORD. In the case of the homopolymers the equation for the computation of the ORD is simplified from (11-5) to the following equation where we still use the same notation

$$[\phi(\lambda)] = 2[\phi_{ii}(\lambda)] - [\phi_i(\lambda)] \tag{11-9}$$

The agreement observed in Fig. 11-2 seems to indicate that the conformation of the bases in the homopolymers is similar to those in the dinucleoside phosphates, and it is supposed in such instances that these homopolymers are in single stranded conformation (see Chapter 9 for further discussion on the subject).

With respect to the application of the method of Cantor and Tinoco to actual RNA samples, comparison of the ORD of tobacco mosaic virus with a computation made on the basis of single-strandedness and a random sequence of bases yields results obtained in Fig. 11-3. Close agreement is found between the measured ORD and the computed ORD at neutral pH

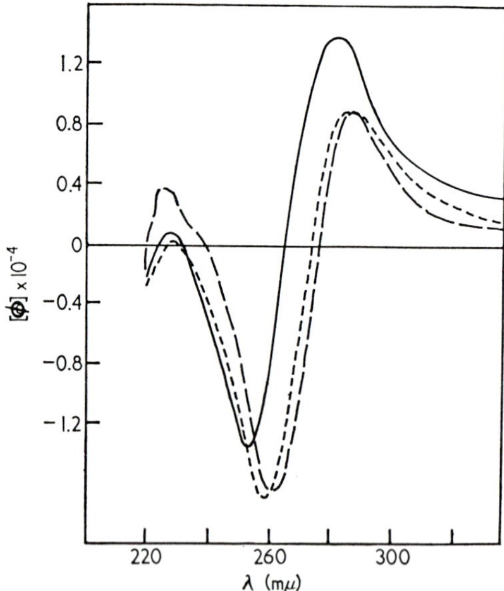


**Fig. 11-3.** Effect of salt concentration on the optical rotatory dispersion of tobacco mosaic virus RNA, at neutral pH. ——, Experimental results in the presence of 0.15 *M* salt; – – –, calculated by nearest-neighbor methods; ----, experimental results in the absence of salt (Cantor *et al.*, 1966).

and at low salt concentration where it is expected that tobacco mosaic virus is in a single stranded conformation (see Chapter 9).

Cantor *et al.* (1966) also attempted to formulate the properties of double stranded RNA molecules from the properties of oligomers. This work was prompted by the fact that at neutral pH and at moderate ionic strength a single stranded model of RNA is not a fair representation of the molecules and there is almost certainly interaction between adenine, uracil, guanine, and cytidine which have not been incorporated in the models now being discussed. To take into consideration the double stranded interactions, a number of equations were derived. First, an equation analogous to (11-5) which can be written as

$$[\phi^{\mathrm{D}}_{\mathrm{RNA}}(\lambda)] = 2 \sum_{ij}\sum \chi_{ij}[\phi^{\mathrm{D}}_{ij}(\lambda)] - \sum \chi_i[\phi_i{}^{\mathrm{D}}(\lambda)] \qquad (11\text{-}10)$$

where $\phi^{\mathrm{D}}_{\mathrm{RNA}}(\lambda)$ is the ORD of the double stranded RNA; $\phi^{\mathrm{D}}_{ij}(\lambda)$ the ORD of the double stranded dimer *ipj* and its antiparallel complement *jpi* or the ORD of a structure of the form

*jpi* Bases in one strand
↕ ↕
*ipj* Bases in complementary strand

$[\phi_i{}^{\mathrm{D}}(\lambda)]$ the ORD per residue of a single base pair, for instance, a structure such as

p*i*p
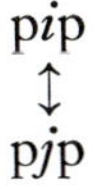
p*j*p

and $\chi_{ij}$ and $\chi_i$ are the mole fraction of the structure shown above. Lastly, the summation in Eq. (11-10) is taken over a single strand only. Various simplifying assumptions are now inserted in the formulations in order to make it workable, first, that the frequency of occurrence of nearest neighbors is random and second the base composition in the different strands is the same. Using those two assumptions the ORD of double stranded RNA could then be written as

$$[\phi^{\mathrm{D}}_{\mathrm{RNA}}(\lambda)] = \{[\phi_1(\lambda)]/2\} + \{[\phi_2(\lambda)]/2\} + 2\chi^2_{\mathrm{AU}}\,\Delta\phi_{\mathrm{AU}}(\lambda) + 2\chi^2_{\mathrm{GC}}\,\Delta\phi_{\mathrm{GC}}(\lambda) + 2\chi_{\mathrm{AU}}\chi_{\mathrm{GC}}\,\Delta\phi_{\mathrm{AUGC}}(\lambda) \qquad (11\text{-}11)$$

where $\Delta\phi_{\mathrm{AU}}(\lambda)$ is the change in ORD when an A–U pair is formed with its fellow pair in the opposite strand, $\Delta\phi_{\mathrm{GC}}(\lambda)$ is the change in ORD when a G–C pair interacts with another G–C pair in the other strand, and $\Delta\phi_{AUGC}(\lambda)$ is the average of the ORD of all other possible interactions. Lastly $\phi_1(\lambda)$ and $\phi_2(\lambda)$ are the molar rotations per residue of the two single strands. These two quantities are different because the sequences of the two strands are not the same although the base composition is. As pointed out by Cantor *et al.* the form of Eq. (11-11) is the same as that of Felsenfeld and Hirschman (1965) [Eq. (8-6)]. In order to use equation (11-11) to compute the ORD of RNA molecules in double stranded conformation estimates of the quantities $\Delta\phi_{AU}(\lambda)$, $\Delta\phi_{CG}(\lambda)$, and, if possible, $\Delta\phi_{AUGC}(\lambda)$, are needed. For $\Delta\phi_{AU}(\lambda)$ Cantor *et al.* took the data of Sarkar and Yang (1965a) on the ORD of poly A, poly U, and poly A–U and used the next equation to estimate $\Delta\phi_{AU}(\lambda)$

$$\Delta\phi_{AU}(\lambda) = [\phi_{\mathrm{poly\,AU}}(\lambda)] \tfrac{1}{2}\{[\phi_{\mathrm{poly\,A}}(\lambda)] + [\phi_{\mathrm{poly\,U}}(\lambda)]\} \qquad (11\text{-}12)$$

A similar equation was written for $\Delta\phi_{GC}(\lambda)$

$$\Delta\phi_{GC}(\lambda) = [\phi_{\mathrm{poly\,GC}}(\lambda)] \tfrac{1}{2}\{[\phi_{\mathrm{poly\,C}}(\lambda)] + [\phi_{\mathrm{poly\,G}}(\lambda)]\} \qquad (11\text{-}13)$$

Since there is no experimental estimate for single-stranded poly G consequently Eq. (11-9) was used to estimate its ORD. The measured rotation for poly G–C and poly C were taken from Sarkar and Yang (1965b). Since the authors could not obtain an experimental estimate of $\Delta\phi_{AUGC}(\lambda)$ (which consists of an average of eight possible interactions) they decided on a rough approximation for $\Delta\phi_{AUGC}(\lambda)$ stating that it would be an average of $\Delta\phi_{AU}(\lambda)$ and $\Delta\phi_{GC}(\lambda)$. This is conceded to be a rough approximation.

With these approximations in hand Cantor *et al.* (1966) proceeded to apply their results to the computation of the ORD of yeast alanine sRNA which was measured by Vournakis and Scheraga (1966). This RNA contains 77 bases, a number of which are absent from the formalism of this chapter (i.e., they are different from the four common bases). The calculation was still made with the following replacements: Inosine, methyl inosine, methyl guanine, and dimethyl guanine were replaced by guanine; pseudo-uracil and thymine were replaced by uracil; and dihydrouracil having no chromophore in the region of interest was assumed to make no contribution. With these replacements a calculation was made of the ORD of this RNA on the basis of a single-stranded conformation using Eq. (11-5). Also made was a calculation of the ORD of the structure shown in Fig. 7 of Cantor *et al.* (1966), the authors noted a number of double-stranded segments and a number of single-stranded segments. In this computation the appropriate equations for single strands and double strands were used. The results of the measured ORD and the computed ORD are shown in Fig. 11-5. There we notice much closer agreement for the computed ORD of the model in Fig. 7 than that found with a calculation based only on a single-stranded RNA. In view of the number of assumptions involved in the computation of the ORD of the model shown in Fig. 7, Cantor *et al.* have cautioned against wholehearted acceptance of the agreement between their computation and the experimental results.

Using the above reasoning Cantor, quoted by Beychok (1967), has computed the ORD of some of the possible conformations of 5S-ribosomal RNA whose sequence was determined by Brownlee *et al.* (1967) and the experimental ORD was found to correspond closely to the model in which all the bases are paired.

It will be observed that all equations in the above material are linear

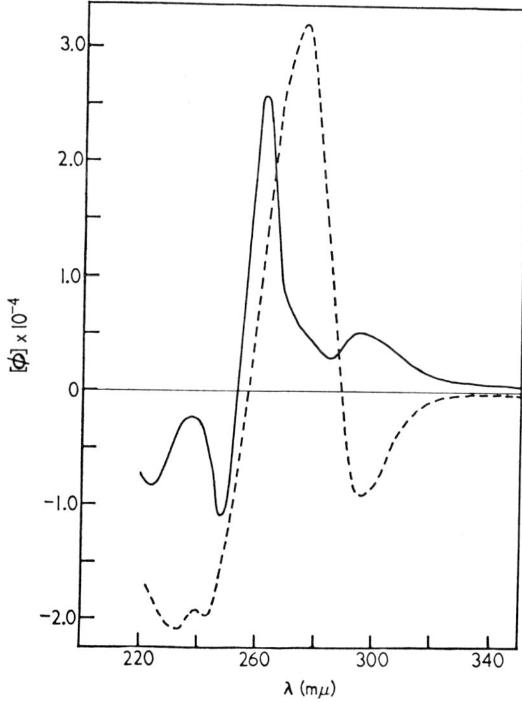


**Fig. 11-4.** ORD difference curves for the formation of base-paired double-stranded RNA from single-stranded RNA. Details are explained in the text. ——, Poly (A + U) − ½ (poly A + poly U); ----, poly (G + C) − ½ (poly C + poly G). Homopolymer spectra were obtained from Sarkar and Yang (1965a,b) (Cantor *et al.*, 1966).

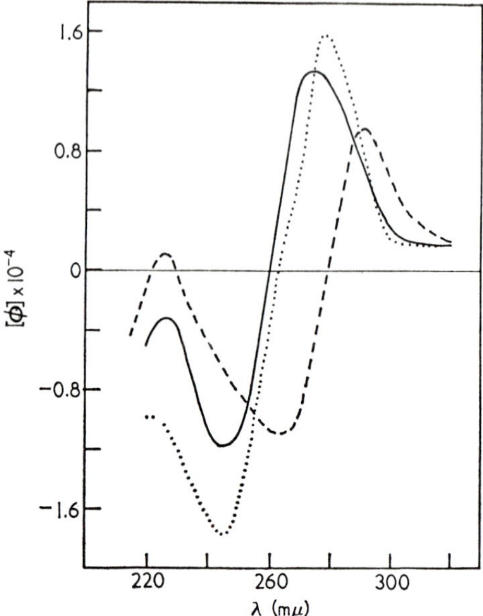


**Fig. 11-5.** Optial rotatory dispersion of the yeast alanine sRNA. ——, Experimental (Vournakis and Scheraga, 1966); ----, calculated ORD of single stranded RNA; · · ·, calculated ORD of the conformation shown in Fig. 7 of Cantor *et al.* (Cantor *et al.*, 1966).

equations and, as before, we can either use them to determine the mole fraction of the various constituents or alternatively knowing the mole fractions and sequence of dimers, we can, *at any given wavelength*, determine the pertinent contribution of any entity at that wavelength. The techniques of function minimization can be used for either purpose and all we need to do here is refer to the pertinent analyses in Chapters 4 and 6 for linear models in terms of parameter determination standard errors and confidence intervals on those parameters.

## 5. Testing Hypotheses about RNA Structure

Since Cantor and Tinoco have postulated that the optical properties of RNA in the single-stranded conformation (at neutral pH and low salt concentration) can be reproduced in terms of the optical properties of the dinucleoside phosphates, it might be worth our while to show how one would test the validity of this hypothesis.

There is no particular reason why we choose this hypothesis for testing, we merely use it as an example. Basically we have two equations, (11-5) and (11-6). If we wish to put Eq. (11-5) to a test we must know the sequence of the various dinucleotides in several RNA molecules. On the other hand a test of the hypothesis underlying Eq. (11-6) requires a random distribution of bases in several RNA molecules. Cantor *et al.* (1966) have stated that they felt that the nearest neighbor frequency of RNA is effectively random. As to whether or not this is the case we will not venture to comment, except to say that for any given RNA one can determine the nearest neighbor frequency and test the hypothesis of whether or not they happen to be random.

In the event that we choose to test the hypothesis of Cantor *et al.* (1966) embodied in Eq. (11-6) and we have a fair number of RNA samples whose nearest neighbor distribution is effectively random, we can proceed as follows. For any given wavelength, say $\lambda_s$, we can write Eq. (11-6) for $m$ RNA samples as follows

$$\begin{aligned} \phi_{\mathrm{RNA}}^{(1)}(\lambda_s) &= \sum_{i=1}^{4} \chi_i^{(1)} \left\{2 \sum_{j=1}^{4} \chi_j^{(1)} [\phi_{ij}(\lambda_s)] - (\phi_i(\lambda_s))]\right\} + \delta_1 \\ &\vdots \\ \phi_{\mathrm{RNA}}^{(m)}(\lambda_s) &= \sum_{i=1}^{4} \chi_i^{(m)} \left\{2 \sum_{j=1} \chi_j^{(m)} [\phi_{ij}(\lambda_s)] - [\phi_i(\lambda_s)]\right\} + \delta_m \end{aligned} \qquad (11\text{-}14)$$

where the superscripts indicate the RNA sample in question. In Eqs. (11-14) we note that the quantities $\phi_{\mathrm{RNA}}^{(k)}(\lambda_s)$, $\chi_i^{(k)}$, $\chi_j^{(k)}$ are all known since we can measure the ORD of the $k$th RNA sample and the mole fractions of its base composition. By suitably rearranging Eqs. (11-14) in order to bring it to a form linear in the $\phi_{ij}(\lambda_s)$ and $\phi_i(\lambda_s)$ and provided the various $\chi_i^{(k)}$ and $\chi_i^{(k)}\chi_j^{(k)}$ are linearly independent, we can solve these equations for the various $\phi_{ij}(\lambda_s)$ and $\phi_i(\lambda_s)$ by minimizing $\sum_{r=1}^{m} \delta_r^2$. Next, it will be noted that we actually *know* the values of $\phi_{ij}(\lambda_s)$ and $\phi_i(\lambda_s)$. Consequently we can proceed to test the hypothesis that the computed $\phi_{ij}(\lambda_s)$ and $\phi_i(\lambda_s)$ obtained from Eq. (11-14) are no different from the experimentally obtained values of those measurements. The procedure for doing this is explained in detail in Chapter 6 where we discuss the linear model of full rank and the testing of hypotheses about it.

For the test of Eq. (11-5) we have to analyze either RNA molecules or oligomers of known sequence. If such molecules are available the procedure outlined above can be easily adapted to these molecules; in such instances the only appreciable difference will be that $\chi_{ij}^{(k)} \neq \chi_i^{(k)}\chi_j^{(k)}$.

For the testing of the hypothesis of Cantor and Tinoco it must be emphasized that the $\phi_{ij}(\lambda_s)$ and $\phi_i(\lambda_s)$ must be computed by minimizing $\sum_{r=1}^{m} \delta_r^2$ and not $\sum_{r=1}^{m} | \delta_r |$ by linear programming or any other method.

In concluding this section on the ORD of RNA, the information we can obtain (provided the criteria of linear independence are satisfied) are the mole fractions of nucleosides or dinucleoside sequences, or if these are known then we could obtain the contribution of the various entities *as they occur in RNA* at any given wavelength. Finally, we can test whether or not the entities as they occur in RNA are any different than when they are separate. Clearly with some imagination we can devise a great many hypotheses and test each one of them. All that is required to generate the ORD of RNA is a linear combination of two entities, a single-stranded entity and a double-stranded entity with the ORD profiles given by the curves shown in Chapter 9. In this case the hypothesis states that all RNA molecules can be represented by an expression of the form

$$\phi_{\mathrm{RNA}}(\lambda) = \alpha_1 f_1(\lambda) + \alpha_2 f_2(\lambda) \tag{11-15}$$

where $f_1(\lambda)$ is the ORD of the single strand and $f_2(\lambda)$ is the ORD of the double strand.

## 6. Optical Rotatory Dispersion of DNA

As we pointed out earlier in this chapter those who followed the arguments on special topics related to the absorption of nucleic acids [based fundamentally on the work of Fresco *et al.* (1963) on RNA, and Felsenfeld and Sandeen (1962) and Felsenfeld and Hirschman (1965) on DNA] will be able to make applications, after making the necessary changes, to either the ORD or CD of DNA. There appears to be, at present, only one more topic on which function minimization can be used and that is the ORD spectra of DNA. In determining the base content of DNA from an ORD measurement Samejima and Yang (1965) discovered a linear relationship between the DNA peak at 290 mμ and the mole per cent of G + C for both the native DNA and heat denatured DNA (Fig. their article). These relationships can be expressed as

$$[\alpha]_{290} = \alpha_1 + \alpha_2 \chi \tag{11-16}$$

where $[\alpha]_{290}$ is the rotation at 290 mμ, and $\alpha_1$, $\alpha_2$ are constants, and $\chi$ is the mole % G + C. The values of $\alpha_1$ and $\alpha_2$ are 26.5 and 850 for the native DNA and 15.4 and 220 for heat denatured DNA. Inasmuch as the linear relationship shown in Eq. (11-16) does not hold for poly dAT (the copolymer of adenine and thymine) it is clear that Eq. (11-16) must only be used for certain values of $\chi$. From the graph of Samejima and Yang it appears that the region of validity is $25 \leq \chi \leq 70$ with the possibility that the higher values of $\chi$ may still fit the relationship. According to Yang† this relationship holds for several wavelengths and consequently we might be able to write for the ORD of DNA

$$\phi_{DNA}(\lambda) = \alpha_1 f_1(\lambda) + \alpha_2 f_2(\lambda) \tag{11-17}$$

where $\alpha_1$ and $\alpha_2$ are the mole % G + C and A + T, respectively, and $f_1(\lambda)$ and $f_2(\lambda)$ are the ORD contributions of G + C and A + T *as found in the DNA*. Thus $f_2(\lambda)$ for instance is not the ORD of dAT since dAT as shown by Samejima and Yang (1965) does not fit the model. Needless to say Eq. (11-17) can now be subjected to the techniques of function minimization to determine $\alpha_1$ and $\alpha_2$, but it must be emphasized that the model depicted by Eq. (11-17) may only hold for values of $\alpha_1$ and $\alpha_2$ in the middle of the interval of those parameters.

In concluding this chapter, various hypotheses put forward by Felsenfeld and his coworkers for the analysis of the spectra of DNA (see Chapter 8) can obviously be tested in a manner similar to that suggested in this chapter for testing the hypotheses put forward by Tinoco and his coworkers on RNA.

## References


Beychok, S. (1967). *Annu. Rev. Biochem.* **37**, 437.

Brownlee, G. G., Sanger, F., and Barrell, B. C. (1967). *Nature* **215**, 735.

Cantor, C. R., and Tinoco, I., Jr. (1965). *J. Mol. Biol.* **13**, 65.

Cantor, C. R., and Tinoco, I., Jr. (1967). *Biopolymers* **5**, 821.

Cantor, C. R., Jaskunas, S. R., and Tinoco, I., Jr. (1966). *J. Mol. Biol.* **20**, 39.

Felsenfeld, G., and Hirschman, S. Z. (1965). *J. Mol. Biol.* **13**, 407.

Felsenfeld, G., and Sandeen, G. (1962). *J. Mol. Biol.* **5**, 587.

Fresco, J. R., Klotz, L. C., and Richards, E. G. (1963). *Cold Spring Harbor Symp. Quant. Biol.* **28**, 83.

Hirschman, S. Z., and Felsenfeld, G. (1965). *J. Mol. Biol.* **13**, 407.

Holcomb, D. N., and Tinoco, I. (1965). Biopolymers **3**, 121.

Samejima, T., and Yang, J. T. (1965). *J. Biol. Chem.* **240**, 2094.

Sarkar, P. K., and Yang, J. T. (1965a). *J. Biol. Chem.* **240**, 2088.

Sarkar, P. K., and Yang, J. T. (1965b). *Biochemistry* **4**, 1238.

Vournakis, M., and Scheraga, H. A. *Abst. Annu. Meet. Biophys. Soc.*, **10***th.*, p. 54, 1966.

Warshaw, M. M., and Tinoco, I., Jr. (1965). *J. Mol. Biol.* **13**, 54.

Warshaw, M. M., and Tinoco, I., Jr. (1966). *J. Mol. Biol.* **20**, 29.

Warshaw, M. M., Bush, C. A., and Tinoco, I. (1965). *Biochem. Biophys. Res. Comm.* **18**, 633.


† Personal communication.

*CHAPTER 12*

# AGGREGATING SYSTEMS

## 1. Introduction

Aggregating systems occur with considerable frequency in biology. Proteins are known to form concentration dependent aggregates, as do purines, pyrimidines, chlorophylls, detergents, nucleotides and several other substances of interest to the biochemist. For references the reader is referred to the article by Adams (1967a). What we consider here is the use of the methods presented in Chapters 4–6 to analyze aggregating systems to determine the association constants of the various species, the molecular weight distribution functions, and the virial coefficients when nonideal systems are being considered. In this chapter we will consider the analysis of data obtained from osmotic coefficient measurements, sedimentation equilibrium light scattering, and gel filtration.

The determination of the association constants of a concentration-dependent aggregating system should yield the molecular weight distribution function, however, it will be found that in a number of cases we do not obtain unequivocal answers and the experimental data will be consistent with several sets of association constants. Such a state of affairs will lead us to consider molecular weight distribution functions and their determination.

## 2. Aggregation Schemes

The general scheme of aggregation which we consider and the equilibrium constants derived therefrom may be written thus

$$\begin{aligned} m_1 + m_1 &\rightleftharpoons m_2, & K_2 &= m_2/(m_1)^2 \\ m_2 + m_1 &\rightleftharpoons m_3, & K_3 &= m_3/(m_1)(m_2) \\ &\vdots & &\vdots \\ m_{n-1} + m_1 &\rightleftharpoons m_n, & K_n &= m_n/(m_1)(m_{n-1}) \end{aligned} \tag{12-1}$$

where $m_i$ is the appropriate concentration of the $i$-mer of the aggregating species. We also consider the aggregating scheme

$$nm_1 \leftrightarrows m_n \qquad (n = 2, 3, \ldots) \tag{12-2}$$

whose equilibrium constants may be written

$$k_2 = m_2/(m_1)^2, \qquad k_3 = m_3(m_1)^3, \qquad \ldots, \qquad k_n = m_n/(m_1)^n \tag{12-3}$$

The two sets of association constants can be related as follows:

$$k_1 = K_1, \qquad k_2 = K_1K_2, \qquad \ldots, \qquad k_n = K_1K_2 \cdots K_n$$

Other schemes of aggregation can also be considered, for example one might have a monomer-dimer-tetramer-hexamer scheme or a monomer-dimer-tetramer-octamer scheme and so on. Schemes of aggregation where there are no intermediates and where we simply get a monomer-polymer equilibrium are possible. In general (except possibly for electrolytes) it may be shown (Steiner, 1952; Adams, 1967a; Ts'o and Chan, 1964) that in the final analysis, most data from aggregating systems, whether ideal or not, can be analyzed in terms of polynomials in the monomer concentration which can be written as

$$f(x) = \alpha_0 + \alpha_1 x + \alpha_2 x^2 + \cdots + a_n x^n \tag{12-4}$$

The coefficients of this polynomial can be equated to the equilibrium constants. The independent variable $x$ represents a quantity dependent on the concentration.

## 3. Number Average Molecular Weights: Ideal Systems

The best biochemical example of aggregating systems employing the use of number average molecular weights is the aggregation of the purines and pyrimidines (Ts'o and Chan, 1964; Chan *et al.*, 1964; Ts'o *et al.*, 1963). Ts'o and Chan (1964) and Kreuzer (1943) have conducted the following analysis for those systems. Using the Gibbs–Duhem relationship they related the molal activity coefficient $\gamma$ to the osmotic coefficient $\phi$ by the following equation:

$$\ln \gamma = (\phi - 1) + \int_0^m (\phi - 1)\, d \ln m \tag{12-5}$$

where $m$ is the molality of the solution. The above authors also showed that

$$\ln(m_1/m) = (\phi - 1) + \int_0^m (\phi - 1)\, d \ln m \tag{12-6}$$

where $m_1$ is the monomer concentration. From Eq. (12-5) and (12-6) we recognize the obvious result that

$$(m_1/m) = \gamma \tag{12-7}$$

In addition, if we note that

$$m = m_1 + 2K_2(m_1)^2 + 3K_2K_3(m_1)^3 + \cdots + nK_2K_3 \cdots K_n(m_1)^n \tag{12-8}$$

then we can divide Eq. (12-8) by $m_1$ to obtain

$$(m/m_1) = 1 + 2K_2m_1 + 3K_2K_3(m_1)^2 + \cdots + nK_2K_3 \cdots K_n(m_1)^{n-1} \tag{12-9}$$

By measuring $\phi$ and representing it as a polynomial in $m$, $\gamma$ can be computed from Eq. (12-5) (Ts'o and Chan, 1964). Representing $\phi$ as a polynomial in $m$ is a device to facilitate the integration required in Eq. (12-5), which employs a technique of function minimization. In this case we obtain the best polynomial fit of $\phi$ in terms of $m$. This was done by using least squares (Ts'o and Chan, 1964). Since we are just fitting data where no physical significance is attached to the parameter, the unrestricted least squares is most suitable. At the end of this chapter more will be said about such data processing.

Knowledge of $m$ and $\gamma$ determines $m_1$ when we make use of Eq. (12-7). When $m_1$ is obtained the only unknowns in Eq. (12-9) are the equilibrium constants and those we determine by the methods of chapter 4. In order to determine the equilibrium constants, Ts'o and Chan (1964) had to make certain assumptions about them. One assumption was $K_1 = K_2 = \cdots = K_n$; another assumption was that only dimers were formed, i.e., $K_3 = K_4 = \cdots = K_n = 0$. Our analysis will be done in terms of different equilibrium constants.

Another, but formally equivalent way of obtaining data from number average molecular weights is due to Steiner (1954, 1968). Steiner derived the following equation in the notation† of the 1968 paper

$$\ln X_{\mathrm{A}} = \int_0^m [(\alpha_n^{-1} - 1)/m]\, dm \tag{12-10}$$

where $X_{\mathrm{A}}$ is the fraction of monomer units, $m$ the total molar concentration for all species, $\alpha_n = M_n/M_{\mathrm{A}}$ the number average degree of association, $M_{\mathrm{A}}$ is the molecular weight of the monomer, and $M_n$ is the number average molecular weight. If the total weight concentration is $c$ then $m = c/M_n$. If we note that

$$m/m_1 = 1 + k_2 m_1 + k_3 m_1{}^2 + \cdots + k_n m_1^{n-1} \tag{12-11}$$

then a series of measurement of $m$ and $m_1$ [which is obtained from Eq. (12-10)] at a given concentration will give us a set of equations like (12-11) which can be solved to determine the association constants. For example, in Eq. (12-9) we can write

$$\begin{aligned} m(c_j)/m_1(c_j) = 1 + 2K_2 m_1(c_j) + 3K_2K_3 m_1{}^2(c_j) + \cdots + nK_2K_3 \\ \cdots k_n m_1^{n-1}(c_j) + \delta_j \qquad (j = 1, 2, \ldots, m) \end{aligned} \tag{12-12}$$

where once again $\delta_1, \delta_2, \ldots, \delta_j, \ldots, \delta_m$ are the errors made in each measurement. To actually compute some of the association constants we use the methods given in Chapter 4 so as to minimize some function of the experimental errors.

Since the association constants must be positive the procedure outlined in Chapter 4 in the application section case must be followed. The weighted least squares or weighted least squares norm is the method of choice if it yields positive coefficients. When it does not yield positive coefficients the method of steepest descent, the Nelder and Mead me-

† Note that $m$ as used by Steiner is different from $m$ as used by T'so and Chan. Also note that $\alpha_n^{-1}$ is equal to the osmotic coefficient.

thod, or linear programming may be attempted. The method for the estimation of the confidence intervals for the association constants obtained from some of the above methods are given in Chapter 6 as well as the statistical testing of hypotheses about the parameters.

The question of how many terms one uses in Eq. (12-12), and similar equations is of great importance. When we are dealing with an aggregating system the chances are that this systems does not aggregate beyond an $n$-mer. In other cases, while the possibility exists of indefinite aggregation, the larger polymers will, in general, exist in insignificant concentration. Consequently, we would like to know how many terms of each of the polynomials shown in the above equations are needed, since Weierstrass's famous approximation theorem (Chapter 3) tells us that the higher the order of the polynomial the closer the fit. Finding out how many terms of a polynomial we need is tantamount to determining the highest aggregating species (in the case of proteins) and the highest aggregating species of significant concentration in the case of a system having indefinite aggregation. In order to do so we must apply certain statistical tests.

*Example* (*Statistical Testing of Hypothesis about Parameters*). To show how the statistical tests work in practice we shall analyze the aggregation of bromouridine, uridine, and cytidine. The least squares or $L_2$ norm approximation was used in all cases. The data for cytidine and uridine are taken from Ts'o *et al.* (1963) and the data for bromouridine are taken from Ts'o and Chan (1964) and are tabulated in Table 12-1. All computations were done on an IBM 1620 computer.

In Chapter 6 we gave the general scheme for the analysis of variance for testing computed coefficients. In what follows we compute the coefficients of polynomials of increasing degree and test for the significance (whether the coefficient is significantly different from zero) of the newest coefficient. Our computations were based on polynomial fits of the data and it is quite possible that orthogonal functions may do a better fitting job.

The coefficients from which the equilibrium constants are to be calculated for our first example, bromouridine, show that for the quadratic fit we have

$$\alpha_1 = 1.7852, \qquad \alpha_2 = 14.8231$$

The analysis of variance for those coefficients is shown in Table 12-2a A value of $F$ of 127.4 is now challenged by comparing it with tabulated

**Table 12-1**

| 5 bromouridine | | Uridine | | Cytidine | |
|---|---|---|---|---|---|
| $m_1$ | $m/m_1$ | $m_1$ | $m/m_1$ | $m_1$ | $m/m_1$ |
| 0.1098 | 1.367 | 0.1268 | 1.18 | 0.1236 | 1.21 |
| 0.1332 | 1.50 | 0.1616 | 1.23 | 0.1552 | 1.29 |
| 0.1533 | 1.63 | 0.1938 | 1.29 | 0.1833 | 1.37 |
| 0.1707 | 1.75 | 0.2232 | 1.35 | 0.2085 | 1.44 |
| 0.1866 | 1.88 | 0.2506 | 1.40 | 0.2314 | 1.51 |
| 0.2008 | 1.92 | 0.2760 | 1.45 | 0.2524 | 1.59 |
| | | 0.2993 | 1.50 | 0.2718 | 1.65 |
| | | 0.3205 | 1.56 | 0.2900 | 1.72 |
| | | 0.3410 | 1.61 | 0.3069 | 1.79 |
| | | 0.3600 | 1.67 | 0.3222 | 1.86 |
| | | 0.3783 | 1.72 | 0.3367 | 1.93 |
| | | 0.3976 | 1.76 | 0.3493 | 2.01 |

values of $F$ for the following degrees of freedom $\nu_1 = 6$, $\nu_2 = 1$. In view of the fact that the 0.01 level of significance is 13.745 for $\nu_1 = 6$ and $\nu_2 = 1$, then we conclude that $\alpha_2$ is certainly significant since $127.4 > 13.745$ and the highest aggregating species of bromouridine is at least a trimer.

The next polynomial is a cubic. On carrying out the computation we obtain

$$\alpha_1 = 0.970344, \qquad \alpha_2 = 26.94895, \qquad \alpha_3 = -41.67380$$

**Table 12-2a**

| SV | *DF* | *SS* | *MS* | *F* |
|---|---|---|---|---|
| Total | 8 | 3.0299 | | |
| Due to $\alpha_1\alpha_2$ | 2 | 3.0272 | | |
| Due to $\alpha_1$ (unadjusted) | 1 | 2.9826 | | |
| Due to $\alpha_2$ (adjusted) | 1 | 0.0446 | 0.0446 | 127.4 |
| Error | 6 | 0.0027 | 0.00035 | |

**Table 12-2b**

| SV | *DF* | *SS* | *MS* | *F* |
|---|---|---|---|---|
| Total | 8 | 3.0299 | | |
| Due to $\alpha_1\alpha_2\alpha_3$ | 3 | 3.0279 | | |
| Due to $\alpha_1\alpha_2$ (unadjusted) | 2 | 3.0272 | | |
| Due to $\alpha_3$ (adjusted) | 1 | 0.0007 | 0.0007 | 1.75 |
| Error | 5 | 0.00200 | 0.0004 | |

The analysis of variance for this set is shown in Table 12-2b. A value of $F$ of 1.75 at $\nu_1 = 5$ is well below the 0.01 significance level and well below even the 0.05 significance level. Consequently, $\alpha_3$ can be considered insignificant. Therefore the data from the aggregation of bromouridine is quite consistent with the fact that the highest aggregating species is a trimer. In the above computation we notice the appearance of the negative coefficient $\alpha_3$. Since we determined that a cubic is not necessary according to our statistical criteria, then the fact that a negative coefficient has appeared should not deter us. The equilibrium constants for the aggregation of bromouridine are $K_2 = 0.89$ and $K_3 = 5.54$.

The next example is uridine. The first computation yields

$$\alpha_1 = 1.0979, \qquad \alpha_2 = 2.0541$$

The analysis of variance for this set is shown in Table 12-3a. This value of $F$ for degrees of freedom $\nu_1 = 12$ and $\nu_2 = 1$ is significant at the 0.01 percent level. Thus uridine is at least a trimer for the next set we have.

$$\alpha_1 = 11.31937, \qquad \alpha_2 = 0.33699, \qquad \alpha_3 = 3.05516$$

**Table 12-3a**

| SV | *DF* | *SS* | *MS* | *F* |
|---|---|---|---|---|
| Total | 14 | 3.15485 | | |
| Due to $\alpha_1\alpha_2$ | 2 | 3.15410 | | |
| Due to $\alpha_1$ (unadjusted) | 1 | 3.13211 | | |
| Due to $\alpha_2$ (unadjusted) | 1 | 0.02199 | 0.02199 | 366.5 |
| Error | 12 | 0.00075 | 0.00006 | |

**Table 12-3b**

| SV | *DF* | *SS* | *MS* | *F* |
|---|---|---|---|---|
| Total | 14 | 3.15485 | | |
| Due to $\alpha_1\alpha_2\alpha_3$ | 3 | 3.15448 | | |
| Due to $\alpha_1\alpha_2$ (unadjusted) | 2 | 3.15410 | | |
| Due to $\alpha_3$ (adjusted) | 1 | 0.00038 | 0.00038 | 11.2 |
| Error | 11 | 0.00037 | 0.000034 | |

The analysis of variance for this set yields Table 12-3b. This value of $F$ is insignificant at the 0.01 level but is significant at the 0.05 level. If we set our acceptance level at 0.01 then the highest aggregating species of uridine is a trimer. On the other hand, if we set the level at 0.05 percent, then we will have to reject this result and say that highest aggregating species is at least a tetramer. Inasmuch as this particular example is interesting, we will carry our analysis a step further. For the fourth degree fit we obtain

$$\alpha_1 = 1.46726, \qquad \alpha_2 = -1.72725, \qquad \alpha_3 = 11.59822, \qquad \alpha_4 = -10.88917$$

The analysis of variance is shown in Table 12-3c. This result shows that $\alpha_4$ is insignificant at the 0.05 level and consequently also at the 0.01 level. The results for uridine can be summarized as follows: If the level of significance is chosen as 0.01, then the highest aggregating species of uridine is a trimer and the equilibrium constants are $K_2 = 0.55$ and $K_3 = 1.25$. On the other hand, if we wish to be more fastidious and choose the level of significance at 0.05, then the highest aggregating species is a tetramer and the equilibrium constants are $K_2 = 0.66$, $K_3 = 0.17$, and $K_4 = 6.80$.

**Table 12-3c**

| SV | *DF* | *SS* | *MS* | *F* |
|---|---|---|---|---|
| Total | 14 | 3.15485 | | |
| Due to $\alpha_1\alpha_2\alpha_3\alpha_4$ | 4 | 3.15453 | | |
| Due to $\alpha_1\alpha_2\alpha_3$ (uadjusted) | 3 | 3.15448 | | |
| Due to $\alpha_4$ (adjusted) | 1 | 0.00005 | 0.00005 | 1.56 |
| Error | 10 | 0.00032 | 0.000032 | |

**Table 12-4a**

| SV | *DF* | *SS* | *MS* | *F* |
|---|---|---|---|---|
| Total | 14 | 5.280724 | | |
| Due to $\alpha_1\alpha_2$ | 2 | 5.279540 | | |
| Due to $\alpha_1$ (unadjusted) | 1 | 5.201100 | | |
| Due to $\alpha_2$ (adjusted) | 1 | 0.078440 | 0.078440 | 8000.4 |
| Error | 12 | 0.001184 | 0.000098 | |

The next example is the aggregation of cytidine. The quadratic fit yields the following coefficients

$$\alpha_1 = 1.05384, \qquad \alpha_2 = 5.06487$$

The analysis of variance is shown in Table 12-4a. This value of $F$ shows that $\alpha_2$ is significant at the 0.01 level. For the cubic fit we have

$$\alpha_1 = 1.39759, \qquad \alpha_2 = 2.07859, \qquad \alpha_3 = 5.97775$$

For the analysis of variance we have Table 12-4b. This value of $F$ shows $\alpha_3$ is significant at the 0.01 level and cytidine is at least a tetramer. The fourth degree fit yields

$$\alpha_1 = 0.982396, \quad \alpha_2 = 8.582916, \quad \alpha_3 = -24.373079, \quad \alpha_4 = 43.725427$$

The analysis of variance for these coefficients is shown in Table 12-4c. This value of $F$ shows that $\alpha_4$ is significant at the 0.01 level and we may conclude that the highest aggregating species is a tetramer and the equilibrium constants are $K_2 = 0.70$, $K_3 = 1.00$, and $K_4 = 2.17$.

**Table 12-4b**

| SV | *DF* | *SS* | *MS* | *F* |
|---|---|---|---|---|
| Total | 14 | 5.280724 | | |
| Due to $\alpha_1\alpha_2\alpha_3$ | 3 | 5.280190 | | |
| Due to $\alpha_1\alpha_2$ (unadjusted) | 2 | 5.27954 | | |
| Due to $\alpha_3$ (adjusted) | 1 | 0.000650 | 0.000650 | 135.4 |
| Error | 11 | 0.000534 | 0.0000485 | |

**Table 12-4c**

| SV | *DF* | *SS* | *MS* | *F* |
|---|---|---|---|---|
| Total | 14 | 5.280724 | | |
| Due to $\alpha_1\alpha_2\alpha_3\alpha_4$ | 4 | 5.280424 | | |
| Due to $\alpha_1\alpha_2\alpha_3$ (unadjusted) | 3 | 5.280190 | | |
| Due to $\alpha_4$ (adjusted) | 1 | 0.000234 | 0.000234 | 7.8 |
| Error | 10 | 0.000300 | 0.000030 | |

In Fig. 12-1 we note how well our computations fit the actual data. (For other relevant discussion on the above examples the reader is referred to Magar, 1969.) In the above examples we concentrated on determining what the highest species of significant concentration is. Actually, statistical testing can be used in connection with other hypotheses.

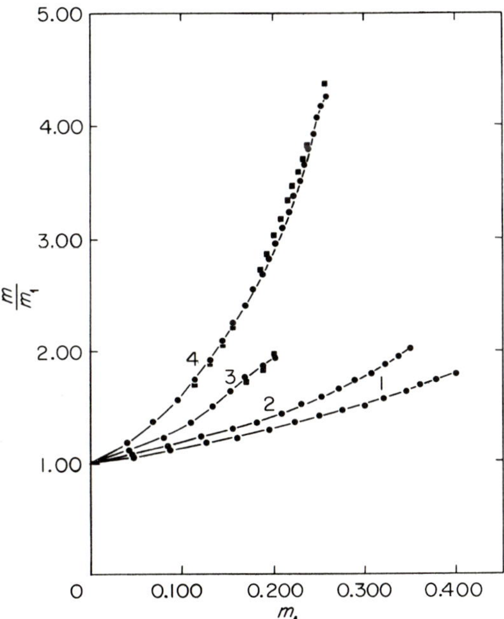


**Fig. 12-1.** The experimental points are shown as round dots and the computed points, when they are distinguishable from the experimental points, are shown as squares. On this scale the experimental points and the computed points for uridine (curve 1) and cytidine (curve 2) are indistinguishable. Curve 3 is bromouridine and curve 4 is purine (Magar, 1969).

For instance, it may well be that the aggregating scheme is not a monomer-dimer-trimer-$\cdots$ scheme but rather a monomer-dimer-tetramer-hexamer- $\cdots$ scheme. To determine whether or not that happens to be the case we could test for the statistical significance of $\alpha_2$. This is done by making our regular fits with $\alpha_2$ included and then by making another fit with $\alpha_2$ excluded, that is to say with a polynomial given by the next equation

$$f(x) = \alpha_0 + \alpha_1 x + \alpha_3 x^3 + \cdots \qquad (12\text{-}13)$$

An analysis of variance is then conducted to determine whether or not $\alpha_2$ is statistically significant.

*Example.* For this example we refer the reader to Magar and Steiner (1970). In this paper some of the data of Solie and Schellman (1968) on purine riboside and deoxyadenosine are analyzed. There an attempt was made to distinguish between models where the association constants are different and models where

$$K_2 = K_3 = K_n = K \qquad (12\text{-}14)$$

## 4. Weight Average Molecular Weights

### 4.1 Ultracentrifugation

The weight average molecular weight is obtained from sedimentation equilibrium studies in the ultracentrifuge. This quantity is also available from light scattering. In general the definition of the various molecular weight averages is given by the following equations

$$\begin{aligned} M_n &= \sum_i (n_i M_i / n_i), & M_w &= \sum_i (n_i M_i^2 / n_i M_i) \\ M_z &= \sum_i (n_i M_i^3 / n_i M_i^2), & M_{z+1} &= \sum_i (n_i M_i^4 / n_i M_i^3) \end{aligned} \qquad (12\text{-}15)$$

where $M_i$ is the molecular weight of the $i$-mer and $n_i$ the number of molecules of the $i$-mer.

#### 4.1.1 *Ideal Systems*

From studies in the ultracentrifuge, assuming the system is ideal in order to characterize the association by determining the constants, we use a treatment due Steiner (1952). Steiner starts by showing that the weight fraction of monomer $\varrho$ can be written as

$$\ln \varrho = \int_0^c \{[(DP)^{-1} - 1]/c\}\, dc \tag{12-16}$$

where $DP$ is the degree of polymerization and $c$ is the concentration of the protein or macromolecule. Steiner (1952) went on to derive the following equation

$$(DP/\varrho) = 1 + 2K_2(c\varrho) + 3K_2K_3(c\varrho)^2 + \cdots + nK_2K_3 \cdots K_n(c\varrho)^{n-1} \tag{12-17}$$

Equation (12-17) could then be used to determine the equilibrium constants. The state of computational aid in 1952 was such that the method by which Steiner went about determining the equilibrium constant left much to be desired. Essentially $(DP/\varrho) - 1$ was plotted against $\varrho c$ and the limiting slope as $c \to 0$ determined $2K_2$. By plotting $(DP/\varrho) - 1 - 2K_2\varrho c$ vs. $(\varrho c)^2$ one went on to obtain $3K_2K_3$ from the limiting slope and so on. Such a process is essentially one of numerical differentiation, with its inherent sources of error compounded as the process is repeated. In addition, since experimental inaccuracies occur at low concentrations, the first $K_i$ (from which subsequent ones are to be determine) are particularly inaccurate.

To over come some of these problems Adams (1967b) proposed the use of the quantity $\sum m_iM_1^2$ to analyze the system. For an ideal system he wrote

$$\sum_i (m_iM_i^2)/M_1^2 = cM_w(c)M_z(c)/M_1^2 = \sum_i i^2k_im_1^i = -\psi \tag{12-18}$$

where $M_i$ is the molecular weight of the $i$-mer and $M_w$ is the weight average molecular weight and $M_z$ the $z$-average molecular weight. The notation in Eq. (12-18) emphasizes that both $M_w$ and $M_z$ are functions of concentration. The important thing to recognize in the formulation of Steiner and Adams is that Eqs. (12-17) and (12-18) are all of the same form as equation (12-8), in the sense that Eq. (12-17), represents $(DP/\varrho)$ as a polynomial in $c\varrho$ and Eq. (12-18) represents the quantity $\psi$ in terms of a polynomial in $m_1$. As such they can be analyzed in a manner identical to our analysis on monouridine, uridine, and cylidine. For example we may write for Eq. (12-17), where the quantity $\varrho$ is evaluated at each concentration, the following $m$ equations

$$\begin{aligned} DP(c_j)/\varrho(c_j) = {} & 1 + 2K_2\varrho(c_j)c_j + 3K_2K_3[\varrho(c_j)(c_j)]^2 + \cdots \\ & + nK_2K_3K_n[\varrho(c_j)(c_j)]^{n-1} + \delta_j \qquad (i = 1, 2, \ldots, m) \end{aligned} \tag{12-19}$$

Next, we may write Eq. (12-18), the ideal case presented by Adams (1967b), as the following equations

$$\psi(c_j) = \frac{1}{2M_1^2} \frac{d}{dc} [c_j M_w(c_j)]^2 = \frac{c_j M_w(c_j) M_z(c_j)}{M_1^2}$$
$$= \sum_{i=1}^{n} i^2 k_i m_1^i(c_j) + \delta_j \qquad (12\text{-}20)$$

The above examples do not exhaust the possibilities of our analyses.

A number of authors have looked at data from nonideal systems and also at data processing methods (Van Holde *et al.*, 1969; Teller *et al.*, 1969; Albright and Williams, 1968; Trautman, 1969; and others). We cannot go through all the contribution of these authors and the reader is referred to the articles on advances in ultracentrifugal analysis in *Ann. N. Y. Acad. Sci.* **164** (1969), for a complete report on this research. Here we treat only a few examples. Many of the analyses presented in that volume can, by suitable rearrangement of the equations, be formulated in a manner amenable to the methods presented in this book.

4.1.2. *Nonideal Systems.*

For nonideal systems certain special considerations must be taken into account in order to carry out meaningful analysis. Adams, Fujita, and Williams (Adams and Fujita, 1963; Adams and Williams, 1964; Adams, 1956a, 1965b) have made a special assumption about these aggregating systems. They assumed that the logarithm of the activity coefficient $y_i$ of each species $i$ can be represented by

$$\ln y_i = iBM_1c + O(c^2) \qquad (12\text{-}21)$$

where $B$ is a virial coefficient and $O(c^2)$ means term of order $c^2$ which are neglected. There is no question about our encountering nonideal systems in practice. The data of Van Holde and Rossetti (1966) on purine and of Van Holde and Rossetti (1967) on adenosine monophosphate indicate definite deviations from ideality. Albright and Williams (1968) working with $\beta$-lactogloulin have also shown deviation from ideality. As shown by Adams (1967a) whenever Eq. (12-21) holds then one can also write, at any concentration, the following

$$M_1/M_{w\text{app}} = (M_1/M_w) + BM_1c \qquad (12\text{-}22)$$

where $M_{w\text{app}}$ is the measured weight average molecular weight and $M_w$ the true weight average molecular weight. Steiner's equation for the nonideal case becomes

$$\ln \varrho_{\text{app}} = \int_0^c \frac{(DP_{\text{app}})^{-1} - 1}{c} \, dc = \ln \varrho + BM_1c \tag{12-23}$$

where $\varrho_{\text{app}}$ is the apparent weight fraction of monomer and $DP_{\text{app}}$ is the apparent degree of polymerization. With these modifications Adams derived the following equation for nonideal systems

$$\frac{d[(M_1/cM_{w\text{app}})]/dc}{[(M_1/cM_{w\text{app}}) - BM_1]^3} = \sum_i i^2K_im_1{}^i = \zeta \tag{12-24}$$

Before proceeding with the analysis of these polynomials to determine the association constants it is necessary to treat the nonideal case to determine $BM_1$ since the analysis for the determination of the association constants can not (if we want our equation to be linear) proceed in absence of the determination of this quantity. Another matter which we discuss is the case considered by Albright and Williams (1968) where more than one virial coefficient enters the nonideal treatment.

#### 4.1.3 *Estimation of Virial Coefficients*

It appears from the work of Adams that the equations containing the virial coefficients differ depending on the mode of aggregation and one cannot write general equations covering all cases. Thus, if we have a monomer-dimer equilibrium we will have to solve an equation for $BM_1$ which is different from the equation one would have to solve for the monomer-dimer-trimer case. It is different in the sense that you cannot obtain the latter case from the former simply by adding another term as one would in most of the previous equations. The fact that we cannot generalize, however, is a hinderance only insofar as it limits our illustrative example. For illustration we have chosen the monomer-dimer-trimer case, but we must emphasize that whenever the equation for a more complicated case is derived it can be analyzed to determine $BM_1$ in precisely the same manner as the monomer-dimer-trimer case. For this case Adams wrote the following equation

$$\frac{6cM_1}{M_{n\text{app}}} - 5c = 2\varrho_{\text{app}} \exp(-BM_1c) + 3BM_1c^2 - \frac{1}{[(M_1/M_{w\text{app}}) - BM_1]} \tag{12-25}$$

In Eq. (12-25) the only unknown is $BM_1$. The apparent number average molecular weight $M_{n\text{app}}$ is obtained at any specific concentration $c_\alpha$ by

$$\int_0^{c_\alpha} (M_1/M_{w\text{app}})\, dc_\alpha = c_\alpha M_1/M_{n\text{app}} \tag{12-26}$$

Albright and Williams (1968), in analyzing a monomer-dimer equilibrium, wrote the following equations for the logarithm of the activity coefficient of the monomer and dimer

$$\begin{aligned} \ln y_1 &= B_1 M_1 c + B_2 M_1 c^2 + \cdots \\ \ln y_2 &= 2B_1 M_1 c + 2B_2 M_1 c^2 + \cdots \end{aligned} \tag{12-27}$$

where $B_2$ is another virial coefficient. Albright and Williams went on to derive the following equation

$$2c = \frac{1}{(M_1/cM_{w\text{app}}) - (\sigma + B_2 M_1 c)} + c\varrho_{\text{app}} \exp(-\sigma c) \tag{12-28}$$

where

$$\sigma = B_1 M_1 + B_2 M_1 c. \tag{12-29}$$

Further, by considering the equation

$$B_2 M_1 c = \frac{(2cM_1/M_{n\text{app}}) - c - c\varrho_{\text{app}} - c^2\sigma}{c^2/3} \tag{12-30}$$

and substituting this value of $B_2 M_1 c$ in Eq. (12-21) they obtained an equation in a single unknown $\sigma$. The respective equations (12-25) and (12-28) were solved by successive approximations† to determine $BM_1$ in the cases analyzed by Adams and to determine $\sigma$ in the cases analyzed by Albright and Williams. In such a procedure Adams uses data from a single concentration only to evaluate $BM_1$. Albright and Williams use the same procedure to determine $\sigma$, again using data from a single concentration. Since $\sigma$ is concentration dependent, solving for $\sigma$ at different concentrations will yield different values of $\sigma$. From Eq. (12-29) we note that the intercept and slope of a plot of $\sigma$ against concentration yield $B_1 M_1$ and $B_2 M_1$ respectively.

As an alternative procedure to successive approximation we advocate the use of some of the nonlinear methods described in Chapter 4 to determine the virial coefficients. The advantage of such a procedure is that *all* data points from the equilibrium curve are used to determine the constants $BM_1$ or $\sigma$ and not just data taken at a single concentration. It is evident, from a statistical point of view, that any parameter determined

† Actually Newton's method for finding roots could have been used.

from a large set of data is more reliable than a parameter determined from a single observation.

If an investigator feels that accuracy is being achieved only in a certain part of the equilibrium curve, then the investigator need use only that part of the curve to compute the virial coefficient. The important thing here is to use as many accurate points as possible to determine the parameter in question. Another advantage of using the methods described in Chapter 4 is that it enables us to determine the confidence interval on the estimated parameter, as shown in Chapter 6.

With respect to the case treated by Albright and Williams we note that we can use nonlinear regression to estimate both $B_1M_1$ and $B_2M_1$ from Eq. (12-28) directly without having to reduce our equation to an equation containing single unknown.

$$2c = \frac{1}{(M_1/cM_{w\text{app}}) - (B_1M_1 + 2B_2M_1c^2)} + c\varrho_{\text{app}} \exp[-(B_1M_1c + B_2M_1c^2)] \tag{12-31}$$

The advantage of such a procedure over that of Albright and Williams is that the entire data set is used at once to determine both $B_1M_1$ and $B_2M_1$ instead of a point by point estimation of $\sigma$. In determining $\sigma$ at any specific point we, in effect, make no allowances for the minimization of the errors, whereas the nonlinear methods described in Chapter 4 are designed specifically to minimize the errors. Using nonlinear regression also allows us to estimate the confidence regions for the parameters $B_1M_1$ and $B_2M_1$. Finally, the use of Eq. (12-31) instead of Eq. (12-30) saves us the trouble of evaluating the integral $\int_0^{c_\alpha} (M_1/M_{w\text{app}})\, dc$ to obtain $M_{n\text{app}}$ since we use $M_{w\text{app}}$ directly in Eq. (12-31). It is pertinent to point out that Albright and Williams (1968), in investigating $\beta$-Lactoglobulin, have found that if they use an equation analogous to their Eq. (23) containing a single virial coefficient, then this parameter, determined at different concentrations, was not constant. This led them to consider equations having more than one virial coefficient. Such considerations give rise to the following important question: Inasmuch as the use of nonlinear regression can yield a single virial coefficient averaged over all the data, do we really need only one virial coefficient or do we need more than one? In the computation performed by Albright and Williams (1968) to obtain a single virial coefficient from an equation analogous to Eq. (23), it is not surprising that different values of the virial coefficient were obtained at different concentrations when successive approximations were used to solve the equation. The question is, if we had solved their equation by nonlinear regression, would the single value of $BM_1$ which we would have obtained be sufficient to describe the data or would we have needed more. To answer this question we should solve, using nonlinear regression, both the equation containing a single virial coefficient and the equation containing two virial coefficients and apply the $F$-test to determine whether the extra virial coefficient is necessary. The difficulties with the $F$-test in nonlinear regression have been taken up in Chapter 6 where it was stated that it is after all better than a naked eye comparison.

We regret that in this section we only treated the case presented by Adams for a monomer-dimer-trimer using one virial coefficient and the case presented by Albright and Williams for a monomer-trimer equilibrium using two virial coefficients. The reason for concentrating on two specific cases is that there does not appear to be an obvious way to generalize the equations of the Adams method to cover the general case of say $n$ aggregating species with $m$ virial coefficients. However we have not encountered to date any specific nonideal aggregating system whose equations cannot be treated by the methods suggested above. Therefore we feel that these methods are are adequate for the determination of the virial coefficients and the confidence intervals on them.

Returning to Adams' equation (12-24), once we determine $BM_1$ by a least squares method, we can then analyze the following $m$ equations which are linear in unknowns $k_1, k_2 \ldots, k_n$.

$$\zeta(c_j) = \frac{d[M_1/cM_{w\text{app}}(c_j)]/dc}{\{[M_1/cM_{w\text{app}}(c_j)] - BM_1\}^3} = \sum_{i=1}^{n} i^2 k_i m_1{}^i(c_j) + \delta_j \qquad (j = 1, 2, \ldots, m) \tag{12-32}$$

4.1.4 *Determination of the Association Constants and Virial Coefficients Simultaneously in Nonideal Cases*

In this section we recognize the fact that Eq. (12-24), the nonideal case with a single virial coefficient given by Adams, can be treated as a nonlinear equation with some modification. In this equation the value of $m_1$ which varies with concentration is not known and can only be determined if $BM_1$ is known. The equation as such cannot be analyzed and $m_1$ has to be substituted for using the following expression

$$m_1 = c\varrho_{\text{app}}/\exp(-BM_1 c) \tag{12-33}$$

When this substitution is made we have no unknown variables only unknown constants and we can solve for $BM_1$ and the $k_i$. Under such circumstances all parameters can be determined at the same time by using steepest descent, Newton–Raphson, or any other nonlinear method mentioned in Chapter 4. Similarly, equations containing more than one virial coefficient and the association constants represented in a manner similar to Eq. (12-24), with suitable modification to exclude unknown variables, can be handled by the nonlinear method.

### 4.1.5 *Determination of Molecular Weight Distributions from Sedimentation Equilibrium Data*

Attempts at determining molecular weight distribution functions from sedimentation equilibrium data have made notable advances recently. Despite this progress a great deal must still be done to ensure the final accuracy of the computed distribution functions. Rinde (1928), Wales (1948), Wales *et al.* (1951), Donnelly (1966), Provencher (1967) and Scholte (1968) are some of authors making fundamental contributions to this subject. The treatment of this subject by most authors starts with the assumptions made by Rinde (1928). These assumptions, as quoted by Fujita (1962) with slight modification in the wording, are the following: The partial specific volumes of all solute components have the same value and may be treated as constant. The density of the solution is approximated by the solvent density $\varrho_0$; this assumes the solution to be sufficiently dilute in all solute components. The components $B_{ik}$, see Eq. (12-34) below, are zero and the solution is so dilute that the effects of terms of the order of the concentration squared are negligible.

The equation for the activity coefficient $y_i$ of species $i$ of a nonelectrolyte in presence of the components 1 to $q$ can be written as

$$\ln y_i = M_i \sum_{k=1}^{q} B_{ik} c_k + \text{terms in higher powers of } c_1, c_2, \ldots, c_q \qquad (12\text{-}34)$$

where $y_i$ is the activity coefficient of species $i$, $M_i$ the molecular weight of species $i$, the $B_{ik}$ the virial coefficients which refer to the limiting slopes of the activity coefficients at $c_1 = c_2 = \cdots = c_q = 0$, and $c_i$ the concentration of species $i$. The implication of some of the above assumptions is that Eq. (12-34) may be written as

$$\ln y_i = M_i c_i \qquad \text{for all } i \qquad (12\text{-}35)$$

When the activity coefficients of all pertinent components are expressed as Eq. (12-35) and provided all the above assumptions hold, the resulting equations at sedimentation equilibrium may be written for all components as the following differential equations

$$dc_i/d\xi = -\lambda M_i c_i \qquad (i = 1, 2, \ldots, q) \tag{12-36}$$

where

$$\lambda = (1 - \bar{v}_{\varrho 0})(r_2{}^2 - r_1{}^2)\mu^2/2RT \tag{12-37}$$

and

$$\xi = (r_2{}^2 - r^2)/(r_2{}^2 - r_1{}^2) \tag{12-38}$$

In the above equation $\mu^2$ is defined as the square of the angular velocity measured in radians per second, $r_1$, $r_2$, and $r$ respectively, are the radical distances from the center of the rotor to the air–liquid meniscus, to the bottom of the cell, and to intermediate points. The differential equations (12-36) are to be solved subject to the boundary conditions dictated by the laws of conservation of mass. The boundary conditions may be written as

$$\int_0^1 c_i(\xi)\, d\xi = c_i{}^0 \qquad (i = 1, 2, \ldots, q) \tag{12-39}$$

where $c_i{}^0$ is the concentration of the solute $i$, before centrifugations. The total concentration $c_0$ of the solution is

$$c_0 = \sum_{i=1}^{q} c_i{}^0 \tag{12-40}$$

Integration of Eqs. (12-36) subject of the above boundary condition results in

$$c_i(\xi) = c_i{}^0 \lambda M_i \exp(-\lambda M_1 \xi)/[1 - \exp(-\lambda M_i)] \qquad (i = 1, 2, \ldots, q) \tag{12-41}$$

On summing all those equations from 1 to $q$ we obtain

$$c(\xi) = c_0 \sum_{i=1}^{q} \lambda M_i f_i \exp(-\lambda M_1 \xi)/[1 - \exp(-\lambda M_i)] \tag{12-42}$$

where we define $f_i$ by

$$f_i = c_i{}^0/c_0 \tag{12-43}$$

For a continuous distribution function of molecular weights Eq. (12-42) becomes

$$c(\xi) = c_0 \int_0^\infty [\lambda M f(M)\, e^{-\lambda M \xi}/(1 - e^{-\lambda M})]\, dM \tag{12-44}$$

and on differentiating Eq. (12-44) with respect to $\xi$ we obtain

$$-(1/c_0)(dc(\xi)/d\xi) = \int_0^\infty [\lambda^2 M^2 f(M)\, e^{-\lambda M\xi}/(1 - e^{-\lambda M})]\, dM \qquad (12\text{-}45)$$

Eqs. (12-44) and (12-45) represent two equivalent integral equations in the unknown function $f(M)$, the molecular weight distribution function. Solution of these integral equations should yield the molecular weight distribution function, and it is to this problem that investigators have addressed themselves.†

Both Wales (1948) and Donnelly (1966) employed the following device for the solution of the equations. By representing the measured quantity $dc(\xi)/d\xi$ in terms of another equation containing a function of $M$ say $\phi(M)$ which can be equated to $f(M)$, the authors were able to determine $f(M)$. The new functions $\phi(M)$ are chosen so that an evaluation of the integral in which they occur is possible. The device amployed by these authors, which can be used in different ways using different functions, may be explained as follows. Either Eq. (12-44) or (12-45) can be written as follows

$$f(c, \xi) = \int_0^\infty \phi(M) U_1(M, \xi, \lambda) U_2(M, \lambda)\, dM \qquad (12\text{-}46)$$

where $U_1(M, \xi, \lambda)$ and $U_2(M, \lambda)$ are known functions. In what follows when we indicate a known function we mean that not only is the function known precisely but it does not present any problems in its integration or subsequent computations. In this equation the function $f(M)$ is replaced by a function $\phi(M)$ so that the integral on the right-hand side can be evaluated. The procedure of Wales (1948) was to rewrite Eq. (12-44) as follows

$$c(\xi)/c_0 = \int_0^\infty \phi(M) \exp MU_3(\xi, \lambda)\, dM \qquad (12\text{-}47)$$

where $U_3(\xi, \lambda)$ is a known function. Next, $\phi(M)$ was related to $f(M)$ by the equation

$$\phi(M) = U_4(M, \xi, \lambda) f(M) \qquad (12\text{-}48)$$

where $U_4(M, \xi, \lambda)$ is a known function. From Eq. (12-48) we see that for every $\phi(M_i)$ we can obtain an $f(M_i)$. The function $\phi(M)$ must now be represented in such a manner as to facilitate the integration on the right-

† Equations (12-44) and (12-45) do not include virial coefficients. It is a simple matter to include one or more virial coefficients in those equations. However, when that is done the virial coefficients will occur nonlinearly. Nevertheless, using methods in Chapter 4 we can solve for both the molecular weight distribution and the virial coefficients.

hand side of Eq. (12-47) and also enable us to account for the experimental data. To do this Wales chose the following representation (where we take as many terms as will fit the data)

$$\phi(M) = e^{-KM} \sum_{j=1}^{n} \theta_j M \tag{12-49}$$

This is a reasonable function since $\phi(M) = 0$ when $M = 0$ and as $M \to \infty$. If we can determine the parameters $\theta_j$ and $K$ we can determine $\phi(M)$ and hence $f(M)$. The parameters $\theta_j$ and $K$ must be related to the experimental data. By substituting the expression for $\phi(M)$ in Eq. (12-49) into Eq. (12-47) and integrating Wales managed to represent the experimental data in terms of the parameters by means of the next equation

$$c(\xi)/c_0 = \sum_{j=1}^{n} j!\theta_j/q^{j+1} \tag{12-50}$$

where $q = K - U_4(\xi, \lambda)$ and the constant $K$ is arbitrary and is such that $q$ must be always positive. By putting

$$p = 1/q \tag{12-51}$$

and dividing both sides of Eq. (12-50) by $p^2$, Wales obtained

$$(1/p^2)[c(\xi)/c_0] = \sum_{j=1}^{n} j!\theta_j p^{j-1} = \psi(p) \tag{12-52}$$

where $\psi(p)$ denotes a polynomial in $p$ written as

$$\psi(p) = \alpha_0 + \alpha_1 p + \alpha_2 p^2 + \alpha_3 p^3 + \cdots \tag{12-53}$$

where $\alpha_0, \alpha_1, \alpha_2, \ldots$ are constants. These constants are readily evaluated by using least squares to fit the experimental quantity $(1/p^2)[dc(\xi)/dc]$ to the polynomial given by Eq. (12-53). To obtain the $\theta_j$ from the $\alpha_j$ we have Eq. (12-54) obtained by equating powers

$$\theta_j \equiv \alpha_{j-1}/j! \tag{12-54}$$

In this manner the $\theta_j$ are evaluated and further, since $K$ has been evaluated previously, we can obtain $\phi(M)$ and hence $f(M)$ by Eq. (12-48).

Donnelly (1966) took a different approach in his treatment of the subject. He substituted a function $\psi(M)$ for $f(M)$ in such a way that the right-hand side of Eq. (12-44) yielded a Laplace transform. This is shown by

$$\int_0^\infty \psi(M) \exp[-U_6(\xi, \lambda)M]\, dM = f(c, \xi) \tag{12-55}$$

where $U_6(\xi, \lambda)$ is a known function. To evaluate the function $\psi(M)$, one has to evaluate the inverse Laplace transform of $f(c, \xi)$. When this is done the function $\psi(M)$ is immediately evaluated and $f(M)$ is also obtained by the relationship established between them by Donnelly.

In order for Donnelly's method to work, the inverse Laplace transform of $f(c, \xi)$ must exist. The function $f(c, \xi)$, which represents the experimental data is fitted to another function with adjustable parameters. This latter function (call it the fitted function) is such that its inverse Laplace transform exists. Donnelly employed two fitted functions to illustrate his method. The first one was essentially a straight line fit. The second fitted function used by Donnelly was a low-order rational function in $\xi$. This allows for a better fit of the experimental data.

Donnelly deplored the fact that one could not compute inverse Laplace transforms of polynomials and consequently one could not use them to fit the experimental data. That, however, is no great loss. From the techniques of function approximation (Rice, 1964; Scheid, 1968) we know that rational functions, by virtue of the fact that they are quotients of polynomials, are a much richer class of functions than polynomials, and consequently it is probable that rational functions will be better able to follow the experimental data. Therefore, the fact that inverse Laplace transforms of higher order polynomial do not exist is not really an obstacle since we can fit the data just as will with higher order rational functions whose inverse Laplace transform does exist. The use of rational functions to fit the experimental data indicates a number of potential uses of Donnelly's method. We shall only mention these briefly since Donnelly's method suffers from the fact that it is only applicable to continuous molecular weight distribution functions and that is rarely the case in practice. As indicated above, if the simple rational function suggested by Donnelly does not fit the data well, then higher order rational functions, which can be defined in general as

$$P(\xi)/Q(\xi) \tag{12-56}$$

where the degree of $P(\xi)$ is less than that of $Q(\xi)$, can be used to fit the experimental data. It can be shown that the inverse Laplace transforms of the class of functions given by Eq. (12-56) always exist (Spiegel, 1965). Further, by the Heaviside expansion theorem (Spiegel, 1965) we can show that the inverse Laplace Transform of any rational function shown above can be written as

$$\sum_{k=1}^{n} [P(\alpha_k)/Q(\alpha_k)] \exp(\alpha_k M) \tag{12-57}$$

where the polynomial $Q(\xi)$ has the $n$ distinct zeros $\alpha_1, \alpha_2, \ldots, \alpha_k, \ldots, \alpha_n$.

From the standpoint of fitting $f(c, \xi)$ to a rational function, we have, in addition to the least squares method used by Donnelly to fit his rational function, the possibility of fitting the data by using an approximation in the $L_\infty$ norm. This latter approximation is also called a min-max approximation in the Chebyshev norm (Rice, 1964) (see Chapter 4 where this fit is discussed further). Since the inverse Laplace transform of rational function exists and rational functions can be fit with the $L_\infty$ norm approximations, the scope of Donnelly's method is greatly expanded. The methods of Wales and Donnelly utilize only a limited amount of information since they use only a single equilibrium run at a single rotor speed. To make the most of the information that one is capable of obtaining Scholte (1968) has suggested the use of Eq. (12-45) at different rotor speeds,† writing

$$f(c, \xi, \lambda) = \sum_i f_i \lambda^2 M_i \exp(-\xi\lambda M_i)/[1 - \exp(-\lambda M_i)] \qquad (12\text{-}58)$$

where

$$f(c, \xi, \lambda) = -(1/c_0)[dc(\xi)/d\xi]_\lambda$$

at the specified rotor speed. Equation (12-102) can be rewritten as

$$f(c, \xi, \lambda)_j = \sum_i f_i K_{ij} + \delta_j \qquad \text{for} \quad j = 1, 2, \ldots, m \qquad (12\text{-}59)$$

for $m$ possible measurements of $f(c, \xi, \lambda)_j$ and where we define

$$K_{ij} = \lambda_j^2 M_i \exp(-\xi_j \lambda_j M_i)/[1 - \exp(-\lambda_j M_i)] \qquad (12\text{-}60)$$

and $\delta_j$ is experimental error in making the $j$th measurement.

To determine the molecular weight distribution function Eq. (12-59) must be solved for the $f_i$ subject to some restriction that the experimental error be minimized. Scholte solved Eq. (12-59) by using linear programming. This method minimized the quantity $\sum_{j=1}^m |\delta_j|$ and solved the problem such that all $f_i$ are positive.

Since Eq. (12-59) is a linear equation we can apply the linear methods described in Chapter 5; in particular, we emphasize the weighted least squares and the constrained least squares as possibilities for its solution.

† This will not work for concentration-dependent aggregating systems.

Some of the problems of least squares methods are the ill-conditioning of equations, although it must be remembered that in principle they ought to work and should be tried.

The problem of whether or not to regard the molecular weight distribution as continuous comes up again here as it will in light scattering. This problem as pointed out previously boils down to the interchange of the integral in Eq. (12-45) with the summation in Eq. (12-58).

In the case of noncontinuous molecular weight distribution the physical situation dictates that we should use summation and Eq. (12-58) is the appropriate equation to use. For continuous molecular weight distribution quadrature must be used to evaluate the integral first.

The linear methods outlined in Chapter 4 will be quite adequate to solve the equation as outlined by Scholte. Further, for the continuous case after the quadrature has been done the methods of Chapter 4 can then be used. To insure the final accuracy of the molecular weight distribution it will prove helpful to know the bounds on the distribution function. We then also discuss practical problems in the solution of equations such as (12-58) in terms of the numerical methods available for its solution and the use of bounds to aid in that solution.

### 4.1.6 *Use of Bounds to Estimate Molecular Distributions*

In general, given the moments of a distribution function its bounds can be estimated. The moments of a molecular weight distribution function can be estimated from sedimentation equilibrium data and they are products of the various molecular weight averages.

Estimation of bounds on the molecular weight distribution function will be helpful for the following reasons. It is possible because of ill-conditioning that data from concentration aggregating systems can give rise to more than one set of association constants (Magar, 1969; Magar and Steiner, 1970; Van Holde *et al.*, 1969). To determine which set is consistent with the data, we compute the molecular weight distribution function from the association constants at a given monomer concentration. Next, by computing the bounds on the distribution function from the moments we can select the distribution function which falls within those bounds and reject the ones that do not. The second reason and more important one for computing bounds on the distribution function is that, in the computation of the distribution function by the methods mentioned in Chapter 4 it is quite possible, because of experimental errors and the numerical procedure employed, that the computed molecular weight distribution function might not coincide with the actual distribution function. Knowing the bounds of the distribution function will indicate whether or not the computed distribution function is consistent with the data. Lastly, knowing the bounds will aid in a small way in the determination of how many terms an investigator requires in Eq. (12-59).

### 4.1.7 *Estimates on the Bounds of a Distribution Function from Its Moments*

The general problem of estimating bounds on a distribution function from its moments was proposed by the famous Russian mathematician Chebyshev in 1874 who stated it in the following manner: Let $F(y)$ be an unknown distribution function defined on the closed interval $[a, b]$ and satisfying the condition

$$F(a) = 0 \tag{12-61}$$

If we define the moments by

$$m_j = \int_a^b y\, F(y)\, dy \qquad (j = 0, 1, 2, \ldots, k) \tag{12-62}$$

and if the moments $m_j$ for $j = 0, 1, \ldots, k$ are known, then for any given value of $y$ such that $a < y < b$, what are the (sharp) upper and lower bounds on $F(y)$. We note that the moments of the molecular weight distribution can be obtained from the various average molecular weights by the following relationships if

$$\int_0^\infty m_j F(m)\, dm = \nu_j \qquad (j = 0, 1, \ldots, k) \tag{12-63}$$

then

$$\begin{gathered} \nu_0 = 1, \qquad \nu_1 = M_w, \qquad \nu_2 = M_w M_z, \qquad \nu_3 = M_w M_z M_{z+1} \\ \nu_4 = M_w M_z M_{z+1} M_{z+2} \end{gathered} \tag{12-64}$$

Since we can only expect to determine with any reasonable accuracy† moments up to $M_{z+2}$ we shall only give the solution to the moment problem (also referred to as the Chebyshev–Markov inequalities) for the case $k = 2, 3,$ and $4$.

In the case where we have two moments $m_1$ and $m_2$ we can, without loss of generality, assume that $m_0 = 1$, $m_1 = 0$ and $m_2 = 1$ since any distribution function can be made to conform to these conditions by a

† Even that is debatable.

linear transformation. It is important to explain this linear transformation since it also illustrates the meaning of the interval $[a, b]$ on which our molecular weight distribution function is defined. The linear transformation brings this about is the following: Every $M_i$ in our molecular weight distribution is transformed to $w_i$ by the equation

$$w_i = (M_i - M_w)/(M_w M_z)^{1/2} \tag{12-65}$$

Now the elements $w_i$ are in the interval $[a, b]$ which is to say that we can set $a = w_1$ and $b = w_n$, where $w_1$ corresponds to $M_1$ the smallest molecular species and $w_n$ corresponds to $M_n$ the largest molecular species. The above assumption makes the moment $m_0 = 1$, $m_1 = 0$, $m_2 = 1$ and implies that all molecular species smaller than $M_w$, have negative values after we perform the transformation. Now if $F(w)$ is the distribution function on $[a, b]$ with moments $m_0 = 1$, $m_1 = 0$, and $m_2 = 1$, then for any given $w_i$ such that $(a < w_i < b)$ we have the following inequalities:

$$0 \leq F(w_i) \leq \frac{1}{1 + w_i^2} \qquad \text{if} \qquad a < w_i \leq -\frac{1}{b} \tag{12-66}$$

$$\frac{1 + bw_i}{(a-b)(a-w_i)} \leq F(w_i) \leq 1 - \frac{1 + aw_i}{(b-a)(b-w_i)} \qquad \text{if} \qquad -\frac{1}{b} \leq w_i \leq -\frac{1}{a} \tag{12-67}$$

$$\frac{w_i^2}{1 + w_i^2} \leq F(w_i) \leq 1 \qquad \text{if} \qquad -\frac{1}{a} \leq w_i < b \tag{12-68}$$

Equations (12-66)–(12-68) determine the bounds when we have two moments only.

In order to determine the bounds using three moments let $F(w)$ be the distribution function and let the moments be $m_0 = 1$, $m_1 = 0$, $m_2 = 1$, and $m_3$, and let

$$h(w_i) = w_i^2 - m_3 w_i - 1 \tag{12-69}$$

defining

$$Z_1 = [m_3 - (a + w_i)]/(1 + aw_i) \tag{12-70}$$

$$Z_2 = [m_3 - (b + w_i)]/(1 + bw_i) \tag{12-71}$$

$$r = [m_3 - (b + w_i)]/(1 + ab) \tag{12-72}$$

$$A(\alpha, \beta, \gamma) = (1 + \alpha\beta)/(\gamma - a)(\gamma - \beta) \tag{12-73}$$

for any given $w_i (a < w_i < b)$ we have the following bounds

$$0 \leq F(w_i) \leq A(b, Z_2, w_i) \quad \text{if} \quad h(w_i) \geq 0; \; w_i < 0 \tag{12-74}$$

$$A(w_i, Z_1, a) \leq F(w_i) \leq A(w_i, Z_1, a) + A(a, Z_1, w_i) \quad \text{if} \quad h(w_i) \leq 0; \; w_i \leq r \tag{12-75}$$

$$A(w_i, b, Z_2) \leq F(w_i) \leq A(w_i, b, Z_2) + A(b, Z_2, w_i) \quad \text{if} \quad h(w_i) \leq 0; \; w_i \geq r \tag{12-76}$$

$$1 - A(a, Z_1, w_i) \leq F(w_i) \leq 1 \quad \text{if} \quad h(w_i) \geq 0; \; w_i > 0 \tag{12-77}$$

For estimating the bounds with four moments we need the molecular weight averages up to and including $M_{z+2}$. If the moments are $m_0 = 1$, $m_1 = 0$, $m_2 = 1$, $m_3$ and $m_4$ we write

$$s(y, \gamma) = y^2 + \frac{\gamma - m_3 + \gamma(m_3 - m_4)}{1 + \gamma(m_3 - \gamma)} y + \frac{\gamma m_3 - m_4 + (m_3 - \gamma)^2}{1 + \gamma(m_3 - \gamma)} \tag{12-78}$$

Next let us set $U(y) = s(y, a)$, $V(y) = s(y, b)$, and $Z(y) = g(y, w_i)$ and let $u_1 < u_2$, $v_1 < v_2$, and $z_1 < z_2$ be the roots of $U(y)$, $V(y)$, and $Z(y)$ respectively, then we have essentially set up the problem so that $a < v_1 < u_1 < v_2 < u_2 < b$. If we define

$$Z_3 = [m_3(a+b+w_i) - m_4 - ab - aw_i - bw_i]/(abw_i + a + b + w_i + m_3) \tag{12-79}$$

$$A = (m_4 - m_3^2 - 1)/[(1 + w_i^2)(m_4 - m_3^2 - 1) + (w_i^2 - m_3 w_i - 1)^2] \tag{12-88}$$

$$B(\alpha, \beta, \gamma) = [m_3 - (\alpha + \beta + Z_3) - \alpha\beta Z_3]/[(\gamma - \alpha)(\gamma - \beta)(\gamma - Z_3)] \tag{12-81}$$

then for any given value $w_i$ such that $a < w_i < b$ we have

$$0 \leq F(w_i) \leq A \quad \text{if} \quad a < w_i \leq v_1 \tag{12-82}$$

$$B(b, w_i, a) \leq F(w_i) \leq B(b, w_i, a) + B(a, b, w_i) \quad \text{if} \quad v_1 \leq w_i \leq u_1 \tag{12-83}$$

$$\frac{1 + w_i z_2}{(z_1 - w_i)(z_1 - z_2)} \leq F(w_i) \leq \frac{1 + w_i z_2}{(z_1 - w_i)(z_1 - z_2)} + A \quad \text{if} \quad u_1 \leq w_i \leq v_2 \tag{12-84}$$

$$1 - b(a, b, w_i) - B(a, w_i, b) \leq F(w_i) \leq 1 - B(a, w_i, b)$$
$$\text{if} \quad v_2 \leq w_i \leq u_2 \tag{12-85}$$

$$1 - A \leq F(w_i) \leq 1$$
$$\text{if} \quad u_2 \leq w_i < b \tag{12-86}$$

Inequalities (12-82), (12-84), and (12-86) also hold for $w_i < z_1$, $z_1 < w_i$, and $z_2 < w_i$ respectively.

The proofs of all the above inequalities are given by Shohat and Tamarkin (1943), Uspensky (1937), and Zelen (1954). As an example of the above inequalities which we take from Zelen (1954) we compute the bounds for a distribution function having the moments $m_0 = 1$, $m_1 = 0$, $m_2 = 1$, $m_3 = 0$, and $m_4 = 3$. We wish determine the bounds of the distribution function at $w_i = 2$, and $w_i = 3$. For bounds using two moments only, we have the following: Since $w_i > 0$, then we have $0.8000 \leq F(2) \leq 1$ and $0.9000 \leq F(3) \leq 1$. For bounds using moments of order four, we must first locate $w_i$ before choosing which inequality to apply. By solving the quadratic equations for $U(y)$ and $V(y)$ we determine $u_1$, $u_2$, $v_1$, and $v_2$ and then find where $w_i$ is located. This is followed by using the appropriate inequality to compute the bounds of $w_i = 2$ and $w_i = 3$. The appropriate inequality is (12-86) so we have

$$1 - \{2/[2(1 + w_i^2) + (w_i^2 - 1)^2]\} \leq F(w_i) \leq 1 \quad \text{for } w_i = 2;\ w_i = 3 \tag{12-87}$$

which makes the bounds

$$0.8947 \leq F(2) \leq 1 \quad \text{and} \quad 0.9777 \leq F(3) \leq 1 \tag{12-88}$$

#### 4.1.8 *Some Practical Consideration in the Determination of Molecular Weight Distributions*

In this section we discuss the practical aspect of the problem. In the analysis of concentration dependent aggregating systems earlier in the chapter, we encountered the possibility of obtaining more than one set of association constants (uridine for example) within the limits of the statistical test utilized. Two ways are available to overcome this problem. In the first case we determine the various molecular weight averages and then compute the bounds on the molecular weight distribution function (MDWF) and see which one of the distribution function computed from the association constants falls within the bounds. Alternatively the molecular weight distribution function can be computed directly using the linear programming method of Scholte (employed at a single rotor speed), or any one or other of the methods described in Section 2 of the theory that are applicable to the linear problem. For concentration-dependent discrete distributions those methods are highly pertinet. In particular, the weighted $L_2$ norm and the constrained $L_2$ norm have never been tried nor has a combination of both.

When the system is concentration-dependent different initial concentrations in any given run will yield different MWD and also different moments. I determine the association constants usually by determining the weight average molecular weight at different concentrations and for this we make several runs at different concentrations. Once the association constants are determined, be they a single set or more, they can be used to compute one or more MWDF as the case may be, at any given initial concentration and weight fraction of the monomer.

To determine the MWD and its bounds, runs are made at a specific initial concentration and at a given speed. The weight fraction of the monomer is then computed using Steiner's equation and the various molecular weight averages. Next, with the association constants obtained previously from a series of different runs but with the *same* initial concentration as the one used to determine the MWD and the weight fraction of the monomer, a MWD can be computed for every set of association constants. The set of association constants giving the MWD which agrees best with the computed MWD (as determined by linear programming or any other method described herein) and which falls within the bounds as determined from the moments can be considered as being most consistent with the data.

Since the determination of the association constants generally involves more than one run and since the MWD and bounds are estimated only from a single run, it is clear that if one makes say three runs to determine the association constants and if the MWD and bounds are computed in each one of those runs the investigator has three independent checks on the ultimate accuracy of the MWD.

Estimation of the bounds of the distribution function might be considered redundant if the distribution function itself is estimated. This of course is true if the distribution function can be accurately estimated. The fact of the matter, however, is that there is experimental error in the estimation of both the distribution function and of the bounds. The general idea advocated here is that *both* sets of information should be estimated to arrive at the most consistent MWD.

One of the more important uses of the bounds will be in the aid they give us in the determination of the distribution functions, particularly, when the distribution is estimated by iteration, since the bounds will give a good starting value for our iterations. Knowing the bounds also helps in the linear programming problem. In determining MWDF by linear programming the problem can be solved not only subject to the constraints that the $f_i$ be all positive as Scholte suggested, but also subject to the constraint that $\sum_i f_i = 1$, and to the constraint that any given $f_i$ must lie within the bounds dictated by the estimate of those quantities from the moments. Only one of the latter two constraints ($\sum_i f_i = 1$) was sometimes incorporated in Scholte's linear programming method. We would like to point out, however, that the inclusion of both constraints is well within the scope of the method. There is no question that linear programming is rendered more consistent with the physical situation when these constraints are taken into consideration. In fact, it is one of the maxims of optimization processes that all constraints consistent with the physical situation must be incorporated in the problem.

A specific problem with the bounds which we must consider is that the estimates on the bounds are based on knowledge of the various molecular weight averages. The averages are not as accurate as one would like them. The reason for this is found in the data processing section of this chapter where we note that we have to differentiate the raw data twice in order to obtain $M_z$ and $M_{z+1}$ and three times in order to obtain $M_{z+2}$. Further, it will be noted that we require the values of the derivatives at points where the accuracy of the data will be the worst. Our suggestions for processing the raw data will enhance the accuracy of the computed molecular weight averages and hence the bounds.

Further refinement in the estimation of the weight average molecular weights were considered in greater detail by Fujita (1962) and by Scholte (1968). Fundamentally the procedure here is to alter the parameter $\lambda$ in Eq. (12-58) by varying the rotor speed or column height. Such a result enables us to write the left-hand side of Eq. (12-58) $f(c, \xi, \lambda)$ as $y(\lambda)$ emphasizing the fact that $\lambda$ is the most important variable for the time being. We also write, after Scholte and Fujita, $y_{1/2}(\lambda), y_{3/4}(\lambda), \ldots$ indicating that we are evaluating $y(\lambda)$ at $\xi = \frac{1}{2}$, $\xi = \frac{3}{4}$, .... Written in these terms we note the following equations:

$$\lim_{\lambda \to 0} y_{1/2}(\lambda) = M_w \tag{12-89}$$

$$M_z = -(5.774/M_w) \int_0^\infty dy_{1/2}(\lambda)\, d(\lambda^2)\, d\lambda \tag{12-90}$$

$$M_{z+1} = -(24/M_w M_z)[dy_{1/2}(\lambda)/d(\lambda^2)]_{\lambda=0} \tag{12-91}$$

$$\lim_{\lambda \to 0} \{[y_{1/4}(\lambda) - y_{3/4}(\lambda)]/2\lambda\} = M_w M_z/4 \tag{12-92}$$

$$\left\{\frac{d}{d\lambda}\left[\frac{y_{1/4}(\lambda) - y_{3/4}(\lambda)}{2}\right]\right\}_{\lambda=0} = \frac{M_w M_z}{4} \tag{12-93}$$

$$\left\{\frac{d}{d(\lambda^2)}\left[\frac{y_{1/4}(\lambda) + y_{3/4}(\lambda)}{2}\right]\right\}_{\lambda=0} = -\frac{M_w M_z M_{z+1}}{96} \tag{12-94}$$

from which we can obtain various molecular weight averages and hence the bounds on the moments. Equations (12-89)–(12-94) were obtained by substituting for the exponential function in Eq. (12-58) by hyperbolic function at the specific values of $\xi$. In this way, the authors could integrate Eq. (12-58) analytically to obtain the desired equations.

The advantages of such equations are that we no longer use the inaccurate data obtained by taking readings at the top and bottom of the cell column, rather we use the data in the middle, or $\frac{1}{4}$ or $\frac{3}{4}$ of the way along the cell. Such data are obtained far more accurately than data obtained at the ends of the cell column. The main disadvantage of these methods is that they require a large number of runs at different speeds and different columns heights. They also require extrapolation to $\lambda = 0$. Scholte has found that from a practical standpoint Eq. (12-92) and (12-93) are to be preferred to Eqs. (12-91) and (12-94) in the estimation of $M_z$ and $M_{z+1}$.

Inasmuch as we have to differentiate and integrate $y(\lambda)$ to obtain the molecular weight averages then what will be said in the data processing section about fitting the raw data applies equally well to fitting the function $y(\lambda)$. In particular, we recommend the unconstrained weighted least squares with polynomials or orthogonal functions to fit $y(\lambda)$. The experimental difficulties in obtaining $y(\lambda)$ over a large range of $\lambda$ have been discussed by Fujita (1962) and Scholte (1968). However, it seems likely that accurate estimates of the various molecular weight averages up to $M_{z+2}$ can be achieved when we use all the available data and a combination of methods as we have advocated here.

From the practical standpoint it seems highly desirable that we should have some kind of standard that will enable us to have access the experimental error made in the reading at the end of the cell column. With such a standard for pinpointing the errors made in the reading, the mathematical methods described herein will go a long way in securing the accuracy of the molecular weight averages and hence the bounds on the distribution functions. Some steps in this direction have been taken by

Richards *et al.* (1968) using ribonuclease as standard and employing the principle of conversation of mass and the labeling of white fringes as an aid to the determination of the averages.

With respect to the determination of the molecular weight distribution we mentioned above the methods of Wales (1948) and Donnelly (1966). Actually Wales (1948) has two methods for computing molecular weight distributions both of which can be rendered more accurate in their estimation of the molecular weight distribution by using some of the nonlinear techniques advocated in Chapter 4. Thus, Eq. (12-50) can be treated by one or other of the nonlinear methods to determine $\theta_j$ and $K$. For this purpose we can use the linearized Taylor series, the method of steepest descent, the Newton–Raphson method, or the Nelder and Mead method to determine these parameters directly. This saves us a number of steps in the computations suggested by Wales and we do not have to determine $K$ by a separate technique. Further, we can determine the standard errors of these parameters and the confidence interval within the confines of our approximations.

Similarly the other method suggested by Wales (1948), and to our knowledge never actually used, can be handled by the nonlinear methods. In this case Eq. (52) of the paper of Wales (1948) is the one used.

With respect to the method of Donnelly (1966), we pointed out earlier that if we wish to consider fits whose inverse Laplace transform exists we can consider rational functions. For this purpose we can extend Donnelly's method and fit the raw data by higher order rational functions using one or other of the nonlinear methods described in Chapter 4 specifically to approximate the raw data by an approximating function in the least squares sense. Alternatively, since there exist convenient ways of approximating the data in terms of rational functions, by using the $L_\infty$ norm or the Chebyshev norm (Scheid, 1968), it may be worth the investigators time to use such an approximation. Thus, if the investigator feels that accurate representation of the raw data can be affected by minimizing the maximum error then an $L_\infty$ fit can be made with rational functions. Once we fit the data by rotational functions, we can determine the inverse Laplace transform of such a function and from that we can determine the molecular weight distribution. The limitation of the method, as recognized by Donnelly, is its applicability only to continuous molecular weight distributions.

In actual application the results obtained by Scholte (1968) were on a hypothetical distribution. However in Fig. 4 of his manuscript we notice that when a random standard error is applied to his data his fits become progressively worse. Scholte partly overcomes this problem by increasing the interval between the molecular weight values or the $M_i$.

The problem of the spacing of molecular weght intervals is common to all methods operating on continuous molecular weight distribution in the sense that we have to decide what the elements $K_{ij}$ of Eq. (12-60) are exactly.† For continuous distribution we resort to quadrature and this leads us naturally into the consideration of the Fredholm integrals equations of the first kind, and these may be solved by a combination of quadrature and least squares which may be written in our notation as

$$f(\zeta) = \int_a^b K(\zeta, M) f(M)\, dM \tag{12-95}$$

where $\zeta$ takes the place of the variables $\xi$ and $\lambda$. To solve this integral equation numerically we have to approximate the integral on the right-hand side of Eq. (12-95) by a summation in the following manner

$$\int_a^b K(\zeta, M) f(M)\, dM = \sum_{i=1}^{n} G_i K(\zeta, M_i) f(M_i) \tag{12-96}$$

Equation (12-96) essentially represents a quadrature scheme. Quadrature schemes, as we observed in Chapter 3, essentially determine the $G_i$ or weights as they are called and the $M_i$ for us. Thus the question of the spacing of $M_i$ is effectively determined for us by the quadrature scheme and is not arbitrarily set as in the method of Scholte. The various $G_i$ and $M_i$ will, of course, have different values depending upon the quadrature scheme used to make the approximation in Eq. (12-96). This is important in practice.

Having performed the approximation in Eq. (12-96) we replace the integral by the quadrature formula and rewrite Eq. (12-95) as

$$f(\zeta_j) = \sum_{i=1}^{n} G_i K(\zeta_j, M_i) f(M_i) + \delta_j \qquad (j = 1, 2, \ldots, m) \tag{12-97}$$

where $\delta_j$ represents both the experimental error made in making the $j$th measurement and the error made in the quadrature approximation. (This latter quantity, which diminishes with $n$, is generally known analytically and depends on the quadrature scheme.) Equation (12-97) is a linear equation in the $f(M_i)$ and can be solved by the usual techniques for those parameters to minimize $\sum_{j=1}^{m} | \delta_j |$ or $\sum_{j=1}^{m} \delta_j^2$. In particular we note that quadrature may be used prior to determining the parameters

† For information on the statistical problems in the choice of the design matrix **k**, write to Grace Wahba of the Department of Statistics, Univ. of Wisconsin at Madison.

by linear programming thus overcoming some of the difficulties encountered by Scholte. For more discussion on Fredholm integral equations of the first kind we refer the reader to Chapter 8 of Kopal (1955).

A combination of least squares and quadrature has been recommended by Provencher (1967). However, Provencher did not use the classical methods of quadrature (Kopal, 1955) to approximate his integral and relied mainly on least squares to solve his problem which essentially was an attempt to fit a Wesslau (1956) distribution with linear combination of Schulz (1939) functions. Provencher was critical of the classical methods of quadrature because he felt that they forced the molecular weight distribution to follow a fixed functional form. This critique does not seem fair, inasmuch as Provencher actually uses fixed functional forms, the Schulz functions, to fit his data. Further, expanding $f(M)$ in a series of orthogonal functions or polynomials as is the case in various quadrature schemes, in principle will reproduce $f(M)$ no matter what its profile looks like. This is a consequence of the Weierstrass approximation theorem (Davis, 1961). Difficulties may be encountered in the numerical extraction of the coefficients of our functions, but that is due to the inherent ill-conditioning of the problem and is not due to restricting $f(M)$ to follow a fixed functional form. For the continuous distributions the classical methods of quadrature followed by least squares analysis or least squares with constraints should work quite well.

Provencher (1967) looked into the problem of using Schulz functions to fit MWDF. For the sake of completeness and to extend Provencher's method by using some of the nonlinear techniques discussed previously we outline his method. In actual application, Provencher set himself the limited problem of synthesizing "experimental data" from a Wesslau (1956) molecular weight distribution written as

$$f(M) = c_0 \exp(-\tau^2/4)/M_0\pi^{1/2} \exp\{-[\ln(M/M_0)]^2/q^2\}, \quad (12\text{-}98)$$

where $c_0$ is the total concentration, $M_0$ the maximum value of $f(M)$ and $\tau$ a positive parameter related to the weight average molecular weight by $M_w = M_0 \exp(3\tau^2/4)$. With this synthetic experimental data Provencher uses a set of Schulz (1939) distribution functions to make his fit. The Schulz functions can be written as

$$f(M) = c_0(\Delta/M_0)^{\Delta+1}[M^\Delta/\Gamma(\Delta + 1)] \exp(-\Delta M/M_0), \quad (12\text{-}99)$$

where $\Delta$ is a positive parameter and $\Gamma$ the gamma function.

When a Schulz function is substituted in Eq. (12-96), the integral in this equation can be solved explicitly to give

$$\frac{2\xi_i c_0}{r_2^2 - r_1^2} \gamma^{\Delta+1}(\Delta + 2)(\Delta + 1)\psi(\Delta + 3, \gamma + \xi_i) = f(\xi, \Delta, \lambda, \gamma, c_0), \tag{12-100}$$

where

$$\gamma = \Delta/\lambda M_0 \tag{12-101}$$

and the function $\psi$ is the generalized Reimann (1859) zeta function and is written as

$$\psi(s, a) = \sum_{n=0}^{\infty} [1/(a + n)^s] \tag{12-102}$$

The technique of using a particular function like the Schulz function in order to evaluate the integral analytically is fairly common in this type of problem and has been used, as we have pointed out, by other authors.

Provencher then attempted to fit his data to a system of equations which can be written as

$$f(\xi_j \lambda_j) = \sum_{i=1}^{n} \alpha_i f_i(\xi_j \lambda_j \Delta_i \gamma_i c_0) + \delta_j \qquad (j = 1, 2, \ldots, m) \tag{12-103}$$

where $n$ was taken to be 2 or 3. By substituting various values for the unknown parameters $\gamma_i$ and $\Delta_i$ (the values were obtained by trial and error), Provencher attempted to solve for the $\alpha_i$ such that $\sum_{j=1}^{m} \delta_j^2$ was minimized. When trial and error values of the $\gamma_i$ and $\Delta_i$ are used, Eq. (12-103) is made linear in the unknowns $\alpha_i$ and it can be solved by the $L_2$ norm methods. If we do not wish to use trial and error to determine the $\gamma_i$ and $\Delta_i$, Eq. (12-103) can be considered to nonlinear in the parammeters $\gamma_i$ and $\Delta_i$ and use any one of the nonlinear methods in Chapter 4 to solve for the $\alpha$, $\gamma_i$, and $\Delta_i$, so as to minimize $\sum_{j=1}^{m} \delta_j^2$. As a matter of fact, Provencher suggested the use of the methods of steepest descent for the solution of this particular problem but did not actually use it.

In general, the Fredholm integral equation of the first kind is extremely difficult to solve for practical cases. The reason for this is pointed out by Tricomi (1957) in his excellent dissertation on integral equations, where he states that such equations admit an infinity of solutions. That this is the case is clear since given that what we determine is a solution vector $f(M_i)$, we can pass an infinity of functions through that set of points which are the elements of this vector. In practice, however, it is highly unlikely that a wildly oscillating function will represent the actual solution of the MWDF. Consequently, with respect to obtaining a solution of a Fredholm integral equation of the first kind we must steer an intermediate course between the infinite solutions prescribed by rigorous mathematical considerations and a single solution we might be prone to accept. The use of bounds will help in such matters. Since we can obtain many solutions, Baker *et al.* (1964) have suggested various "smoothing" methods for the solutions of such equations. In their paper they discuss the many ways of selecting smoothness.

With respect to what quadrature scheme one may use, it will be remembered that different quadrature schemes give rise to different values for the spacing of the molecular weights, $M_i$. Some quadrature schemes tend to concentrate the $M_i$ only in certain parts of the molecular weight interval. If the experimental data obtained in that region are inaccurate, difficulties can arise and they must be avoided by using a more suitable quadrature scheme.

### 4.1.9 *Data Processing Aspects of Osmotic Pressure and Sedimentation Equilibrium Measurements.*

The quantity that is experimentally determined in the osmotic coefficient is the quantity $\phi$ in Eq. (12-5). As stated in that section, in order to determine $\gamma$ we have to represent $\phi$ as a polynomial in $m$. Thus

$$\phi(m) = \alpha_0 + \alpha_1 m + \alpha_2 m^2 + \cdots + \alpha_n m^n \qquad (12\text{-}104)$$

to facilitate integration.

To determine the $\alpha_i$ we measure $\phi$ at a number of $m$ and solve for the $\alpha_i$ by least squares. Ts'o and Chan (1964) fitted $\phi(m)$ by a sixth degree polynomial in $m$. Since there are no constraints on the $\alpha$ a straightforward application of least squares is in order. Once again we mention that high degree polynomials may be ill-conditioned and orthogonal functions may be appropriate.

For sedimentation equilibrium the raw data determined are essentially the concentration of the solute versus the distance along the cell. The raw data obtained from sedimentation equilibrium curves must be processed with their future use in mind. If we wish to determine bounds on the molecular distribution function we must determine the molecular weight averages. Fujita (1962) wrote the following equations for the molecular weight averages

$$M_z = (1/\lambda^2 c_0 M_w)[(dc/d\xi)_{\xi=1} - (dc/d\xi)_{\xi=0}] \qquad (12\text{-}105)$$

$$M_{z+1} = (1/\lambda^3 c_0 M_w M_z)[(d^2c/d\xi^2)_{\xi=0} - (d^2c/d\xi^2)_{\xi=1}] \qquad (12\text{-}106)$$

and for higher averages

$$M_{z+p} = \frac{1}{\lambda^{p+2} c_0 M_w M_z \cdots M_{z+p-1}} \left[ \left( \frac{d^{p+1} c}{d\xi^{p+1}} \right)_{\xi=0} - \left( \frac{d^{p+1} c}{d\xi^{p+1}} \right)_{\xi=1} \right] \quad (12\text{-}107)$$

when $p$ is an odd integer and

$$\frac{1}{\lambda^{p+2} c_0 M_w M_z \cdots M_{z+p-1}} \left[ \left( \frac{d^{p+1} c}{d\xi^{p+1}} \right)_{\xi=1} - \left( \frac{d^{p+1} c}{d\xi^{p+1}} \right)_{\xi=0} \right] \quad (12\text{-}108)$$

when $p$ is an even integer. From the above equation the data we seek to analyze must undergo differentiation several times. In certain instances integration also must be performed. Fundamentally this is a problem of data smoothing complicated in some cases by discontinuities in the molecular weight versus concentration curves.

According to the theory of Adams and Fujita (1963) these discontinuities should not exist, but they have been observed by Adams (1962), Squire and Li (1961), Jeffrey and Coates (1966), and Kakiuchi (1965). Albright and Williams (1968) felt that the result they obtained for $\beta$-lactoglobulin, while merging quite well at low concentration, showed some evidence of discontinuities at higher concentrations. These authors did not think this effect pronounced. A number of reasons have been given by various authors as to why this might happen but none are very convincing. To quote Albright and Williams (1968), "It is difficult to think of reasons why the theory of Adams and Fujita is sometimes not observed in practice." In view of the fact that the discontinuities are found after the processing of the raw data has occurred (to obtain the $M_{w\text{app}}$ versus concentration curve one must differentiate first), it may well be possible to eliminate these discontinuities if we are more efficient in this aspect of the work.

In suggesting that the raw data is processed by using the methods in this section it is supposed throughout that the raw data will in fact give a continuous curve when $M_{w\text{app}}$ is plotted against concentration. The possibility that the raw data does not in fact do that and that the discontinuities are not due to experimental error but are real has been suggested by Squire (personal communication). Thus, various computations by Squire and Li (1961) have shown that higher aggregates *not* in equilibrium with the monomer can give rise to these discontinuities. Whatever the case may be, it is clear that Squire's hypothesis must be ruled out if we are to use the data processing methods described in this section for overlapping concentration runs. It must be emphasized here that we are mainly concerned with the numerical analysis aspect of the problem.

Consequently, we will not say anything in regard to the suggestions by various authors as to how to overcome this problem for the experimental standpoint.

Jeffrey and Coates (1966) have suggested that we represent the logarithm of the concentration by the following equation:

$$\ln c = a + br^2 + dr^4 \tag{12-109}$$

where $a$, $b$, and $d$ are constants. To obtain these constants Jeffrey and Coates used a least squares approximation (Chapter 4), and then computed the weight average molecular weight. Using this method these authors found discontinuities in the $M_{w\text{app}}$ versus concentration curve. To avoid these difficulties they decided to fit their experimental data by means of a Fourier series, and they came to tentative conclusion that the discontinuities observed in the $M_{w\text{app}}$ versus concentration curve were due to the least squares fit (presumably because the Fourier series fit yielded less pronounced discontinuities).

In actual application Jeffrey and Coates used the least squares polynomial to fit most of the raw data used in their paper. In opting to use the least squares polynomial fit of ln $c$, despite their misgivings, the authors were effectively choosing mathematical convenience, since a polynomial is more rapidly differentiated than a Fourier series. Inasmuch as polynomial approximations are really the best representation of the data in terms of future analysis, it is pertinent to inquire as to the origin of the negative conclusion of Jeffrey and Coates. As we have pointed out in Chapter 4, a least squares polynomial solution, despite the fact that it defines the coefficients uniquely, gives rise to problems in extricating them because of the structure of the matrices involved. The higher the order of the polynomial the more difficult it is to determine the coefficients because of the ill-conditioning of the equations.

Since the least squares polynomial with which Jeffrey and Coates were dealing with was of low degree, it is a little difficult to see that the least squares polynomial was the source of the discontinuities. Further, if we remember that any kind of fit is only as good as the data that is being fitted, then the errors encountered in making the concentration versus distance measurement must be significant, at least in certain parts of the cell. It is a well known fact that it becomes increasingly difficult to measure the concentration at the ends of the column, especially at the bottom.

In the light of the above we feel that a weighted least squares (Chapter 4) may well be the best method for processing the data, since such a procedure will take into consideration the readings at the ends of the cell, weighting them in an inverse proportion to the error made in their measurement. Since the higher the order of polynomial the better it is able to follow any given data curve, it might be felt that a higher order polynomal is required to fit the data. However, as we pointed out above, while this could be done it is sometimes done at the risk of producing ill-conditioned equations, and it would be better to use orthogonal polynomials to fit the data.

In Chapter 4 we mentioned the use of the $L_\infty$ norm. In sedimentation equilibrium min-max approximations may be used when the investigator is supposing equal error throughout the solution column. This might be the case in the instances when one desires to neglect the top and bottom of the columns or in the instance one seeks specifically to minimize the error at the ends of the column since it is at the ends that the maximum error will occur.

It will be recalled that if one is to use Donnelly method to determine molecular weight distributions, the raw data must be processed in such a manner that the inverse Laplace transform of the fitting function must exist. Since the inverse Laplace transform or rational functions exist and since there exists nice min-max approximations with rational functions, it may be worthwhile for the investigators to use min-max fits with rational functions as a starting point for determining the molecular weight distribution function by Donnelly's method. Certainly it would be interesting to see how least squares fits of the raw data with rational functions compared with a min-max fit in the final computation of the molecular weight distribution.

## 4.2 Light Scattering and Low-Angle X-Ray Scattering

Procedures can be used for analyzing light scattering and low-angle X-ray scattering data that are quite similar to those applied to sedimentation equilibrium. Using light scattering and X-ray scattering techniques one may determine molecular weight distributions as well as particle scattering factors which are indicative of the shape of the particle.

### 4.2.1 *Molecular Weight Distributions*

In light scattering or X-ray scattering, whenever the molecular dimensions of the scattering particles exceed $\lambda/10$, where $\lambda$ is the wavelength of the incident radiation, interference between the various scattering centers in the molecule results. These interference effects are best described by means of a function $P(\theta)$, which depends on the angle of observation of the scattered radiation $\theta$, and is called the particle scattering factor. The equation for the particle scattering factor was first derived by Debye (1915) and may be written as

$$P(\theta) = (1/n^2) \sum_{i=1}^{n} \sum_{j=1}^{n} (\sin \mu r_{ij})/\mu r_{ij} \tag{12-110}$$

where $\mu = (4\pi/\lambda) \sin (\theta/2)$, $n$ is the number of scattering centers and $r_{ij}$ the distance between the $i$th and $j$th scattering centers. The function $P(\theta)$ depends on the shape of the particle.

As pointed out by Zimm, (1948), in a heterogeneous system, at infinite dilution, the total light scattering will be the sum of the scattering from each of the individual components. The same is true of X-ray scattering (see for example Guinier and Fournet 1955). For a polydisperse sample, Eq. (11) of Zimm's paper can be considered as an integral equation in the molecular weight distribution for the various aggregates function, provided $P(\theta)$ is known. This equation is written as

$$\langle P(\theta) \rangle = \int_a^b P(\theta, M) f(M) \, dM \tag{12-111}$$

where both $\langle P(\theta) \rangle$ and $P(\theta, M)$ are quantities which are known or can be determined and $f(M)$ is the molecular weight distribution. Rewriting Eq. (12-111) with a slight change of notation and replacing the integration by summation, we obtain

$$\langle P(\theta) \rangle = \Big( \sum_i P_i(\theta) c_i M_i \Big) \Big/ \Big( \sum_i c_i M_i \Big)$$

or (12-112)

$$c_0 \langle P(\theta) \rangle = (1/M_w) \sum_i P_i(\theta) c_i M_i$$

where $P_i(\theta)$ is the particle scattering factor of the $i$th particle, $c_i$ the concentration of the $i$th species, and $M_i$ its molecular weight. The quantity $M_w$ is the weight-average molecular weight and $c_0$ is the total concentration.

To obtain some useful information on the molecular weight distribution Zimm used the following distribution function written in terms of the degree of polymerization $\gamma$

$$f(\gamma) = (y^{z+1}/z!) \gamma^z e^{-y\gamma} \tag{12-113}$$

where $y$ and $z$ are the adjustable parameters. The width of the distribution is characterized by the parameter $z$. To our knowledge treatment of this subject has not been extended beyond the work of Zimm. Although Zimm derived his equation while considering the subject of light scattering, what is said of light scattering is also true of X-ray scattering; consequently what is said of light scattering here can be said, after making necessary changes, for X-ray scattering. We consider Eqs. (12-111) and (12-112) for the determination of the molecular weight distribution and we assume for the time being that $P_i(\theta)$ can be computed for all contributing species. The quantity $P_i(\theta)$ has in fact been computed for several different shapes and we leave to the discussion how these computation affect our results.

Since we wish to concentrate on biological systems, it is important to point out that these systems tend to be concentration-dependent aggregating systems and they tend to concentrate in integer multiples of the monomer. This distinction is not an idle one, from the point of view of the numerical analyst because in the case of the discretely aggregating biological system we will, either know, or have a fairly good idea, of how to effect the summation in Eq. (12-112) exactly. This is the case because we know the values of the $M_i$, whereas in Eq. (12-111) we must somehow approximate the integral by quadrature.

Using Eq. (12-112) we can measure the light scattered at as many different angles or $\theta$ as we please; consequently we can write Eq. (12-111) as many different times as we make measurements, say, $m$ times:

$$c_0\langle P(\theta_j)\rangle = (1/M_w) \sum_i P_i(\theta_j)c_iM_i + \delta_j \qquad (j = 1, 2, \ldots, m) \qquad (12\text{-}114)$$

where $\langle P(\theta_j)\rangle$ is $\langle P(\theta)\rangle$ determined at the $j$th measurement and $\delta_j$ is the experimental error encountered in making that measurement. If the $P_i(\theta)$ are known, then Eq. (12-114) could be considered as a linear equation in the unknown products $c_iM_i$. Further, if we note that when the distributions are discrete we have

$$M_i = iM_1 \qquad (12\text{-}115)$$

we also have the constraint

$$\sum_i c_i = c_0 \qquad (12\text{-}116)$$

where $c_0$ is the total concentration. If we substitute for the $M_i$ in Eq. (12-114) by using Eq. (12-115), then Eq. (12-114) can be written in terms of $M_1$ only, the molecular weight of the monomer or the smallest species. This is generally known in concentration-dependent aggregating systems (proteins, nucleotides for instance). Consequently, we write

$$c_0\langle P(\theta_j)\rangle = (1/M_w) \sum_i iP_i(\theta_j)M_1c_i + \delta_j \qquad (j = 1, 2, \ldots, m) \quad (12\text{-}117)$$

We can solve Eq. (12-117) to determine the unknowns $c_i$ such that $\sum_j^m |\delta_j|$ or $\sum_j^m \delta_i^2$ is a minimum or some function of the experimental error is minimized. When that is done, we can use the values of $c_i$ to determine the molecular weight distribution. Our problem, therefore, is to solve Eq. (12-117) in such a way that $\sum_{j=1}^m |\delta_j|$ or $\sum_{j=1}^m \delta_j^2$ is a minimum and that can be done by several methods. Before considering such solutions, consider the case of a continuous distribution.

Zimm's equation in the continuous case is essentially a Fredholm integral equation of the first kind. The numerical methods available for its solution include quadrature where we approximate the integral on the right-hand side of Eq. (12-111) by a summation in the following manner

$$\int_a^b P(\theta, M)f(M)dM = \sum_{R=1}^q G_R P(\theta, M_R)f(M_R) \qquad (12\text{-}118)$$

Equation (12-118) represents a quadrature scheme which determines for us the $G_R$ or weights as they are called and the $M_R$ (see Chapter 3). Consequently, the question of the spacing of the $M_R$ is effectively determined for us by quadrature, and we do not have to arbitrarily space them. The $G_R$ and the $M_R$ will have different values depending on the quadrature scheme used to approximate the integral. Once the integral is properly approximated, we can rewrite Eq. (12-111) $m$ times as follows:

$$\langle P(\theta_j)\rangle = \sum_{R=1}^q G_R P(\theta, M_R)f(M_R) + \delta_j \qquad (j = 1, 2, \ldots, m) \quad (12\text{-}119)$$

where $\delta_j$ now incorporates both the experimental error and the error made in the quadrature. (This latter error is generally small and tends to zero as $q$ increases.)

The similarity between Eq. (12-117) and Eq. (12-119) is now evident in the sense that they both represent simultaneous linear equations in the unknowns $c_i$ and $f(M_R)$. Equation (12-119) can be solved so as to minimize either $\sum_{j=1}^m |\delta_j|$ or $\sum_{j=1}^m \delta_j^2$.

As before in the instance that we seek to minimize $\sum_{j=1}^m |\delta_j|$ we use linear programming. Linear programming must be used, because Eqs.

(12-117) and (12-119) must be solved subject to the constraints $c_i \geq 0$ for all $i$ or $f(M_R) \geq 0$ for all $R$. If, on the other hand, we wish to minimize $\sum_j^m \delta_j^2$ then the following methods of chapter 4 may be suggested for solving Eqs. (12-117) and (12-119): The least squares, the weighted least squares the constrained least squares; the method of steepest descent, and the Nelder and Mead method. In the use of each one of these methods, we must obtain an answer such that $c_i \geq 0$ for all $i$ and $f(_RM) \geq 0$ for all $R$. Other methods are outlined in Chapter 4.

### 4.2.2 *Determination of the Particle Scattering Factors*

If we assume we have an independent method for the determination of the molecular weight distribution, such as sedimentation equilibrium, and if the aggregating system is concentration dependent which is to say that at various concentrations we will have different molecular weight distributions, then we can think of the $\alpha_i$(which are now functions of concentration) as known quantities and proceed to write Eq. (12-114) for $m$ different concentrations at a given single scattering angle $\theta$ as follows

$$\langle P(\theta, c_j)\rangle = \sum_i \alpha_{ij} P_i(\theta) + \delta_j \qquad (j = 1, 2, \ldots, m) \quad (12\text{-}120)$$

The above $m$ equations which now represent the measurement of $P(\theta)$ for $m$ solution at different concentrations at a *given* $\theta$ can now be solved for $P_i(\theta)$ to determine those values, since the $\alpha_{ij}$ are known from sedimentation equilibrium studies. By determining the entire particle scattering factor for each one of the $m$ solution we can determine all the quantities $P_i(\theta_k)$ and they are the particle scattering factor of the various aggregates. $P_1(\theta)$ for the monomer, $P_2(\theta)$ for the dimer and so on.

Equation (12-120) is identical in algebraic form to Eq. (12-117) and is solved to determine the $P_i(\theta)$ by the techniques outlined in Chapter 4 to minimize $\sum_j^m | \delta_j |$ or $\sum_j^m \delta_j^2$, as the case may be. To obtain the standard error on the parameters we refer the reader to Chapter 6.

The question of scattering factors $P_i(\theta)$ to be used in Eq. (12-117) or $P(\theta, M)$ to be used in Eq. (12-119) should be considered. These particle scattering factors are assumed computable from theoretical principles using Eq. (12-110) as opposed to particle scattering factors experimentally determined from Eq. (12-120).

Geidushek and Holtzer (1958) have tabulated the particle scattering factor for a number of shapes. Kratky (1963) in a later review, increased that number. For linear polymers such as Gaussian coils, rods, etc., as well as particles whose shapes do not change with increasing molecular weights the tabulated $P_i(\theta)$ can be used. For discrete systems, unless $P_1(\theta)$ is known it does not seem possible that we can compute $P_2(\theta)$, $P_3(\theta)$, etc. There is no reason, however, why certain assumptions cannot be made as regards $P_1(\theta)$ followed by a comparison of the experimental results to see how well such assumptions fit. Obviously, there are possible assumptions that could be made about the system either intuitively or on the basis of other independent evidence. For continuous molecular weight distributions of particles whose shape does alter with increasing size the problem of determining $P(\theta, M)$ is less formidable since quadrature will effectively approximate $P(\theta, M)$.

For discrete aggregating systems where the particle shape changes with size (and this is probably true of proteins), we must make a number of assumptions about $P_2(\theta)$, $P_3(\theta)$, . . . . Proteins tend to be ellipsoids and assumptions as to how these ellipsoids are put together will be called for. Such procedures for computing particle scattering factors are not new and various particle scattering factors for a very large number of possible shapes of proteins in single or multiple units have appeared in the literature from time to time (Bloomfield, 1965; Kratky and Kreutz, 1960; Luzzatti *et al.*, 1961; Cleeman and Kratky, 1960; Kratky *et al.*, 1955; Maomon, 1957; Mittelbach and Porod, 1961, 1962; Kratky and Porod, 1948).

With the present high-speed computers it appears that the use of (12-110) will enable us to compute the scattering curves for any shape and that greatly enhances the possibilities of the method, especially when it comes to comparing the computed scattering curves with those derived from a solution to Eq. (12-120).

It is necessary to emphasize that the methods we propose for computing molecular weight distributions require very accurate data. However, if the data is not very accurate the solution of the equations can still be effected and the resulting parameters will in general provide useful information about the distribution. Certainly, for the purpose of comparing two or more samples, there is no reason why the polydispersity of such samples cannot be characterized in terms of only a few (five or six) parameters.

On comparing light scattering to sedimentation equilibrium for the determination of molecular weight distribution, at the present level of experimental accuracy, sedimentation equilibrium is probably the more accurate method. However, given the same level of experimental accuracy, light scattering, by virtue of the fact that the particle scattering factors have more features (curvature) than their equivalent counterparts in sedimentation equilibrium, will permit a more unequivocal determination of molecular weight distributions. Unlike Zimm, who cleverly suggested that we use the inverse of the particle scattering factor to facilitate extrapolation in the determination of $M_w$ and the mean-square radius, the use of particle scattering factors with all its features will be quite beneficial in such analyses.

If the accuracy of the light-scattering experiments does not exceed that of sedimentation equilibrium we can neglect the use of light-scattering data for determining molecular weight distributions and concentrate on determining the shapes of the various aggregating species from the evaluation of the particle scattering factors of systems already characterized (as regards molecular weight distributions) by sedimentation equilibrium. We believe for the present it is in that respect that light scattering will be most useful.

#### 4.2.3 *Comment on Other Possible Ways of Determining Molecular Weight Distributions Using Similar Integral Equations*

Examples of other applications, specifically with respect to the determination of molecular weight distributions, may be given as follows: If $F(\theta, M)$ is a well-defined quantity of a polymer solution dependent on some experimental quantity $\theta$ and the polymer molecular weight $M$, then the average quantity $\langle F(\theta)\rangle$ which is measured for a solution of a polymer whose molecular weight distribution is given by $W(M)$ will be given by an integral equation such as Eq. (12-111) and solved by the methods outlined. In such instances $\langle F(\theta)\rangle$ may be a shear rate or a frequency dependent viscoelastic quantity such as the viscosity or storage modulus (Peticolas, 1961, 1962).

## 5. Gel Filtration Methods in the Determination of Molecular Weight Distributions

In recent years gel filtration has come to play an important part in the analysis of aggregating systems. Two methods have been proposed to handle data from gel filtration. Because of the analogy the methods show to transport phenomena they have been labeled integral boundary and differential boundary analysis. The differential boundary method has been used by Chiancone *et al.* (1968).

## 5.1 Integral Boundary Method for Gel Filtration

The method described by Ackers (1968) is a method for determining the molecular size[†] distribution. It depends on the use of molecular sieving columns of different porosity. Ackers based his method on the determination of the empirical quantity $\sigma_x$. The subscript $x$ pertains to the porosity of the column and the quantity $\sigma_x$ may be determined from the equation

$$\bar{V} = V_0 + \bar{\sigma}_x V_i \tag{12-121}$$

where $V_0$ is the void volume of the column, $V_i$ the internal gel volume, and $\bar{V}$ the elution volume of the species of interest. It might be noted that Ackers in his paper writes the quantity $\bar{\sigma}_x$ as $\bar{\sigma}_{sx}$. In such cases the first subscript refers to the total concentration of the reaction mixture put on the column.[‡] The quantity $\sigma_x$ has been shown be Ackers to be of the form

$$\sigma_x = \sum_{i=1}^{k} \sigma_{ix} c_i \Big/ \sum_{i=1}^{k} c_i = \sum_{i=1}^{k} f_i \sigma_{ix} \tag{12-122}$$

where $c_i$ is the concentration of the $i$th species, $k$ the number of species present in the system, $f_i$ the weight fraction of the $i$th species, and $\sigma_{ix}$ the molecular sieve coefficient of the $i$th species on a column $x$ where $x$ is the characteristic porosity. The molecular sieve coefficient $\sigma$ for the $i$th species on a column $x$ is a function of the molecular radius $a_i$ of the $i$th species and a set of calibration constants $u_{xi}$ (Porath, 1963; Squire, 1964; Laurent and Killander, 1964; Ackers, 1964). In general we may express $\sigma_{ix}$ as follows

$$\sigma_{ix} = \sigma(a_i, u_{xi}) \tag{12-123}$$

to indicate its dependence on $a_i$ and $u_{xi}$. An exact expression for $\sigma_{ix}$ derived by Laurent and Killander (1964) can be written as

$$\sigma_{ix} = P_x \exp[-\pi L_x(a_i + Q_x)^2] \tag{12-124}$$

† The phrase molecular size distribution is preferred to molecular weight distribution since it is function of molecular radius and is only indirectly related to molecular weight.

‡ We really do not have any need for the first subscript to carry out the analysis for the simple reason that in concentration-dependent aggregating systems a different concentration will yield a different molecular size distribution. The analysis, as will be seen, needs only the application of a *single* concentration to several columns of *different* porosity to determine the molecular size distribution. However, to determine stoichiometry and equilibrium constants one must know the absolute concentration of the monomers or any other polymer.

where $P_x$, $L_x$ and $Q_x$ are calibration constants for a given gel. With this expression for $\sigma_{ix}$ we can now express Equation (12-122) in the form

$$\sigma_x = \sum_{i=1}^{k} f_i\{P_x \exp[-\pi L_x(a_i + Q_x)^2]\} \tag{12-125}$$

Since the only unknowns in Eq. (12-125) are $f_i$ and $a_i$, Ackers observes that by determining $\sigma_x$ for $2k$-1 columns of different porosity and bearing in mind that

$$\sum_{i=1}^{k} f_i = 1 \tag{12-126}$$

we can determine the unknowns $f_i$ and $a_i$. Ackers does not provide a solution for this problem but goes on to solve a similar problem for two components.

Before indicating the solution to a system of $2k$-1 or more nonlinear equations such as Eq. (12-125) to determine the $f_i$ and the $a_i$, we shall explain how Ackers solved a similar problem for a simulated experiment involving only two components. In solving this problem Ackers does not use the calibrating function given in Eq. (12-124) but introduces another calibrating function (Ackers, 1967) written as

$$\sigma = \operatorname{erfc}[(a - \alpha_0)/\beta_0] = 1 - (2/\sqrt{\pi} \int_0^{(a-\alpha_0)/\beta_0} \exp(-a^2)\, da \tag{12-127}$$

or

$$a = \alpha_0 + \beta_0 \operatorname{erfc}^{-1}\sigma \tag{12-128}$$

where $a$ is the molecular radius of the component in question, $\alpha_0$ and $\beta_0$ constants characteristic of the column, and $\operatorname{erfc}^{-1}$ the inverse error function complement of $\sigma$. The tables of these functions are found in most mathematical handbooks. Since Ackers is interested in the molecular radius as one of the quantities he finally decides on Eq. (12-127) as opposed to Eq. (12-125) for the calibrating function for obvious reasons. In the first place, the molecular radius does not occur in a quadratic form as it does in Eq. (12-124) and in the second only two constants $\alpha_0$ and $\beta_0$ are required for the calibration of the gel instead of three in the case of the Eq. (12-124).

Ackers' experimental procedure for a two-component system existing as a monomer-dimer equilibrium is to choose three columns with different porosity. Then using a number of molecules of known molecular radius he proceeds to make a plot of the molecular radius versus $\operatorname{erfc}^{-1} \sigma$.

This can easily be done once one determines $\sigma$ for the proteins in question and consults the mathematical tables. According to Eq. (12-128) the plot of molecular radius versus $\text{erfc}^{-1}\,\sigma$ should yield $\alpha_0$ and $\beta_0$ the caliberation constants. The reaction mixture of interacting components is then put on each of his three columns to determine $\sigma_1$, $\sigma_2$, and $\sigma_3$ for this mixture. Subsequently, by using the following equations

$$\begin{aligned} \sigma_1 &= f_1 \,\text{erfc}[(a_1 - \alpha_1)/\beta_1] + (1 - f_1)\,\text{erfc}[(a_2 - \alpha_1)/\beta_1] \\ \sigma_2 &= f_1 \,\text{erfc}[(a_1 - \alpha_2)/\beta_2] + (1 - f_1)\,\text{erfc}[(a_2 - \alpha_2)/\beta_2] \\ \sigma_3 &= f_1 \,\text{erfc}[(a_1 - \alpha_3)/\beta_3] + (1 - f_1)\,\text{erfc}[(a_2 - \alpha_3)/\beta_3] \end{aligned} \qquad (12\text{-}129)$$

and recalling that $\alpha_1$, $\alpha_2$, $\alpha_3$, $\beta_1$, $\beta_2$, and $\beta_3$ are known caliberation constants for each of the three columns, then the unknown quantities are $f_1$, $a_1$, and $a_2$. Ackers solved the above equations graphically for $f_1$, $a_1$, and $a_2$ and Table II of Ackers (1968) shows the results of his simulated experiments where he assumed that the radii of his molecular components to be $a_1 = 24.4$ Å and $a_2 = 31.0$ Å and solved his equation for various concentrations of his reaction mixture assuming an experimental error of 0.1 ml in the centroid volume.

The column labeled second calculations in Ackers' paper results from averaging all the values of $a_1$ and $a_2$ obtained from the first calculation and inserting those averaged values back in Eqs. (12-129). When that is done the author has three equations for any given $f_1$ and consequently solves three times for $f_1$. The average of those three values for the $f_1$ are tabulated in the second column.

### 5.2 Solutions to the Problem of Determining the Molecular Weight Distribution Function in the General Case

In general, the system of equations we have to solve depends on the caliberation function used. For the caliberation function given by Laurent and Killander we have the following equations

$$\begin{aligned} \sigma_1 &= \sum_{i=1}^{k} f_i\{P_1 \exp[-\pi L_1(a_i + Q_1)^2]\} + \delta_1 \\ \sigma_2 &= \sum_{i=1}^{k} f_i\{P_2 \exp[-\pi L_2(a_i + Q_2)^2]\} + \delta_2 \\ &\vdots \\ \sigma_m &= \sum_{i=1}^{k} f_i[P_m \exp[-\pi L_m(a_i + Q_m)^2]\} + \delta_m \end{aligned} \qquad (12\text{-}130)$$

to be solved for $f_i$ and $a_i$, the $\delta_j$ being errors incurred in the measurements. Equation (12-130) is clearly a nonlinear equation and if we seek to minimize $\sum_{j=1}^{m} \delta_j^2$ the available methods of solution are linearizing a Taylor series, steepest descent, Marquart method, Newton–Raphson method, Nelder and Mead, as well as other methods discussed in Chapter 4. We have not tried any of these methods so our suggestions in Chapter 4 hold.

The other calibration function presents us with the following equations

$$\begin{aligned}
\sigma_1 &= \sum_{i=1}^{k} f_i\left[1 - (2/\sqrt{\pi}) \int_0^{(a_i-\alpha_1)/\beta_1} \exp(-a^2{}_i)\, da_i\right] + \delta_1 \\
\sigma_2 &= \sum_{i=1}^{k} f_i\left[1 - (2/\sqrt{\pi}) \int_0^{(a_i-\alpha_2)/\beta_2} \exp(-a_i^2)\, da_i\right] + \delta_2 \\
&\vdots \\
\sigma_m &= \sum_{i=1}^{k} f_i\left[1 - (2/\sqrt{\pi}) \int_0^{(a_i-\alpha_m)/\beta_m} \exp(-a_i^2)\, da_i\right] + \delta_m
\end{aligned} \tag{12-131}$$

where the $\alpha_j$ and $\beta_j$ are the calibration constants of the $j$th column. Once again Eqs. (12-131) are nonlinear and what was said of Eqs. (12-130) applies, after necessary charges are made, to Eqs. (12-131).

To determine the confidence interval on the parameters we refer the reader to Chapter 6. To determine how many terms we require for the analysis or the highest aggregating species we may use the $F$-test but it must be remembered that our problem here is a nonlinear one.

The solution of the nonlinear equation presented above essentially determines the molecular weight distribution and from that point of view the problem is solved. There are, however, a number of further considerations in relation to the work of Ackers which we may look at now. Ackers wrote the following equations for the partition coefficient $\sigma_{sx}$ of a reaction mixture. (Note that the partition coefficient now depends on the concentration of the reaction mixture and $x$ the porosity of the column on which it was put.)

$$\begin{aligned}
\sigma_{sx} = \sigma_{1x} &+ (2K_2/M_1)(2_x - \sigma_{1x})C_s + [3K_3(\sigma_{3x} - \sigma_{1x}) \\
&- 8K_2^2(\sigma_{2x} - \sigma_{1x})](C_s^2/M_1^2) + \cdots
\end{aligned} \tag{12-132}$$

where $C_s$ is the concentration of the $s$th mixture.

Ackers suggested that we use this equation to solve for $M_1$ the molecular weight of the monomer, since all the quantities in this equation other than $M_1$ are known (the equilibrium constants $K_2$, $K_3$, etc., are known if we know the molecular weight distribution function). Although Ackers suggested we solve Eq. (12-132) by iterations to find $M_1$, it is clear that nonlinear methods described in Chapter 4 will do a better job in solving for $M_1$. Actually, determination of $M_1$ is rarely a problem especially in the analysis of proteins since $M_1$ generally is known from some other experiment. When $M_1$ is known then Eq. (12-132) can be used to determine the equilibrium constants $K_2$, $K_3$, etc., if we regard them as unknowns. There is considerable merit in determining the equilibrium constants in that way inasmuch as the problem is now linear and all the linear methods and statistical techniques pertinent to the linear methods can now be brought to bear on the problem. In particular, the statistical techniques will be most useful in the determination of the highest aggregating species as we have shown earlier.

The question as to how many components are in the aggregating or what is the highest aggregating species was also considered by Ackers (1968). A matrix rank analysis was suggested to determine the number of linearly independent components. If we define $O_{sx} = C_s \bar{\sigma}_{sx}$, where $C_s$ as before is the total concentration of reaction mixture, we can write $O_{sx}$ as

$$O_{sx} = \sum_{i=1}^{k} C_{si} \sigma_{ix} \tag{12-133}$$

or in matrix notation we can write

$$\mathbf{O} = \mathbf{C}\boldsymbol{\sigma} = \begin{bmatrix} O_{11} & O_{12} & \cdots & O_{1n} \\ O_{21} & O_{22} & \cdots & O_{2n} \\ \vdots & \vdots & & \vdots \\ O_{m1} & O_{m2} & \cdots & O_{mn} \end{bmatrix}$$

$$= \begin{bmatrix} C_{11} & C_{12} & \cdots & C_{1k} \\ C_{21} & C_{22} & \cdots & C_{2k} \\ \vdots & \vdots & & \vdots \\ C_{m1} & C_{m2} & \cdots & C_{mk} \end{bmatrix} \begin{bmatrix} \sigma_{11} & \sigma_{12} & \cdots & \sigma_{1n} \\ \sigma_{21} & \sigma_{22} & \cdots & \sigma_{2n} \\ \vdots & \vdots & & \vdots \\ \sigma_{k1} & \sigma_{k2} & \cdots & \sigma_{kn} \end{bmatrix} \tag{12-139}$$

Effectively we will know how many components are present if we can know how many terms to use in Eq. (12-133). This means how many columns of the matrix **C** are required to completely define this system. Since the elements of **C** are not known Ackers suggests that the rank of the matrix **O** should tell us the rank of **C** since **σ** is nonsingular (see theorem on matrices in Chapter 2). Ackers did not perform any matrix rank analysis, consequently what criteria one should use to terminate the analysis is not mentioned. We presume the criteria given by Wallace and Katz (1964) should be adequate (H. Warshaw, personal communication). The use of matrix rank analysis and statistical analysis to determine how many terms are required have been compared in Chapter 9, where we mentioned the advantage of statistical analysis. Warshaw has communicated the following to the author.

We currently use two methods of matrix rank analysis. One is the Wallace and Katz (1964) method. The other is an unpublished method developed by Warshaw and W. Godschalk in which at the *k*th stage of matrix reduction, we minimize

$$\sum_{i=l+1}^{m} \sum_{j=k+1}^{n} (S'_{ij}/S_{ij})^2$$

where

$$S'_{ij} = [S_{ij}^2 + S_{kj}^2(\sigma_{ik}/\sigma_{kk})^2 + S_{ik}^2(\sigma_{kj}/\sigma_{kk})^2 + S_{kk}^2(\sigma_{ik}\sigma_{kj}/\sigma_{kk}^2)^2]^{1/2}$$

where $S_{ij}$ is the error on matrix (data) element $\sigma_{ij}$ and $m$ and $n$ are the number of rows and columns. That is, we try pivoting *each* (data) submatrix element to the upper left corner of the (data) submatrix, and we choose the one which propagates the least error. We terminate the analysis when all elements of the next (data) submatrix are less than two or three times (depending on the size of the submatrix) their corresponding elements in the error matrix.

In discussing matrix rank analysis to determine the number of components in a system Ackers noted that the bilinear form in Eq. (12-133) is completely determined whenever $mn \geq k(m + n)$ so we could in principle solve for the elements of the matrices $\mathbf{C}$ and $\boldsymbol{\sigma}$ without having to calibrate the columns. The fact that a bilinear form is completely determined whenever $mn \geq k(m + n)$ leads us to consider the following bilinear form which may be used to determine the molecular weight distribution. Ackers wrote

$$\bar{\sigma}_{sx} = \sum_{i=1}^{k} f_{si}\sigma_{ix} \tag{12-135}$$

From this bilinear form we can determine in principle the molecular weight distribution.

Solving a bilinear form or using the other method proposed by Ackers requires a great deal of experimental labor, a fact recognized by Ackers who proposed to overcome some of his problems by constructing a column with a continuous gradient of porosity. Theoretical considerations lead to the integral equation†

$$\sigma_t(x) = \int_0^\infty f(a)\sigma(a, x)\, da \tag{12-136}$$

where $f(a)$ represents the molecular weight distribution function. In order to solve this equation for $f(a)$ the function $\sigma_t(x)$ and kernel $\sigma(a, x)$ must be determined. Ackers proposes to determine both quantities by novel experimental techniques. For $\sigma_t(x)$ a direct densitometric scan of a saturated column is required. The kernel of the equation is determined from the calibration of the column with standards of known $a$. With these functions in hand, the molecular weight distribution function $f(a)$ may be determined by solving Eq. (12-136). It will be interesting to see how well these new experimental techniques work.‡

Some recent results from Ackers' laboratory concerning the scanning methods and the solution of a system of simulated equations of the form (12-131) have kindly been supplied by H. Warshaw. The scanning system (Brumbaugh and Ackers, 1968) greatly reduces the number of experiments required to solve for the molecular radii, stoichiometry, and equilibrium constants. One can stack a number of gels into a single column and do a saturation experiment of the type described by Brumbaugh and Ackers. The gels can easily be accomodated in only 2 columns. Not only that, but within a gel region, 50–100 values of $\bar{\sigma}_x$ can be averaged, vastly improving the accuracy of the data.

With respect to the simulation of Eq. (12-131), the following was stated by Warshaw and Ackers (1969) using the Nelder and Mead method.

> We successfully solved a simulation of a three-component system (monomer-dimer-tetramer) on ten gels. In solving for the necessary parameters (the $a_i$'s and $f_i$'s) in the 10 non-linear equations, we made extensive use of a Nelder and Mead minimization computer program written by Peter Spragg. We found this procedure to be quite effective. In solving for either 2 or 3 parameters, our solution was *independent of the initial guess.* This is of course very valuable. With 4 or more parameters, the solutions would generally depend on the initial guess, due to the existence of several local minima. In these cases it was necessary to make several initial guesses, and choose the solution at the true minimum. With more than 5 parameters, the solutions generally are unreliable, because of the errors we have put on the data.

† A great deal about the solution of such integral equations has been stated in the text of this chapter.

‡ If they work well, and it looks like they might, columns might end up doing a major part of the work of the ultracentrifuge.

## 5.3 Differential Boundary Methods in Gel Filtration

This approach elaborated by Chiancone *et al.* (1968) may be labeled the difference boundary method because of its close analogy to other techniques (in sedimentation and electrophoresis) using the same mathematical ideas. The equations pertaining to this method are best derived by using Fig. 12-2. In this figure the continuous heavy lines are the concentration of the effluent as a function of time, and the broken line the corresponding concentration of the solution flowing into the column. In these experiments the rate of flow is kept constant, and therefore the

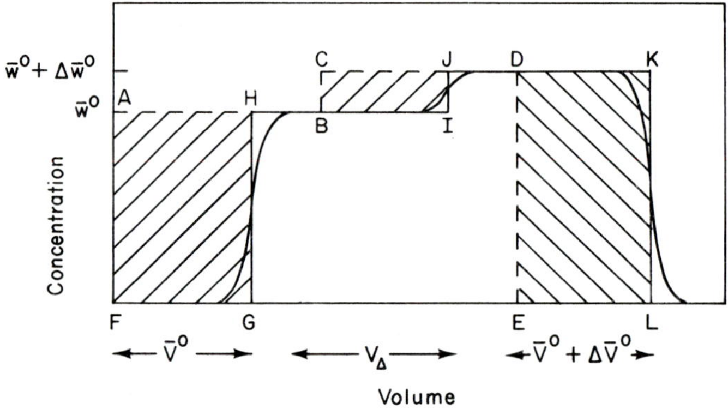


**Fig. 12-2.** The relationship between the elution volume, $V_\Delta$, of a finite difference boundary formed between two solutions of plateau constituent concentrations $\bar{w}^0$ and $\bar{w}^0 + \Delta\bar{w}^0$, and their constituent elution volumes, $\bar{V}^0$ and $\bar{V}^0 + \Delta\bar{V}^0$, respectively. ----, inflowing solution; ——, outflowing solution. *Vertical solid lines* represent the position of the appropriate equivalent sharp boundary (Chiancone *et al.*, 1968).

abscissa represents either the time or the volume of liquid that has entered or left the column. The vertical solid lines in Fig. 12-2 have been constructed in close analogy with the Longsworth (1943) method for electrophoresis experiments that shows the positions of the equivalent sharp boundaries corresponding to the leading integral boundary *GH*, the trailing integral boundary *LK*, and the finite difference boundary *IJ*. The path of the inflowing solution on the figure follows the letters *A*, *B*, *C*, *D*, *E*, and *M*, and that of the effluent *F*, *G*, *H*, *I*, *J*, *K*, *L*, and *M*. At any instant the total amount of solute held by the column is equal to the difference between the amounts that have entered and left the column, and this can be read from the figure as the area encompassed by the broken and solid lines. The areas above the solid line count as positive and those below it as negative. The total amount of solute that can be included at equilibrium in the column will be termed and is a function of the total concentration $\bar{w}$. The quantity $\bar{w}^0$ is the initial concentration of the solute. The corresponding value of $\mathscr{I}^0$ is given by the area $AHGF$. At point $B$ we increase the concentration of the incoming solution by $\Delta\bar{w}^0$. The corresponding increase in $\mathscr{I}^0$ is $\Delta\mathscr{I}^0$ and is given by the area $CJIB$. At point $D$ the inflowing solution is replaced by solvent and all the included solute $\mathscr{I}^0 + \Delta\mathscr{I}^0$ given by the area $DKLE$ is swept out of the column. From the diagram we have

$$\mathscr{I}^0 = \bar{w}^0\bar{V}^0 \tag{12-137}$$

and

$$\begin{aligned}\mathscr{I} + \Delta\mathscr{I}^0 &= \bar{w}^0\bar{V}^0 + \Delta(\bar{w}^0\bar{V}^0)\\ &= (\bar{w}^0 + \Delta\bar{w}^0)(\bar{V}^0 + \Delta\bar{V}^0)\end{aligned} \tag{12-138}$$

In order to have mass conservation we must have the following

$$\text{area}(AHGF) + \text{area}(CJIB) = \text{area}(DKLE) \tag{12-139}$$

This means that

$$\bar{w}^0\bar{V}^0 + (\Delta\bar{w}^0)V_\Delta = \mathscr{I}^0 + \Delta\mathscr{I}^0 \tag{12-140}$$

Now in the limit as $\Delta\bar{w}^0 \to 0$, for a true differential boundary, we must have

$$\lim_{\Delta w\to 0} V_\Delta = V = d\mathscr{I}^0/d\bar{w}^0 = d(\bar{w}^0\bar{V}^0)/d\bar{w}^0 \tag{12-141}$$

Consequently to obtain the quantity of interest $V$ we take tangents on the finite difference boundaries. In the actual experiments $\bar{V}^0$ and $V_\Delta$ are measured for a series of values of $\bar{w}^0$. Chiancone *et al.* (1968) recommended that we set $\Delta\bar{w}^0$ at about 10% of $\bar{w}^0$. Making use of these measurements and Eqs. (12-138) and (12-141) we can obtain $\mathscr{I}^0$ and $\Delta\mathscr{I}^0/\Delta\bar{w}^0$ as a function of $\bar{w}^0$. This enables us to compute $V$. To obtain the equilibrium constants from differential boundary gel filtration data Chiancone and coworkers who were analyzing hemoglobin supposed that their aggregating system was monomer-dimer-tetramer and wrote

$$w_2 = K_{1,2}w_1^2, \qquad w_3 = K_{1,3}w_1^3, \qquad w_4 = K_{1,4}w_1^4 \tag{12-142}$$

and

$$\bar{w} = w_1 + w_2 + w_3 + w_4 \tag{12-143}$$

where $w_1$, $w_2$, $w_3$, and $w_4$ are the concentrations of monomer, dimer, trimer, and tetramer respectively, and $\bar{w}$ the total concentration of the solute, and $K_{1,2}$ $K_{1,3}$, and $K_{1,4}$ association constants. This model Chiancone and coworkers simplified further by rewriting Eq. (12-140) as follows:

$$w_2 = K_{1,2}w_1^2, \qquad w_3 = K_{2,4}w_1w_2, \qquad w_4 = K_{2,4}w_2^2 \tag{12-144}$$

where

$$\varrho = (K_{1,2}K_{1,3}/K_{1,4}) \tag{12-145}$$

and

$$K_{2,4} = [K_{1,4}/(K_{1,2})^2]. \tag{12-146}$$

With these definitions the authors went on to analyze Eq. (12-150) (as shown below) to determine the equilibrium constants. Before we proceed with consideration of Eq. (12-150), we emphasize that most aggregating schemes can be analyzed by this method provided the pertinent equations for the system under consideration are properly derived. Further, in Eq. (12-150) we require that a number quantity corresponding to the elution volume of the various aggregating species must be determined. This quantity is determined by means of a calibrating equation given by a number of authors (Whitaker, 1963; Andrews, 1964; Determann and Michel, 1966) and can be written as

$$\bar{V} = A + B \log M_w \tag{12-147}$$

where $A$ and $B$ are constants. Strictly speaking $\bar{V}$ depends on the concentration of the solute and consequently the authors wrote the following equation for $V$ to take the effect concentration into account

$$V_n = (V_n)_0(1 - g\bar{w}) \tag{12-148}$$

where $g$ is a constants and $(V_n)_0$ is the elution volume of the $n$-mer at zero concentration. As stated above we require the quantities $(V_1)_0$, $(V_2)_0$, $(V_3)_0$ and $(V_4)_0$ for further analysis. For their purposes Chiancone and coworkers obtained the quantity $(V_1)_0$ from the elution volume of myoglobin at zero concentration. To obtain $(V_4)_0$ we take the elution volume of hemoglobin. Since we know the molecular weights of myoglobin and hemoglobin, then we can use Eq. (12-147) to determine the constants $A$ and $B$ and that, in turn, will help us to determine $(V_2)_0$ and $(V_3)_0$ the elution volume of the dimer and trimer. Finally, to derive the equation we want to analyze we make use of the equation

$$\bar{w}\bar{V}^0 = \mathcal{T}^0 = \bar{w}_1{}^0V_1 + w_2{}^0V_2 + w_3{}^0V_3 + w_4{}^0V_4 \tag{12-149}$$

and if we substitute for the $V$ using Eq. (12-148) we obtain

$$\bar{w}^0\bar{V}^0 = (1 - g\bar{w}^0)[(V_1)_0 w_1{}^0 + K_{1,2}(V_2)_0(w_1{}^0)^2 + \varrho K_{1,2}K_{2,4}(w_1{}^0)/(V_3)_0 + K_{1,2}^2 K_{2,4}(w_1{}^0)/(V_4)_0] \quad (12\text{-}150)$$

in terms of $w_1{}^0$ only. This latter equation is the equation required to determine the equilibrium constants. Before doing so, let us note one particular assumption implicit in this derivation, that the parameter $g$ which is inserted in Eq. (12-148) to take into account the concentration dependence of the evolution volume is assumed the same for all species. The authors have discussed this assumption at some length in their paper and conclude that it had little effect on their final results.

Chiancone *et al.* (1968) applied the Nelder and Mead and the Newton–Raphson methods to Eq. (12-150) to evaluate the constants $(V_4)_0$, $K_{1,2}$, $K_{2,4}$, and $g$ by using a series of experimental values for $\bar{w}^0$ and $\bar{V}^0$. It is not at all clear to this author how this was done without one knowing the values of $\omega_1{}^0$ and $\varrho$, but we presume from Table I of Chiancone *et al.* that $\varrho$ ($\delta$ in the notation of the authors) at least was set equal to 1 in the computation. Possibly $\omega_1{}^0$ was also computed in the process so the authors in effect obtained $(V_4)_0$, $K_{1,2}$, $K_{2,4}$, $g$ and $\omega_1{}^0$ using the Nelder and Mead and Newton–Raphson methods. The results of these authors are shown in Table I of that paper.

## 5.4 Winzor's Method for Determing Association Constants

Proceeding in complete analogy to Steiner's procedure, Winzor (1969) stated that the weight fraction of the monomer $\alpha_A$ at a concentration $c_0$ applied to the gel could be obtained from the following equation:

$$\ln \alpha_A = \int_0^{c_0} [(V_A/V_w) - 1]\, d \ln c \quad (12\text{-}151)$$

From Eq. (12-151) we note the $\alpha_A$ can be determined from a series of measurements of $V_w$ at different applied concentrations of the protein $c_0$. Here $V_w$ is equal to first moment of a boundary in an elution profile and $V_A$ is obtained by extrapolating a $V_w$ vs. $c_0$ plot to zero solute concentration. Once $\alpha_A$ is known at different $c_0$ then we can write the following equation $m$ times with our usual notation:

$$c_{0j}V_{wj} = V_A[\alpha_{Aj}c_{0j} + 2K_2(\alpha_{Aj}c_{0j})^2 + 3K_2K_3(\alpha_{Aj}c_{0j})_3 + \cdots] + \delta_j \quad (j = 1, 2, \ldots, m) \quad (12\text{-}152)$$

Since all the quantities except the association constants are known then we can solve for them to minimize $\sum_{j=1}^{m} \delta_j^2$ or some other function of the experimental error.

## 6. Mixed Aggregation

The theory of mixed aggregation using number average molecular weights has been dealt with by Steiner (1968). For an example of the application of such a theory to the mixed aggregation of inosine and cytidine, where the association constants of cytidine and inosine are determined in the presence of the self aggregation of those nucleosides, the reader is referred to Magar *et al.* (1971). In that paper other theories and methods on the same subject are discussed.

### References

Ackers, G. K. (1964). *Biochemistry* **3**, 723.
Ackers, G. K. (1967). *J. Biol. Chem.* **242**, 3237.
Ackers, G. K. (1968). *J. Biol. Chem.* **243**, 2056.
Adams, E. T., Jr. (1956a). *Biochemistry* **4**, 1646.
Adams, E. T., Jr. (1956b). *Biochemistry* **4**, 1655.
Adams, E. T., Jr. (1962). Ph. D. Thesis. Univ. of Wisconsin, Madison, Wisconsin.
Adams, E. T., Jr. (1967a). *In* "Fractions" (W. Hauck, ed.), Vol. 3. McGraw-Hill, New York.
Adams, E. T., Jr. (1967b). *Biochemistry* **6**, 1864.
Adams, E. T., Jr., and Williams, J. W. (1964). *Amer. Chem. Soc.* **86**, 3454.
Adams, E. T., Jr., and Fujita, H. (1963). *In* "Ultracentrifugal Analysis in Theory and Experiment" (J. W. Williams, ed.). Academic Press, New York.
Albright, D. A., and Williams, J. W. (1968). *Biochemistry* **7**, 67.
Andrews, P. (1964). *Biochemistry* **91**, 222.
Baker, C. T. H., Fox, L., Mayers, D. F., and Wright, K. (1964). *Comput. J.* **7**, 141.
Bloomfield, V. (1965). *Biochemistry* **5**, 684.
Brumbaugh, E. E. and Ackers, G. K. (1968).
Chan, S. I., Schweitzer, M. P., Ts'o, P. O. P., and Helmkamp, O. (1964). *J. Amer. Chem. Soc.* **86**, 4182.
Chebyshev, P. (1874). *J. Math. Pure Appl.* **19**, 157.
Chiancone, E., Gilbert, L. M., Gilbert, G. A., and Kellett, G. L. (1968). *J. Biol. Chem.* **243**, 1212.
Cleeman, J. C., and Kiatky, O. (1960). *Naturforscher* **156**, 525.
Davis, P. J. (1961). "Interpolation and Approximation." Ginn (Blaisdell), Boston, Massachusetts.
Debye, P. (1915). *Ann. Phys.* (*Leipzig*) **56**, 809.
Determann, H., and Michel, W. (1966). *J. Chromatogr.* **25**, 303.

Donnelly, T. H. (1966). *J. Phys. Chem.* **70**, 1862.
Fujita, H. (1962). "Mathematical Theory of Sedimentation Analysis." Academic Press, New York.
Geidushek, P., and Holtzer, A. (1958). *Advan. Biol. Med. Phys.* **6**, 431.
Guinier, A., and Fournet, G. (1955). "Small Angle X-Ray Scattering." Wiley, New York.
Jeffrey, P. D., and Coates, H. H. (1966). *Biochemistry* **5**, 489.
Kakiuchi, K. (1965). *J. Phys. Chem.* **69**, 1869.
Kopal, Z. (1955). "Numerical Analysis." Wiley, New York.
Kratky, O. (1963). *Progr. Biophys. Mol. Biol.* **13**, 105.
Kratky, O., and Kreutz, W. (1960). *Z. Elektrochem.* **64**, 880.
Kratky, O., and Porod, G. (1948). *Acta Phys. Austr.* **2**, 133.
Kratky, O., Porod, G., Sekora, A., and Paletta, R. (1955). *J. Polymer Sci.* **16**, 163.
Kreuzer (1943). *Z. Phys. Chem.* (*Leipzig*) **B53**, 213.
Laurent, T. C., Killander, J. (1964). *J. Chromatogr.* **14**, 317.
Luzzatti, V., Witz, J., and Nicolaieff, A. (1961). *J. Mol. Biol.* **3**, 379.
Magar, M. E. (1969). *J. Theor. Biol.* **22**, 291.
Magar, M. E. (1970). *J. Theor. Biol.* **27**, 127.
Magar, M. E., and Steiner, R. F. (1970). *Biochim. Biophys. Acta* **224**, 80.
Magar, M. E., Steiner, R. F., and Kolinski, R. (1971). *Biophys. J.* **11**, 387.
Maomon, G. E. (1957). *Biochim. Biophys. Acta* **26**, 233.
Mittelbach, P., and Porod, G. (1961). *Acta Phys. Austr.* **14**, 185.
Mittelbach, P., and Porod, G. (1962). *Acta Phys. Austr.* **15**, 122.
Peticolas, W. L. (1961). *J. Chem. Phys.* **35**, 2128.
Peticolas, W. L. (1962). *J. Chem. Phys.* **37**, 2331.
Porath, J. (1963). *Pure Appl. Chem.* **6**, 233.
Provencher, S. W. (1967). *J. Chem. Phys.* **46**, 3229.
Roark, E., and Yphantis, D. (1969). *Ann. N.Y. Acad. Sci.* **164**, 245.
Rice, J. R. (1964). "The Approximation of Functions." Addison-Wesley, Reading, Massachusetts.
Richards, G. E., Teller, D. C., and Schachman, H. K. (1968). *Biochemistry* **7**, 1054.
Rinde, H. (1928). Ph. D. Thesis, Uppsala Univ., Uppsala, Sweden.
Reimann, G. (1859). *Monatsber. Deut. Akad. Wiss. Berlin* 671–680.
Scheid, L. (1968). "Numerical Analysis." McGraw-Hill, New York.
Scholte, Th. G. (1968). *J. Polym. Sci. Ser. A*-2 **6**, 91, 191.
Schulz, Z. (1939). *Z. Phys. Chem.* (*Leipzig*) **B43**, 25.
Shohat, J. A., and Tamarkin, J. D. (1943). *Amer. Math. Soc.*
Solie, T. N., and Schellman, J. (1968). *J. Mol. Biol.* **33**, 61.
Spiegel, M. (1965). "Laplace Transforms." McGraw-Hill, New York.
Squire, P. G. (1964). *Arch. Biochem. Biophys.* **107**, 471.
Squire, P. G., and Li, C. H. (1961). *J. Amer. Chem. Soc.* **83**, 3521.
Steiner, R. F. (1952). *Arch. Biochem. Biophys.* **39**, 333.
Steiner, R. F. (1954). *Arch. Biochem. Biophys.* **49**, 400.
Steiner, R. F. (1968). *Biochemistry* **7**, 2201.
Teller, D. C., Horbett, T. A., Richards, E. G., and Schachman, H. K. (1969). *Ann. N.Y. Acad. Sci.* **164**, 66.
Trautman, R. (1969). *Ann. N.Y. Acad. Sci.* **164**, 52.
Tricomi, F. G. (1957). "Integral Equations." Wiley, New York.
Ts'o, P. O. P., and Chan, S. I. (1964). *J. Amer. Chem. Soc.* **86**, 4176.

Ts'o, P. O. P., Melvin, I. S., and Olson, A. C. (1963). *J. Amer. Chem. Soc.*
Uspensky, J. V. (1937). "Introduction to Mathematical Probability." McGraw-Hill, New York.
Van Holde, K. E., and Rossetti, C. P. (1967). *Biochemistry* **6**, 2189.
Van Holde, K. E., Rossetti, G. P., and Dyson, R. D. (1969). *Ann. N.Y. Acad. Sci.* **164**, 279.
Wales, M. (1948). *J. Phys. Colloid Chem.* **52**, 235.
Wales, M., Adler, F. T., and Van Holde, K. E. (1951). *J. Phys. Chem.* **55**, 145.
Wallace, R., and Katz, S. M. (1964). *J. Phys. Chem.* **68**, 3890.
Warshaw, H., and Ackers, G. K. (1969). *Congress Int. Union Pure Appl. Biophys., 3rd*, p. 166.
Wesslau, H. (1956). *Makromol. Chem.* **20**, 111.
Whitaker, R. (1963). *Anal. Chem.* **35**, 1950.
Williams, J. M., Van Holde, K. E., Baldwin, R. L., and Fujita, H. (1958). *Chem. Rev.* **58**, 715.
Winzor, D. J. (1969). *In* "Physical-Chemical Techniques in Protein Chemistry" (E. J. Leach, ed.). Academic Press, New York.
Zelen, M. (1954). *J. Res. Nat. Bur. Stand.* **53**, 377.
Zimm, B. H. (1948). *J. Chem. Phys.* **16**, 1099.

*CHAPTER 13*

# ALLOSTERIC EFFECTS AND OTHER COOPERATIVE PHENOMENA IN PROTEIN–LIGAND EQUILIBRIA[†]

## 1. Introduction

Protein–ligand interaction plays a fundamental role in the regulation of enzyme and protein action. Such interactions are of fundamental physiological importance (Grisolia, 1964; Gerhardt and Pardee, 1962; Koshland, 1963). The term allosteric effect (Monod *et al.*, 1963) was coined to define specific and indirect interactions between distinct binding sites, and it is believed that such effects are finally responsible for the performance of regulatory functions.

## 2. Description of Some Models in the Literature

Several models have been proposed to explain how protein action is modified by ligands. In view of the fact that to a very large extent proteins

[†] This chapter owes a great deal to the contributions of R. F. Steiner and J. Fletcher.

are made of subunits, Monod *et al.* (1965) proposed a general interpretation of allosteric effects in terms of the quaternary structure of proteins. This model supposes that there are two forms of a protein, one in which all the subunits have one conformation and the other in which the subunits are in another conformation. Atkinson *et al.* (1965), discussing DPN-linked isocitrate dehydrogenase and phosphofructokinase, suggested that a substrate or modifier might first combine with one site and alter the binding at still another site. Koshland *et al.* (1966) proposed several models to describe the possible conformation changes that proteins undergo when binding substrates.

Frieden (1967) discussed some of the allosteric models, enumerated some of the assumptions involved in them, and suggested methods for plotting the data. He also pointed out that, strictly speaking, the model of Monod *et al.* (1965) is more applicable to binding studies than to kinetic studies. To overcome this and to extend Monod's model to deal with kinetic phenomena, Frieden adopted the approach that a modifier may affect the maximum velocity as the equilibrium constant. He derived an equation taking those factors into consideration. A more general treatment of such behavior is given by Whitehead (1970).

This chapter is principally concerned with the determination of the parameters characterizing the binding of ligands to proteins and the equilibria between different protein species for all the important models proposed in the literature. In addition, since the possibility exists that a protein not having subunits may show preferential binding to more than one molecule of the same ligand, we shall also consider this phenomena.

In what follows we review the methods that have been employed previously to determine the parameters. We then consider the equations representing the models and we show that most of them can be linearized. Next we discuss the methods of analysis for those models. Lastly we apply the methods to the analysis of a number of cases considered in detail by Magar *et al.* (1971).

Previously suggested methods for obtaining the *differing* equilibrium constants between protein and ligand are quite unsatisfactory, and they invariably involve a process of trial and error. In the case where we have a single protein chain binding to several small molecules, methods for determining an equilibrium constant when it is *identical* for the small molecules have been previously described by Scatchard (1949) and Klotz *et al.* (1946). When the equilibrium constants are different, a series of limiting slopes and intercepts are used as a basis of calculating those constants. These methods have been discussed by Edsall and Wyman (1958) and applied to the specific case of the binding of zinc and copper ions to imidazoles. In the final analysis, even these authors resort to trial and error to determine the equilibrium constants.

With respect to the model of Monod *et al.* (1965), once again trial and error has been used to determine the parameter of this model. Rubin and Changeux (1966) have proposed graphical methods to determine the extent of shifting of equilibria between various protein species in the presence of modifying reagents. Similar methods have been proposed to determine the extent of cooperativity due to the presence of a ligand. Changeux and Rubin (1968) actually applied their methods to the data of Changeux *et al.* (1968) and Gerhardt and Schachman (1968) on aspartic transcarbamylase. These authors obtained only limits to certain parameters (e.g., in their notation parameters $L'$ and $c$) and then used these roughly defined parameters to obtain the other parameters to the model.

Koshland *et al.* (1966) employed nomograms and graphs to determine some of the parameters using a limited number of points. Frieden did not address himself to the problem of parameter determination except in a very minor way. The same is true of Whitehead, the main thrust of whose paper is directed toward determining whether or not modifiers fit in a single site on the protein or on two sites.

Unlike authors who were using trial and error to estimate parameters or monograms that used a few specific points, the method of function minimization will use all the data to determine the pertinent parameters. Problems will arise during the determination of the parameters because of ill-conditioning and these will be pointed out in the practical examples we work out.

From the standpoint of the statistician and numerical analyst who ultimately wish to extract the parameters of the model a great deal of similarity between the models is evident. These will be pointed out. As a matter of fact Magar and Steiner (1971) have shown that the model first proposed by Adair (1925) to explain the binding of oxygen to hemoglobin is quite general and that most models recently proposed are special mathematical cases of Adair's original model. Thus the tetrahedral, square, linear and concerted models of Koshland *et al.* (1966), the model of Monod *et al.* (1965), and Frieden's Eq. (3) and the Scatchard model are all special cases of Adair's equation as will be demonstrated.

## 3. Sequential Binding or Adair's Equation—Proteins without Any Subunits

The equilibria for the case of a protein without any subunits but with differing affinity to more than one molecule of a ligand may be written as follows:

$$\begin{array}{ll} P + S \rightleftharpoons PS_1, & K_1 = [PS_1]/[P][S] \\ PS_1 + S \rightleftharpoons PS_2, & K_2 = [PS_2]/[PS_1][S] \\ \vdots & \vdots \\ PS_{n-1} + S \rightleftharpoons PS_n, & K_n = [PS_n]/[PS_{n-1}][S] \end{array} \tag{13-1}$$

where $P$ is the protein and $S$ the ligand. The equation to be analyzed is shown below

$$J = \frac{K_1[S] + 2K_1K_2[S]^2 + \cdots + nK_1K_2 \cdots K_n[S]^n}{1 + K_1[S] + K_1K_2[S]^2 + \cdots + K_1K_2 \cdots K_n[S]^n} \tag{13-2}$$

where $J =$ (moles of $S$ combined moles of $P$). By a number of rearrangements beginning with multiplication of the left-hand side of the equation by the denominator of the right-hand side we can write (13-2) as follows:

$$J = \sum_{i=1}^{n} \beta_i \phi(S) \tag{13-3}$$

where we have put

$$\beta_1 = K_1, \qquad \beta_2 = K_1K_2, \qquad \ldots, \qquad \beta_n = K_1K_2 \cdots K_n \tag{13-4}$$

and

$$\begin{array}{l} \phi_1(S) = (1 - J)[S], \\ \phi_2(S) = (2 - J)[S]^2, \qquad \ldots, \qquad \phi_n(S) = (n - J)[S]^n \end{array} \tag{13-5}$$

In this manner the equation is linearized. Another method of linearizing Eq. (13-2) is due to Steiner (1953). This can be demonstrated as follows by dividing Eq. (13-2) by $[S]$

$$\frac{J}{[S]} = \frac{K_1 + 2K_1K_2[S] + \cdots + nK_1K_2 \cdots K_n[S]^{n-1}}{1 + K_1[S] + K_1K_2[S]^2 + \cdots + K_1K_2 \cdots K_n[S]^n} \tag{13-6}$$

In Eq. (13-6) the numerator of the right-hand side is equal to the first derivative of the denominator. From any table of integrals we can ascertain that

$$\left[\int f'(x)\,dx\right]\Big/ f(x) = \ln f(x)$$

so that

$$\int_0^{[S]_{\max}} (J/[S])\,d[S] = 1 + K_1[S] + K_1K_2[S]^2 + \cdots + K_1K_2\cdots K_n[S]^n \tag{13-7}$$

where $[S]_{\max}$ is the highest ligand concentration used. By carrying out the integral on the right-hand side of (13-7) at different values of $[S]$, we effectively get a set of data to fit to the polynomial in (13-7).

In actual application, as we demonstrate later on in the chapter either Eq. (13-2) may be treated directly as a nonlinear problem or the linearized form of equations such as (13-3) or (13-7) may be used to determine the parameters. At this stage the reader might note that Eq. (13-2) can be written as

$$J = \frac{\beta_1[S] + 2\beta_2[S]^2 + \cdots + n\beta_n[S]^n}{1 + \beta_1[S] + \beta_2[S]^2 + \cdots + \beta_n[S]^n} \tag{13-8}$$

and Eq. (13-7) can be written as

$$\lambda = \int_0^{[S]_{\max}} (J/[S])\,d[S] = 1 + \beta_1[S] + \beta_2[S]^2 + \cdots + \beta_n[S]^n \tag{13-9}$$

Equations (13-8) and (13-9) will be used to show the equivalence between the Adair model, as we shall call it, and the other models.

## 4. Models of Koshland, Nemethy, and Filmer

Comparison of Eq. (13-8) with those given by Koshland *et al.* (1966) for the "linear," "square," and "tetrahedral" models where the subunits are identical, reveals that Eq. (13KNF)† for the tetrahedral model, Eqs. (18KNF) and (19KNF) for the square model, and Eq. (24KNF) for the linear model have the same form as Eq. (13-2) and can be linearized in the same manner as shown above, except that the constants $\beta_1, \beta_2, \ldots, \beta_4$ have different meanings. Kirtley and Koshland (1967) recognized that the square model could be written in the above manner. In the cases dealt

† Equations of other authors will be labelled by the number of the equation in the original paper followed by the first letters of the authors' surnames. For example, Eq. (3) in Frieden's paper will be labelled as (3F).

with by Koshland *et al.* where the subunits are not identical, Eq. (33KNF) for the square case and Eq. (38KNF) for the "rectangular" case can also be written as Eq. (13-8). Here again the constants have different meanings but the form of the equation remains the same.

In view of the basic equivalence in form of the equations corresponding to the various explicit models treated in the theory of Koshland *et al.* (1966) it will not be invariably possible to make the distinction between these models. In general, it will be possible to compute the parameters characterizing the association in terms of a particular model only if the number of (identical) subunits $n$ is equal to, or greater than, the number of parameters to be determined. A choice between models is theoretically possible only if $n$ exceeds the number of parameters.

It is convenient to group the models according to whether all subunits of a given protein molecule are constrained to adopt the same conformation, or whether subunits of different conformation are allowed in the same molecule. The latter case corresponds to the tetrahedral, square, and linear models of Koshland *et al.* (1966) while the former includes the "concerted" model of these authors, as well as the model of Monod *et al.* (1965).

These models assume a system of identical subunits, each of which may exist in either of two conformations, $A$ or $B$. Only form $B$ binds substrate, with an association constant $K_s$.

$$K_s = [BS]/[B][S] \tag{13-10}$$

where $[S]$, $[B]$, and $[BS]$ are the concentrations of free substrate, subunits in form $B$, and substrate-form $B$ complexes, respectively. We also have we equilibrium constant for $A \rightleftharpoons B$ conversion

$$K_t = [B]/[A] \tag{13-11}$$

and the quasi-chemical equilibrium constant for nearest neighbor interactions between $A$ and $B$ forms

$$K_{AB} = [AB][A]/[AA][B] \tag{13-12}$$

where $[AB]$ and $[AA]$ are the concentrations of nearest neighbor $A$–$B$ and $A$–$A$ pairs, respectively, while $[A]$ and $[B]$ refer to noninteracting subunits.

$$K_{BB} = [BB][A][A]/[AA][B][B] \tag{13-13}$$

The analogous quantity $K_{AA}$ is taken to be equal to unity.

### 4.1 Tetrahedral Model

For this case, in which the binding of ligand is described by Eq. (13KNF), the parameters of Eqs. (13-2) and (13-3) have the following meanings:

$$\beta_1 = 4K_{AB}^3 K_s K_t, \qquad \beta_2 = 6K_{AB}^4 K_{BB}(K_s K_t)^2$$
$$\beta_3 = 4K_{AB}^3 K_{BB}^3 (K_s K_t)^3, \qquad \beta_4 = K_{BB}^6 (K_s K_t)^4 \tag{13-14}$$

Since $K_s$ and $K_t$ occur only as the product, they cannot be evaluated separately and must be regarded as collectively comprising a single variable $K_s K_t$. In addition, the following relationships exist:

$$\ln \beta_4 = 2 \ln \beta_3 - 2 \ln \beta_1 \qquad \text{or} \qquad \beta_4 = (\beta_3/\beta_1)^2$$
$$3 \ln(\beta_2/6) - \ln(\beta_3/4) = 3 \ln(\beta_1/4) \tag{13-15}$$

These relations provide a test of whether a particular set of data obeys the tetrahedral model. Since there are effectively only two independent equations relating $K_{AB}$, $K_{BB}$, and $K_s K_t$, it is not possible to obtain a unique solution for these unknowns. If a value is assumed for $K_{AB}$ (or either of the other two variables), then values may be computed for the other unknowns.

### 4.2 Square Model

This model assumes four subunits in a square pattern with no interactions across the diagonal. From Eq. (18KNF) we note that

$$\beta_1 = 4K_{AB}^2 K_s K_t, \qquad \beta_2 = (2K_{AB}^4 + 4K_{AB}K_{BB})(K_s K_t)^2$$
$$\beta_3 = 4K_{AB}^2 K_{BB}^2 (K_s K_t)^3, \qquad \beta_4 = K_{BB}^4 (K_s K_t)^4 \tag{13-16}$$

In this case, we have

$$\ln \beta_4 = 2 \ln \beta_3 - 2 \ln \beta_1 \qquad \text{or} \qquad \beta_4 = (\beta_3/\beta_1)^2$$
$$\beta_2 = (\beta_1^2/8) + (\beta_1\beta_3)^{1/2} \tag{13-17}$$

These equations may be used to test the adherence of data to the square model.

### 4.3 Linear Model

In this case the four subunits have a linear arrangement, so that the two interior subunits have two nearest neighbors each, while the terminal subunits have one. For this model, from (24KNF),

$$\begin{aligned}\beta_1 &= 2(K_{AB} + K_{AB}^2)K_s K_t, \\ \beta_2 &= (K_{AB}^2 K_{BB} + 2K_{AB}^3 + K_{AB}^2 + 2K_{AB}K_{BB})(K_s K_t)^2 \\ \beta_3 &= 2(K_{AB}^2 K_{BB} + K_{AB}K_{BB}^2)(K_s K_t)^3 \\ \beta_4 &= K_{BB}^3 (K_s K_t)^4\end{aligned} \tag{13-18}$$

In contrast to the preceding two cases, Eqs. 13-18 are independent, so that it is possible to solve for $K_{AB}$, $K_{BB}$, and $K_s K_t$. By eliminating $K_{BB}$ and $K_s K_t$, we have

$$\beta_3 - \frac{4\beta_1^3\left[(1+K_{AB})^4 \dfrac{\beta_2^2}{\beta_1^4} - 3K_{AB}(1+K_{AB})^2 \dfrac{\beta_2}{\beta_1^2} + (K_{AB}^2)/2 - K_{AB} - 1\right]}{(K_{AB}+2)^2(1+K_{AB})^3} = 0 \tag{13-19}$$

Equation (13-19) is a nonlinear equation in the unknown $K_{AB}$ and can be solved by Newton's method. By substitution of this value in the first of Eqs. (13-18), $K_s K_t$ may be obtained. Substitution of these values in the second of Eqs. (13-18) yields $K_{BB}$. Anticipating our dealing with the subject of the distinction between models we might say that to test for the consistency of the data with the model, the computed values of $K_{BB}$ and $K_s K_t$ may be inserted in the last of Eqs. (13-18) and compared to the value of $\beta_4$ obtained from a least squares procedure.

### 4.4 Concerted Model

This model assumes that the conformations of all subunits change simultaneously, so that the protein is either all $A$ or all $B$. The predicted binding isotherm for this case has a somewhat different form from those of the preceding models (30KNF),

$$J = \frac{4K_s[S](1 + K_s[S])^3}{K_w^{-4} + (1 + K_s[S])^4} \tag{13-20}$$

Here $K_w$, whose value is independent of the number of bound ligands, is the equilibrium constant for the interconversion of an $A$ and $B$ unit and includes the nearest neighbor interaction terms. For the interconversion of $A_4$ and $B_4$,

$$[B_4] = K_w^4[A_4] \tag{13-21}$$

Defining

$$\lambda' = K_w^{-4} + (1 + K_s[S])^4 \tag{13-22}$$

we have

$$J = d \ln \lambda'/d \ln[S] \tag{13-23}$$

and

$$\ln \lambda'/\lambda_0' = \int_0^{[S]} (J/[S])\, d[S] \tag{13-24}$$

where

$$\begin{aligned}\lambda_0' &= \text{value of } \lambda \text{ when } [S] = 0\\ &= K_w^{-4} + 1\end{aligned}$$

Also

$$\begin{aligned}\lambda'/\lambda_0' = 1 &+ 4[K_s/(1 + K_w^{-4})][S] + 6[K_s^2/(1 + K_w^{-4})][S]^2\\ &+ 4[K_s^3/(1 + K_w^{-4})][S]^3 + [K_s^4/(1 + K_w^{-4})][S_4]^4\end{aligned} \tag{13-25}$$

Thus

$$\begin{aligned}\beta_1 &= 4K_s/(1 + K_w^{-4}), \qquad \beta_2 = 6K_s^2/(1 + K_w^{-4})\\ \beta_3 &= 4K_s^3/(1 + K_w^{-4}), \qquad \beta_4 = K_s^4/(1 + K_w^{-4})\end{aligned} \tag{13-26}$$

In this case

$$K_s = \tfrac{2}{3}(\beta_2/\beta_1) \tag{13-27}$$

and $K_w$ may then be computed from any one of Eqs. (13-26).

### 4.5 Nonidentical Subunits

In the event that the subunits are not identical, and where the two subunits have different intrinsic binding constants with respect to the substrates, these binding constants may be defined as

$$K_{s\alpha} = [B_\alpha S]/[B_\alpha][S], \qquad K_{s\beta} = [B_\beta S]/[B_\beta][S] \tag{13-28}$$

where $K_{s\alpha}$ refers to the binding constant of the $A$ subunit and $K_{s\beta}$ refers to the binding constant of the $B$ subunit. The equation which we seek to analyze is now Eq. (KNF33) of Koshland *et al.* (1966) and it takes the form

$$J = \frac{\begin{array}{c} 2(K_{s\alpha}K_t + K_{s\beta}K_t)[S] + 2[(K_{s\alpha}K_t)^2 + (K_{s\beta}K_t)^2 \\ + 4K_{BB}(K_{s\alpha}K_t)(K_{s\beta}K_t)][S]^2 + 6K_{BB}^2(K_{s\alpha}K_t)(K_{s\beta}K_t)(K_{s\alpha}K_t \\ + K_{s\beta}K_t)[S]^3 + 4[(K_{s\alpha}K_t)(K_{s\beta}K_t)]^2K_{BB}^4[S]^4 \end{array}}{\begin{array}{c} 1 + 2(K_{s\alpha}K_t + K_{s\beta}K_t)[S] + [(K_{s\alpha}K_t)^2 + (K_{s\beta}K_t)^2 \\ + 4K_{BB}(K_{s\alpha}K_t)(K_{s\beta}K_t)][S]^2 + 2K_{BB}^2(K_{s\alpha}K_t)(K_{s\beta}K_t)[(K_{s\alpha}K_t) \\ + (K_{s\beta}K_t)][S]^3 + K_{BB}^4[(K_{s\alpha}K_t)(K_{s\beta}K_t)]^2[S]^4 \end{array}} \tag{13-29}$$

In this equation it will be noted that we assume that $K_t$ is constant. This means we do not include the possibility that $K_t$ may take two forms, $K_{t\alpha}$ and $K_{t\beta}$, characterizing the possible isomerization between the active and inactive forms of the two subunits. Equation (13-29) can be recognized as having the same form as (13-8) where

$$\begin{aligned} \beta_1 &= K_{s\alpha}K_t + K_{s\beta}K_t, \\ \beta_2 &= (K_{s\alpha}K_t)^2 + (K_{s\beta}K_t)^2 + 4K_{BB}(K_{s\alpha}K_t)(K_{s\beta}K_t) \\ \beta_3 &= K_{BB}^2(K_{s\alpha}K_t)(K_{s\beta}K_t)(K_{s\alpha}K_t + K_{s\beta}K_t) \\ \beta_4 &= [(K_{s\alpha}K_t)(K_{s\beta}K_t)]^2K_{BB}^4 \end{aligned} \tag{13-30}$$

## 5. The Model of Monod, Wyman, and Changeux

The equilibrium equations considered by Monod *et al.* (1965), in their notation, may be written as follows:

$$\begin{array}{cc} \multicolumn{2}{c}{R_0 \rightleftharpoons T_0} \\ R_0 + F \rightleftharpoons R_1, & T_0 + F \rightleftharpoons T_1 \\ R_1 + F \rightleftharpoons R_2, & T_1 + F \rightleftharpoons T_2 \\ \vdots & \vdots \\ R_{n-1} + F \rightleftharpoons R_n, & T_{n-1} + F \rightleftharpoons T_n \end{array} \tag{13-31}$$

If we put

$$(F/K_R) = A, \qquad (K_R/K_T) = C, \qquad L = (T_0/R_0) \tag{13-32}$$

we obtain

$$\bar{Y}_F = \frac{LCA(1 + CA)^{n-1} + A(1 + A)^{n-1}}{L(1 + AC)^n + (1 + A)^n} \tag{13-33}$$

The meaning of the symbols in Eqs. (13-31)–(13-33) are as follows: $F$ is the ligand concentration and, therefore, $A$ may be called the "normalized" ligand concentration, $R_0$ and $T_0$ represent two states of the protein, $K_R$ and $K_T$ are the microscopic dissociation constants of the ligand and the two states of the protein, $L$ is the equilibrium constant between the two states of the protein and $\bar{Y}_F$ is the saturation function or what is equivalent to $J$ in the previous case.

For illustrative reasons we shall treat the above case for $n = 3$, showing how we may linearize. All other values of $n$ follow in the same pattern as the illustrative example. For $n = 3$, rearranging Eq. (13-12), we obtain:

$$\bar{Y}_F = \sum_{i=1}^{3} \gamma_i \phi_i(A) \tag{13-34}$$

where

$$\gamma_1 = 1 + LC, \qquad \gamma_2 = 1 + LC^2, \qquad \gamma_3 = 1 + LC^3 \tag{13-35}$$

and

$$\begin{gathered} \phi_1(A) = (A - 3AY_F), \qquad \phi_2(A) = (A^2 - 3A^2Y_F), \\ \phi_3(A) = (A^3 - A^3Y_F) \end{gathered} \tag{13-36}$$

Or, if we prefer to write $Y_F$ specifically as a function of the ligand concentration instead of the normalized ligand concentration, we may write

$$Y_F = \sum_{i=1}^{3} \varrho_i \phi_i(F) \tag{13-37}$$

where

$$\begin{gathered} \varrho_1 = 1 + (LC/K_R), \qquad \varrho_2 = 1 + (LC^2/K_R{}^2), \\ \varrho_3 = 1 + (LC^3/K_R{}^3) \end{gathered} \tag{13-38}$$

and

$$\begin{gathered} \phi_1(F) = F - 3FY_F, \qquad \phi_2(F) = F^2 - 3F^2Y_F, \\ \phi_3(F) = F^3 - F^3Y_F \end{gathered} \tag{13-39}$$

Here again we note that we can either treat Eq. (13-37) as a nonlinear equation in the parameters or use a linearized form.

A number of things may be noted about the model of Monod *et al.* (1965). This model is similar to the concerted model of Koshland *et al.* (1966), except that the ligand can be bound by either the $A$ or $B$ form. A derivation similar to that given for the concerted model of Koshland is given for comparison. This derivation yields for four sites (using the KNF terminology):

$$\ln \frac{\lambda''}{\lambda''_0} = \int_0^{[S]} J \frac{d[S]}{[S]} \tag{13-40}$$

$$\begin{aligned} \frac{\lambda''}{\lambda''_0} = 1 = {} & \frac{4[S]}{(1 + K_W^{-4})} (K_A + K_W^{-4} K_B) + \frac{6[S]^2}{(1 + K^{-4})} (K_A^{\;2} + K_W^{-4} K_B^{\;2}) \\ & + \frac{4[S]}{(1 + K_W^{-4})} (K_A^{\;3} + K_W^{-4} K_B^{\;3}) \\ & + \frac{[S]^4}{(1 + K_W^{-4})} (K_A^{\;4} + K_W^{-4} K_B^{\;4}) \end{aligned} \tag{13-41}$$

where $K_A$ and $K_B$ are the association constants of ligand with a subunit in the $A$ and $B$ conformation, respectively.

In this case,

$$\beta_1 = \frac{4}{(1 + K_W^{-4})} (K_A + K_W^{-4} K_B), \qquad \beta_2 = \frac{4}{(1 + K_W^{-4})} (K_A^{\;2} + K_W^{-4} K_B^{\;2})$$

$$\beta_3 = \frac{4}{(1 + K_W^{-4})} (K_A^{\;3} + K_W^{-4} K_B^{\;3}), \qquad \beta_4 = \frac{1}{(1 + K_W^{-4})} (K_A^{\;4} + K_W^{-4} K_B^{\;4}) \tag{13-42}$$

No simple analytical expression can be obtained in this case relating the $\beta$ and the association constants. However, given any three $\beta$ computed by our previous method, any three of the above equations can be considered as nonlinear equations in the three unknowns $K_W$, $K_A$, and $K_B$ and solved by Newton's procedure for simultaneous nonlinear equations or some variation thereof (Schied, 1968).

To generalize the Monod, Wyman, and Changeux model we do not assume that each conformer binds the substrate with the same equilibrium constant. Thus the equilibria in the notation of (MWC) can be written

$$\begin{array}{llll} & R_0 \rightleftharpoons T_0, & \gamma = T_0/R_0 & \\ R_0 + S \rightleftharpoons R_1, & \psi_1 = R_1/R_0 S, & T_0 + S \rightleftharpoons T_1, & \phi_1 = T_1/T_0 S \\ R_1 + S \rightleftharpoons R_2, & \psi_2 = R_2/R_1 S, & T_1 + S \rightleftharpoons T_2, & \phi_2 = T_2/T_1 S \\ R_2 + S \rightleftharpoons R_3, & \psi_3 = R_3/R_2 S, & T_2 + S \rightleftharpoons T_3, & \phi_3 = T_3/T_2 S \\ R_3 + S \rightleftharpoons R_4, & \psi_4 = R_4/R_3 S, & T_3 + S \rightleftharpoons T_4, & \phi_4 = T_4/T_3 S \end{array} \tag{13-43}$$

The equation for $J$ for four sites in its most general form for the unrestricted model of (MWC) is

$$J = \frac{\beta_1 S + \beta_2 S^2 + \beta_3 S^3 + \beta_4 S^4}{1 + \gamma + \varepsilon_1 S + \varepsilon_2 S^2 + \varepsilon_3 S^3 + \varepsilon_4 S^4} \tag{13-44}$$

where $\beta_1$, $\beta_2$, $\beta_3$, $\beta_4$, $\varepsilon_1$, $\varepsilon_2$, $\varepsilon_3$, and $\varepsilon_4$ are functions of the equilibrium $\psi_1$, $\psi_2$, $\psi_3$, $\psi_4$, $\phi_1$, $\phi_2$, $\phi_3$, $\phi_4$. More pertinent, $\varepsilon_1$, $\varepsilon_2$, $\varepsilon_3$, and $\varepsilon_4$ are simple multiples of $\beta_1$, $\beta_2$, $\beta_3$, and $\beta_4$. Therefore Eq. (13-44) while having nine parameters can be readily characterized by only five parameters.

Comparison of Eq. (13-8) with Eq. (13-44) reveals the presence of the parameter $\gamma$ (the equilibrium constant between $R_0$ and $T_0$) in the denominator of $\gamma$. The questions as to whether we have two forms of a protein binding a substate or only one form can be settled when we determine if the data will fit an equation such as (13-8) with four parameters of Eq. (13-44) which contains the fifth parameter $\gamma$. A statistical test can be applied to determine if $\gamma = 0$.

## 6. Expansion of Models

Beyond allowing for two different conformers to combine with a single substrate, one can conceive of a substrate and a modifier combining with one or both forms of protein conformers and so on.

If $[M]$ represents the concentration of free modifier, then the possible equilibria may be represented by

$$\begin{aligned}
P + S &\rightleftharpoons PS, & \beta_{10} &= [PS]/[P][S] \\
P + M &\rightleftharpoons PM, & \beta_{01} &= [PM]/[P][M] \\
PS + M &\rightleftharpoons PSM, & \beta_{11} &= (PSM]/[P][S][M] \\
PSM + M &\rightleftharpoons PSM_2, & \beta_{12} &= [PSM_2]/[P][S][M]^2 \\
&\vdots \\
PS_{i-1}M_j + S &\rightleftharpoons PS_iM_j, & \beta_{ij} &= [PS_iM_j]/[P][S]^i[M]
\end{aligned} \tag{13-45}$$

In this case, if we define

$$\lambda_S = 1 + \beta_{10}[S] + \beta_{11}[S][M] + \beta_{12}[S][M]^2 + \cdots + \beta_{ij}[S]^i[M]^j + \cdots \tag{13-46}$$

we then have

$$J_S = (\partial \ln \lambda_s / \partial \ln[S])_{[M]} \tag{13-47}$$

where $J_S$ is the number of moles of ligand $S$ bound per mole of protein and

$$\ln(\lambda_S/\lambda_0) = \int_0^{[S]} (J_S/[S])\, d[S] \qquad ([M] = \text{constant})$$

where the integration is carried out at a constant value of $[M]$, and where $\lambda_0 = 1 + \sum_{j\geq 1} \beta_{0j} M^j$ and is obtained by application of Eq. (13-7) in absence of $S$. This is readily achieved if the modifier is present in excess, so that $[M]$ is approximately equal to the total modifier concentration. If this is not the case then binding measurements must be carried out for both $S$ and $M$ and $J_S$ obtained as a function of $[S]$ for a series of (constant) values of $[M]$ by suitable interpolation.

We may then write

$$\lambda_S = 1 + \alpha_1[S] + \alpha_2[S]^2 + \cdots \tag{13-48}$$

where

$$\alpha_1 = \sum_{j>0} \beta_{1j}[M]^j \qquad \text{and} \qquad \alpha_2 = \sum_{j>0} \beta_{2j}[M]^j \tag{13-49}$$

If $\alpha_1$ is determined as a function of $[M]$, then by least squares polynomial fitting of

$$\alpha_1 = \beta_{10} + \beta_{11}[M] + \beta_{12}[M]^2 + \cdots \tag{13-50}$$

the consecutive constants $\beta_{10}$, $\beta_{11}$, etc., may be obtained. However, see comment at the end of Section 8 of this chapter.

## 7. Cases Considered by Frieden

In general, the models considered by Frieden (1967) are applicable to kinetic data. Frieden (uses the same notation and) adds several assumptions to those of Monod *et al.* (1965) in order to modify their equation for kinetic data. For details the reader should consult Frieden's paper. With these new assumptions incorporated, the equation of Monod *et al.* becomes

$$\frac{v}{nV_{\max}} = \frac{LCA(1+A)^{n-1} + A(1+A)^{n-1}}{L(1+CA)^n + (1+A)^n} \tag{13-51}$$

where $v$ is the initial velocity and $V_{\max}$ is the maximum velocity of the reaction. In absence of modifiers Frieden's model proposed that once a ligand binds to one site it alters the dissociation constants of the other sites by a factor $i$. The addition of a second ligand alters the dissociation constant of unoccupied sites by a factor of $j$ and so on. The expression for an enzyme with four sites for the initial velocity is given by

$$\frac{v}{nV_{\max}} = \frac{4(A/K_A)[1 + (3iA/K_A) + 3i^2j(A/K_A)^2 + i^3j^2k(A/K_A)^3]}{1 + (4A/K_A) + 6i(A/K_A)^2 + 4i^2j(A/K_A)^3 + i^3j^2k(A/K_A)^4} \tag{13-52}$$

Here again $V_{\max}$ is the maximum velocity of the reaction and $K_A$ is the microscopic association constant for the ligand when no sites are occupied. Equation (13-52) can be linearized by both methods outlined previously.

Lastly Frieden derived a more general equation for the initial velocity given by

$$\frac{v}{nV_{\max}} = \frac{(k_5 + k_6M)(1 + A)^{n-1}(1 + M)^{n-1} + (k_5' + k_6'dM)LCA(1 + CA)^{n-1}(1 + dM)^{n-1}}{(1 + M)^n(1 + A)^n + (1 + dM)^nL(1 + CA)^n} \tag{13-53}$$

where $M$ is the modifier concentration and $k_5$, $k_6$, $k_5'$, and $k_6'$ the rate constants for the rate determining steps. These steps involve the breakdown of the enzyme substrate complex in one or the other conformational form $k_5$ and $k_5'$ and the enzyme substrate modifier complex in one or the other forms $k_6$ or $k_6'$. It is not difficult to see that by manipulation of Eq. (13-53) in a manner similar to all the previous equations, we can finally equate $v/nV_{\max}$ to the sum of constants terms and variable terms. The constants terms contain all the constants in Eq. (13-53) including the rate constants. The variable terms are functions of *both* the substrate and modifier concentration.

## 8. Scatchard's Model

We pay brief attention to this model since Fletcher *et al.* (1970) dealt with it in detail and showed that it can be considered as a special case of Eq. (13-8). Scatchard's (1949) equation is derived on the basis of the assumption that there exist $m$ classes of different and independent sites. The equation for $J$ is given by

$$J = \sum_{i=1}^{m} [n_ik_i[S]/(1 + k_i[S])] \tag{13-54}$$

where $n_i$ represents the number of idntical sites of type $i$ and $k_i$ the microscopic binding constant of this site. Fletcher *et al.* (1970) have shown how one can represent the $k_i$ of Eq. (13-54) in terms of the $\beta_i$ of Eq. (13-8). See Fletcher's papers for details.

Before proceeding to the subject of distinguishing between models, let us note that the number of possible equilibria and hence of parameters multiply rapidly as we consider these additional assumptions. Further, if as Frieden (1967) did we consider kinetic phenomena, rate constants will enter our equations, thereby further multiplying the number of parameters involved [see (6F) and Whitehead (1970)]. Such a multitude of constants is likely to give rise to ill-conditioned equations. From the standpoint of the statistician and numerical analyst it is fruitless to discuss or to attempt to make distinctions between models for such cases since there is no way of obtaining good and unique solutions (Magar *et al.*, 1971).

## 9. Distinguishing between Different Models

The fact that all the equations cited above have the same basic form as Eq. (13-9) (in the sense that $\lambda$, $\lambda'/\lambda_0'$, and $\lambda''/\lambda_0''$ may in each case be expressed as a four-term polynomial in $[S]$) complicates the problem of distinguishing between models.

In principle, the relative magnitudes of the $\beta_i$ can be made the basis for testing the adherence of data to a particular model. For the tetrahedral, square, and concerted models of Koshland *et al.* (1966), there are in each case two equations relating the $\beta_i$. In particular, the predicted relationship between $\beta_1$, $\beta_2$, and $\beta_3$ is very different for these three cases [Eqs. (13-15), (13-17), and (13-27)]. If the data are consistent with any one of these models, then the corresponding relationships between the $\beta_i$ should hold, and conversely, the adherence to one of these relationships is an indication that the corresponding model is applicable.

In the actual situation, however, experimental error can be expected to nullify partially the equality given by Eqs. (13-15), (13-17), (13-27). It is thus important to devise a statistical test or criterion which will predict the degree of probability with which any given set of $\beta$ can be expected to satisfy these relationships.

When, as in the present case, parameters are estimated by a linear procedure, a confidence interval can be placed on each parameter. One of the simplest statistical tests which could be used is to compute the 95% confidence interval for a particular $\beta$ and then to determine whether the value of this $\beta$ computed from the other $\beta$ using Eqs. (13-15), (13-17), or (13-27) falls within this confidence interval.

For example, if Eq. (13-15) were used, one would determine whether $(\hat{\beta}_3/\hat{\beta}_1)^2$ falls in the 95% confidence interval of $\hat{\beta}_4$ and whether $(3/2^{5/3})\hat{\beta}_1\hat{\beta}_3^{1/3}$ falls in the confidence interval of $\hat{\beta}_2$ (where $\hat{\beta}_1$, $\hat{\beta}_2$, and $\hat{\beta}_3$ are the values of $\beta_1$, $\beta_2$, and $\beta_3$ estimated by a least squares procedure). If these relationships are satisfied, it is possible to state that the data are consistent with the tetrahedral model.

Another possible test may be based upon the likelihood ratio principle (Chapter 6). The likelihood ratio is formed by minimizing the pertinent quadratic forms with respect to the parameters in the unrestricted space of the parameters and with respect to the parameters in a restricted space consistent with any special case of the general equation. An analysis of variance table may be set up (Chapter 6) and the hypothesis that the data restricted case of the general model may be tested at a level of significance of 0.05 or 0.01.

In the linear model of Koshland *et al.* (1966), as well as in the model of Monod *et al.* (1965) no simple relationship exists between the $\beta$. However, once the $\beta$ are obtained, the physical parameters characteristic of the model can be computed as described earlier. To test for consistency with the model, one may compare the observed and computed values of the $\beta$ which was not used in determining the parameters of the model.

For example, in the Monod, Wyman, and Changeaux case the first three of equations (13-42) may be used to determine $K_w$, $K_A$, and $K_B$. Then, using these estimated values of $K_w$, $K_A$, and $K_B$ in the fourth of equation, we determine whether or not the quantity $[1/(1 + K_w^{-4})](K_A{}^4 + K_w^{-4}K_B{}^4)$ falls within the 95% confidence interval of $\beta_4$.

## 10. Comments on Procedures

In the cases of the tetrahedral, square, and concerted models of Koshland *et al.* (1966) it is clearly not possible, even in principle, to evaluate the characteristic parameters of the models, since the number of independent equations is less than the number of unknowns. However, if a value is assumed for one parameter, it is possible to compute values for the remaining parameters consistent with this assumption. Only in the cases of the linear model of Koshland, Nemethy, and Filmer and the model of Monod *et al.* (1965) is it possible, in principle, to assign unique values to the parameters.

In the preceding sections we have demonstrated that a very large number of the models discussed may be regarded as special cases of the general equation (13-8) and that the linearization procedures developed for the general equation can be applied to the analysis in terms of any specific model. It is important to emphasize the truism that any set of data fitting a special case will also fit the general case. From this we note that *Adair's equation first derived in 1925 can account for most of the allosteric phenomena. The main assumption of Adair's derivation is that the equilibrium constants are unequal.*

If the various models proposed can be regarded as reasonably accurate representations of actual situations, the linearization approach outlined here has a number of advantages. These include the fact that the statistical tests suggested for distinguishing between explicit models become relatively rigorous when the relevant parameters are estimated from a linear model. There is in addition the practical advantage that estimation of the $\beta$ requires only limited computer facilities, although for the Monod, Wyman, and Changeux model and the linear model of Koshland, Nemethy, and Filmer nonlinear techniques must be used to determine the parameters characterizing the models after the $\beta$ have been determined by a linear method. Standard procedures are available for this purpose (Chapter 4).

As mentioned earlier, it should be stressed that the criteria proposed above for distinguishing between models cannot do more than decide which of a set of equations best fits the data. The fact that an equation derived for a particular model can adequately represent the data does not constitute proof that the model corresponds to reality.

## 11. Analysis of Data

Two kinds of data were subjected to the analytical methods. Data from the literature on the binding by bovine serum albumin of acetyl tryptophan and thyroxine are analyzed, as well as "synthetic" data generated by assigning arbitrary values of the equilibrium constants and the ligand concentration and computing the number of ligands bound per mole of protein. This artificial data is then perturbed by applying normally distributed random errors with a given fixed standard deviation, and the methods of model fitting described herein are then applied to this data.

While the real data analyzed here do not display cooperativity, the simulated data represent a case of cooperative phenomena. The synthetic data and random noise were produced in the computational facilities of DCRT-NIH. The computers used were the models 50 and 65 of the

IBM 360 system. The programs used were written ˙n FORTRAN IV and all computations were made in double precision, giving 64 bits, or about 17 decimal places of computational accuracy.

The random errors were produced by the McLaren–Marsaglia method (McLaren and Marsaglia, 1964). Effectively, the random numbers were generated from a uniform distribution and transformed to points on a normal distribution by the Box–Mueller transformation. The mean value of this distribution at a given value of concentration is the exact value of $J$ (the number of moles of bound ligand per protein molecule). The generated values of $J$ fall randomly within an area around $J$ whose extent is indicated by the standard deviation, $\sigma$. The specific versions of these algorithms are standard NIH program library routines known as NIHV10-11-12. These subroutines are written in the 360 assembly language and are recalled by FORTRAN programs.

The nonlinear least squares programs NIHH22-23 were used to recover parameters from the synthetic data with added noise. These programs are described in detail elsewhere (Fletcher and Shrager, 1968) and will not be discussed further here.

For linear programming, multiple regression analysis, and linear least squares polynomial fitting, the General Electric time sharing facilities were used. The linear program used was LINE1 $1*** with some modification. The multiple regression program used was MREGI $***; the least squares polynomial program was POLFIT***.

## 11.1 RESULTS: SIMULATED DATA

For the simulated data the chosen range of ligand concentration (in arbitrary units) was 0.05 to 10.0. Four consecutive association steps were postulated, corresponding to the following stepwise binding constants: $K_1 = 0.10$, $K_2 = 0.25$, $K_3 = 0.20$, $K_4 = 0.12$.

The above range for the ligand concentration was chosen primarily to avoid troublesome computer round-off error. For many real data the free ligand concentration may be of the order of $10^{-4}$ $M$ or less. Such low numbers may lead to round-off error during machine computation and it is desirable to change the concentration scale by multiplying by an appropriate power of ten. This transformation is especially recommended when using standard package programs that may not use double precision arithmetic. Using the above association constants and range of ligand concentrations together with Eq. (13-8), values of $J$ at equally spaced intervals of $[S]$ were generated and a random error of a given standard deviation $\sigma$ was added to the data. Whenever the added random

**Table 13-1**

SIMULATED DATA

| Method uses | Program | Number of terms | Values of stepwise constants | |
|---|---|---|---|---|
| | | *Random error* $\sigma = 0.005$ | | |
| Equation (13-8) | Marquardt | 4 | $K_1 = 0.096$,<br>$K_3 = 0.18$, | $K_2 = 0.272$<br>$K_4 = 0.13$ |
| Equation (13-3) | Linear programming (LINEI$) | 4 | $K_1 = 0.092$,<br>$K_3 = 0.16$, | $K_2 = 0.31$<br>$K_4 = 0.12$ |
| Equation (13-3) | Multiple regression (MREGI$) | 4 | $K_1 = 0.083$,<br>$K_3 = 0.16$, | $K_2 = 0.34$<br>$K_4 = 0.13$ |
| Equation (13-9) | Linear programming (LINEI$) | 4 | $K_1 = 0.10$,<br>$K_3 = 0.20$, | $K_2 = 0.25$<br>$K_4 = 0.12$ |
| Equation (13-9) | Least squares polynomial (POLFIT) | 4 | $K_1 = 0.10$,<br>$K_3 = 0.25$, | $K_2 = 0.21$<br>$K_4 = 0.10$ |
| | | *Random error* $\sigma = 0.01$ | | |
| Equation (13-8) | Marquardt | 4 | $K_1 = 0.093$,<br>$K_3 = 0.175$, | $K_2 = 0.284$<br>$K_4 = 0.13$ |
| Equation (13-3) | Multiple regression | 4 | $K_1 = 0.045$,<br>$K_3 = 0.092$, | $K_2 = 0.794$<br>$K_4 = 0.186$ |
| Equation (13-3) | Least squares polynomial | 4 | $K_1 = 0.10$,<br>$K_3 = 0.25$, | $K_3 = 0.22$<br>$K_4 = 0.11$ |
| | Linear programming | 4 | $K_1 = 0.083$,<br>$K_3 = 0.13$, | $K_2 = 0.37$<br>$K_4 = 0.16$ |
| | | *Random error* $\sigma = 0.0125$ | | |
| Equation (13-3) | Multiple regression | 4 | $K_1 = 0.018$,<br>$K_3 = 0.062$, | $K_2 = 2.137$<br>$K_4 = 0.244$ |
| | | *Random error* $\sigma = 0.015$ | | |
| Equation (13-8) | Marquardt | 4 | $K_1 = 0.09$,<br>$K_3 = 0.16$, | $K_2 = 0.30$<br>$K_4 = 0.135$ |

**Table 13-1** (*continued*)

| Method uses | Program | Number of terms | Values of stepwise constants | |
|---|---|---|---|---|
| Equation (13-9) | Linear programming | 4 | $K_1 = 0.089$, $K_3 = 0.157$, | $K_2 = 0.315$ $K_4 = 0.14$ |
| Equation (13-9) | Least squares polynomial | 4 | $K_1 = 0.091$, $K_3 = 0.174$, | $K_3 = 0.294$ $K_4 = 0.13$ |
| Equation (13-9) | Least squares polynomial | 5 | $K_1 = 0.092$, $K_3 = 0.157$, $(K_5 = 0.013)^a$ | $K_2 = 0.315$ $K_4 = 0.127$ |
| Equation (13-9) | Multiple regression | 4 | $\beta_1 = 0.089 \pm 0.185 \times 10^{-3}$ $\beta_2 = 0.028 \pm 0.752 \times 10^{-4}$ $\beta_3 = 0.0044 \pm 0.144 \times 10^{-4}$ $\beta_4 = 0.616 \times 10^{-3} \pm 0.571 \times 10^{-6}$ | |
| Equation (13-9) | Multiple regression | 4 | $K_1 = 0.089$ $K_2 = 0.314$ $K_3 = 0.157$ $K_4 = 0.140$ | |

[a] Statistically insignificant values are given in parentheses.

error produced a negative value for $J$ this random error was set equal to zero in the subsequent analysis. When the simulated data were used *without* addition of random error all methods discussed here recovered the association constants from the synthetic data exactly.

The results on the simulated data at different levels of random error are all listed in Table 13-1.

## 11.2 Results: Real Data

We turn now to the analysis of actual data. In dealing with the analysis of scattered real data, the actual data giving $J[S]$ as a function of $[S]$ were expressed as an 11th power polynomial in $[S]$, using a least squares fit. This function was then used in the integration of Eq. (13-9) to determine $\lambda$ as a function of $[S]$. The function $\lambda$ was then used for the computation of macroscopic binding constants from Eq. (13-9) by linear programming or least squares regression analysis. We shall consider first the binding of thyroxine by bovine serum albumin. The data are those of Tabachnik (1967).

Table 13-2 cites the macroscopic association constants obtained by least squares polynomial, linear programming, and multiple regression

**Table 13-2**[a]

| Method uses | Program | Number of terms | Values of constants |
|---|---|---|---|
| Equation (13-9) | Least squares polynomial | 4 | $\beta_1 = 15.1 \times 10^5$<br>$\beta_2 = 68.2 \times 10^{10}$<br>$\beta_3 = 27.0 \times 10^{15}$<br>$\beta_4 = 11.8 \times 10^{20}$ |
| | Linear programming | 5 | $\beta_1 = 18.6 \times 10^5$<br>$\beta_2 = 64.8 \times 10^{10}$<br>$\beta_3 = 69.2 \times 10^{15}$<br>$\beta_4 = 0$<br>$\beta_5 = 80.0 \times 10^{25}$ |
| Equation (13-8) | Marquardt | 5 | $\beta_1 = 13.4 \times 10^5$<br>$\beta_2 = 62.7 \times 10^{10}$<br>$\beta_3 = 28.3 \times 10^{15}$<br>$\beta_4 = 73.0 \times 10^{20}$<br>$\beta_5 = 39.5 \times 10^{25}$ |
| | | 6 | $\beta_1 = 13.5 \times 10^5$<br>$\beta_2 = 62.8 \times 10^{10}$<br>$\beta_3 = 40.9 \times 10^{15}$<br>$\beta_4 = 70.3 \times 10^{20}$<br>$\beta_5 = 40.3 \times 10^{25}$<br>$(\beta_6 = 35 \times 10^{27})$ |
| Equation (13-9) | Multiple regression | 3 | $\beta_1 = 53.1 \times 10^5$<br>$\beta_2 = -92.73 \times 10^{10}$<br>$\beta_3 = 268.01 \times 10^{15}$ |
| Equation (13-9) | Multiple regression | 4 | $\beta_1 = 1.43 \times 10^5$<br>$\beta_2 = 154.89 \times 10^{10}$<br>$\beta_3 = -123.68 \times 10^{15}$<br>$\beta_4 = 199.84 \times 10^{20}$ |
| Equation (13-9) | Multiple regression | 5 | $\beta_1 = 7.3 \times 10^5 \pm 2.9 \times 10^5$<br>$\beta_2 = 93.8 \times 10^{10} \pm 17.9 \times 10^{10}$<br>$\beta_3 = 41.2 \times 10^{15} \pm 46.1 \times 10^{15}$<br>$\beta_4 = 11.05 \times 10^{20} \pm 51.8 \times 10^{20}$<br>$\beta_5 = 77.05 \times 10^{25} \pm 21.0 \times 10^{25}$ |

[a] Data from Tabachnik (1967). 0.06 *M* phosphate, pH 7.4, 30°C.

**Table 13-3**

BINDING CONSTANTS FOR THE ACETYL TRYPTOPHAN–SERUM ALBUMIN SYSTEM[a]

| Method uses | Program | Number of terms | Values of constants |
|---|---|---|---|
| Equation (13-9) | Least squares polynomial | 2 | $\beta_1 = 3.22 \times 10^4$<br>$\beta_2 = 0.396 \times 10^8$ |
| Equation (13-9) | Least squares polynomial | 3 | $\beta_1 = 3.25 \times 10^4$<br>$\beta_2 = 0.213 \times 10^8$<br>$\beta_3 = 0.190 \times 10^{12}$ |
| Equation (13-9) | Least squares multiple regression | 2 | $\beta_1 = 3.18 \pm 0.127 \times 10^4$<br>$\beta_2 = 0.445 \pm 0.126 \times 10^8$ |
| Equation (13-9) | Least squares multiple regression | 3 | $\beta_1 = 3.04 \times 10^4$<br>$\beta_2 = 0.8 \times 10^8$<br>$\beta_3 = 0.241 \times 10^{12}$ |
| Equation (13-9) | Linear programming | 3 | $\beta_1 = 3.20 \times 10^4$<br>$\beta_2 = 0.435 \times 10^8$<br>$\beta_3 = 0$ |
| Equation (13-8) | Marquardt | 2 | $\beta_1 = 3.448 \times 10^4$<br>$\beta_2 = 0.091 \times 10^8$ |

[a] Data from McMenamy (1963).

fits of Eq. (13-9), as well as by the use of Marquardt's method in conjunction with Eq. (13-8).

Marquardt's method may also be applied to the same data using the Scatchard model (1949). This yields the conclusion that there is a single strong site with $k_1$ (the microscopic association constant) equal to $9.54 \times 10^5$ and five weaker sites with $k_i$ equal to $0.85 \times 10^5$.

The second set of experimental data analyzed are those for the binding of acetyl tryptophan by serum albumin (McMenamy, 1963). The results obtained by application of the various procedures are cited in Table 13-3. All methods agree as to both the magnitude of $K_1$ (or $\beta_1$) and its large magnitude relative to the higher stepwise binding constants.

## 11.3 COMMENTS ON ANALYZED DATA

The simulated data which were analyzed here could correspond to the case of four equivalent sites with some degree of cooperativity of binding.

The use of the Scatchard model may be inappropriate for a system of this kind, which displays significant cooperativity, since the values of the individual microscopic binding constants are in this case strongly dependent upon the number of bound ligands. Strictly speaking, the Scatchard model is only applicable to systems where the binding sites are completely independent.

The following general conclusions may be derived from the results with simulated data. Both Marquardt's method using Eq. (13-8) and use of (13-9) to fit $\lambda$ permit extraction of the original constants with reasonable accuracy. Indeed, the numerical results cited in Table 13-1 do not indicate a decisive superiority of either approach. Thus, at $\sigma = 0.005$ Marquardt's method gives values for all four stepwise constants which were slightly closer to the original values than those yielded by Eq. (13-9) (using the least squares polynomial fit); at $\sigma = 0.015$ the reverse was true. In both cases the difference was small. If five terms were included, the use of Eq. (13-9) yielded a value for $\beta_5$ which was either negative or not statistically significant, indicating that four terms are sufficient.

Similarly, no clearcut choice is possible between linear programming and least squares polynomial fits on the basis of the analysis of the simulated data, using Eq. (13-9). Linear programming gave slightly more accurate values at $\sigma = 0.005$; at $\sigma = 0.015$ the reverse was true.

As judged by reproduction of the original constants, both Marquardt's method and the approach of the linear methods may be regarded as reasonably adequate and there appears to be little to choose between them, except that the latter approach requires much less elaborate computer facilities.

Although linear programming, used in conjunction with Eq. (13-9) reproduced the original constants with fidelity comparable to that attained with least squares polynomial fits, the latter method is probably intrinsically preferable because of the availability of suitable statistical criteria for termination of the series. However, for the simulated data considered here, attempts at a five term fit using linear programming invariably gave a value of zero for $\beta_5$, thereby establishing the termination point of the series without any necessary for further analysis.

The use of the linearization technique of Eq. (13-3) can be clearly rejected, in view of the wide discrepancies between the computed constants and the original values for $\sigma \leq 0.01$. The use of such implicit regression which is suggested in one or two other parts of the book (there is an example in enzyme kinetics) must be used with great caution. In general, when there is a choice between implicit regression and use of a nonlinear equation, we would advise using the nonlinear equation.

At present we have not explored the reasons for the failure of this method. Two possibilities are suggested for this failure. First, inasmuch as the quantity $J$, with the error associated with it, occurs on the right-hand side of Eq. (13-3) it is possible that this could cause a progressive worsening of the fit as the error increased. The second reason could be that this particular linearizing transformation gives rise to ill-conditioned equations.

It should be mentioned that we have not attempted to establish the maximum degree of error which can be tolerated before all methods yield unacceptable results. A rigorous treatment of this aspect would require a Monte Carlo approach to simulation.

While the analysis of real data, for which the binding parameters are not know *a priori*, does not provide a valid assessment of the reliabilities of the different approaches, a comparison of the results obtained is nevertheless of interest. We shall consider first the relatively simple case of the binding of acetyl tryptophan by serum albumin.

Analysis of the data of McMenamy (1963) using Eq. (13-9) and by Marquardt's method gave similar results (Table 13-3). All methods agree in assigning a value close to $3.2 \times 10^4$ for the first binding constant. The second constant is of much lower magnitude. It should be stressed that only the macroscopic binding constants can be reliably derived from data of this kind.

The procedure of associating microscopic constants unambiguously with various sites in the sense that a given site has a given fixed microscopic constant is often not feasible. Such classification is not meaningful in the absence of detailed information about the topography of the sites and the tertiary structure of the protein, although it may often be possible to make reasonable surmises on the basis of the relative magnitudes of the consecutive stepwise constants.

The physical situation in the acetyl tryptophan case could, for example, correspond either to two different sites, one strong and the other weak, or to two equivalent sites with strong mutual interference. Since acetyl tryptophan is charged, the latter would be the case if the two sites were closely spaced, because of electrostatic repulsion. Alternatively, the latter situation might reflect a conformational change, resulting from the binding of the first ligand.

We shall consider finally the more complicated case of the binding of thyroxine by serum albumin for which the experimental scatter was quite large. The data of Tabachnik were originally analyzed in terms of the Scatchard model using Eq. (13-54). Tabachnik, like Steiner *et al.* (1966) and Sterling (1964) concluded that human serum albumin contains one strong binding site for thyroxine and several weaker ones. Specifically Tabachnik (1967) found the single strong site to have an intrinsic association constant of $14.1\times10^5 \pm 2.0\times10^5$ and the five or six weaker sites to have association constants of about $0.5\times10^5$.

The original analysis of Tabachnik, as well as those of Steiner, Roth, and Robbins and of Sterling, are subject to the limitations implicit in the Scatchard model. In addition, since they are based upon a visual fit of the data, there is an inevitable element of subjectivity. The use of the methods described herein eliminates the subjectivity. However, the problem is ill-conditioned. The data of Tabachnik show that we must start with at least four terms since for its maximum value $J > 3$. Fitting Eq. (13-9) with least squares polynomials with four terms gave all positive constants. On the other hand five term fits had a negative term.

The various computer programs employed to analyze the data in terms of Eq. (13-9) gave results in reasonable agreement for the first two binding constants, with one exception (Table 13-2). The general pattern was that the first thyroxine is bound much more strongly than those bound subsequently, is agreement with the original conclusions of Tabachnik and of Steiner and coworkers. This could arise from the presence of one strong site and several weak sites, or of several equivalent sites with strong mutual interference. To illustrate the ambiguities which can occasionally arise in the computer analysis of this kind of problem we have included an anomalous result in Table 13-2. The fit of Eq. (13-9), using the least squares multiple regression program (MREGI $), gave five terms with $\beta_1 = 7.3\times10^5$ and $\beta_2 = 93.8\times10^{10}$. In terms of microscopic constants this would mean that the second thyroxine is bound more strongly than the first. In contrast, a fit of Eq. (13-9) with the same data using the least squares polynomial program (POLFIT) gave $\beta_1 = 15.1\times10^5$ and $\beta_2 = 68.2\times10^{10}$, in reasonable agreement with the other methods, all of which gave values of $\beta_1$ between $13\times10^5$ and $18\times10^5$. We will discuss the origin of the divergent result obtained with MREGI $ later.

The linear programming solution gave unacceptable values for the two higher constants since it is very hard to account for a physical situation where $\beta_4 = 0$. Marquardt's method as described by Fletcher and Spector (1968) gives six sites with $K_1 = 13.4\times10^5$ and $K_2 = 62.7\times10^5$ (Table 13-2). An advantage of Marquardt's method is that it presents, at each solution attempt, a complete accounting of the fitting conditions. That is, it indicates the condition of the fitting process and if ill-conditioned indicates the offending parameters. It gives directly which parameters can be modified and by how much, without altering the goodness of visual fit (or sum of squares) to the data.

The results of the above computations for thyroxine show that only the first two constants are reasonably well determined and, even for these, anomalous results were obtained with one program. Attempts to extract more than three parameters from a $J$ vs. $A$ curve such as the one given by thyroxine are quite hazardous. It is important to emphasize here that even if the data were extremely accurate one could not extract more than a limited number of parameters from an essentially featureless curve.

One final remark may be made concerning the differences between the least squares polynomial fit and multiple regression for this particular example. In principle these techniques employ exactly the same numbers and invert the same matrix. Yet we obtain different results in this case. In practice, however, to avoid computer round-off error, the matrix $X'X$ is orthogonalized in the least squares polynomial program (POLFIT). In the multiple regression program (MREGI \$) the matrix $\mathbf{X'X}$ is transformed to the correlation matrix, the values of whose elements lie between $-1$ and $+1$. In either case these transformations prevent a great deal of round-off error. They also may account for the slight differences in the values of the computed parameters when the problem is well-conditioned and probably for the major differences in the present ill-conditioned case, since round-off error may persist to unequal extents. A second contributing factor is that, when using POLFIT, the function fitted is $(\lambda - 1)/[S]$, while, when using MREGI \$, the function fitted is $\lambda - 1$.

## References

Adair, G. S. (1925). *J. Biol. Chem.* **63**, 529.

Atkinson, D. E., Hathaway, J. A., and Smith, E. C. (1965). *J. Biol. Chem.* **240**, 2682.

Box, G. E. P., and Muller, M. E. (1958). *Ann. Math. Stat.* **29**, 610.

Changeux, J. P., Gerhardt, J. C., and Schachman, H. K. (1968). *Biochemistry* **7**, 521.

Changeux, J. P., and Rubin, M. M. (1968). *Biochemistry* **7**, 533.

Edsall, J. T., and Wyman, J. (1958). "Biophysical Chemistry." Academic Press, New York.

Fletcher, J. E., Spector, A. A., and Ashbrook, J. D. (1970). *Biochemistry* **9**, 2580.

Fletcher, J. E., and Spector, A. A. (1968). *Comput. Biomed. Res.* **2**, 164.

Fletcher, J. E., and Shrager, R. I. (1968). *Tech. Rept.* 1 DCRT, National Institute of Health, Bethesda, Maryland.

Frieden, E. (1967). *J. Biol. Chem.* **242**, 2045.
Gerhardt, J. C., and Pardee, A. (1962). *J. Biol. Chem.* **237**, 891.
Gerhardt, J. C., and Schachman, H. K. (1968). *Biochemistry* **7**, 538.
Grisolia, S. (1964). *Physiol. Rev.* **44**, 657.
Klotz, I. M., Walker, F. M., and Pivan, R. B. (1946). *J. Amer. Chem. Soc.* **68**, 1486.
Kirtley, M. E., and Koshland, D. E., Jr. (1967). *J. Biol. Chem.* **242**, 4192.
Koshland, D. E., Jr. (1963). *Science* **142**, 1583.
Koshland, D. E., Jr., Nemethy, G., and Filmer, D. (1966). *Biochemistry* **5**, 364.
Kowalik, A., and Morrison, J. F. (1968). Math. Biosci. **2**, 57.
McLaren, M. D., and Marsaglia, G. (1964). *J. Ass. Comput. Mach.* **12**, 83.
McMenamy, R. (1963). *Arch. Biochem. Biophys.* **103**, 409.
Magar, M. E., and Steiner, R. F. (1971). *J. Theor. Biol.* **32**, 495.
Magar, M. E., Steiner, R. F., and Fletcher, J. E. (1971). *J. Theor. Biol.* **32, 59**.
Monod, J., Changeux, J. P., and Jacob, F. (1963). *J. Mol. Biol.* **6**, 306.
Monod, J., Wyman, J., and Changeux, J. P. (1965). *J. Mol. Biol.* **12**, 88.
Rubin, M. M., and Changeux, J. P. (1966). *J. Mol. Biol.*
Scatchard, G. (1949). *Ann. N.Y. Acad. Sci.*, **51**, 660.
Schied, F. (1968). "Numerical Analysis." McGraw-Hill, New York.
Steiner, R. F. (1953). *Arch. Biochem. Biophys.* **46**, 291.
Steiner, R. F., Roth, J., and Robbins, J. (1966). *J. Biol. Chem.* **241**, 560.
Sterling, K. (1964). *J. Clin. Invest.* **43**, 1721.
Tabachnik, M. (1967). *J. Biol. Chem.* **242**, 1646.
Whitehead, E. (1970). *Biochemistry* **9**, 1440.

*CHAPTER 14*

# ENZYME KINETICS

## 1. Introduction

In this chapter we are mainly concerned with the determination of the kinetic parameters of enzyme reactions. Given the rate equations pertaining to any specific mechanism, we wish to show how the kinetic parameters are determined in the best possible way. We do not derive any rate equations in this chapter, consequently a reader who does not have a rate equation to analyze will benefit little from this chapter. It is, of course, possible that the reader may want to analyze the Michealis–Menten equation to determine $K_m$ and $V_{max}$, which we will demonstrate.

As a matter of fact, we gave the Michaelis–Menten equation as an illustrative example in the introductory chapter. If it is thought by the investigator that it is too troublesome to go through a prolonged analysis, that is to say the investigator is not interested in the precise value of the kinetic parameters, a great deal can still be learnt about reaction mechanism and inhibition patterns simply by inspecting the kinetic data. For a detailed exposition of this subject we refer the reader to Cleland (1963b).

In principle, the derivation of a rate equation is a straightforward algebraic procedure. In practice, however, as the reactions get more complex the amount of labor involved grows very rapidly. In the last dozen years or so a great deal of progress has been achieved in cutting the amount of work involved in the derivation of rate equations. The diagramatic procedure of King and Altman (1956) is used to a very large degree by most practicing kineticists (Cleland, 1967). Volkenstein and and Goldstein (1966) have further reduced the labor by suggesting certain shortcuts in the procedure of King and Altman. The method of King and Altman has been presented in varying degrees of detail in several text books: Dixon and Webb (1964), Amdur and Hammes (1966), and Mahler and Cordes (1966). Using the method of King and Altman all kinds of rate equations describing different kinetic schemes can be derived. Once the assumed rate equations are in hand we can proceed to analyze them to determine the parameters. Most rate equations can be linearized, that is to say the parameters in the rate equations occur linearly (see Chapter 1), and in presenting some typical examples we show how that can be done. The fact that some rate equations are not readily linearized should not present any problems since the methods indicated in Chapter 4 are capable of handling the nonlinear problems also.[†]

Inasmuch as the rate equation will be either linear or nonlinear, the methods can handle *all* rate equations regardless of their complexity. In analyzing the assumed rate equation, an important thing to note (especially in regard to the subject of enzyme kinetics) is the word of caution found at the end of the introductory chapter. There we emphasized that the methods which were going to be presented in Chapter 4 would ultimately yield kinetic parameters that would fit the data with great precision. If the investigator obtains values for the parameters that bear little resemblance to what he believes to be the physical situation, then one of two things could be wrong. Either there is excessive experimental error or the model used to analyze the data does not reflect the realities of the situation. The mathematical methods presented in Chapter 4 cannot, in the last analysis, be faulted. In obtaining the answer to any specific problem they are only as good as the precision of the accumulated data and the model used to analyze it.

In gathering data for enzyme kinetics it is the experience of this author that statistical weighting of the data is almost essential if we are to draw

[†] There is evidence to show that the results are better if the equations analyzed here are in their nonlinear form (see Colquhoun, 1971).

any meaningful inferences from the computed parameters. To quote a practicing kineticist regarding the use of statistics in the treatment of enzyme kinetics, "Data processing of this sort should be considered mandatory for serious kinetic studies" Cleland (1967). In view of the fact that there is really no limit to the rate equation that can be analyzed, we will have to be selective in our choice of the models to be discussed.

What we discuss here is the determination of the kinetic parameters and linearization of a complex rate equation of interest which arises from a synthetase reaction. What shall be said of this specific case can also be said of any other case so it is an illustrative example for the entire problem. Analysis of the synthetase reaction follows a brief discussion of two papers by Wilkinson (1961) and Cleland (1963a) who analyzed some of the more familiar equations. These authors emphasized the statistical aspects and the available computer programs for processing enzyme kinetic data. The paper by Wilkinson (1961) amply demonstrates, by practical example, the use of some of the methods we have been talking about. Wilkinson analyzed the raw data of Atkinson *et al.* (1961) on the kinetics of the enzyme nicotinamide mononucleotide adenylytransferase.

## 2. Two Cases Treated by Wilkinson

Wilkinson (1961) discusses a number of topics which we treat in Chapters 4–6. Among these we find the definition of a number statistical terms, linear regression analysis, and a word or two about nonlinear regression analysis. He also discusses applications of these methods to two specific examples to determine kinetic parameters. These two examples are the Michaelis–Menten case to determine $V_{max}$ and $K_m$ and the estimation of dissociation constants from a plot of $K_m$ vs. pH. The equation required for the latter case is due to Dixon (1953). The Michaelis–Menten equation can be written in the linear form as follows

$$1/v = (1/V_{max}) + (K_m/V_{max})(1/A) \tag{14-1}$$

where $v$ is the velocity of the reaction and $A$ the substrate concentration. In our notation this is transformed into

$$f(x) = \alpha_1 f_1(x) + \alpha_2 f_2(x) + \delta \tag{14-2}$$

where $\delta$ is the error made in a measurement and where we have put

$$\begin{aligned} f(x) &= 1/v, \qquad f_1(x) = 1, \qquad f_2(x) = 1/A \\ \alpha_1 &= 1/V_{max}, \qquad \alpha_2 = K_m/V_{max} \end{aligned} \tag{14-3}$$

Equation (14-2) is essentially that of a straight line, and is the simplest case treated in Chapter 4. The question of weighting the data which is essential in enzyme kinetics shows us that the weighting of the velocities must be proportional to the fourth power of the velocity. For details of actual application see Wilkinson's paper.

The second problem treated by Wilkinson was the determination of certain dissociation constants from a $K_m$ vs. pH curve. The original equation was given by Dixon (1953) as follows

$$K_m = K_m^*[1 + ([H^+]/K_1)][1 + (K_2/[H_+])] \tag{14-4}$$

where $K_m^*$ is the asymptotic value of $K_m$ and $K_1$ and $K_2$ are the dissociation constants of two ionizable groups. Equation (14-4) can now be written as

$$K_m = K_m^* + (K_m^*[H^+]/K_1) + (K_m^*K_2/[H^+]) + (K_m^*K_2/K_1) \tag{14-5}$$

or if we wish to write it in our familiar notation we can write

$$f(x) = \alpha_1 f_1(x) + \alpha_2 f_2(x) + \alpha_3 f_3(x) + \delta \tag{14-6}$$

where $\delta$ is the error made in measurement and where

$$\begin{aligned} \cdot(x) &= K_m, \quad f_1(x) = 1, \quad f_2(x) = [H^+], \quad f_3(x) = 1/[H^+] \\ \alpha_1 &= K_m + (K_m^*K_2/K_1), \quad \alpha_2 = K_m^*/K_1, \quad \alpha_3 = K_m^*K_2 \end{aligned} \tag{14-7}$$

Determination of the parameters $\alpha_1$, $\alpha_2$, and $\alpha_3$ results in our determining the dissociation constants. Again we refer the reader to Chapter 4 for the solution of Eq. (14-6). The dissociation constants $K_m^*$, $K_1$, and $K_2$, can be determined graphically and the reader is referred to the paper by Atkinson *et al.* (1961) Fig. 1A to see how that is done.

## 3. The Effects of pH on $V_{max}$

Although Wilkinson did not treat the effect of pH on $V_{max}$ we mention it here since it gives rise to an equation quite similar to Eq. (14-6). The equation for the variation of $V_{max}$ with pH is given by Dixon and Webb (1964) as

$$V_{max} = k_{+2}^*E/[1 + ([H^+]/K_{es_1}) + (K_{es_2}/[H^+])] \tag{14-8}$$

Here $E$ is the enzyme concentration, $k_{+2}^*$ is the asymptotic value of the rate constant pertinent to the breakdown of the enzyme substrate complex, and $K_{es_1}$ and $K_{es_2}$ are the ionization constants of the complex which affect the acid and the alkaline side of the pH curve respectively. Inverting Eq. (14-8).

$$1/V_{\max} = (1/k_{+2}^{*}E) + ([H^{+}]/K_{\mathrm{es}_1}k_{+2}^{*}E) + (K_{\mathrm{es}_1}/k_{+2}^{*}E[H_{+}]) \quad (14\text{-}9)$$

As we can rewrite Eq. (14-9) as

$$f(x) = \alpha_1 f_1(x) + \alpha_2 f_2(x) + \alpha_3 f_3(x) + \delta \quad (14\text{-}10)$$

where

$$\begin{aligned} f(x) &= 1/V_{\max}, \quad f_1(x) = 1, \quad f_2(x) = [H^{+}], \quad f_3(x) = 1/[H^{+}] \\ \alpha_1 &= 1/k_{+2}^{*}E, \quad \alpha_2 = 1/K_{\mathrm{es}_1}k_{+2}^{*}E, \quad \alpha_3 = K_{\mathrm{es}_2}/k_{+2}^{*}E \end{aligned} \quad (14\text{-}11)$$

The reader will note that (14-10) is identical to (14-6) and what has been said of (14-6) can be said of (14-10).

The paper by Wilkinson as stated before analyzes the data of Atkinson *et al.* (1961) on nicotinamide mononucleotide adenylytransferase showing, very nicely, how the methods are used and the results we can expect to obtain. It is a good example of what we can do in practice for the Michealis–Menten case and the variation of $K_m$ with pH and is easily extended to more complex cases.

## 4. Cases Treated by Cleland

Cleland (1963a) presented several more rate equations for analysis including the two treated by Wilkinson. Some of the other equations presented by Cleland are the following

$$v = (V_{\max}A^2)/(a + 2bA + A^2) \quad (14\text{-}12)$$

where $a$, $b$, and $V_{\max}$ are kinetic parameters and $A$ the substrate concentration. If Eq. (14-22) is left in its present form it can be treated as a nonlinear equation. Alternatively, we linearize as follows

$$1/v = (a/V_{\max})(1/A^2) + (2b/V_{\max})(1/A) + (1/V_{\max}) \quad (14\text{-}13)$$

This equation can be recast in our notation thus

$$f(x) = \alpha_1 f_1(x) + \alpha_2 f_2(x) + \alpha_3 f_3(x) + \delta \quad (14\text{-}14)$$

where $\delta$ is the error made at the measurment and where

$$\begin{aligned} f(x) &= 1/v, \quad f_1(x) = 1/A^2, \quad f_2(x) = 1/A, \quad f_3(x) = 1 \\ \alpha_1 &= a/V_{\max}, \quad \alpha_2 = 2b/V_{\max}, \quad \alpha_3 = 1/V_{\max} \end{aligned} \quad (14\text{-}15)$$

The next equation presented by Cleland was on the mechanism of a bisubstrate reaction written as

$$v = (V_{\max}AB)/(K_{ia}K_b + K_aB + K_bA + AB) \tag{14-16}$$

where $A$ and $B$ are the substrate and $K_{ia}$, $K_a$, $K_b$, and $V_{\max}$ are the parameters to be determined. Since Cleland states that the solution for the kinetic parameters were arrived at by iteration, he probably was treating Eq. (14-16) as a nonlinear equation. In point of fact, this equation is readily linearized in the following manner

$$(1/v) = \phi_0 + (\phi_1/A) + (\phi_2/B) + (\phi_{12}/AB) \tag{14-17}$$

where $\phi_0$, $\phi_1$, $\phi_2$, and $\phi_{12}$ are kinetic parameters. This equation will be recognized as one previously presented by Dalziel (1957) whose notation we use. When the constants are suitably defined this equation can be cast in our notation as follows

$$f(x) = \alpha_1 f_1(x) + \alpha_2 f_2(x) + a_3 f_3(x) + \alpha_4 f_4(x) + \delta \tag{14-18}$$

where $\delta$ is the experimental error made in the measurement and where we define

$$f(x) = (1/v),\quad f_1(x) = 1,\quad f_2(x) = 1/A,\quad f_3(x) = 1/B,\quad f_4(x) = 1/AB$$
$$\alpha_1 = \phi_0,\qquad \alpha_2 = \phi_1,\qquad \alpha_3 = \phi_2,\qquad \alpha_4 = \phi_{12} \tag{14-19}$$

It will be noted that we can write a similar equation to (14-17) for the reverse reaction, thus

$$(1/v) = \phi_0' + (\phi_1'/A') + (\phi_2'/B') + (\phi_{12}'/A'B') \tag{14-20}$$

where $A'$ and $B'$ are the products of the reaction and $\phi_0'$, $\phi_1'$, $\phi_2'$ and $\phi_{12}'$ are the kinetic parameters of the reverse reaction.

By solving for Eqs. (14-17) and (14-20) in the manner indicated in Chapter 5 we can determine the parameters $\phi_0$, $\phi_1$, $\phi_2$, $\phi_{12}$, $\phi_0'$, $\phi_1'$, $\phi_2'$, and $\phi_{12}'$. Dalziel (1957) has shown that when certain relationships between these parameters hold we can make inferences about the mechanisms, whether the order of addition of the reactants is random, sequential, etc. The reader is referred to Dalziel (1957), Cleland (1963c), and Cleland (1967), for further reading on the meaning and relationships of these constants.

In addition to the above equations presented by Cleland, we can give the following equations for enzyme reaction in presence of inhibitors.

$$v = (V_{max}A)/\{K_m[1 + (I/K_i)] + A\} \tag{14-21}$$

$$v = (V_{max}A)/\{K_m + A[1 + (I/K_i)]\} \tag{14-22}$$

$$v = (V_{max}A)/\{K_m[1 + (I/K_{is})] + A[1 + (I/K_{ii})]\} \tag{14-23}$$

(14-24)

In the above equations $V_{max}$, $K_m$, $K_{is}$, $K_i$, and $K_{ii}$ are all kinetic parameters. Equation (14-21) represents linear competitive inhibition, Eq. (14-22) is a case of linear uncompetitive inhibition, Eq. (14-23) represents a case of linear noncompetitive inhibition and Eq. (14-24) represents in the nomenclature of Cleland (1963a) *S*-parabolic, *I*-linear noncompetitive inhibition and *S*-linear and *I*-parabolic noncompetitive inhibition respectively when the linear and parabolic factors are reversed in the denominator of Eq. (14-24). A glance at Eqs. (14-21)–(14-24) shows that some of them can be readily linearized if necessary. Cleland[†] (1963a) has stated that several programs in FORTRAN are available for handling these equations and the two cases treated by Wilkinson. From the general trend of Cleland's discussion we have drawn the conclusion that Cleland analyzed most of the above equations by treating them as nonlinear problems. We do not know which, if any, of the nonlinear methods described in Chapter 4 he used to analyze his data or what method he used to obtain the standard errors. The discussion in Chapters 5 and 6, however, will serve to point out to the reader precisely what is involved in these problems.

Inasmuch as most of the equations can be linearized, we recommend using linear forms because of the advantages this offers from a statistical point of view. For instance, in using Eq. (14-17) for a two substrate reaction, one can gather all the $m \times n$ data points at $m$ different concentrations of substrate $A$ and $n$ different concentration of substrate $B$, feed them to the computer together with the weights for the velocity and obtain the four kinetic parameters $\phi_0$, $\phi_1$, $\phi_2$, and $\phi_{12}$ together with the standard error of each of these parameters. This saves the time of drawing several straight lines and computing their intercepts and slopes. More important the labor saving device is the elimination of any naked eye bias an investigator may happen to have. It will also be recalled that the

[†] Present address: Department of Biochemistry, University of Wisconsin, Madison, Wisconsin.

computer will give the investigator the "best" fit in the least squares sense. Further, if it is desired the computer can be programmed to test the parameters $\phi_0$, $\phi_1$, $\phi_2$, $\phi_{12}$, $\phi_0'$, $\phi_1'$, $\phi_2'$, and $\phi_{12}'$ to tell the investigator what reaction mechanism is being followed and with what degree of probability.

## 5. A Complex Case and Ways of Handling It

The equation which we treat is the generalized rate equation for the kinetic scheme (14-25),

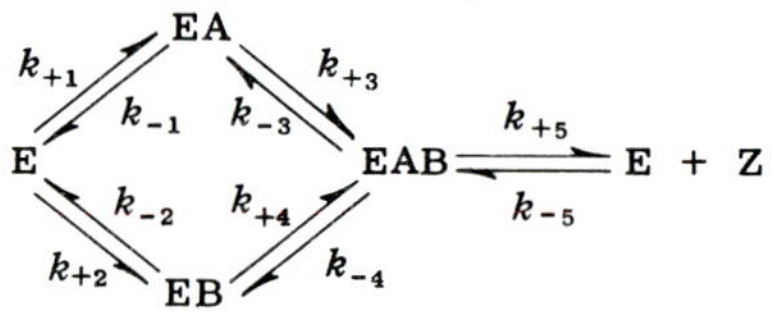

$$\begin{array}{ccccc} & & \mathrm{EA} & & \\ & k_{+1} \nearrow \swarrow k_{-1} & & k_{+3} \searrow \nwarrow k_{-3} & \\ \mathrm{E} & & & & \mathrm{EAB} \underset{k_{-5}}{\overset{k_{+5}}{\rightleftharpoons}} \mathrm{E} + \mathrm{Z} \\ & k_{+2} \searrow \nwarrow k_{-2} & & k_{+4} \nearrow \swarrow k_{-4} & \\ & & \mathrm{EB} & & \end{array}$$

where $A$ and $B$ are substrates and $Z$ is the product. Dixon and Webb (1964) have determined the rate equation for this reversible system and have shown it to be

$$v = \frac{(c_1AB + c_2A^2B + c_3AB^2 - c_4Z - c_5AZ - c_6BZ)E}{\begin{array}{c} c_7 + c_8A + c_9B + c_{10}AB + c_{11}A^2 + c_{12}B^2 + c_{13}A^2B \\ + c_{14}B^2A + c_{15}Z + c_{16}AZ + c_{17}BZ + c_{18}ABZ \end{array}} \tag{14-26}$$

where

$$\begin{aligned}
c_1 &= k_{+1}k_{-2}k_{+3}k_{+5} + k_{-1}k_{+2}k_{+4}k_{+5}, & c_2 &= k_{+1}k_{+3}k_{+4}k_{+5} \\
c_3 &= k_{+2}k_{+3}k_{+4}k_{+5}, & c_4 &= k_{-1}k_{-2}k_{-3}k_{-5} + k_{-1}k_{-2}k_{-4}k_{-5} \\
c_5 &= k_{-1}k_{-3}k_{+4}k_{-5}, & c_6 &= k_{-2}k_{+3}k_{-4}k_{-5} \\
c_7 &= k_{-1}k_{-2}k_{-3} + k_{-1}k_{-2}k_{-4} + k_{-1}k_{-2}k_{+5} & & \qquad (14\text{-}27) \\
c_8 &= k_{+1}k_{-2}k_{-3} + k_{+1}k_{-2}k_{-4} + k_{+1}k_{-2}k_{+5} + k_{-1}k_{-3}k_{+4} + k_{-1}k_{+4}k_{+5} \\
c_9 &= k_{-1}k_{+2}k_{-3} + k_{-1}k_{+2}k_{-4} + k_{-1}k_{+2}k_{+5} + k_{-2}k_{+3}k_{-4} + k_{-2}k_{+3}k_{+5} \\
c_{10} &= k_{+1}k_{-2}k_{+3} + k_{-1}k_{+2}k_{+4} + k_{+1}k_{+3}k_{-4} + k_{+2}k_{-3}k_{+4} + k_{+3}k_{+4}k_{+5} \\
c_{11} &= k_{+1}k_{-3}k_{+4} + k_{+1}k_{+4}k_{+5}, & c_{12} &= k_{+2}k_{+3}k_{-4} + k_{+2}k_{+3}k_{+5} \\
c_{13} &= k_{+1}k_{+3}k_{+4}, & c_{14} &= k_{+2}k_{+3}k_{+4} \\
c_{15} &= k_{-1}k_{-2}k_{-5} + k_{-1}k_{-4}k_{-5} + k_{-2}k_{-3}k_{-5}, & c_{16} &= k_{-1}k_{+4}k_{-5} + k_{-3}k_{+4}k_{-5} \\
c_{17} &= k_{-2}k_{+3}k_{-5} + k_{+3}k_{-4}k_{-5}, & c_{18} &= k_{+3}k_{+4}k_{-5}
\end{aligned}$$

Various simplifications can occur in Eq. (14-26), for instance, the forward rate at $Z = 0$ is given by

$$v = \frac{(c_1AB + c_2A^2B + c_3AB^2)E}{c_7 + c_8A + c_9B + c_{10}AB + c_{11}A^2 + c_{12}B^2 + c_{13}A^2B + c_{14}AB^2} \tag{14-28}$$

and putting $A = B = 0$ we obtain the backward rate:

$$v = (c_4ZE)/(c_7 + c_{15}Z) \tag{14-29}$$

Dixon and Webb observed that this equation is of the same form as the Michaelis–Menten equation, but with more complex expression for $V_{\max}$ and $K_{\mathrm{m}}$ which can now be written as

$$V_{\max} = (c_4E)/c_{15}, \qquad K_{\mathrm{m}} = (c_7/c_{15}) \tag{14-30}$$

In the event that one of the two initial reactants, say $B$, is kept constant, we obtain for the forward rate

$$v = (iA^2 + jA)E/(k + lA^2 + mA) \tag{14-31}$$

where

$$\begin{array}{lll} i = c_2B, & j = c_1B + c_3B^2, & k = c_7 + c_9B + c_{12}B^2 \\ l = c_{11} + c_{13}B, & m = c_8 + c_{10}B + c_{14}B^2 & \end{array} \tag{14-32}$$

A similar and somewhat symmetrical equation is obtained when $A$ is held constant.

Our objectives in treating this example are to determine all the rate constants $k_{+1}, k_{-1}, \ldots, k_{+5}, k_{-5}$ in the reaction sequence depicted in (14-25). To do so we could linearize Eq. (14-26) by writing

$$\sum_{i=1}^{17} \alpha_i f_i(x) + \delta \tag{14-33}$$

In Eq. (14-33) we have separated the constant from the variables and have represented the experimental error by $\delta$. The constants are the $\alpha_i$ and the variables are the $f_i(x)$ and they are defined below:

$$\begin{array}{llll} \alpha_1 = (c_1E/c_7), & \alpha_2 = (c_2E/c_7), & \alpha_3 = (c_3E/c_7), & \alpha_4 = (-c_4E/c_7) \\ \alpha_5 = (-c_4E/c_7), & \alpha_6 = (-c_6E/c_7), & \alpha_7 = (-c_8/c_7), & \alpha_8 = (-c_9/c_7) \\ \alpha_9 = (-c_{10}/c_7), & \alpha_{10} = (-c_{11}/c_7), & \alpha_{11} = (-c_{12}/c_7), & \alpha_{12} = (-c_{13}/c_7) \\ \alpha_{13} = (-c_{14}/c_7), & \alpha_{14} = (-c_{15}/c_7), & \alpha_{15} = (-c_{15}/c_7), & \alpha_{16} = (-c_{17}/c_7) \\ \alpha_{17} = (-c_{18}/c_7) & & & \end{array} \tag{14-34}$$

and

$$\begin{array}{llll} f_1(x) = AB, & f_2(x) = A^2B, & f_3(x) = AB^2, & f_4(x) = Z \\ f_5(x) = AZ, & f_6(x) = BZ, & f_7(x) = Av, & f_8(x) = Bv \\ f_9(x) = ABv, & f_{10}(x) = A^2v, & f_{11}(x) = B^2v, & f_{12}(x) = A^2Bv \\ f_{13}(x) = AB^2v, & f_{14}(x) = Zv, & f_{15}(x) = AZv, & f_{16}(x) = BZv \\ f_{17}(x) = ABZv & & & \end{array} \tag{14-35}$$

Whether or not the linearized expression (14-33) will work well in actual practice we cannot say since we have not tried it. It is an implicit regression problem because the velocity occurs on both sides of the equation. Our practical results in using similar methods of linearization in protein ligand equilibria indicate that this may not be a good method. Treating Eq. (14-26) as a nonlinear problem may be a better way.

It is, of course, possible to solve Eq. (14-32) by the techniques in Chapter 4 provided we have linear independence of the $f_i(x)$. The experimental data that are required for both the linear and non linear cases are the measurements of the initial velocity at a variety of concentrations of substrates and product. From a single experiment, one can get a number of points provided one knows the exact initial velocity at the exact concentration of substrates and product. The problem as presented above should determine 17 coefficients. However, we have only ten kinetic parameters, consequently our problem is overdetermined and some of the $c_i$ are expressible in terms of each other. This fact must be taken into consideration and used to our advantage. The fact that we have constraints makes linear programming highly advisable if the linearized equation is to be used.

In general, we must make use of every available piece of information to obtain our final results. In the event that we think, or have experimental evidence, that certain parameters should be constrained to lie within certain values, then these facts must be taken into consideration. For instance, we can using a stopped flow apparatus determine $k_{+1}$ or $k_{+2}$. If an equilibrium dialysis experiment is done on the reaction $E + A \rightleftharpoons EA$ then the dissociation constant for this reaction coupled with the stopped flow experiment should yield $k_{+1}$ and $k_{-1}$. If these values of $k_{+1}$ and $k_{-1}$ or any other rate constants are known precisely or approximately the problem is greatly simplified because our methods permit us to constrain the parameters.

From a computational standpoint and given the possible experimental error we do not think linear least squares will work very well. However, the use of nonlinear least squares or any other method may be attempted by breaking up the problem and treating some of the equations obtained by putting $A = B = 0$ or $Z = 0$ and varying $A$ (or $B$) keeping $B$ (or $A$) constant. Breaking up the problem in this manner and analyzing the less complicated equation will yield more accurate answers. Whenever possible, therefore, we recommend this procedure.

In the case we are treating we have four "sub-equations" namely Eqs. (14-28), (14-29), (14-31), and its symmetrical counterpart written in terms of $B$. Equation (14-29) is essentially the Michaelis–Menten case where we have $A = B = 0$ and we measure the velocity as a function of the product concentration. This equation is used to solve for $V_{\max}$ and $K_m$ which gives us $c_4/c_{15}$ and $c_7/c_{15}$ respectively. Next we take Eq. (14-31) where we set $B$ constant and $Z = 0$ and solve for $i$, $j$, $k$, $l$, and $m$. Either using the equation directly and treating it as a nonlinear problem or linearizing thus

$$v(k + lA^2 + mA) = iA^2E + jAE$$

or (14-36)

$$v = (iA^2E/k) + (jAE/k) - (lA^2v/k) - (mAv/k)$$

by measuring the initial velocity at various concentration of $A$ and at constant $B$ we would be able to obtain the parameters $i/k$, $j/k$, and $m/k$ which can be written in terms of the $c'$ as follows

$$\begin{aligned} i/k &= c_2B/(c_7 + c_9B + c_{12}B^2) \\ j/k &= (c_1B + c_3B^2)/(c_7 + c_9B + c_{12}B^2) \\ l/k &= (c_{11} + c_{13}B)/(c_7 + c_9B + c_{12}B^2) \\ m/k &= (c_8 + c_{10}B + C_{14}B^2)/(c_7 + c_9B + c_{12}B^2) \end{aligned} \tag{14-37}$$

By determining $i/k$, $j/k$, $l/k$, and $m/k$ at different concentrations of $B$ we solve for the $c$ by treating Eq. (14-37) as nonlinear or we could linearize (14-37) and solve for it to obtain the parameters. We also note that if we had retained $B$ as the variable in Eq. (14-31) and kept $A$ constant we could have obtained a similar set of relations for $i$, $j$, $k$, $l$, and $m$ call them $i'$, $j'$, $k'$, $l'$, and $m'$

$$\begin{aligned} i' = c_3A, \quad j' = c_1A + c_2A^2, \quad k' = c_7 + c_8A + c_{11}A^2 \\ l' = c_{12} + c_{14}A^2, \quad m' = c_9 + c_{18}A + c_{13}A^2 \end{aligned} \tag{14-38}$$

Consequently from a set of experiments at $m$ concentrations of $A$ and $n$ concentration of $B$ we can obtain all the following parameters $c_1, c_2, \ldots,$

$c_{15}$. For $c_{16}$, $c_{17}$, and $c_{18}$ the entire original Eq. (14-27) is then used after the values of the parameters previously determined from the simpler cases are substituted; that is to say, Eq. (14-27) is used with all the values of the constants $c_1, c_2, \ldots, c_{15}$ inserted in their respective places. Further, with the results one obtains from measuring the velocity at $m$ concentrations of $A$ and $n$ concentration of $B$ we could have used Eq. (14-29) to determine $c_1$, $c_2$, $c_3$, $c_7$, $c_8$, $c_9$, $c_{10}$, $c_{11}$, $c_{12}$, $c_{13}$, and $c_{14}$. Once again this equation could have been used as is or it could have been linearized.

For the solution of the equations presented throughout this chapter there are several methods indicated in Chapter 4. Of these methods, linear programming will probably adequately handle the complex cases especially when there are constraints on the parameters. Unfortunately we cannot find the standard error of the parameters using linear programming, and therefore a wise alternative is to break up the problem and deal with it as we have indicated here by using least squares or one or the other of the methods mentioned in Chapter 4 that will enable us to estimate the standard errors of the parameters.

In Chapter 6 we show how that is done, and we also show how we may test the hypotheses about the kinetic parameters. In testing hypotheses about parameters we determine, given the accumulated data, whether or not and with what degree of probability a certain rate constant or equilibrium constant is equal to zero or some other special quantity. Such testing will be very valuable in discussions about intermediary metabolism.

In concluding this chapter about enzyme kinetics, we would like to mention that a large number of rate equations and development of special topics such as the effect of pH on $K_m$ or $V_{max}$ have not been included. We believe, that all of these cases can be adequately handled by the methods of Chapter 4. The reader whose rate equation did not appear here, or who is making a pH-dependent measurement will probably be able to analyze his kinetic equation whatever its form to obtain the kinetic parameters and the standard error of these parameters and to test hypotheses about them. If the rate equation of interest has not appeared in this chapter, that does not mean it is not amenable to treatment, it very probably is, and we recommend that you see your programmer about how to analyze it.

In this chapter also, we were not concerned with the meaning of the parameters obtained and their relation to the actual mechanism. These questions have been dealt with in several textbooks and articles indicated throughout this chapter. For the handling of equations occurring in rapid reactions and transient enzyme kinetics the reader is referred to the next chapter.

We cannot close this chapter without one final word of advice to enzyme kineticists. We strongly recommend that after collecting the data such as concentrations and the velocity as a function of substrate, inhibitor, and you then let the computer do the statistical weighting, the determination of the parameters and their standard errors, the testing of hypothesis about the parameter, and so forth. In this way you can maintain the proper caution and objectivity.

## References


Amdur, I., and Hammes, G. G. (1966). "Chemical Kinetics, Principles and Selected Topics." MacGraw-Hill, New York.

Atkinson, M. R., Jackson, J. F., and Morton, R. K. (1961). *Biochem. J.* **80**, 318.

Cleland, W. W. (1963a). *Nature* **198**, 463.

Cleland, W. W. (1963b). *Biochim. Biophys. Acta* **67**, 188.

Cleland, W. W. (1963c). *Biochim. Biophys. Acta.*

Cleland, W. W. (1967). *Annu. Rev. Biochem.* **35**, 77.

Colquhoun, D. (1971). "Lectures in Biostatistics." Oxford Univ. Press (Clarendon), London and New York.

Dalziel, K. (1957). *Acta Chem. Scand.* **11**, 1076.

Dixon, M. (1953). *Biochem. J.* **55**, 161.

Dixon, M., and Webb, E. C. (1964). "Enzymes." Academic Press, New York.

King, E. L., and Altman, C. (1956). *J. Phys. Chem.* **60**, 1375.

Mahler, H. R., and Cordes, E. H. (1966). "Biological Chemistry," Harper, New York.

Volkenstein, M. W., and Goldstein, B. N. (1966). *Biochim. Biophys. Acta* **115**, 471.

Wilkinson, G. N. (1961). *Biochem. J.* **80**, 324.

*CHAPTER 15*

# RAPID REACTIONS: TRANSIENT ENZYME KINETICS AND OTHER RAPID BIOLOGICAL REACTIONS

## 1. Introduction

In this chapter we concentrate on the determination of kinetic parameters by techniques generally employed for the study of rapid reactions, namely, flow techniques and relaxation methods. To borrow a few words from Bernhard (1968), we are talking here of transient kinetics because the turnover number of enzyme catalyzed reactions are rapid (the rate constant for the breakdown of the final enzyme substrate complex before the liberation of products is generally of the order of $10^2$ sec$^{-1}$ or less). To investigate what happens during that short time we must employ the special techniques mentioned above. Once we determine the four kinetic parameters [from Eq. (14-17)] for a two substrate reaction studied in the steady state we have exhausted the information available unless we resort to experimental techniques specifically designed to probe the rapid initial phase of the reaction (Gibson, 1968). Consequently, if one is done with the steady-state kinetics of a reaction and one still wants to do kinetics, then there is no choice but to resort to the measurement of rapid reactions.

## 2. Techniques of Measuring Rapid Reactions

At the heart of the matter, insofar as the techniques for measuring rapid reactions are concerned, lies the measurement of the concentration of various reacting species in as short a time as possible. To attain this end various investigators have devised different methods. For a survey of the techniques and methods used in chemistry the reader is referred to the volume by Friess *et al.* (1963) which includes articles by a number of original contributors on the subject: Chance, Roughton, Eigen, DeMayer, Porter, and others.

In this volume the authors describe the various instruments (their own and others') and the methods of measuring rapid reactions. In biochemistry the techniques most commonly used for following rapid reactions are flow techniques (both stopped flow and continuous flow) and relaxation methods.

In a stopped flow apparatus two reactants (enzyme and substrate usually) are pushed by means of syringes into a chamber where they are rapidly mixed and then pushed into an observation cell where the flow is stopped by another syringe. This is immediately followed by recording (usually by a spectrophotometer) of the contents of the observation cell to determine the concentration of some chemical species as a function of time. The time course of the appearance or disappearance of that chemical species is followed only after the mixing and the stoppage of flow. Gibson extended the scope of the stopped flow method (Gibson and Greenwood, 1965) by coupling it with a flashphotolysis devise. Essentially what is being done here is that after the flow has stopped a flash is directed at the solution which causes the photodissociation of the reactants. When the flash is removed the reaction resulting from the recombination of the reactants is followed. This procedure enables one to reduce the time involved before making the first observation.

In a continuous flow apparatus the solutions are rapidly mixed and then pushed along an observation tube where the concentration of a chemical species is recorded. The time scale for a continuous flow apparatus is obtained in following manner. If the observation point is

$x$ cm away from the mixing chamber and the flow velocity is $v$ cm/sec then we are essentially recording the concentration of the chemical species at a time $t = x/v$ sec. By varying $x$ or $v$ we can change $t$ and thus record the concentration of the chemical species of interest at various times. In general, flow techniques can record at times varying from 1–5000 msec. If longer times are required then one mixes manually and places the observation chamber in the recording device.

For time resolution faster than a millisecond down to the microsecond, range relaxation methods are used. These methods, first introduced by Manfred Eigen (Eigen, and DeMaeyer, 1963), depend on taking a system already in *equilibrium* and perturbing this equilibrium by means of a change in temperature or pressure for examples. The system is then observed as it shifts or relaxes to its new equilibrium by monitoring the change in concentration of one or more (usually one) of the reactants as the equilibrium is shifted.

In this chapter we examine and analyze a number of equations that arise during the course of working with rapid reactions. Once again, since we can not handle all cases in this text, we shall have to be selective. In the demonstrations of cases given in the next section we must bear in mind that the methods advocated will work for reactions more or less complex that the ones given.

## 3. Combination of Hemoglobin with Ligands

The combination of hemoglobin or any other protein for that matter, with four molecules of the same ligand is the example we use. It is, of course, no accident that we select this example, inasmuch as this reaction has been continuously investigated over the last forty years or so. Next we investigate the general equations for multiple intermediates in enzyme substrate reactions. Lastly, we look at the equations produced in the consideration of a complex reaction when investigated by relaxation methods where it will be shown that the resulting equations are essentially no different from the previous case except for the fact that concentrations (in the previous case) are now replaced by equilibrium concentrations. Needless to say the analysis of the equations presented for the above cases will also be applicable to nonenzymic reactions.

Most of the information in this section is taken from the article by Gibson and Roughton (1957) on the combination of carbon monoxide whit sheep hemoglobin. The general reaction scheme according to Adair

(1925) may be written as

$$
\begin{aligned}
P + A_1 &\underset{k_1}{\overset{k_1'}{\rightleftharpoons}} PA_1, &\quad k_1'/k_1 &= K_1 \\
PA_1 + A_2 &\underset{k_2}{\overset{k_2'}{\rightleftharpoons}} PA_2, &\quad k_2'/k_2 &= K_2 \\
PA_2 + A_3 &\underset{k_3}{\overset{k_3'}{\rightleftharpoons}} PA_3, &\quad k_3'/k_3 &= K_3 \\
PA_3 + A_1 &\underset{k_4}{\overset{k_4'}{\rightleftharpoons}} PA_4, &\quad k_4'/k_4 &= K_4
\end{aligned}
\tag{15-1}
$$

where $P$ is a protein or any reacting species for that matter (in the case discussed by Adair, as stated above, the species was hemoglobin), and $A$ the ligand and the $k$ and $K$ rate constants and equilibrium constants respectively. Adair went on to derive the following equation for the combination of oxygen with hemoglobin

$$
y = \frac{K_1 p + 2K_1K_2p^2 + 3K_1K_2K_3p^3 + 4K_1K_2K_3K_4p^4}{4(1 + K_1p + K_1K_2p^2 + K_1K_2K_3p^3 + K_1K_2K_3K_4p^4)} \tag{15-2}
$$

where $y$ is the fractional saturation of hemoglobin with oxygen and $p$ the oxygen pressure. Equation (15-2) is readily recognized as Equation (13-2) when $n = 4$ and can be handled in the same manner to obtain the dissociation constants $K_1$, $K_2$, $K_3$, and $K_4$.

Roughton *et al.* (1955), analyzing the reaction of sheep hemoglobin, obtained the four intermediate equilibrium constants $K_1$ to $K_4$. The method used involved the analysis of accurate data on the combination of hemoglobin at 0–2% saturation to obtain an accurate $K_1$ and at 98–100% saturation to obtain an accurate $K_4$ by extrapolation. Obtaining accurate data at the extremes of the saturation curve is not unlike the methods advocated by Nozaki *et al.* (1957) in the analysis of copper–imidazole and zinc–imidazole binding.

For the determination of the kinetic parameters in Eq. (15-1) we proceed as follows. Suppose we are dealing with the combination of hemoglobin and carbon monoxide. If we let

$$
[CO] = q_0, \quad [Hb_4] = r_0, \quad [Hb_4CO] = [Hb_4(CO)_2]
$$
$$
= [Hb_4(CO)_3] = [Hb_4(CO)_4] = 0
$$

at the start of the reaction when $t = 0$, and if we let

$$[CO] = q, \quad [Hb_4] = r, \quad [Hb_4(CO)] = s$$
$$[Hb_4(CO)_2] = u, \quad [Hb_4(CO)_3] = v, \quad [Hb_4(CO)_4] = w$$

at time $t$, we obtain the following differential equations

$$\begin{aligned} dr/dt &= -k_1'qr, & ds/dt &= k_1'qr - k_2'qs \\ du/dt &= k_2'qr - k_3'qu, & dw/dt &= k_3'qu - k_4'qv \end{aligned} \tag{15-3}$$

Further, we have the following constraints for the total hemoglobin and carbon monoxide

$$\begin{aligned} r_0 &= r + s + u + v + w \\ q_0 &= q + s + 2u + 3v + 4w = q + r_0 z \end{aligned}$$

where $z$ is the fractional saturation and can be written as

$$z = (s + 2u + 3v + 4w)/r_0 \tag{15-4}$$

The solution of the differential equations (15-3) arrived at by Bateman (1910), *on the assumption that $q_0$ is very large and remains so for the duration of the reaction*, is

$$\begin{aligned}
r &= r_0 \exp(-\lambda_1't) \\
s_0 &= r_0\lambda_1'\left[\frac{\exp(-\lambda_1't)}{(\lambda_2' - \lambda_1')} + \frac{\exp(-\lambda_2't)}{(\lambda_1' - \lambda_2')}\right] \\
u_0 &= r_0\lambda_1'\lambda_2'\left[\frac{\exp(-\lambda_1't)}{(\lambda_2' - \lambda_1')(\lambda_3' - \lambda_1')}\right. \\
&\quad \left. + \frac{\exp(-\lambda_2't)}{(\lambda_1' - \lambda_2')(\lambda_3' - \lambda_2')} + \frac{\exp(-\lambda_3't)}{(\lambda_1' - \lambda_3')(\lambda_2' - \lambda_3')}\right] \\
v_0 &= r_0\lambda_1'\lambda_2'\lambda_3'\left[\frac{\exp(-\lambda_1't)}{(\lambda_2 - \lambda_1)(\lambda_3' - \lambda_1')(\lambda_4' - \lambda_1')}\right. \\
&\quad + \frac{\exp(-\lambda_2't)}{(\lambda_1' - \lambda_2')(\lambda_3' - \lambda_2')(\lambda_4' - \lambda_2')} \\
&\quad + \frac{\exp(-\lambda_3't)}{(\lambda_1' - \lambda_3')(\lambda_2' - \lambda_3')(\lambda_4' - \lambda_3')} \\
&\quad \left. + \frac{\exp(-\lambda_4't)}{(\lambda_1' - \lambda_4')(\lambda_2' - \lambda_4')(\lambda_3' - \lambda_4')}\right] \\
w_0 &= r_0 - r - s_0 - u_0 - v_0
\end{aligned} \tag{15-5}$$

In the above equation we have put

$$\lambda_1' = q_0 k_1', \qquad \lambda_2' = q_0 k_2', \qquad \lambda_3' = q_0 k_3', \qquad \lambda_4' = q_0 k_4' \tag{15-6}$$

Gibson and Roughton used the above equation to make preliminary estimates of the parameters $k_1'$, $k_2'$, $k_3'$, and $k_4'$ in the following manner: By assuming certain values for $\lambda_1'$, $\lambda_2'$, $\lambda_3'$, and $\lambda_4'$ and inserting them in Eq. (15-5) the quantities $s_0$, $u_0$, $v_0$, and $w_0$ were computed. Next by substituting in Eq. (15-4)

$$z = (s_0 + 2u_0 + 3v_0 + 4w_0)/r$$

the authors computed a value of $z$ and compared this value with the experimentally determined $z$. In this manner preliminary estimates were obtained for $k_1'$, $k_2'$, $k_3'$, and $k_4'$. The preliminary estimates were refined to make the estimates as close as possible to the computed data. The authors did not describe how they went about making these refinements, presumably they used one or other of the nonlinear methods described in Chapter 4.

The solution of the actual differential equations without the restrictive assumption that $q_0$ is very large and remains so for the duration of the reaction may be written as

$$r = r_0 \exp \int_0^t -k_1' q \, dt \tag{15-7a}$$

$$s = k_1' r \left[ \frac{1}{(k_2' - k_1')} + \frac{(r/r_0)^{(k_2' - k_1')/k_1'}}{(k_1' - k_2')} \right] \tag{15-7b}$$

$$u = k_1' k_2' r \left[ \frac{1}{(k_2' - k_1')(k_3' - k_1')} + \frac{(r/r_0)^{(k_2' - k_1')/k_1'}}{(k_1' - k_2')(k_3' - k_2')} + \frac{(r/r_0)^{(k_3' - k_1')/k_1'}}{(k_1' - k_3')(k_2' - k_3')} \right] \tag{15-7c}$$

$$\begin{aligned} v = k_1' k_2' k_3' r \Bigg[ & \frac{1}{(k_2' - k_1')(k_3' - k_1')(k_4' - k_1')} \\ & + \frac{(r/r_0)^{(k_2' - k_1')/k_1'}}{(k_1' - k_2')(k_3' - k_2')(k_4' - k_2')} \\ & + \frac{(r/r_0)^{(k_3' - k_1')/k_1'}}{(k_1' - k_3')(k_2' - k_3')(k_4' - k_3')} \\ & + \frac{(r/r_0)^{(k_4' - k_1')/k_1'}}{(k_1' - k_4')(k_2' - k_4')(k_2' - k_4')} \Bigg] \end{aligned} \tag{15-7d}$$

We recognize that the above equations are nonlinear in the parameters. The same can be said of Bateman's equations and consequently the nonlinear iterative techniques described in Chapter 4 can be used to solve them both. With respect to the actual solutions two things can be done in our attempts to estimate the rate constants. We can either solve Bateman's equation to obtain $\lambda_1'$, $\lambda_2'$, $\lambda_3'$, and $\lambda_4'$ which can provide preliminary estimates for $k_1'$, $k_2'$, $k_3'$, and $k_4'$ by using Eq. (15-6) or we can solve Eqs. (15-7) directly for $k_1'$, $k_2'$, $k_3'$, and $k_4'$. In the first instance we set down the following equation

$$z_j = [s_0(t_j) + 2u_0(t_j) + 3v_0(t_j) + 4w_0(t_j)] + \delta_j \qquad (j = 1, 2, \ldots, m) \tag{15-8}$$

where $z_j$ is the measured fractional saturation at time $t_j$; $s_0(t_j)$, $u_0(t_j)$, $v_0(t_j)$, and $w_0(t_j)$ the values of $s_0$, $u_0$, $v_0$, and $w_0$ at time $t_j$; and $\delta_j$ the experimental error made in the $j$th measurement. In order to satisfy the condition of Bateman's equations $s_0$, $u_0$, $v_0$, must be small. Equations (15-8) is now solved for $\lambda_1'$, $\lambda_2'$, $\lambda_3'$, and $\lambda_4'$ by some of the methods described in Chapter 4, e.g., the linearized Taylor series, the Newton–Raphson, steepest descent, etc. To clarify how one obtains the solutions within the formalism of Chapter 4 let us recall that $z_j$ is a known quantity and is the response of the system, and $s_0(t_j)$, $u_0(t_j)$, $3v_0(t_j)$ and $w_0(t_j)$ are all functions of $t_j$ and $r_0$, both of which are known. The only unknowns in Eq. (15-8) are the parameters $\lambda_1'$, $\lambda_2'$, $\lambda_3'$, $\lambda_4'$, and the equation is solved for them.

The values of $\lambda_1'$, $\lambda_2'$, $\lambda_3'$, and $\lambda_4'$ thus obtained are then used to obtain preliminary estimates of $k_1'$, $k_2'$, $k_3'$, and $k_4'$ using (15-6). With this knowledge of $k_1'$, $k_2'$, $k_3'$, and $k_4'$ we can set out to solve Eqs. (15-7a)–(15-7d) derived by Gibson and Roughton, by using the values of the $k$ as the starting point of our iterative procedures. The equations derived by Gibson and Roughton may now be written as the following $m$ equations

$$z_j = \{[s(t_j) + 2u(t_j) + 3v(t_j) + 4w(t_j)]/r_0\} + \delta_j \qquad (j = 1, 2, \ldots, m) \tag{15-9}$$

where $z_j$ as before represents the experimentally measured fractional saturation at time $t_j$ and $s(t_j)$, $u(t_j)$, $v(t_j)$, and $w(t_j)$ the values of $s$, $u$, $v$, and $w$ obtained from Eqs. (15-7a)–(15-7d) at time $t_j$. It will be noted that these functions are written in terms of the rate constants and in terms of $r$. Now $r$ can be obtained from Eq. (15-7a) if we know the values of $k_1'$. Since all we have is a rough estimate of $k_1'$ then the value of $r$ obtained from Eq. (15-7a) is not an accurate one. To overcome this problem we may replace $r$ by $r_0 \exp \int_0^{t_j} - k_1' q \, dt$ instead of the $r$ in Eq. (15-9) without complicating the nonlinear problem to any great extent, since the integral can be easily executed leaving us with $k_1'$ as the only unknown.

The nonlinear techniques for solving (15-7) directly (without having to obtain our initial guess from the Bateman equations) by inserting suitable initial guesses for $k_1'$, $k_2'$, $k_3'$, and $k_4'$ can handle the problem without any difficulty. We believe that the direct solution of Eqs. (15-7a)–(15-7d) is probably a better procedure than solving Bateman's equations first, in view of the shorter number of steps.

Sometimes it is valuable to obtain an estimate of $k_1'$ to use in the above equations. Gibson and Roughton employed the following device to obtain such an estimate. At low values of time the rate of combination of Hemoglobin with carbon monoxide is given by the equation

$$[1/(q_0 - r_0)] \ln[(q_0 - z)/(r_0 - z)] = (-k_1't/4) + \text{constant} \qquad (15\text{-}10)$$

where $q_0$ and $r_0$ have been identified before as the starting concentrations of the ligand and the starting concentration of the protein respectively. A plot of

$$[1/(q_0 - r_0)] \ln[(q_0 - z)/(r_0 - z)]$$

versus time should yield an asymptotic slope (as $t \to 0$) equal to $k_1'/4$. Gibson and Roughton have stated that for the case of combination of carbon monoxide with hemoglobin one generally obtains a $k_1'$ accurate to within 1% provided $k_2'/k_1'$ is greater than unity. This reasoning, presumably, should hold true for other reactions as well.

Once we determine $k_1'$, $k_2'$, $k_3'$, and $k_4'$ and, if we note we can determine $K_1$, $K_2$, $K_3$, and $K_4$ as outlined in Chapter 13, then $k_1$, $k_2$, $k_3$, and $k_4$ can also be determined. The standard error and the confidence intervals on these parameters can be estimated as outlined in Chapter 6. The problem of such estimates in the nonlinear cases has already been discussed in that chapter.

## 4. A Potential Method for the Determination of Spectra of Intermediates

The experiments of Gibson and Roughton point to some interesting consideration in connection with the possible determination of the spectral properties of the intermediates. The fact that the spectral properties of a protein changes on combination with the ligand leads one to the following question: If the protein combines with more than one ligand, then the resulting spectrum is probably that of all the components of the reaction mixture namely $P$, $PA_1$, $PA_2$, $PA_3$, etc. Now, if that happens to be the case, then it is pertinent to inquire after the spectral properties of all the components in the reaction mixture. When the intermediates have different spectral properties then it may be possible to determine the concentration of each intermediate in the presence of others. This will enable us to determine the quantities $r(t)$, $s(t)$, $u(t)$, etc. directly.

This topic has been dealt with to some extent in Chapter 9 when we described the use of the matrix rank analysis by Ainsworth and Bingham (1968) for the hemoglobin–carbon monoxide reaction. To conduct this analysis we suppose that the equilibrium constants $K_1, K_2, K_3, \ldots$ for any system we choose to analyze are known or can be determined as outlined in Chapter 13. When such is the case we can measure the absorption at wavelength $\lambda_1$ of $j$ reaction mixtures at differing concentration of the protein and ligand. This sets up the following $m$ equations at wavelength $\lambda_1$

$$A_{j\lambda_1} = \varepsilon_{1\lambda_1}c_{j1} + \varepsilon_{2\lambda_1}c_{j2} + \varepsilon_{3\lambda_1}c_{j3} + \varepsilon_{4\lambda_1}c_{j4} + \delta_j \qquad (j = 1, 2, \ldots, m) \tag{15-11}$$

where $A_{j\lambda_1}$ is the absorption of the $j$th reaction mixture at wavelength $\lambda_1$ and $\varepsilon_{i\lambda_1}$ is the extinction coefficient of the $i$th compound at $\lambda_1$. When the equilibrium constant are known and the concentrations of the initial species are known then the $c_{ji}$ (which are the concentrations of the $i$th species in the $j$th reaction mixture) can be computed for each of $j$ reaction mixtures. Solution of the $m$ equations (15-11) for the $\varepsilon_{i\lambda_1}$ by using least squares or any other method for linear problems described is Chapter 4 may be used. The process is then repeated for as many wavelengths as we please to produce the spectrum of all the intermediates. Once this is done, we can, by recording the spectrum of the reaction mixture at time $t$, determine the concentration of all $n$ intermediates as we have shown in Chapter 7. Needless to say, in such cases $m$ must be greater than or equal to $n$.

## 5. Determination of the Number of Intermediates and Rate Constants in Enzyme Catalyzed Reactions

In this section we determine the number of intermediates and rate constants at the early initial phase of the enzyme reaction. This type of kinetics has been referred to as presteady-state kinetics, and the actual cases analyzed have pertained to reactions containing a single intermediate.

An interesting treatment of the general problem has been recently presented by Darvey *et al.* (1966). These authors treat the case given by the following kinetic scheme

$$E + A \underset{k_2}{\overset{k_1}{\rightleftharpoons}} X_1 \underset{k_3}{\overset{k_3}{\rightleftharpoons}} X_2 \underset{k_6}{\overset{k_5}{\rightleftharpoons}} \cdots \underset{k_{2i}}{\overset{k_{2i-1}}{\rightleftharpoons}} X_i \underset{k_{2i+2}}{\overset{k_{2i+1}}{\rightleftharpoons}} \cdots X_m \underset{k_{2m+2}}{\overset{k_{2m+1}}{\rightleftharpoons}} E + P \qquad (15\text{-}12)$$

where $A$ is the substrate, $P$ the product and $X_1, X_2, \ldots, X_m$ the $m$ intermediates. The $k_i$ are the rate constants. The steady-state equations for such mechanisms cannot determine the number of intermediates. Peller and Alberty (1959), have pointed out that from the four kinetic parameters in the steady-state equation (14-17) one can provide certain estimates for the lower limits of two bimolecular constants $k_1$ and $k_{2m+2}$ as well as two other constants. The other constants may be any one of the $k_{2i+1}$ for the forward rate and any one for the $k_{2i+2}$ for the reverse reaction.

Darvey *et al.* (1966) treated the general scheme depicted in Eq. (15-12) for any number of intermediates. Prior to this treatment Roughton (1954) and Ouellet and Laidler (1956) treated the specific cases with one and two intermediates respectively. Actual experiments were done by Gutfreund (1954, 1955) who used a number of measurements on the hydrolysis of synthetic amino acid esters by trypsin and chymotrypsin as raw data.

What we hope to determine by using the treatment of Darvey *et al.* (1966) is the number of intermediates and some rate constants. The rate of change with respect to time of the substrate, intermediates products, and enzyme can be written as a set of simultaneous differential equations

$$\begin{aligned}
dA/dt &= -k_1EA + k_2X_1 \\
dX_1/dt &= k_1EA - k_2X_1 - k_3X_1 + k_4X_2 \\
dX_2/dt &= k_3X_1 - k_4X_2 - k_5X_2 + k_6X_3 \\
&\vdots \\
dX_i/dt &= k_{2i-1}X_{i-1} - (k_{2i} + k_{2i+1})X_i + k_{2i+2}X_{i+1} \qquad (15\text{-}13) \\
&\vdots \\
dX_m/dt &= k_{2m-1}X_{m-1} - (k_{2m} + k_{2m+1})X_m + k_{2m+2}EP \\
dP/dt &= k_{2m+1}X_m - k_{2m+2}EP \\
dE/dt &= -k_1EA + k_2X_1 + k_{2m+1}X_m - k_{2m+2}EP
\end{aligned}$$

The initial concentration of all components at time $t = 0$ can be written as

$$\begin{aligned} A(0) &= A^0, \qquad E(0) = E^0, \qquad P(0) = 0 \\ X_i(0) &= 0, \qquad i = 1, 2, \ldots, m \end{aligned} \tag{15-14}$$

A solution of the above simultaneous differential equations should tell us how the substrates, products, and intermediates should behave with respect to time. In other words the solution should produce the functions $A(t)$, $X_i(t)$, and $P(t)$ in Taylor series (see Chapter 3) about $t = 0$

$$\begin{aligned} A(t) &= A(0) + A'(0)t + (1/2!)A''(0)t^2 + \cdots \\ X_i(t) &= X_i(0) + X_i'(0)t + (1/2!)X_i''(0)t^2 + \cdots \\ P(t) &= P(0) + P'(0)t + (1/2!)P''(0)t^2 + \cdots \end{aligned} \tag{15-15}$$

The solution for these equations are

$$A(t) = A^0 - k_1E^0A^0t + (1/2!)k_1E^0A^0[k_1A^0 + k_1E^0 + k_2]t^2 + \cdots \tag{15-16}$$

$$X_i(t) = (1/i!)\prod_{j=0}^{i-1} k_{2j+1}E^0A^0t^i - [1/(i+1)!]\prod_{j=0}^{i-1} k_{2j+1}E^0A^0 \times \left[k_1A^0 + k_1E^0 + \sum_{j=2}^{2i+1} k_j\right]t^i + \cdots \quad (i = 1, 2, \ldots, m) \tag{15-17}$$

$$P(t) = [1/(m+1)!]\prod_{j=0}^{m} k_{2j+1}E^0A^0t^{m+1} - [1/(m+2)!]\prod_{j=0}^{m} k_{2j+1}E^0A^0 \times \left[k_1A^0 + (k_1 + k_{2m+2})E^0 + \sum_{j=2}^{2m+1} k_j\right]t^{m+2} + \cdots \tag{15-18}$$

where $A(t)$, $X_i(t)$, and $P(t)$ are the concentration of the substrate, the $i$th intermediate, and the product at time $t$, respectively, and $E^0$ and $A^0$ are initial concentrations of the enzyme and substrate. In order to determine the sum of the rate constants and the number of intermediates, Darvey *et al.* (1966) divided Eq. (15-18) by $t^{m+1}$ to obtain

$$P(t)/t^{m+1} = [1/(m+1)!]\prod_{j=0}^{m} k_{2j+1}E^0A^0 - [1/(m+2)!]\prod_{j=0}^{m} k_{2j+1}E^0A^0 \times \left[k_1A^0 + (k_1 + k_{2m+2})E^0 + \sum_{j=2}^{2m+1} k_j\right]t + \cdots \tag{15-19}$$

By giving $m$ the value it can take in Eq. (15-19), i.e., 0, 1, 2, . . . , $m$ the authors proposed a sequence of plots of $P/t$, $P/t^2$, $P/t^3$, vs. $t$. The power of

$t$ in the last plot of this sequence giving a nonnegative initial slope is taken as the number of intermediates. The authors also noted that a number of rate constants as well as products and sums of rate constants can be obtained from a variety of linear plots. The method suggested for determining rate constants determines only two rate constants $k_1$ and $k_{2m+2}$. In what follows we suggest a method for determining all the rate constants as well as the number of intermediates.

If we recognize that Eq. (15-16) is a polynomial in time then it is readily seen (using the notation employed in this book) that by equating

$$f(x) = A(t) - A^0, \qquad f_1(x) = t, \qquad f_2(x) = t^2, \qquad f_3(x) = t^3), \ldots$$

and (15-20)

$$\alpha_1 = -k_1 A^0 E^0, \qquad \alpha_2 = (1/2!) k_1 E^0 A^0 [k_1 A^0 + k_1 E^0 + k_2], \ldots$$

etc., we can solve for $\alpha_1$ by our usual linear methods to determine all the rate constants since in $\alpha_1$, $k_1$ is the only unknown and in $\alpha_2$, $k_2$ is the only unknown after $k_1$ has been determined. To determine the number of intermediates we use the analysis of variance techniques demonstrated in Chapter 6 and 12. However, unlike concentration-dependent aggregating systems two coefficients are tested statistically at the same time (see Chapter 6), since for every intermediate there is a forward and backward rate constant. If the forward rate constant is not significantly different from zero that does not mean that the intermediate does not exist because the other rate constant pertaining to that reaction may have a large value. The equation for product formation as a function of time, Eq. (15-18) can also be handled in the same manner as Eq. (15-16) and the values of the rate constants compared for consistency.

The methods suggested for determining the rate constants use all data points representing either $A(t)$ or $P(t)$. Both least squares or linear programming can be used to determine parameters. As indicated earlier only least squares will allow us precise statistical testing (Magar, 1969).

The procedures described here and in other chapters require accurate data. Data for presteady-state kinetics will generally be generated by flow techniques since we require an initial mixing. This is in contrast to the relaxation methods where the system is in equilibrium. The equations for such systems, which are similar to the equations in this section, will be considered in the next Section. We must us emphasize the necessary of extremely accurate data over a very short period of time. Taking an example from the work of Gutfreund (1954, 1955), on the hydrolysis of synthetic amino acid esters by trypsin and chymotrypsin, it is evident from his data that we can usually determine the presence of only a single intermediate. To determine more than one intermediate, if more than one exists, requires a detailed study in the 0–3000-μsec region.

## 6. Relaxation Methods

In what follows we consider the general case of several intermediates, *in equilibrium*, for the reaction scheme given by

$$A + B + \cdots \underset{k_2}{\overset{k_1}{\rightleftharpoons}} Y_1 \underset{k_4}{\overset{k_3}{\rightleftharpoons}} Y_2 \rightleftharpoons \cdots \underset{k_{2i}}{\overset{k_{2i-1}}{\rightleftharpoons}} Y_i \underset{k_{2i+2}}{\overset{k_{2i+1}}{\rightleftharpoons}} \cdots \rightleftharpoons Y_m \underset{k_{2m+2}}{\overset{k_{2m+1}}{\rightleftharpoons}} Q + P + \cdots \tag{15-21}$$

where $A$, $B$, ... are the reactants, $Q$, $P$, ... the products, and $Y_i$ the $i$th intermediate. If we assume (Eigen and DeMaeyer, 1963) that when the system is perturbed the deviation in concentration of any given species from its equilibrium concentration is small and that the deviation is proportional to the difference in concentration between the equilibrium value and the actual value being measured, then we can set up the following $n$ equations

$$-d(\Delta y_i)/dt = \sum_{q=1}^{m} A_{iq}\, \Delta y_q \qquad (i = 1, 2, \ldots) \tag{15-22}$$

where $\Delta y_i$ is the deviation of the species $Y_i$ from its equilibrium concentration and the $A_{iq}$ are known functions of the rate constants and the equilibrium concentrations. It is easy to see the similarity between Eqs. (15-13) and (15-21). In Eq. (15-21) instead of having concentrations as in Eq. (15-13) we have differences in equilibrium concentrations. The problem of analyzing Eq. (15-21) is dealt with in exactly the same manner as the analysis of Eqs. (15-13), which can be outlined briefly as follows. The first of Eqs. (15-13) can be written as

$$(dA/dt) = -k_1 EA(t_j) + k_2 X_1(t_j) + \delta_j \qquad (j = 1, 2, \ldots, n) \tag{15-13$'$}$$

Now, *if* we can make those $n$ measurements for $dA/dt$, $EA$, and $X$ at $n$ different times, then Eq. (15-13) can be solved by the usual techniques described in Chapter 4 so as to minimize $\sum_{j=1}^{n} \delta_1{}^2$ and to determine $k_1$ and $k_2$. The other equations of (15-13) can be solved in a similar manner. Likewise, if all the $\Delta y_i$ are known then any one of the Eqs. (15-22) can be set up as follows

$$(-d\,\Delta y_i/dt)_j = \sum_{q=1}^{m} A_{iq}\,\Delta y_q(t_j) + \delta_j \qquad (j = 1, 2, \ldots, n) \quad (15\text{-}23)$$

and solved for all the $A_{iq}$. It will be recalled that we know (or should know) the equilibrium concentrations of the reactants, intermediates, and products otherwise some of the $A_{iq}$ cannot be determined. Determining all the $A_{iq}$ provides a check on the consistency of the system since any $A_{iq}$ or combination of $A_{iq}$ obtained from any one of Eqs. (15-23) should be set equal to some of the other $A_{iq}$ obtained from any other of Eqs. (15-23). In fact, if we are inclined we can make use of this fact as an aid in solution of Eqs. (15-23). A most important step in the practical application of the differential form of equations arising in this work, such as Eqs. (15-13) and (15-23), is determining the concentration of the intermediates. If these do not have different physical properties and cannot be monitored independently of each other with time, then the differential form of the equations are not very helpful. The problems of using the differential forms has been discussed in Chapter 16 where a similar problem arises for discussion (see Hartley, 1948).

The general solution of Eqs. (15-22) have been given by Eigen and DeMaeyer as

$$\Delta y_i = \sum_{q=1}^{m} A_{iq} \exp(t/\tau_q) \qquad (15\text{-}24)$$

where $A_{iq}$ are again functions of the equilibrium concentrations and rate constants and $\tau_q$ the $q$th relacation time. To obtain the $A_{iq}$ explicitly in terms of rate constants (a feat achieved only in certain specific situations) we can proceed in a manner identical to Darvey *et al.* (1966) and expand the $\Delta y_i(t)$ as a Taylor series about the point $t = 0$ in the same way as we have done for Eqs. (15-13). Then by substituting for the various derivatives at zero time we can obtain a solution for the $\Delta y_i$ as polynomials in time. This is essentially Eq. (15-24) when the exponential is expanded and the relaxation times lumped in the coefficients of the polynomial in $t$. This is as it should be, since the relaxation times are functions of the rate constants and the equilibrium concentrations. We leave it as an exercise to the interested reader to derive the explicit expression for the $A_{iq}$. When that is done, the $\Delta y_i(t)$ appears as explicit functions of the rate constants, equilibrium concentrations, and time. With such a representation of the $\Delta y_i(t)$ we can proceed to analyze any given $\Delta y_i(t)$ for the rate constants.

Investigators working with relaxation methods have had a tendency to concentrate only on the determination of relaxation times (Eigen and De Maeyer, 1963; Hammes and Schimmel, 1966). This tendency makes for an easier mathematical analysis because it ignores the explicit structure of the $A_{iq}$. In such instances Eq. (15-24) is analyzed directly. If we can measure

$$\Delta y_i(t_j) = \sum_{q=1}^{m} A_{iq} \exp(-t_j/\tau_q) + \delta_j \qquad (j = 1, 2, \ldots, n) \quad (15\text{-}25)$$

where $\delta_j$ is the experimental error in a measurement at $t_j$. Equation (15-25), a sum of exponentials, is a nonlinear equation in the unknowns $A_{iq}$ and $\tau_q$ and can be solved for the parameters such that $\sum_{j=1}^{n} \delta_j^2$ is a minimum. As we noted in Chapter 4 there is quite a bit of literature on the fitting of data with sums of exponentials and for those readers who are interested we refer them to Chapter 4 for a brief discussion on these methods and a listing of relevant references. The confidence interval on the determined parameters and their standard errors as well as how many relaxation times are involved can be found in Chapter 6. Methods of statistical testing about hypotheses on the determined parameters can also be found in Chapter 6 but these pertain to the linear equations only. In Eqs. (15-25) the fact that the $A_{iq}$ are not written out explicitly does not matter, because the relaxation times are the quantities that are generally required. As a matter of fact, in an inordinate number of instances, even if the $A_{iq}$ were written out explicitly the fact that we may not know the equilibrium concentration of some of the participating species makes it impossible to extract any information regarding rate constants from the $A_{iq}$. Consequently, for those who are interested in relaxation times only, the explicit expressions for the $A_{iq}$ are best left alone. On the other hand, if one wants to extract maximum information about the system there is no reason why an explicit expression for $\Delta y_i(t)$ cannot be analysed in an identical manner to the method we proposed for the equations of Darvey *et al.* (1966). To our knowledge no systematic analysis of Eq. (15-25) by the techniques mentioned in Chapter 4 has been undertaken.

## 7. A Word of Caution

In concluding this chapter we would like to emphasize what Eigen and DeMaeyer (1963) have said about analysis of relaxation spectra, "It is obvious that application of such mathematical procedures requires very accurate measurements." We would like to reiterate this sentence for the methods we suggest. It is interesting to note that Eigen and DeMaeyer (1963) were making the statement in connection with the analysis of the following equation (in their notation) containing two relaxation times.

$$\text{response } (t) = A_{\text{I}} \exp(-t/\tau_{\text{I}}) + A_{\text{II}} \exp(-t/\tau_{\text{II}}) \qquad (15\text{-}26)$$

These authors suggested that if $\tau_{\text{I}}$ is slightly smaller that $\tau_{\text{II}}$, then repeated differentiation with respect to $t$ and a plot of the quantity

$$[d^n x(t)/dt^n]/[d^{n+1} x(t)/dt^{n+1}]$$

vs. $n$ should yield the relaxation time by extrapolating to large $n$. With all due respect to Professors Eigen and DeMaeyer and with our wholehearted support for their advocacy of extremely accurate data we must state that from the point of view of the numerical analyst, obtaining a reliable relaxation time $\tau_{\text{I}}$ using the above procedure is extremely hazardous simply because the above procedure depends on repeated numerical differentiation of experimental data, a notoriously unreliable process as was stated several times before. In fact there are some numerical analysts who feel that a case could be made for using some numerical differentiation results as random number generators.

Returning to the point about accurate data, let us once again emphasize that it will not do the investigator much good to use inaccurate data as a basis for fitting a series of exponentials or for any manner of analysis advocated herein. With inaccurate data the investigator can obtain as precise a fit as possible but the validity of the conclusions drawn from such data will continue to remain doubtful. Consequently we feel that much time and money would be saved if the investigator would probe into the question of the precision of the expected result from any given technique before venturing very far with his commitments to any given experiment.

## References

Adair, G. S. (1925). *J. Biol. Chem.* **63**, 529.

Ainsworth, S., and Bingham, W. S. (1968). *Biochem. Biophys. Acta* **160**, 10.

Bateman, H. (1910). *Proc. Cambridge Phil. Soc.* **15**, 423.

Bernhard, S. (1968). "Enzymes: Structure and Function." Benjamin, New York.

Darvey, I. G., Prokhovnik, S. J., and Williams, J. F. (1966). *J. Theor. Biol.* **13**, 90.

Eigen, M., and DeMaeyer, L. (1963). *In* "Technique of Organic Chemistry" (Friess, Lewis, and A. Weissberger, eds.), Vol. 8, Part II. Wiley (Interscience), New York.

Friess, S., Lewis, N., and Weissberger, A., eds. (1963). "Technique of Organic Chemistry," Vol. 8, Part II. Wiley (Interscience), New York.
Gibson, Q. H. (1968). *Pacific Slope Biochem. Conf. Symp. Santa Barbara*, 1968.
Gibson, H. Q., and Greenwood, C. (1965). *J. Biol. Chem.* **240**, 2694.
Gibson, Q. H., and Roughton, F. W. J. (1957). *Proc. Royal Soc. Ser. B* **146**, 206.
Gutfreund, H. (1954). *Discuss. Faraday Soc.* **17**, 220.
Gutfreund, H. (1955). *Discuss. Faraday Soc.* **20**, 167.
Hammes, G. G., and Schimmell, P. (1966). *J. Phys. Chem.*
Hartley, H. O. (1948) *Biometrika* **51**, 347.
Magar, M. E. (1969). *J. Theor. Biol.* **25**, 345.
Nozaki, Y., Gurd, F. R. N., Chen, R., and Edsall, J. T. (1957). *J. Amer. Chem. Soc.* **79**, 2123.
Ouellet, L., and Laidler, K. J. (1956). *Can. J. Chem.* **34**, 146.
Peller, L., and Alberty, R. A. (1959). *J. Amer. Chem. Soc.* **81**, 5907.
Roughton, F. W. J. (1954). *Discuss. Faraday Soc.* **17**, 116.
Roughton, F. W. J., Otis, A. B., and Lyster, R. L. J. (1955). *Proc. Royal Soc. Ser. B* **144**, 29.

*CHAPTER 16*

# TRACER TECHNIQUES IN COMPARTMENTALIZED SYSTEMS

## 1. Introduction

Tracers are generally used to observe a population of things by labeling some of them. The mixing of the labeled species with the population as a whole, should, provided there is through mixing, yield information on the population. There is really no restriction as to what kind of label we attach to any species. Thus the label could be a dye, a radioactive atom, or whatever. Various information can be obtained using tracer techniques. For example, we might want to find the rate of exchange of some molecule between compartments or the half life of a given population undergoing metabolism. To repeat a common theme from other chapters: Whatever system we are analysing, for whatever purpose, we will ultimately encounter equations, linear or nonlinear in the parameters, we can solve these equations to determine the parameters (which generally have some physical significance) and their standard error, and we can test various hypotheses about them.

In this chapter we shall only examine a limited number of topics as examples. Undoubtedly, there are many more complicated models that could be analyzed, but those follow the general schemes for linear and nonlinear models. Consequently, we advise the reader with any tracer model where parameters are to be determined to consult a statistician or programmer if we do not deal with the model of interest. The chances are that the techniques of function minimization will yield an answer to any problem encountered.

## 2. Two-Compartment Systems

The main objectives of subsequent analyses is to determine rates of transport of material from one compartment into another. We will deal with the generalized multicompartment system presently, however, to give the reader some idea of what is happening let us consider a two-compartment system. In a two-compartment system the rate of change (with time) of the absolute amount of tracer is equal to the amount coming in minus the amount going out. If the absolute amounts of tracer in compartment 1 and compartment 2 are given by $R_1$ and $R_2$, then we have

$$dR_1/dt = S_1(da_1/dt) = \beta(a_2 - a_1) \tag{16-1}$$

and similarly for the rate of change in compartment 2 we have

$$dR_2/dt = S_2(da_2/dt) = \beta(a_1 - a_2) \tag{16-2}$$

where $\beta$ is the rate of exchange, $S_1$ and $S_2$ amounts of substance $S$ in compartments 1 and 2, $a_1$ and $a_2$ the specific activities in compartments 1 and 2, and $t$ the time. We also note that the specific activities $a_1$ and $a_2$ can be defined by the following equations

$$a_1 = (R_1/S_1), \qquad a_2 = (R_2/S_2) \tag{16-3}$$

As mentioned earlier we seek to determine the rate of exchange $\beta$. To treat this matter as a linear problem we can use either Eq. (16-1) or (16-2). By determining $R_1$, $R_2$, $a_2$, and $a_1$ at various times we can proced to use function minimization to determine the rate constant. This case involves simply a least squares fit of the straight line $(dR_1/dt)$ vs. $a_2 - a_1$. The slope of this line will be $\beta$. Note that in obtaining the quantity $(dR_1/dt)$ from the raw data at various times we do not obtain the first derivative as dictated by Eqs. (16-1) or (16-2), but a series of finite differences which are in essence a numerical differentiation of the data. If we analyze (16-1) on (16-2) we are treating the *differential form* of the equations.

It is more rigorous to determine $\beta$ from a nonlinear equation. We can solve the simultaneous differential equations (16-1) and (16-2) for $a_1$ and $a_2$ and write for the solution of $a_1$

$$a_1 = [a(0)/S][S_1 + S_2 \exp(-\beta St/S_1S_2)] \tag{16-4}$$

The condition imposed in this solution is that all activity is in compartment 1 and the specified activity at the start of the experiment in this compartment is $a(0)$. If we want to determine $\beta$ so as to minimize the experimental error we write the following $m$ equations at $m$ different times $t_1, t_2, \ldots, t_m$

$$\begin{aligned} a_1(t_1) &= [a(0)/S]\{S_1(t_1) + S_2(t_1)\exp[-\beta St_1/S_1(t_1)S_2(t_2)]\} + \delta_1 \\ &\vdots \\ a_1(t_m) &= [a(0)/S]\{S_1(t_m) + S_2(t_m)\exp[-\beta St_m/S_1(t_m)S_2(t_m)]\} + \delta_m \end{aligned} \tag{16-5}$$

The $m$ equations (16-5) can be solved for $\beta$ such that $\sum_{j=1}^{m} \delta_j^2$ is a minimum by most of the nonlinear techniques in Chapter 4. As an alternative to the use of Eq. (16-5) we can use, for instance, the equation for the solution for $a_2$, which can be written as

$$a_2 = [a(0)S_1/S][1 - \exp(-\beta St/S_1S_2)] \tag{16-6}$$

This equation can be handled in a similar manner to Eqs. (16-5) to solve for $\beta$. Still another alternative would be the use of the difference between $a_1$ and $a_2$ as follows

$$a_1 - a_2 = a(0)\exp\{-\beta[(1/S_1) + (1/S_2)]t\} \tag{16-7}$$

and this can be written as the following $m$ equations at $m$ different times as shown below

$$\begin{aligned} a_1(t_1) - a_2(t_1) &= a(0)\exp\{-\beta[(1/S_1(t_1)) + (1/S_2(t_2))]t_1\} \\ &\vdots \\ a_1(t_m) - a_2(t_m) &= a(0)\exp\{-\beta[(1/S_1(t_m)) + (1/S_2(t_m))]t_m\} \end{aligned} \tag{16-8}$$

which can be solved by the nonlinear techniques in Chapter 4 to determine $\beta$ such that $\sum_{j=1}^{m} \delta_j^2$ is a minimum.

The above ways of solving for $\beta$ may appear to complicate things unnecessarily. This is partially true since Eq. (16-7) can, by taking logarithms, be linearized as follows

$$\ln(a_1 - a_2) = \ln a(0) - \beta[(1/S_1) + (1/S_2)]t \tag{16-9}$$

On the other hand we can look at Eqs. (16-4), (16-6), and (16-7) as some of the simplest forms of nonlinear equations, being nonlinear equations in a single unknown. It will be noted that in solving these equations such that the error is minimized we are obtaining the best fit to the raw experimental data. We are doing almost the same thing with these equations in their differential form when we use Eqs. (16-1) or (16-2), but in these instances we have to differentiate the data first before we can make a least squares fit for the straight line given by those equations. In considering which kind of fit we would like to make, we must weigh the advantage of the fact that we do not differentiate in the nonlinear case against the advantage that we can employ statistical tests in the linear case (Chapter 6 and below).

## 3. Closed Systems and Open Systems

The systems of compartments which we were analyzing are closed systems, which means we do not permit communication between our compartments and spaces which are not considered part of our compartment system. In other words the total amount of tracer in our compartment system remains constant. On the other hand an open system permits exchange with the environment or spaces which are not part of our compartment systems. The case of a one-compartment open system is of interest since it shows how we can readily convert from a closed compartment system to an open compartment system. To convert from a closed two-compartment system to an open one-compartment system we allow $S_2$ or the quantity of material in the second compartment (which now becomes the exterior of our system) to approach infinity. Under such circumstances $a_2$ in Eq. (16-7) will tend to zero since the activity will be diluted with nonlabeled material [see Eq. (16-3)]. This transforms Eq. (16-7) for $a_1$ to

$$a_1 = a(0) \exp(-\beta t/S_1) \tag{16-10}$$

Consequently to go from closed to open systems we simply pursue the analogy given here for any system of equations. As more compartments are added the problem becomes more complicated.

## 4. Three-Compartment Systems

What has been said of the two-compartment system in terms of the form of the equation analyzed can be said, in an expanded form, of multicompartment systems. For any system, the investigator can either use the differential form of the equations, in which case the exchange rates appear linearly in the equations, or integrate these equations, in which case the parameters will be seen to occur nonlinearly. In what follows we show how a three-compartment system can be analyzed first in its differential form and then after integration.

The three-compartment system can be represented as in Fig. 16-1. The mathematical formulation runs as follows, where we have assumed that $\beta_{ij} = \beta_{ij}$

$$\begin{aligned} S_1(da_1/dt) &= \beta_{12}(a_2 - a_1) + \beta_{13}(a_3 - a_1) \\ S_2(da_2/dt) &= \beta_{12}(a_1 - a_2) + \beta_{23}(a_3 - a_2) \\ S_3(da_3/dt) &= \beta_{13}(a_1 - a_3) + \beta_{23}(a_2 - a_3) \end{aligned} \tag{16-11}$$

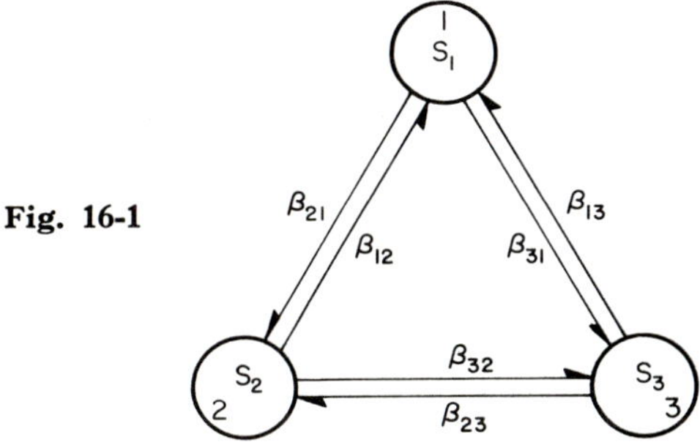


**Fig. 16-1**

Equations (16-11) are three equations in the unknowns $\beta_{12}, \beta_{13}, \beta_{23}$ and we might consider solving them for these parameters. The problem of linear dependence prevents such a solution in the above form because if we add the first two equations and examine the third we arrive at the relationship

$$S_1(da_1/dt) + S_2(da_2/dt) + S_3(da_3/dt) = 0 \tag{16-12}$$

which expresses the fact that the total activity in the system is constant. From the linear dependence exhibited by Eqs. (16-11) we can only solve for two rates. If we represent Eqs. (16-11) taken at different distributions so that the $a_i$ are different, then we can eliminate the linear dependence, calling $a_{11}$ the specific activity of $a_1$ in the first distribution of radioactivity and generalizing to all the $a_i$ such that $a_{ij}$ is the specific activity of $a_i$ in the $j$th distribution. We can write Eqs. (16-11) as follows

$$
\begin{aligned}
S_1(da_{11}/dt) &= \beta_{12}(a_{21} - a_{11}) + \beta_{13}(a_{31} - a_{11}) + \delta_1 \\
S_2(da_{21}/dt) &= \beta_{12}(a_{11} - a_{21}) + \beta_{23}(a_{31} - a_{21}) + \delta_2 \\
S_3(da_{31}/dt) &= \beta_{13}(a_{11} - a_{31}) + \beta_{23}(a_{23} - a_{31}) + \delta_3 \\
S_1(da_{12}/dt) &= \beta_{12}(a_{22} - a_{12}) + \beta_{13}(a_{32} - a_{12}) + \delta_4 \\
&\vdots \\
S_3(da_{3n}/dt) &= \beta_{13}(a_{1n} - a_{3n}) + \beta_{23}(a_{2n} - a_{3n}) + \delta_{3n}
\end{aligned}
\tag{16-13}
$$

The above Eqs. (16-13) give us $3 \times n$ equations in the three unknowns $\beta_{12}$, $\beta_{23}$, and $\beta_{13}$ and these can be solved such that $\sum_{j=1}^{3n} \delta_j^2$ is a minimum by the techniques outlined in Chapter 4. For the determination of confidence intervals on the estimated parameters and the testing of hypotheses about these parameters the reader is referred to Chapter 6.

To show the form of the equations we get when we integrate equations such as (16-11), and to deal with a more complicated system we study the work of Skinner *et al.* (1959) on a three-compartment system depicted in Fig. 16-2. In this figure the $q_i$ represent the quantity of radio-

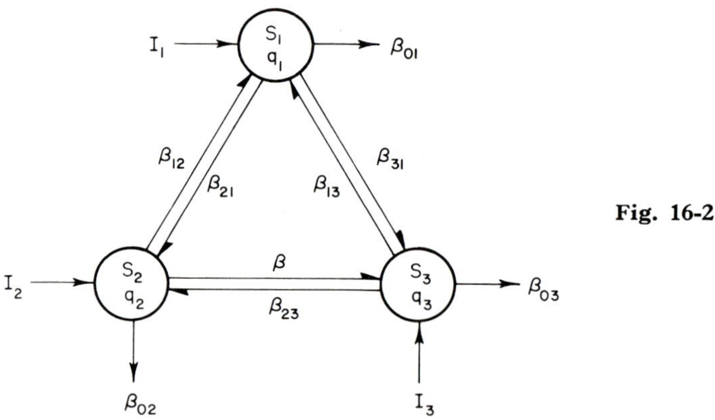


**Fig. 16-2**

active material at any instant in time in compartment $i$. The $I_i$ are diluting material with zero activity and the $S_i$ and $\beta_{ij}$ have the same meaning as before. The differential equations for this system can be written as

$$
\begin{aligned}
dq_1/dt &= -\beta_{11}q_1 + \beta_{12}q_2 + \beta_{13}q_3 \\
dq_2/dt &= -\beta_{22}q_2 + \beta_{21}q_1 + \beta_{23}q_3 \\
dq_3/dt &= -\beta_{33}q_3 + \beta_{31}q_1 + \beta_{22}q_2
\end{aligned}
\tag{16-14}
$$

In the above equations we have

$$\beta_{11} = \beta_{01} + \beta_{21} + \beta_{31}, \quad \beta_{22} = \beta_{02} + \beta_{12} + \beta_{32}, \quad \beta_{33} = \beta_{03} + \beta_{13} + \beta_{23} \tag{16-15}$$

The solution of the equation subject to the boundary conditions

$$q_1 = q_{10}, \qquad q_2 = q_3 = 0 \tag{16-16}$$

where $q_{10}$ is the radioactivity injected into the system at $t = 0$ (this radioactivity is injected into compartment 1, only) is

$$q_1/q_{10} = A_1 \exp(-\alpha_1 t) + A_2 \exp(-\alpha_2 t) + A_3 \exp(-\alpha_3 t) \tag{16-17a}$$

$$q_2/q_{10} = B_1 \exp(-\alpha_1 t) + B_2 \exp(-\alpha_2 t) + B_3 \exp(-\alpha_3 t) \tag{16-17b}$$

$$q_3/q_{10} = C_1 \exp(-\alpha_1 t) + C_2 \exp(-\alpha_2 t) + C_3 \exp(-\alpha_3 t) \tag{16-17c}$$

where

$$\begin{aligned} A_i &= [(-\alpha_i + \beta_{22})(-\alpha_i + \beta_{33}) - \beta_{23}\beta_{32}]/\Delta_i \\ B_i &= [(-\alpha_i + \beta_{33})\beta_{12} + \beta_{23}\beta_{31}]/\Delta_i \\ C_i &= [(-\alpha_i + \beta_{22})\beta_{31} + \beta_{21}\beta_{23}]/\Delta_i \end{aligned} \tag{16-18}$$

$$\begin{aligned} \Delta_1 &= (-\alpha_1 + \alpha_2)(-\alpha_1 + \alpha_3) \\ \Delta_2 &= (-\alpha_2 + \alpha_3)(-\alpha_2 + \alpha_1) \\ \Delta_3 &= (-\alpha_3 + \alpha_1)(-\alpha_3 + \alpha_2) \end{aligned} \tag{16-19}$$

Equation (16-17a) is a nonlinear equation in the unknown $A_1$, $A_2$, $A_3$, $\alpha_1$, $\alpha_2$, and $\alpha_3$. Equation (16-17b) is nonlinear in the unknown $B_1$, $B_2$, $B_3$, $\alpha_1$, $\alpha_2$, and $\alpha_3$, and lastly Eq. (16-17c) is nonlinear in the unknown $C_1$, $C_2$, $C_3$, $\alpha_1$, $\alpha_2$, and $\alpha_3$. One can take any one of those equations and treat it as a nonlinear equation and attempt to determine the parameters from it. For the sake of illustration we can write Eq. (16-17a) as the following $m$ equations

$$\begin{aligned} (q_1/q_{10})_{t_1} &= \sum_{i=1}^{3} A_i \exp(-\alpha_i t_1) + \delta_1 \\ (q_1/q_{10})_{t_2} &= \sum_{i=1}^{3} A_i \exp(-\alpha_i t_2) + \delta_2 \\ &\vdots \\ (q_1/q_{10})_{t_m} &= \sum_{i=1}^{3} A_i \exp(-\alpha_i t_m) + \delta_m \end{aligned} \tag{16-20}$$

where the quantities $(q_1/q_{10})_{t_j}$ means the ratio of the radioactivity in compartment 1 at time $t_j$ to the amount of radioactivity injected at the start of the experiments. Once again we note that the equations are nonlinear equations in the unknowns $A_i$ and $\alpha_i$, and we can solve for them by the techniques of Chapter 4 using either the general nonlinear techniques or one of the specialized methods for the analysis of a sum of exponentials. Once these parameters are determined we can then use Eqs. (16-19) to obtain the rates of exchange. The difficulties in analyzing a sum of exponentials have been mentioned in Chapter 4. For the determination of confidence intervals on the parameters the reader is referred to Chapter 6 on the analysis of nonlinear models.

Skinner *et al.* (1959) having solved for the general system containing three compartments simplified matters by equating various rates to zero. Thus, if there was no exchange between compartment 1 and compartment 3 then $\beta_{13} = \beta_{31} = 0$; if compartment 2 did not exchange with the exterior of the system then $\beta_{02} = 0$; or if the backward rate of exchange was equal to the forward rate of exchange in any given compartment then $\beta_{ij} = \beta_{ji}$ as we assumed when we set up Eqs. (16-11); etc.

If we solve for the parameters when they occur linearly, then as we have shown in Chapter 6, we can test hypotheses about them subject to the limitations of our use of the differential forms of the equations. In the event that we have estimated the parameters using nonlinear methods, hypothesis testing, as we pointed out in Chapter 6 is no longer an easy matter.

## 5. Problems with the Use of Simultaneous Linear Differential Equations

As we pointed out in the chapter on rapid reactions, when we have the differential coefficients of measured quantities we run into some problems. Equations (16-13) and some subsequent equations in this chapter have these quantities in them. In order to work with the unintegrated equations to keep the problem linear in the parameters one must ask how does one evaluate $da_{ij}/dt$ given that what is measured is a set of $a_{ij}$ vs. $t$. Hartley (1948) stated that it would be difficult to fit to an empirical series of observations (in our case $a_{ij}$) a condition involving its differential coefficient. For Hartley's exposition of this and related problems we refer the reader to the original paper. It is possible to fit the $a_{ij}$ vs. $t$ curve to a cubic or fourth-power polynomial and to use the derivative of the polynomial evaluated at the appropriate $t$. This procedure which facilitates our task in practice will, however, not give a true confidence interval on the parameters as determined from Eq. (16-13). Any such confidence interval will always be much smaller than the true confidence interval.

## 6. Multicompartment Systems

Turning to the general case of a multicompartment system which has been treated by Berman and Schoenfeld (1956) we hasten to point out that this is really no different from the previous cases insofar as our analyses are concerned. For the treatment of the differential form the parameters occur linearly in the equations, and for the treatment of the integrated equations the parameters occur nonlinearly as parts of terms in the sum of exponentials. For the sake of completeness we show analyzed equations. The differential equations for the case of $n$ compartments using our previous notation is

$$dq_i/dt = -\beta_{ii}q_i + \sum_{k=1}^{n} \beta_{ik}q_k \qquad (i = 1, 2, \ldots, n) \qquad (16\text{-}21)$$

and the integrated equations are

$$q_i(t) = \sum_{j=1}^{n} A_j \exp(-\alpha_j t) \qquad (i = 1, 2, \ldots, n) \qquad (16\text{-}22)$$

where the $A_j$ and $\alpha_j$ are complex functions of the exchange rates and isotope concentrations. [See Berman and Schoenfeld (1956) for the use of Laplace transforms to integrate (16-21).] Both sets of equations (16-21) and (16-22) can be handled by the techniques of Chapter 4, the former being linear in the parameters and the latter nonlinear in the parameters.

## 7. Simulation of Tracer Data

Another way of looking at the problem of multicompartment systems with the objective of determining whether or not we can determine the parameters is to simulate the experiments. As in Chapter 13 where we simulated results on protein–ligand equilibria we use the same type of procedure to generate data and the potential error associated with this data to determine what experimental accuracy is required for a given accuracy of a model's parameters.

Myhill and coworkers have approached this problem (Myhill *et al.*, 1965; Myhill, 1968). The equation analyzed by Myhill (1968) was Eq. (16-22) with three terms. With this equation a large amount of possible data can be simulated with the following situations varied: (a) Time interval over which data was collected, (b) number of data points taken, (c) spacing of data points, (d) magnitude of random error applied to data, (e) variation this error causes in the time interval of data collection, (f) statistical distribution of error in each data point, and (g) values of the $\alpha_j$ and $A_j$. In this manner Myhill was able to simulate data which closely resembled the experimental situation one was likely to encounter in practice. By using methods of function minimization such as expanding in a Taylor series (see Chapter 4) Myhill could compute the parameters $\alpha_j$ and $A_j$ and could determine how close they were to the parameters before the error was added. Such a procedure enabled one to determine the extent of accuracy in the determination of parameters to be expected in a given experiment. Although some of the variations Myhill puts the process of simulating the data can be handled analytically to determine the extent of error, it is clear that the practical approach give in Myhill (1968) is readily comprehensible to most investigators working in this field and allows them the flexibility of simulating data to suit their own particular experimental situation.

The idea of simulating the kind of data one expects to obtain in an experiment to determine how much error one can tolerate in order to determine parameters with a given accuracy is clearly exploitable for any model. This technique has been used by Magar *et al.* (1971) in protein–ligand equilibria.

## 8. Modification of the Multicompartiment System Model and Other Models

The general treatment of a multicompartment system can be modified in various ways in the same manner as a three compartment system. This can be done by lumping compartments, allowing or disallowing exchange between specific compartments, and so on. Further, one can, as in the three compartment system, test all kinds of hypotheses about the computed exchange rates.

Beyond the treatment of multicompartment systems there are various models of varying degrees of complexity that arise in working with tracers. For instance, the treatment of the circulatory system gives rise to models where there is exchange between a blood vessel and its surroundings, or flow into perfused organs, and so on. For a discussion of number of these models we refer the reader to Sheppard (1962). In all of these cases there can be obtained either linear or nonlinear equations that can be analyzed according to the methods discussed here in.

## References

Berman, M. (1965). *In* "Computers in Biomedical Research" (R. W. Stacy and B. D. Waxman, eds.), Vol. 2, Academic Press, New York.

Berman, M., and Schoenfeld, R. (1956). *J. Appl. Phys.* **27**, 1361.

Branson, H. (1952). *Arch. Biochem. Biophys.* **36**, 48.

Hart, H. E. (1955). *Bull. Math. Biophys.* **17**, 87.

Hartley, H. O. (1948). *Biometrika* **51**, 347.

Hearon, J. F. (1953). *Bull. Math. Biophys.* **15**, 121.

Magar, M. E., Steiner, F., and Fletcher, J. E. (1971). *J. Theor. Biol.* **32**, 59.

Myhill, J. (1968). *J. Theor. Biol.* **23**, 218.

Myhill, J., Wadsworth, G. P., and Brownell, G. L. (1965). *Biophys. J.* **5**, 89.

Sheppard, C. W. (1962). "Basic Principles of Tracer Methods." Wiley, New York.

Sheppard, C. W., and Householder, A. S. (1951). *J. Appl. Phys.* **22**, 510.

Skinner, S. M., Clark, R. E., Baker, N., and Shipley, R. A. (1959). *Amer. J. Physiol.* **196**, 238.

Solomon, A. K. (1949). *J. Clin. Invest.* **28**, 1297.

# *APPENDIX*

# STATISTICAL TABLES

**Table A-1**

AREA OF THE STANDARD NORMAL DISTRIBUTION[a]

| $z$ | $A(z)$ | $z$ | $A(z)$ | $z$ | $A(z)$ | $z$ | $A(z)$ |
|---|---|---|---|---|---|---|---|
| .00 | .00000 | .50 | .19146 | 1.00 | .34134 | 1.50 | .43319 |
| .01 | .00399 | .51 | .19497 | 1.01 | .34375 | 1.51 | .43448 |
| .02 | .00798 | .52 | .19847 | 1.02 | .34614 | 1.52 | .43574 |
| .03 | .01197 | .53 | .20194 | 1.03 | .34849 | 1.53 | .43699 |
| .04 | .01595 | .54 | .20540 | 1.04 | .35083 | 1.54 | .43822 |
| .05 | .01994 | .55 | .20884 | 1.05 | .35314 | 1.55 | .43943 |
| .06 | .02392 | .56 | .21226 | 1.06 | .35543 | 1.56 | .44062 |
| .07 | .02790 | .57 | .21566 | 1.07 | .35769 | 1.57 | .44179 |
| .08 | .03188 | .58 | .21904 | 1.08 | .35993 | 1.58 | .44295 |
| .09 | .03586 | .59 | .22240 | 1.09 | .36214 | 1.59 | .44408 |
| .10 | .03983 | .60 | .22575 | 1.10 | .36433 | 1.60 | .44520 |
| .11 | .04380 | .61 | .22907 | 1.11 | .36650 | 1.61 | .44630 |
| .12 | .04776 | .62 | .23237 | 1.12 | .36864 | 1.62 | .44738 |
| .13 | .05172 | .63 | .23565 | 1.13 | .37076 | 1.63 | .44845 |
| .14 | .05567 | .64 | .23891 | 1.14 | .37286 | 1.64 | .44950 |
| .15 | .05962 | .65 | .24215 | 1.15 | .37493 | 1.65 | .45053 |
| .16 | .06356 | .66 | .24537 | 1.16 | .37698 | 1.66 | .45154 |
| .17 | .06750 | .67 | .24857 | 1.17 | .37900 | 1.67 | .45254 |
| .18 | .07142 | .68 | .25175 | 1.18 | .38100 | 1.68 | .45352 |
| .19 | .07535 | .69 | .25490 | 1.19 | .38298 | 1.69 | .45449 |
| .20 | .07926 | .70 | .25804 | 1.20 | .38493 | 1.70 | .45543 |
| .21 | .08317 | .71 | .26115 | 1.21 | .38686 | 1.71 | .45637 |
| .22 | .08706 | .72 | .26424 | 1.22 | .38877 | 1.72 | .45728 |
| .23 | .09095 | .73 | .26730 | 1.23 | .39065 | 1.73 | .45818 |
| .24 | .09483 | .74 | .27035 | 1.24 | .39251 | 1.74 | .45907 |
| .25 | .09871 | .75 | .27337 | 1.25 | .39435 | 1.75 | .45994 |
| .26 | .10257 | .76 | .27637 | 1.26 | .39617 | 1.76 | .46080 |
| .27 | .10642 | .77 | .27935 | 1.27 | .39796 | 1.77 | .46164 |
| .28 | .11026 | .78 | .28230 | 1.28 | .39973 | 1.78 | .46246 |
| .29 | .11409 | .79 | .28524 | 1.29 | .40147 | 1.79 | .46327 |
| .30 | .11791 | .80 | .28814 | 1.30 | .40320 | 1.80 | .46407 |
| .31 | .12172 | .81 | .29103 | 1.31 | .40490 | 1.81 | .46485 |
| .32 | .12552 | .82 | .29389 | 1.32 | .40658 | 1.82 | .46562 |
| .33 | .12930 | .83 | .29673 | 1.33 | .40824 | 1.83 | .46638 |
| .34 | .13307 | .84 | .29955 | 1.34 | .40988 | 1.84 | .46712 |
| .35 | .13683 | .85 | .30234 | 1.35 | .41149 | 1.85 | .46784 |
| .36 | .14058 | .86 | .30511 | 1.36 | .41309 | 1.86 | .46856 |
| .37 | .14431 | .87 | .30785 | 1.37 | .41466 | 1.87 | .46926 |
| .38 | .14803 | .88 | .31057 | 1.38 | .41621 | 1.88 | .46995 |
| .39 | .15173 | .89 | .31327 | 1.39 | .41774 | 1.89 | .47062 |
| .40 | .15542 | .90 | .31594 | 1.40 | .41924 | 1.90 | .47128 |
| .41 | .15910 | .91 | .31859 | 1.41 | .42073 | 1.91 | .47193 |
| .42 | .16276 | .92 | .32121 | 1.42 | .42220 | 1.92 | .47257 |
| .43 | .16640 | .93 | .32381 | 1.43 | .42364 | 1.93 | .47320 |
| .44 | .17003 | .94 | .32639 | 1.44 | .42507 | 1.94 | .47381 |
| .45 | .17364 | .95 | .32894 | 1.45 | .42647 | 1.95 | .47441 |
| .46 | .17724 | .96 | .33147 | 1.46 | .42785 | 1.96 | .47500 |
| .47 | .18082 | .97 | .33398 | 1.47 | .42922 | 1.97 | .47558 |
| .48 | .18439 | .98 | .33646 | 1.48 | .43056 | 1.98 | .47615 |
| .49 | .18793 | .99 | .33891 | 1.49 | .43189 | 1.99 | .47670 |

[a] Hayslett (1968).

Table A-1 (*continued*)

| $z$ | $A(z)$ | $z$ | $A(z)$ | $z$ | $A(z)$ | $z$ | $(Az)$ |
|---|---|---|---|---|---|---|---|
| 2.00 | .47725 | 2.50 | .49379 | 3.00 | .49865 | 3.50 | .49977 |
| 2.01 | .47778 | 2.51 | .49396 | 3.01 | .49869 | 3.51 | .49978 |
| 2.02 | .47831 | 2.52 | .49413 | 3.02 | .49874 | 3.52 | .49978 |
| 2.03 | .47882 | 2.53 | .49430 | 3.03 | .49878 | 3.53 | .49979 |
| 2.04 | .47932 | 2.54 | .49446 | 3.04 | .49882 | 3.54 | .49980 |
| 2.05 | .47982 | 2.55 | .49461 | 3.05 | .49886 | 3.55 | .49981 |
| 2.06 | .48030 | 2.56 | .49477 | 3.06 | .49889 | 3.56 | .49981 |
| 2.07 | .48077 | 2.57 | .49492 | 3.07 | .49893 | 3.57 | .49982 |
| 2.08 | .48124 | 2.58 | .49506 | 3.08 | .49896 | 3.58 | .49983 |
| 2.09 | .48169 | 2.59 | .49520 | 3.09 | .49900 | 3.59 | .49983 |
| 2.10 | .48214 | 2.60 | .49534 | 3.10 | .49903 | 3.60 | .49984 |
| 2.11 | .48257 | 2.61 | .49547 | 3.11 | .49906 | 3.61 | .49985 |
| 2.12 | .48300 | 2.62 | .49560 | 3.12 | .49910 | 3.62 | .49985 |
| 2.13 | .48341 | 2.63 | .49573 | 3.13 | .49913 | 3.63 | .49986 |
| 2.14 | .48382 | 2.64 | .49585 | 3.14 | .49916 | 3.64 | .49986 |
| 2.15 | .48422 | 2.65 | .49598 | 3.15 | .49918 | 3.65 | .49987 |
| 2.16 | .48461 | 2.66 | .49609 | 3.16 | .49921 | 3.66 | .49987 |
| 2.17 | .48500 | 2.67 | .49621 | 3.17 | .49924 | 3.67 | .49988 |
| 2.18 | .48537 | 2.68 | .49632 | 3.18 | .49926 | 3.68 | .49988 |
| 2.19 | .48574 | 2.69 | .49643 | 3.19 | .49929 | 3.69 | .49989 |
| 2.20 | .48610 | 2.70 | .49653 | 3.20 | .49931 | 3.70 | .49989 |
| 2.21 | .48645 | 2.71 | .49664 | 3.21 | .49934 | 3.71 | .49990 |
| 2.22 | .48679 | 2.72 | .49674 | 3.22 | .49936 | 3.72 | .49990 |
| 2.23 | .48713 | 2.73 | .49683 | 3.23 | .49938 | 3.73 | .49990 |
| 2.24 | .48745 | 2.74 | .49693 | 3.24 | .49940 | 3.74 | .49991 |
| 2.25 | .48778 | 2.75 | .49702 | 3.25 | .49942 | 3.75 | .49991 |
| 2.26 | .48809 | 2.76 | .49711 | 3.26 | .49944 | 3.76 | .49992 |
| 2.27 | .48840 | 2.77 | .49720 | 3.27 | .49946 | 3.77 | .49992 |
| 2.28 | .48870 | 2.78 | .49728 | 3.28 | .49948 | 3.78 | .49992 |
| 2.29 | .48899 | 2.79 | .49736 | 3.29 | .49950 | 3.79 | .49992 |
| 2.30 | .48928 | 2.80 | .49744 | 3.30 | .49952 | 3.80 | .49993 |
| 2.31 | .48956 | 2.81 | .49752 | 3.31 | .49953 | 3.81 | .49993 |
| 2.32 | .48983 | 2.82 | .49760 | 3.32 | .49955 | 3.82 | .49993 |
| 2.33 | .49010 | 2.83 | .49767 | 3.33 | .49957 | 3.83 | .49994 |
| 2.34 | .49036 | 2.84 | .49774 | 3.34 | .49958 | 3.84 | .49994 |
| 2.35 | .49061 | 2.85 | .49781 | 3.35 | .49960 | 3.85 | .49994 |
| 2.36 | .49086 | 2.86 | .49788 | 3.36 | .49961 | 3.86 | .49994 |
| 2.37 | .49111 | 2.87 | .49795 | 3.37 | .49962 | 3.87 | .49995 |
| 2.38 | .49134 | 2.88 | .49801 | 3.38 | .49964 | 3.88 | .49995 |
| 2.39 | .49158 | 2.89 | .49807 | 3.39 | .49965 | 3.89 | .49995 |
| 2.40 | .49180 | 2.90 | .49813 | 3.40 | .49966 | 3.90 | .49995 |
| 2.41 | .49202 | 2.91 | .49819 | 3.41 | .49968 | 3.91 | .49995 |
| 2.42 | .49224 | 2.92 | .49825 | 3.42 | .49969 | 3.92 | .49996 |
| 2.43 | .49245 | 2.93 | .49831 | 3.43 | .49970 | 3.93 | .49996 |
| 2.44 | .49266 | 2.94 | .49836 | 3.44 | .49971 | 3.94 | .49996 |
| 2.45 | .49286 | 2.95 | .49841 | 3.45 | .49972 | 3.95 | .49996 |
| 2.46 | .49305 | 2.96 | .49846 | 3.46 | .49973 | 3.96 | .49996 |
| 2.47 | .49324 | 2.97 | .49851 | 3.47 | .49974 | 3.97 | .49996 |
| 2.48 | .49343 | 2.98 | .49856 | 3.48 | .49975 | 3.98 | .49997 |
| 2.49 | .49361 | 2.99 | .49861 | 3.49 | .49976 | 3.99 | .49997 |
| | | | | | | 4.00 | .49997 |

## Table A-2

$t$-DISTRIBUTION[a]

| Degrees of Freedom | $\gamma = .10$ | $\gamma = .05$ | $\gamma = .025$ | $\gamma = .01$ | $\gamma = .005$ |
|---|---|---|---|---|---|
| 1 | 3.078 | 6.314 | 12.706 | 31.821 | 63.657 |
| 2 | 1.886 | 2.920 | 4.303 | 6.965 | 9.925 |
| 3 | 1.638 | 2.353 | 3.182 | 4.541 | 5.841 |
| 4 | 1.533 | 2.132 | 2.776 | 3.747 | 4.604 |
| 5 | 1.476 | 2.015 | 2.571 | 3.365 | 4.032 |
| 6 | 1.440 | 1.943 | 2.447 | 3.143 | 3.707 |
| 7 | 1.415 | 1.895 | 2.365 | 2.998 | 3.499 |
| 8 | 1.397 | 1.860 | 2.306 | 2.896 | 3.355 |
| 9 | 1.383 | 1.833 | 2.262 | 2.821 | 3.250 |
| 10 | 1.372 | 1.812 | 2.228 | 2.764 | 3.169 |
| 11 | 1.363 | 1.796 | 2.201 | 2.718 | 3.106 |
| 12 | 1.356 | 1.782 | 2.179 | 2.681 | 3.055 |
| 13 | 1.350 | 1.771 | 2.160 | 2.650 | 3.012 |
| 14 | 1.345 | 1.761 | 2.145 | 2.624 | 2.977 |
| 15 | 1.341 | 1.753 | 2.131 | 2.602 | 2.947 |
| 16 | 1.337 | 1.746 | 2.120 | 2.583 | 2.921 |
| 17 | 1.333 | 1.740 | 2.110 | 2.567 | 2.898 |
| 18 | 1.330 | 1.734 | 2.101 | 2.552 | 2.878 |
| 19 | 1.328 | 1.729 | 2.093 | 2.539 | 2.861 |
| 20 | 1.325 | 1.725 | 2.086 | 2.528 | 2.845 |
| 21 | 1.323 | 1.721 | 2.080 | 2.518 | 2.831 |
| 22 | 1.321 | 1.717 | 2.074 | 2.508 | 2.819 |
| 23 | 1.319 | 1.714 | 2.069 | 2.500 | 2.807 |
| 24 | 1.318 | 1.711 | 2.064 | 2.492 | 2.797 |
| 25 | 1.316 | 1.708 | 2.060 | 2.485 | 2.787 |
| 26 | 1.315 | 1.706 | 2.056 | 2.479 | 2.779 |
| 27 | 1.314 | 1.703 | 2.052 | 2.473 | 2.771 |
| 28 | 1.313 | 1.701 | 2.048 | 2.467 | 2.763 |
| 29 | 1.311 | 1.699 | 2.045 | 2.462 | 2.756 |
| 30 | 1.310 | 1.697 | 2.042 | 2.457 | 2.750 |
| 40 | 1.303 | 1.684 | 2.021 | 2.423 | 2.704 |
| 60 | 1.296 | 1.671 | 2.000 | 2.390 | 2.660 |
| 120 | 1.289 | 1.658 | 1.980 | 2.358 | 2.617 |
| $\infty$ | 1.282 | 1.645 | 1.960 | 2.326 | 2.576 |

[a] Hayslett (1968).

**Table A-3**

$\chi^2$-DISTRIBUTION[a]

| $n$ | $\gamma = .995$ | $\gamma = .99$ | $\gamma = .975$ | $\gamma = .95$ | $\gamma = .05$ | $\gamma = .025$ | $\gamma = .01$ | $\gamma = .005$ | $n$ |
|---|---|---|---|---|---|---|---|---|---|
| 1 | .0000393 | .000157 | .000982 | .00393 | 3.841 | 5.024 | 6.635 | 7.879 | 1 |
| 2 | .0100 | .0201 | .0506 | .103 | 5.991 | 7.378 | 9.210 | 10.597 | 2 |
| 3 | .00717 | .115 | .216 | .352 | 7.815 | 9.348 | 11.345 | 12.838 | 3 |
| 4 | .207 | .297 | .484 | .711 | 9.488 | 11.143 | 13.277 | 14.860 | 4 |
| 5 | .412 | .554 | .831 | 1.145 | 11.070 | 12.832 | 15.086 | 16.750 | 5 |
| 6 | .676 | .872 | 1.237 | 1.635 | 12.592 | 14.449 | 16.812 | 18.548 | 6 |
| 7 | .989 | 1.239 | 1.690 | 2.167 | 14.067 | 16.013 | 18.475 | 20.278 | 7 |
| 8 | 1.344 | 1.646 | 2.180 | 2.733 | 15.507 | 17.535 | 20.090 | 21.955 | 8 |
| 9 | 1.735 | 2.088 | 2.700 | 3.325 | 16.919 | 19.023 | 21.666 | 23.589 | 9 |
| 10 | 2.156 | 2.558 | 3.247 | 3.940 | 18.307 | 20.483 | 23.209 | 25.188 | 10 |
| 11 | 2.603 | 3.053 | 3.816 | 4.575 | 19.675 | 21.920 | 24.725 | 26.757 | 11 |
| 12 | 3.074 | 3.571 | 4.404 | 5.226 | 21.026 | 23.337 | 26.217 | 28.300 | 12 |
| 13 | 3.565 | 4.107 | 5.009 | 5.892 | 22.362 | 24.736 | 27.688 | 29.819 | 13 |
| 14 | 4.075 | 4.660 | 5.629 | 6.571 | 23.685 | 26.119 | 29.141 | 31.319 | 14 |
| 15 | 4.601 | 5.229 | 6.262 | 7.261 | 24.996 | 27.488 | 30.578 | 32.801 | 15 |
| 16 | 5.142 | 5.812 | 6.908 | 7.962 | 26.296 | 28.845 | 32.000 | 34.267 | 16 |
| 17 | 5.697 | 6.408 | 7.564 | 8.672 | 27.587 | 30.191 | 33.409 | 35.718 | 17 |
| 18 | 6.265 | 7.015 | 8.231 | 9.390 | 28.869 | 31.526 | 34.805 | 37.156 | 18 |
| 19 | 6.844 | 7.633 | 8.907 | 10.117 | 30.144 | 32.852 | 36.191 | 38.582 | 19 |
| 20 | 7.434 | 8.260 | 9.591 | 10.851 | 31.410 | 34.170 | 37.566 | 39.997 | 20 |
| 21 | 8.034 | 8.897 | 10.283 | 11.591 | 32.671 | 35.479 | 38.932 | 41.401 | 21 |
| 22 | 8.643 | 9.542 | 10.982 | 12.338 | 33.924 | 36.781 | 40.289 | 42.796 | 22 |
| 23 | 9.260 | 10.196 | 11.689 | 13.091 | 35.172 | 38.076 | 41.638 | 44.181 | 23 |
| 24 | 9.886 | 10.856 | 12.401 | 13.848 | 36.415 | 39.364 | 42.980 | 45.558 | 24 |
| 25 | 10.520 | 11.524 | 13.120 | 14.611 | 37.652 | 40.646 | 44.314 | 46.928 | 25 |
| 26 | 11.160 | 12.198 | 13.844 | 15.379 | 38.885 | 41.923 | 45.642 | 48.290 | 26 |
| 27 | 11.808 | 12.879 | 14.573 | 16.151 | 40.113 | 43.194 | 46.963 | 49.645 | 27 |
| 28 | 12.461 | 13.565 | 15.308 | 16.928 | 41.337 | 44.461 | 48.278 | 50.993 | 28 |
| 29 | 13.121 | 14.256 | 16.047 | 17.708 | 42.557 | 45.722 | 49.588 | 52.336 | 29 |
| 30 | 13.787 | 14.953 | 16.791 | 18.493 | 43.773 | 46.979 | 50.892 | 53.672 | 30 |

[a] Hayslett (1968).

**Table A-4**

*F*-Distribution $\gamma = 0.05$[a]

| $\nu_2$ \ $\nu_1$ | 1 | 2 | 3 | 4 | 5 | 6 | 7 | 8 | 9 |
|---|---|---|---|---|---|---|---|---|---|
| 1 | 161.45 | 199.50 | 215.71 | 224.58 | 230.16 | 233.99 | 236.77 | 238.88 | 240.54 |
| 2 | 18.513 | 19.000 | 19.164 | 19.247 | 19.296 | 19.330 | 19.353 | 19.371 | 19.385 |
| 3 | 10.128 | 9.5521 | 9.2766 | 9.1172 | 9.0135 | 8.9406 | 8.8868 | 8.8452 | 8.8123 |
| 4 | 7.7086 | 6.9443 | 6.5914 | 6.3883 | 6.2560 | 6.1631 | 6.0942 | 6.0410 | 5.9988 |
| 5 | 6.6079 | 5.7861 | 5.4095 | 5.1922 | 5.0503 | 4.9503 | 4.8759 | 4.8183 | 4.7725 |
| 6 | 5.9874 | 5.1433 | 4.7571 | 4.5337 | 4.3874 | 4.2839 | 4.2066 | 4.1468 | 4.0990 |
| 7 | 5.5914 | 4.7374 | 4.3468 | 4 1203 | 3.9715 | 3.8660 | 3.7870 | 3.7257 | 3.6767 |
| 8 | 5.3177 | 4.4590 | 4.0662 | 3.8378 | 3.6875 | 3.5806 | 3.5005 | 3.4381 | 3.3881 |
| 9 | 5.1174 | 4.2565 | 3.8626 | 3.6331 | 3.4817 | 3.3738 | 3.2927 | 3.2296 | 3.1789 |
| 10 | 4.9646 | 4.1028 | 3.7083 | 3.4780 | 3.3258 | 3.2172 | 3.1355 | 3.0717 | 3.0204 |
| 11 | 4 8443 | 3.9823 | 3.5874 | 3.3567 | 3.2039 | 3.0946 | 3.0123 | 2.9480 | 2.8962 |
| 12 | 4.7472 | 3.8853 | 3.4903 | 3.2592 | 3.1059 | 2.9961 | 2.9134 | 2.8486 | 2.7964 |
| 13 | 4.6672 | 3.8056 | 3.4105 | 3.1791 | 3.0254 | 2.9153 | 2.8321 | 2.7669 | 2.7144 |
| 14 | 4.6001 | 3.7389 | 3.3439 | 3.1122 | 2.9582 | 2.8477 | 2.7642 | 2.6987 | 2.6458 |
| 15 | 4.5431 | 3.6823 | 3.2874 | 3.0556 | 2.9013 | 2.7905 | 2.7066 | 2.6408 | 2.5876 |
| 16 | 4.4940 | 3.6337 | 3.2389 | 3.0069 | 2.8524 | 2.7413 | 2.6572 | 2.5911 | 2.5377 |
| 17 | 4.4513 | 3.5915 | 3.1968 | 2.9647 | 2.8100 | 2.6987 | 2.6143 | 2.5480 | 2.4943 |
| 18 | 4.4139 | 3.5546 | 3.1599 | 2.9277 | 2.7729 | 2.6613 | 2.5767 | 2.5102 | 2.4563 |
| 19 | 4.3808 | 3.5219 | 3.1274 | 2.8951 | 2.7401 | 2.6283 | 2.5435 | 2.4768 | 2.4227 |
| 20 | 4.3513 | 3.4928 | 3.0984 | 2.8661 | 2.7109 | 2.5990 | 2.5140 | 2.4471 | 2.3928 |
| 21 | 4.3248 | 3.4668 | 3.0725 | 2.8401 | 2.6848 | 2.5757 | 2.4876 | 2.4205 | 2.3661 |
| 22 | 4.3009 | 3.4434 | 3.0491 | 2.8167 | 2.6613 | 2.5491 | 2.4638 | 2.3965 | 2.3419 |
| 23 | 4.2793 | 3.4221 | 3.0280 | 2.7955 | 2.6400 | 2.5277 | 2.4422 | 2.3748 | 2.3201 |
| 24 | 4.2597 | 3.4028 | 3.0088 | 2.7763 | 2.6207 | 2.5082 | 2.4226 | 2.3551 | 2.3002 |
| 25 | 4.2417 | 3.3852 | 2.9912 | 2.7587 | 2.6030 | 2.4904 | 2.4047 | 2.3371 | 2.2821 |
| 26 | 4.2252 | 3.3690 | 2.9751 | 2.7426 | 2.5868 | 2.4741 | 2.3883 | 2.3205 | 2.2655 |
| 27 | 4.2100 | 3.3541 | 2.9604 | 2.7278 | 2.5719 | 2.4591 | 2.3732 | 2.3053 | 2.2501 |
| 28 | 4.1960 | 3.3404 | 2.9467 | 2.7141 | 2.5581 | 2.4453 | 2.3593 | 2.2913 | 2.2360 |
| 29 | 4.1830 | 3.3277 | 2.9340 | 2.7014 | 2.5454 | 2.4324 | 2.3463 | 2.2782 | 2.2229 |
| 30 | 4.1709 | 3.3158 | 2.9223 | 2.6896 | 2.5336 | 2.4205 | 2.3343 | 2.2662 | 2.2107 |
| 40 | 4.0848 | 3.2317 | 2.8387 | 2.6060 | 2.4495 | 2.3359 | 2.2490 | 2.1802 | 2.1240 |
| 60 | 4.0012 | 3.1504 | 2.7581 | 2.5252 | 2.3683 | 2.2540 | 2.1665 | 2.0970 | 2.0401 |
| 120 | 3.9201 | 3.0718 | 2.6802 | 2.4472 | 2.2900 | 2.1750 | 2.0867 | 2.0164 | 1.9588 |
| ∞ | 3.8415 | 2.9957 | 2.6049 | 2.3719 | 2.2141 | 2.0986 | 2.0096 | 1.9384 | 1.8799 |

[a] Hayslett (1968).

**Table A-4** *(continued)*

| 10 | 12 | 15 | 20 | 24 | 30 | 40 | 60 | 120 | ∞ |
|---|---|---|---|---|---|---|---|---|---|
| 241.88 | 243.91 | 245.95 | 248.01 | 249.05 | 250.09 | 251.14 | 252.20 | 253.25 | 254.32 |
| 19.396 | 19.413 | 19.429 | 19.446 | 19.454 | 19.462 | 19.471 | 19.479 | 19.487 | 19.496 |
| 8.7855 | 8.7446 | 8.7029 | 8.6602 | 8.6385 | 8.6166 | 8.5944 | 8.5720 | 8.5494 | 8.5265 |
| 5.9644 | 5.9117 | 5.8578 | 5.8025 | 5.7744 | 5.7459 | 5.7170 | 5.6878 | 5.6581 | 5.6281 |
| 4.7351 | 4.6777 | 4.6188 | 4.5581 | 4.5272 | 4.4957 | 4.4638 | 4.4314 | 4.3984 | 4.3650 |
| 4.0600 | 3.9999 | 3.9381 | 3.8742 | 3.8415 | 3.8082 | 3.7743 | 3.7398 | 3.7047 | 3.6688 |
| 3.6365 | 3.5747 | 3.5108 | 3.4445 | 3.4105 | 3.3758 | 3.3404 | 3.3043 | 3.2674 | 3.2298 |
| 3.3472 | 3.2840 | 3.2184 | 3.1503 | 3.1152 | 3.0794 | 3.0428 | 3.0053 | 2.9669 | 2.9276 |
| 3.1373 | 3.0729 | 3.0061 | 2.9365 | 2.9005 | 2.8637 | 2.8259 | 2.7872 | 2.7475 | 2.7067 |
| 2.9782 | 2.9130 | 2.8450 | 2.7740 | 2.7372 | 2.6996 | 2.6609 | 2.6211 | 2.5801 | 2.5379 |
| 2.8536 | 2.7876 | 2.7186 | 2.6464 | 2.6090 | 2.5705 | 2.5309 | 2.4901 | 2.4480 | 2.4045 |
| 2.7534 | 2.6866 | 2.6169 | 2.5436 | 2.5055 | 2.4663 | 2.4259 | 2.3842 | 2.3410 | 2.2962 |
| 2.6710 | 2.6037 | 2.5331 | 2.4539 | 2.4202 | 2.3803 | 2.3392 | 2.2966 | 2.2524 | 2.2064 |
| 2.6021 | 2.5342 | 2.4630 | 2.3879 | 2.3487 | 2.3082 | 2.2664 | 2.2230 | 2.1778 | 2.1307 |
| 2.5437 | 2.4753 | 2.4035 | 2.3275 | 2.2878 | 2.2468 | 2.2043 | 2.1601 | 2.1141 | 2.0658 |
| 2.4935 | 2.4247 | 2.3522 | 2.2756 | 2.2354 | 2.1938 | 2.1507 | 2.1058 | 2.0589 | 2.0096 |
| 2.4499 | 2.3807 | 2.3077 | 2.2304 | 2.1898 | 2.1477 | 2.1040 | 2.0584 | 2.0107 | 1.9604 |
| 2.4117 | 2.3421 | 2.2686 | 2.1906 | 2.1497 | 2.1071 | 2.0629 | 2.0166 | 1.9681 | 1.9168 |
| 2.3779 | 2.3080 | 2.2341 | 2.1555 | 2.1141 | 2.0712 | 2.0264 | 1.9796 | 1.9302 | 1.8780 |
| 2.3479 | 2.2776 | 2.2033 | 2.1242 | 2.0825 | 2.0391 | 1.9938 | 1.9464 | 1.8963 | 1.8432 |
| 2.3210 | 2.2504 | 2.1757 | 2.0960 | 2.0540 | 2.0102 | 1.9645 | 1.9165 | 1.8657 | 1.8117 |
| 2.2967 | 2.2258 | 2.1508 | 2.0707 | 2.0283 | 1.9842 | 1.9380 | 1.8895 | 1.8380 | 1.7831 |
| 2.2747 | 2.2036 | 2.1282 | 2.0476 | 2.0050 | 1.9605 | 1.9139 | 1.8649 | 1.8128 | 1.7570 |
| 2.2547 | 2.1834 | 2.1077 | 2.0267 | 1.9838 | 1.9390 | 1.8920 | 1.8424 | 1.7897 | 1.7331 |
| 2.2365 | 2.1649 | 2.0889 | 2.0075 | 1.9643 | 1.9192 | 1.8718 | 1.8217 | 1.7684 | 1.7110 |
| 2.2197 | 2.1479 | 2.0716 | 1.9898 | 1.9464 | 1.9010 | 1.8533 | 1.8027 | 1.7488 | 1.6906 |
| 2.2043 | 2.1323 | 2.0558 | 1.9736 | 1.9299 | 1.8842 | 1.8361 | 1.7851 | 1.7307 | 1.6717 |
| 2.1900 | 2.1179 | 2.0411 | 1.9586 | 1.9147 | 1.8687 | 1.8203 | 1.7689 | 1.7138 | 1.6541 |
| 2.1768 | 2.1045 | 2.0275 | 1.9446 | 1.9005 | 1.8543 | 1.8055 | 1.7537 | 1.6981 | 1.6377 |
| 2.1646 | 2.0921 | 2.0148 | 1.9317 | 1.8874 | 1.8409 | 1.7918 | 1.7396 | 1.6835 | 1.6223 |
| 2.0772 | 2.0035 | 1.9245 | 1.8389 | 1.7929 | 1.7444 | 1.6928 | 1.6373 | 1.5766 | 1.5089 |
| 1.9926 | 1.9174 | 1.8364 | 1.7480 | 1.7001 | 1.6491 | 1.5943 | 1.5343 | 1.4673 | 1.3893 |
| 1.9105 | 1.8337 | 1.7505 | 1.6587 | 1.6084 | 1.5543 | 1.4952 | 1.4290 | 1.3519 | 1.2539 |
| 1.8307 | 1.7522 | 1.6664 | 1.5705 | 1.5173 | 1.4591 | 1.3940 | 1.3180 | 1.2214 | 1.0000 |

**Table A-5**

$F$-DISTRIBUTION $\gamma = 0.025^{a}$

| $\nu_2$ \ $\nu_1$ | 1 | 2 | 3 | 4 | 5 | 6 | 7 | 8 | 9 |
|---|---|---|---|---|---|---|---|---|---|
| 1 | 647.79 | 799.50 | 864.16 | 899.58 | 921.85 | 937.11 | 948.22 | 956.66 | 963.28 |
| 2 | 38.506 | 39.000 | 39.165 | 39.248 | 29.298 | 39.331 | 39.355 | 39.373 | 39.387 |
| 3 | 17.443 | 16.044 | 15.439 | 15.101 | 14.885 | 14.735 | 14.624 | 14.540 | 14.473 |
| 4 | 12.218 | 10.649 | 9.9792 | 9.6045 | 9.3645 | 9.1973 | 9.0741 | 8.9796 | 8.9047 |
| 5 | 10.007 | 8.4336 | 7.7636 | 7.3879 | 7.1464 | 6.9777 | 6.8531 | 6.7572 | 6.6810 |
| 6 | 8.8131 | 7.2598 | 6.5988 | 6.2272 | 5.9876 | 5.8197 | 5.6955 | 5.5996 | 5.5234 |
| 7 | 8.0727 | 6.5415 | 5.8898 | 5.5226 | 5.2852 | 5.1186 | 4.9949 | 4.8994 | 4.8232 |
| 8 | 7.5709 | 6.0595 | 5.4160 | 5.0526 | 4.8173 | 4.6517 | 4.5286 | 4.4332 | 4.3572 |
| 9 | 7.2093 | 5.7147 | 5.0781 | 4.7181 | 4.4844 | 4.3197 | 4.1971 | 4.1020 | 4.0260 |
| 10 | 6.9367 | 5.4564 | 4.8256 | 4.4683 | 4.2361 | 4.0721 | 3.9498 | 3.8549 | 3.7790 |
| 11 | 6.7241 | 5.2559 | 4.6300 | 4.2751 | 4.0440 | 3.8807 | 3.7586 | 3.6638 | 3.5879 |
| 12 | 6.5538 | 5.0959 | 4.4742 | 4.1212 | 3.8911 | 3.7283 | 3.6065 | 3.5118 | 3.4358 |
| 13 | 6.4143 | 4.9653 | 4.3472 | 3.9959 | 3.7667 | 3.6043 | 3.4827 | 3.3880 | 3.3120 |
| 14 | 6.2979 | 4.8567 | 4.2417 | 3.8919 | 3.6634 | 3.5014 | 3.3799 | 3.2853 | 3.2093 |
| 15 | 6.1995 | 4.7650 | 4.1528 | 3.8043 | 3.5764 | 3.4147 | 3.2934 | 3.1987 | 3.1227 |
| 16 | 6.1151 | 4.6867 | 4.0768 | 3.7294 | 3.5021 | 3.3406 | 3.2194 | 3.1248 | 3.0488 |
| 17 | 6.0420 | 4.6189 | 4.0112 | 3.6648 | 3.4379 | 3.2767 | 3.1556 | 3.0610 | 2.9849 |
| 18 | 5.9781 | 4.5597 | 3.9539 | 3.6083 | 3.3820 | 3.2209 | 3.0999 | 3.0053 | 2.9291 |
| 19 | 5.9216 | 4.5075 | 3.9034 | 3.5587 | 3.3327 | 3.1718 | 3.0509 | 2.9563 | 2.8800 |
| 20 | 5.8715 | 4.4613 | 3.8587 | 3.5147 | 3.2891 | 3.1283 | 3.0074 | 2.9128 | 2.8365 |
| 21 | 5.8266 | 4.4199 | 3.8188 | 3.4754 | 3.2501 | 3.0895 | 2.9686 | 2.8740 | 2.7977 |
| 22 | 5.7863 | 4.3828 | 3.7829 | 3.4401 | 3.2151 | 3.0546 | 2.9338 | 2.8392 | 2.7628 |
| 23 | 5.7498 | 4.3492 | 3.7505 | 3.4083 | 3.1835 | 3.0232 | 2.9024 | 2.8077 | 2.7313 |
| 24 | 5.7167 | 4.3187 | 3.7211 | 3.3794 | 3.1548 | 2.9946 | 2.8738 | 2.7791 | 2.7027 |
| 25 | 5.6864 | 4.2909 | 3.6943 | 3.3530 | 3.1287 | 2.9685 | 2.8478 | 2.7531 | 2.6766 |
| 26 | 5.6586 | 4.2655 | 3.6697 | 3.3289 | 3.1048 | 2.9447 | 2.8240 | 2.7293 | 2.6528 |
| 27 | 5.6331 | 4.2421 | 3.6472 | 3.3067 | 3.0828 | 2.9228 | 2.8021 | 2.7074 | 2.6309 |
| 28 | 5.6096 | 4.2205 | 3.6264 | 3.2863 | 3.0625 | 2.9027 | 2.7820 | 2.6872 | 2.6106 |
| 29 | 5.5878 | 4.2006 | 3.6072 | 3.2674 | 3.0438 | 2.8840 | 2.7633 | 2.6686 | 2.5919 |
| 30 | 5.5675 | 4.1821 | 3.5894 | 3.2499 | 3.0265 | 2.8667 | 2.7460 | 2.6513 | 2.5746 |
| 40 | 5.4239 | 4.0510 | 3.4633 | 3.1261 | 2.9037 | 2.7444 | 2.6238 | 2.5289 | 2.4519 |
| 60 | 5.2857 | 3.9253 | 3.3425 | 3.0077 | 2.7863 | 2.6274 | 2.5068 | 2.4117 | 2.3344 |
| 120 | 5.1524 | 3.8046 | 3.2270 | 2.8943 | 2.6740 | 2.5154 | 2.3948 | 2.2994 | 2.2217 |
| $\infty$ | 5.0239 | 3.6889 | 3.1161 | 2.7858 | 2.5665 | 2.4082 | 2.2875 | 2.1918 | 2.1136 |

[a] Hayslett (1968).

**Table A-5** (*continued*)

| 10 | 12 | 15 | 20 | 24 | 30 | 40 | 60 | 120 | $\infty$ |
|---|---|---|---|---|---|---|---|---|---|
| 968.63 | 976.71 | 984.87 | 993.10 | 997.25 | 1001.4 | 1005.6 | 1009.8 | 1014.0 | 1018.3 |
| 39.398 | 39.415 | 39.431 | 39.448 | 39.456 | 39.465 | 39.473 | 39.481 | 39.490 | 39.498 |
| 14.419 | 14.337 | 14.253 | 14.167 | 14.124 | 14.081 | 14.037 | 13.992 | 13.947 | 13.902 |
| 8.8439 | 8.7512 | 8.6565 | 8.5599 | 8.5109 | 8.4613 | 8.4111 | 8.3604 | 8.3092 | 8.2573 |
| 6.6192 | 6.5246 | 6.4277 | 6.3285 | 6.2780 | 6.2269 | 6.1751 | 6.1225 | 6.0693 | 6.0153 |
| 5.4613 | 5.3662 | 5.2687 | 5.1684 | 5.1172 | 5.0652 | 5.0125 | 4.9589 | 4.9045 | 4.8491 |
| 4.7611 | 4.6658 | 4.5678 | 4.4667 | 4.4150 | 4.3624 | 4.3089 | 4.2544 | 4.1989 | 4.1423 |
| 4.2951 | 4.1997 | 4.1012 | 3.9995 | 3.9472 | 3.8940 | 3.8398 | 3.7844 | 3.7279 | 3.6702 |
| 3.9639 | 3.8682 | 3.7694 | 3.6669 | 3.6142 | 3.5604 | 3.5055 | 3.4493 | 3.3918 | 3.3329 |
| 3.7168 | 3.6209 | 3.5217 | 3.4186 | 3.3654 | 3.3110 | 3.2554 | 3.1984 | 3.1399 | 3.0798 |
| 3.5257 | 3.4296 | 3.3299 | 3.2261 | 3.1725 | 3.1176 | 3.0613 | 3.0035 | 2.9441 | 2.8828 |
| 3.3736 | 3.2773 | 3.1772 | 3.0728 | 3.0187 | 2.9633 | 2.9063 | 2.8478 | 2.7874 | 2.7249 |
| 3.2497 | 3.1532 | 3.0527 | 2.9477 | 2.8932 | 2.8373 | 2.7797 | 2.7204 | 2.6590 | 2.5955 |
| 3.1469 | 3.0501 | 2.9493 | 2.8437 | 2.7888 | 2.7324 | 2.6742 | 2.6142 | 2.5519 | 2.4872 |
| 3.0602 | 2.9633 | 2.8621 | 2.7559 | 2.7006 | 2.6437 | 2.5850 | 2.5242 | 2.4611 | 2.3953 |
| 2.9862 | 2.8890 | 2.7875 | 2.6808 | 2.6252 | 2.5678 | 2.5085 | 2.4471 | 2.3831 | 2.3163 |
| 2.9222 | 2.8249 | 2.7230 | 2.6158 | 2.5598 | 2.5021 | 2.4422 | 2.3801 | 2.3153 | 2.2474 |
| 2.8664 | 2.7689 | 2.6667 | 2.5590 | 2.5027 | 2.4445 | 2.3842 | 2.3214 | 2.2558 | 2.1869 |
| 2.8173 | 2.7196 | 2.6171 | 2.5089 | 2.4523 | 2.3937 | 2.3329 | 2.2695 | 2.2032 | 2.1333 |
| 2.7737 | 2.6758 | 2.5731 | 2.4645 | 2.4076 | 2.3486 | 2.2873 | 2.2234 | 2.1562 | 2.0853 |
| 2.7348 | 2.6368 | 2.5338 | 2.4247 | 2.3675 | 2.3082 | 2.2465 | 2.1819 | 2.1141 | 2.0422 |
| 2.6998 | 2.6017 | 2.4984 | 2.3890 | 2.3315 | 2.2718 | 2.2097 | 2.1446 | 2.0760 | 2.0032 |
| 2.6682 | 2.5699 | 2.4665 | 2.3567 | 2.2989 | 2.2389 | 2.1763 | 2.1107 | 2.0415 | 1.9677 |
| 2.6396 | 2.5412 | 2.4374 | 2.3273 | 2.2693 | 2.2090 | 2.1460 | 2.0799 | 2.0099 | 1.9353 |
| 2.6135 | 2.5149 | 2.4110 | 2.3005 | 2.2422 | 2.1816 | 2.1183 | 2.0517 | 1.9811 | 1.9055 |
| 2.5895 | 2.4909 | 2.3867 | 2.2759 | 2.2174 | 2.1565 | 2.0928 | 2.0257 | 1.9545 | 1.8781 |
| 2.5676 | 2.4688 | 2.3644 | 2.2533 | 2.1946 | 2.1334 | 2.0693 | 2.0018 | 1.9299 | 1.8527 |
| 2.5473 | 2.4484 | 2.3438 | 2.2324 | 2.1735 | 2.1121 | 2.0477 | 1.9796 | 1.9072 | 1.8291 |
| 2.5286 | 2.4295 | 2.3248 | 2.2131 | 2.1540 | 2.0923 | 2.0276 | 1.9591 | 1.8861 | 1.8072 |
| 2.5112 | 2.4120 | 2.3072 | 2.1952 | 2.1359 | 2.0739 | 2.0089 | 1.9400 | 1.8664 | 1.7867 |
| 2.3882 | 2.2882 | 2.1819 | 2.0677 | 2.0069 | 1.9429 | 1.8752 | 1.8028 | 1.7242 | 1.6371 |
| 2.2702 | 2.1692 | 2.0613 | 1.9445 | 1.8817 | 1.8152 | 1.7440 | 1.6668 | 1.5810 | 1.4822 |
| 2.1570 | 2.0548 | 1.9450 | 1.8249 | 1.7597 | 1.6899 | 1.6141 | 1.5299 | 1.4327 | 1.3104 |
| 2.0483 | 1.9447 | 1.8326 | 1.7085 | 1.6402 | 1.5660 | 1.4835 | 1.3883 | 1.2684 | 1.0000 |

**Table A-6**

*F*-DISTRIBUTION $\gamma = 0.01$[a]

| $\nu_2$ \ $\nu_1$ | 1 | 2 | 3 | 4 | 5 | 6 | 7 | 8 | 9 |
|---|---|---|---|---|---|---|---|---|---|
| 1 | 4052.2 | 4999.5 | 5403.3 | 5624.6 | 5763.7 | 5859.0 | 5928.3 | 5981.6 | 6022.5 |
| 2 | 98.503 | 99.000 | 99.166 | 99.249 | 99.299 | 99.332 | 99.356 | 99.374 | 99.388 |
| 3 | 34.116 | 30.817 | 29.457 | 28.710 | 28.237 | 27.911 | 27.672 | 27.489 | 27.345 |
| 4 | 21.198 | 18.000 | 16.694 | 15.977 | 15.522 | 15.207 | 14.976 | 14.799 | 14.659 |
| 5 | 16.258 | 13.274 | 12.060 | 11.392 | 10.967 | 10.672 | 10.456 | 10.289 | 10.158 |
| 6 | 13.745 | 10.925 | 9.7795 | 9.1483 | 8.7459 | 8.4661 | 8.2600 | 8.1016 | 7.9761 |
| 7 | 12.246 | 9.5466 | 8.4513 | 7.8467 | 7.4604 | 7.1914 | 6.9928 | 6.8401 | 6.7188 |
| 8 | 11.259 | 8.6491 | 7.5910 | 7.0060 | 6.6318 | 6.3707 | 6.1776 | 6.0289 | 5.9106 |
| 9 | 10.561 | 8.0215 | 6.9919 | 6.4221 | 6.0569 | 5.8018 | 5.6129 | 5.4671 | 5.3511 |
| 10 | 10.044 | 7.5594 | 6.5523 | 5.9943 | 5.6363 | 5.3858 | 5.2001 | 5.0567 | 4.9424 |
| 11 | 9.6460 | 7.2057 | 6.2167 | 5.6683 | 5.3160 | 5.0692 | 4.8861 | 4.7445 | 4.6315 |
| 12 | 9.3302 | 6.9266 | 5.9526 | 5.4119 | 5.0643 | 4.8206 | 4.6395 | 4.4994 | 4.3875 |
| 13 | 9.0738 | 6.7010 | 5.7394 | 5.2053 | 4.8616 | 4.6204 | 4.4410 | 4.3021 | 4.1911 |
| 14 | 8.8616 | 6.5149 | 5.5639 | 5.0354 | 4.6950 | 4.4558 | 4.2779 | 4.1399 | 4.0297 |
| 15 | 8.6831 | 6.3589 | 5.4170 | 4.8932 | 4.5556 | 4.3183 | 4.1415 | 4.0045 | 3.8948 |
| 16 | 8.5310 | 6.2262 | 5.2922 | 4.7726 | 4.4374 | 4.2016 | 4.0259 | 3.8896 | 3.7804 |
| 17 | 8.3997 | 6.1121 | 5.1850 | 4.6690 | 4.3359 | 4.1015 | 3.9267 | 3.7910 | 3.6822 |
| 18 | 8.2854 | 6.0129 | 5.0919 | 4.5790 | 4.2479 | 4.0146 | 3.8406 | 3.7054 | 3.5971 |
| 19 | 8.1850 | 5.9259 | 5.0103 | 4.5003 | 4.1708 | 3.9386 | 3.7653 | 3.6305 | 3.5225 |
| 20 | 8.0960 | 5.8489 | 4.9382 | 4.4307 | 4.1027 | 3.8714 | 3.6987 | 3.5644 | 3.4567 |
| 21 | 8.0166 | 5.7804 | 4.8740 | 4.3688 | 4.0421 | 3.8117 | 3.6396 | 3.5056 | 3.3981 |
| 22 | 7.9454 | 5.7190 | 4.8166 | 4.3134 | 3.9880 | 3.7583 | 3.5867 | 3.4530 | 3.3458 |
| 23 | 7.8811 | 5.6637 | 4.7649 | 4.2635 | 3.9392 | 3.7102 | 3.5390 | 3.4057 | 3.2986 |
| 24 | 7.8229 | 5.6136 | 4.7181 | 4.2184 | 3.8951 | 3.6667 | 3.4959 | 3.3629 | 3.2560 |
| 25 | 7.7698 | 5.5680 | 4.6755 | 4.1774 | 3.8550 | 3.6272 | 3.4568 | 3.3239 | 3.2172 |
| 26 | 7.7213 | 5.5263 | 4.6366 | 4.1400 | 3.8183 | 3.5911 | 3.4210 | 3.2884 | 3.1818 |
| 27 | 7.6767 | 5.4881 | 4.6009 | 4.1056 | 3.7848 | 3.5580 | 3.3882 | 3.2558 | 3.1494 |
| 28 | 7.6356 | 5.4529 | 4.5681 | 4.0740 | 3.7539 | 3.5276 | 3.3581 | 3.2259 | 3.1195 |
| 29 | 7.5976 | 5.4205 | 4.5378 | 4.0449 | 3.7254 | 3.4995 | 3.3302 | 3.1982 | 3.0920 |
| 30 | 7.5625 | 5.3904 | 4.5097 | 4.0179 | 3.6990 | 3.4735 | 3.3045 | 3.1726 | 3.0665 |
| 40 | 7.3141 | 5.1785 | 4.3126 | 3.8283 | 3.5138 | 3.2910 | 3.1238 | 2.9930 | 2.8876 |
| 60 | 7.0771 | 4.9774 | 4.1259 | 3.6491 | 3.3389 | 3.1187 | 2.9530 | 2.8233 | 2.7185 |
| 120 | 6.8510 | 4.7865 | 3.9493 | 3.4796 | 3.1735 | 2.9559 | 2.7918 | 2.6629 | 2.5586 |
| ∞ | 6.6349 | 4.6052 | 3.7816 | 3.3129 | 3.0173 | 2.8020 | 2.6393 | 2.5113 | 2.4073 |

[a] Hayslett (1968).

**Table A-6** *(continued)*

| 10 | 12 | 15 | 20 | 24 | 30 | 40 | 60 | 120 | ∞ |
|---|---|---|---|---|---|---|---|---|---|
| 6055.8 | 6106.3 | 6157.3 | 6208.7 | 6234.6 | 6260.7 | 6286.8 | 6313.0 | 6339.4 | 6366.0 |
| 99.399 | 99.416 | 99.432 | 99.449 | 99.458 | 99.466 | 99.474 | 99.483 | 99.491 | 99.501 |
| 27.229 | 27.052 | 26.872 | 26.690 | 26.598 | 26.505 | 26.411 | 26.316 | 26.221 | 26.125 |
| 14.546 | 14.374 | 14.198 | 14.020 | 13.929 | 13.838 | 13.745 | 13.652 | 13.558 | 13.463 |
| 10.051 | 9.8883 | 9.7222 | 9.5527 | 9.4665 | 9.3793 | 9.2912 | 9.2020 | 9.1118 | 9.0204 |
| 7.8741 | 7.7183 | 7.5590 | 7.3958 | 7.3127 | 7.2285 | 7.1432 | 7.0568 | 6.9690 | 6.8801 |
| 6.6201 | 6.4691 | 6.3143 | 6.1554 | 6.0743 | 5.9921 | 5.9084 | 5.8236 | 5.7372 | 5.6495 |
| 5.8143 | 5.6668 | 5.5151 | 5.3591 | 5.2793 | 5.1981 | 5.1156 | 5.0316 | 4.9460 | 4.8588 |
| 5.2565 | 5.1114 | 4.9621 | 4.8080 | 4.7290 | 4.6486 | 4.5667 | 4.4831 | 4.3978 | 4.3105 |
| 4.8492 | 4.7059 | 4.5582 | 4.4054 | 4.3269 | 4.2469 | 4.1653 | 4.0819 | 3.9965 | 3.9090 |
| 4.5393 | 4.3974 | 4.2509 | 4.0990 | 4.0209 | 3.9411 | 3.8596 | 3.7761 | 3.6904 | 3.6025 |
| 4.2961 | 4.1553 | 4.0096 | 3.8584 | 3.7805 | 3.7008 | 3.6192 | 3.5355 | 3.4494 | 3.3608 |
| 4.1003 | 3.9603 | 3.8154 | 3.6646 | 3.5868 | 3.5070 | 3.4253 | 3.3413 | 3.2548 | 3.1654 |
| 3.9394 | 3.8001 | 3.6557 | 3.5052 | 3.4274 | 3.3476 | 3.2656 | 3.1813 | 3.0942 | 3.0040 |
| 3.8049 | 3.6662 | 3.5222 | 3.3719 | 3.2940 | 3.2141 | 3.1319 | 3.0471 | 2.9595 | 2.8684 |
| 3.6909 | 3.5527 | 3.4089 | 3.2588 | 3.1808 | 3.1007 | 3.0182 | 2.9330 | 2.8447 | 2.7528 |
| 3.5931 | 3.4552 | 3.3117 | 3.1615 | 3.0835 | 3.0032 | 2.9205 | 2.8348 | 2.7459 | 2.6530 |
| 3.5082 | 3.3706 | 3.2273 | 3.0771 | 2.9990 | 2.9185 | 2.8354 | 2.7493 | 2.6597 | 2.5660 |
| 3.4338 | 3.2965 | 3.1533 | 3.0031 | 2.9249 | 2.8422 | 2.7608 | 2.6742 | 2.5839 | 2.4893 |
| 3.3682 | 3.2311 | 3.0880 | 2.9377 | 2.8594 | 2.7785 | 2.6947 | 2.6077 | 2.5168 | 2.4212 |
| 3.3098 | 3.1729 | 3.0299 | 2.8796 | 2.8011 | 2.7200 | 2.6359 | 2.5484 | 2.4568 | 2.3603 |
| 3.2576 | 3.1209 | 2.9780 | 2.8274 | 2.7488 | 2.6675 | 2.5831 | 2.4951 | 2.4029 | 2.3055 |
| 3.2106 | 3.0740 | 2.9311 | 2.7805 | 2.7017 | 2.6202 | 2.5355 | 2.4471 | 2.3542 | 2.2559 |
| 3.1681 | 3.0316 | 2.8887 | 2.7380 | 2.6591 | 2.5773 | 2.4923 | 2.4035 | 2.3099 | 2.2107 |
| 3.1294 | 2.9931 | 2.8502 | 2.6993 | 2.6203 | 2.5383 | 2.4530 | 2.3637 | 2.2695 | 2.1694 |
| 3.0941 | 2.9579 | 2.8150 | 2.6640 | 2.5848 | 2.5026 | 2.4170 | 2.3273 | 2.2325 | 2.1315 |
| 3.0618 | 2.9256 | 2.7827 | 2.6316 | 2.5522 | 2.4699 | 2.3840 | 2.2938 | 2.1984 | 2.0965 |
| 3.0320 | 2.8959 | 2.7530 | 2.6017 | 2.5223 | 2.4397 | 2.3535 | 2.2629 | 2.1670 | 2.0642 |
| 3.0045 | 2.8685 | 2.7256 | 2.5742 | 2.4946 | 2.4118 | 2.3253 | 2.2344 | 2.1378 | 2.0342 |
| 2.9791 | 2.8431 | 2.7002 | 2.5487 | 2.4689 | 2.3860 | 2.2992 | 2.2079 | 2.1107 | 2.0062 |
| 2.8005 | 2.6648 | 2.5216 | 2.3689 | 2.2880 | 2.2034 | 2.1142 | 2.0194 | 1.9172 | 1.8047 |
| 2.6318 | 2.4961 | 2.3523 | 2.1978 | 2.1154 | 2.0285 | 1.9360 | 1.8363 | 1.7263 | 1.6006 |
| 2.4721 | 2.3363 | 2.1915 | 2.0346 | 1.9500 | 1.8600 | 1.7628 | 1.6557 | 1.5330 | 1.3805 |
| 2.3209 | 2.1848 | 2.0385 | 1.8783 | 1.7908 | 1.6964 | 1.5923 | 1.4730 | 1.3246 | 1.0000 |

**Table A-7**

*F*-DISTRIBUTION $\gamma = 0.005$[a]

| $\nu_2$ \ $\nu_1$ | 1 | 2 | 3 | 4 | 5 | 6 | 7 | 8 | 9 |
|---|---|---|---|---|---|---|---|---|---|
| 1 | 16211 | 20000 | 21615 | 22500 | 23056 | 23437 | 23715 | 23925 | 24091 |
| 2 | 198.50 | 199.00 | 199.17 | 199.25 | 199.30 | 199.33 | 199.36 | 199.37 | 199.39 |
| 3 | 55.552 | 49.799 | 47.467 | 46.195 | 45.392 | 44.838 | 44.434 | 44.126 | 43.882 |
| 4 | 31.333 | 26.284 | 24.259 | 23.155 | 22.456 | 21.975 | 21.622 | 21.352 | 21.139 |
| 5 | 22.785 | 18.314 | 16.530 | 15.556 | 14.940 | 14.513 | 14.200 | 13.961 | 13.772 |
| 6 | 18.635 | 14.544 | 12.917 | 12.028 | 11.464 | 11.073 | 10.786 | 10.566 | 10.391 |
| 7 | 16.236 | 12.404 | 10.882 | 10.050 | 9.5221 | 9.1554 | 8.8854 | 8.6781 | 8.5138 |
| 8 | 14.688 | 11.042 | 9.5965 | 8.8051 | 8.3018 | 7.9520 | 7.6942 | 7.4960 | 7.3386 |
| 9 | 13.614 | 10.107 | 8.7171 | 7.9559 | 7.4711 | 7.1338 | 6.8849 | 6.6933 | 6.5411 |
| 10 | 12.826 | 9.4270 | 8.0807 | 7.3428 | 6.8723 | 6.5446 | 6.3025 | 6.1159 | 5.9676 |
| 11 | 12.226 | 8.9122 | 7.6004 | 6.8809 | 6.4217 | 6.1015 | 5.8648 | 5.6821 | 5.5368 |
| 12 | 11.754 | 8.5096 | 7.2258 | 6.5211 | 6.0711 | 5.7570 | 5.5245 | 5.3451 | 5.2021 |
| 13 | 11.374 | 8.1865 | 6.9257 | 6.2335 | 5.7910 | 5.4819 | 5.2529 | 5.0761 | 4.9351 |
| 14 | 11.060 | 7.9217 | 6.6803 | 5.9984 | 5.5623 | 5.2574 | 5.0313 | 4.8566 | 4.7173 |
| 15 | 10.798 | 7.7008 | 6.4760 | 5.8029 | 5.3721 | 5.0708 | 4.8473 | 4.6743 | 4.5364 |
| 16 | 10.575 | 7.5138 | 6.3034 | 5.6378 | 5.2117 | 4.9134 | 4.6920 | 4.5207 | 4.3838 |
| 17 | 10.384 | 7.3536 | 6.1556 | 5.4967 | 5.0746 | 4.7789 | 4.5594 | 4.3893 | 4.2535 |
| 18 | 10.218 | 7.2148 | 6.0277 | 5.3746 | 4.9560 | 4.6627 | 4.4448 | 4.2759 | 4.1410 |
| 19 | 10.073 | 7.0935 | 5.9161 | 5.2681 | 4.8526 | 4.5614 | 4.3448 | 4.1770 | 4.0428 |
| 20 | 9.9439 | 6.9865 | 5.8177 | 5.1743 | 4.7616 | 4.4721 | 4.2569 | 4.0900 | 3.9564 |
| 21 | 9.8295 | 6.8914 | 5.7304 | 5.0911 | 4.6808 | 4.3931 | 4.1789 | 4.0128 | 3.8799 |
| 22 | 9.7271 | 6.8064 | 5.6524 | 5.0168 | 4.6088 | 4.3225 | 4.1094 | 3.9440 | 3.8116 |
| 23 | 9.6348 | 6.7300 | 5.5823 | 4.9500 | 4.5441 | 4.2591 | 4.0469 | 3.8822 | 3.7502 |
| 24 | 9.5513 | 6.6610 | 5.5190 | 4.8898 | 4.4857 | 4.2019 | 3.9905 | 3.8264 | 3.6949 |
| 25 | 9.4753 | 6.5982 | 5.4615 | 4.8351 | 4.4327 | 4.1500 | 3.9394 | 3.7758 | 3.6447 |
| 26 | 9.4059 | 6.5409 | 5.4091 | 4.7852 | 4.3844 | 4.1027 | 3.8928 | 3.7297 | 3.5989 |
| 27 | 9.3423 | 6.4885 | 5.3611 | 4.7396 | 4.3402 | 4.0594 | 3.8501 | 3.6875 | 3.5571 |
| 28 | 9.2838 | 6.4403 | 5.3170 | 4.6977 | 4.2996 | 4.0197 | 3.8110 | 3.6487 | 3.5186 |
| 29 | 9.2297 | 6.3958 | 5.2764 | 4.6591 | 4.2622 | 3.9830 | 3.7749 | 3.6130 | 3.4832 |
| 30 | 9.1797 | 6.3547 | 5.2388 | 4.6233 | 4.2276 | 3.9492 | 3.7416 | 3.5801 | 3.4505 |
| 40 | 8.8278 | 6.0664 | 4.9759 | 4.3738 | 3.9860 | 3.7129 | 3.5088 | 3.3498 | 3.2220 |
| 60 | 8.4946 | 5.7950 | 4.7290 | 4.1399 | 3.7600 | 3.4918 | 3.2911 | 3.1344 | 3.0083 |
| 120 | 8.1790 | 5.5393 | 4.4973 | 3.9207 | 3.5482 | 3.2849 | 3.0874 | 2.9330 | 2.8083 |
| ∞ | 7.8794 | 5.2983 | 4.2794 | 3.7151 | 3.3499 | 3.0913 | 2.8968 | 2.7444 | 2.6210 |

[a] Hayslett (1968).

**Table A-7** (*continued*)

| 10 | 12 | 15 | 20 | 24 | 30 | 40 | 60 | 120 | ∞ |
|---|---|---|---|---|---|---|---|---|---|
| 24224 | 24426 | 24630 | 24836 | 24940 | 25044 | 25148 | 25253 | 25359 | 25465 |
| 199.40 | 199.42 | 199.43 | 199.45 | 199.46 | 199.47 | 199.47 | 199.48 | 199.49 | 199.51 |
| 43.686 | 43.387 | 43.085 | 42.778 | 42.622 | 42.466 | 42.308 | 42.149 | 41.989 | 41.829 |
| 20.967 | 20.705 | 20.438 | 20.167 | 20.030 | 19.892 | 19.752 | 19.611 | 19.468 | 19.325 |
| 13.618 | 13.384 | 13.146 | 12.903 | 12.780 | 12.656 | 12.530 | 12.402 | 12.274 | 12.144 |
| 10.250 | 10.034 | 9.8140 | 9.5888 | 9.4741 | 9.3583 | 9.2408 | 9.1219 | 9.0015 | 8.8793 |
| 8.3803 | 8.1764 | 7.9678 | 7.7540 | 7.6450 | 7.5345 | 7.4225 | 7.3088 | 7.1933 | 7.0760 |
| 7.2107 | 7.0149 | 6.8143 | 6.6082 | 6.5029 | 6.3961 | 6.2875 | 6.1772 | 6.0649 | 5.9505 |
| 6.4171 | 6.2274 | 6.0325 | 5.8318 | 5.7292 | 5.6248 | 5.5186 | 5.4104 | 5.3001 | 5.1875 |
| 5.8467 | 5.6613 | 5.4707 | 5.2740 | 5.1732 | 5.0705 | 4.9659 | 4.8592 | 4.7501 | 4.6385 |
| 5.4182 | 5.2363 | 5.0489 | 4.8552 | 4.7557 | 4.6543 | 4.5508 | 4.4450 | 4.3367 | 4.2256 |
| 5.0855 | 4.9063 | 4.7214 | 4.5299 | 4.4315 | 4.3309 | 4.2282 | 4.1229 | 4.0149 | 3.9039 |
| 4.8199 | 4.6429 | 4.4600 | 4.2703 | 4.1726 | 4.0727 | 3.9704 | 3.8655 | 3.7577 | 3.6465 |
| 4.6034 | 4.4281 | 4.2468 | 4.0585 | 3.9614 | 3.8619 | 3.7600 | 3.6553 | 3.5473 | 3.4359 |
| 4.4236 | 4.2498 | 4.0698 | 3.8826 | 3.7859 | 3.6867 | 3.5850 | 3.4803 | 3.3722 | 3.2602 |
| 4.2719 | 4.0994 | 3.9205 | 3.7342 | 3.6378 | 3.5388 | 3.4372 | 3.3324 | 3.2240 | 3.1115 |
| 4.1423 | 3.9709 | 3.7929 | 3.6073 | 3.5112 | 3.4124 | 3.3107 | 3.2058 | 3.0971 | 2.9839 |
| 4.0305 | 3.8599 | 3.6827 | 3.4977 | 3.4017 | 3.3030 | 3.2014 | 3.0962 | 2.9871 | 2.8732 |
| 3.9329 | 3.7631 | 3.5866 | 3.4020 | 3.3062 | 3.2075 | 3.1058 | 3.0004 | 2.8908 | 2.7762 |
| 3.8470 | 3.6779 | 3.5020 | 3.3178 | 3.2220 | 3.1234 | 3.0215 | 2.9159 | 2.8058 | 2.6904 |
| 3.7709 | 3.6024 | 3.4270 | 3.2431 | 3.1474 | 3.0488 | 2.9467 | 2.8408 | 2.7302 | 2.6140 |
| 3.7030 | 3.5350 | 3.3600 | 3.1764 | 3.0807 | 2.9821 | 2.8799 | 2.7736 | 2.6625 | 2.5455 |
| 3.6420 | 3.4745 | 3.2999 | 3.1165 | 3.0208 | 2.9221 | 2.8198 | 2.7132 | 2.6016 | 2.4837 |
| 3.5870 | 3.4199 | 3.2456 | 3.0624 | 2.9667 | 2.8679 | 2.7654 | 2.6585 | 2.5463 | 2.4276 |
| 3.5370 | 3.3704 | 3.1963 | 3.0133 | 2.9176 | 2.8187 | 2.7160 | 2.6088 | 2.4960 | 2.3765 |
| 3.4916 | 3.3252 | 3.1515 | 2.9685 | 2.8728 | 2.7738 | 2.6709 | 2.5633 | 2.4501 | 2.3297 |
| 3.4499 | 3.2839 | 3.1104 | 2.9275 | 2.8318 | 2.7327 | 2.6296 | 2.5217 | 2.4078 | 2.2867 |
| 3.4117 | 3.2460 | 3.0727 | 2.8899 | 2.7941 | 2.6949 | 2.5916 | 2.4834 | 2.3689 | 2.2469 |
| 3.3765 | 3.2111 | 3.0379 | 2.8551 | 2.7594 | 2.6601 | 2.5565 | 2.4479 | 2.3330 | 2.2102 |
| 3.3440 | 3.1787 | 3.0057 | 2.8230 | 2.7272 | 2.6278 | 2.5241 | 2.4151 | 2.2997 | 2.1760 |
| 3.1167 | 2.9531 | 2.7811 | 2.5984 | 2.5020 | 2.4015 | 2.2958 | 2.1838 | 2.0635 | 1.9318 |
| 2.9042 | 2.7419 | 2.5705 | 2.3872 | 2.2898 | 2.1874 | 2.0789 | 1.9622 | 1.8341 | 1.6885 |
| 2.7052 | 2.5439 | 2.3727 | 2.1881 | 2.0890 | 1.9839 | 1.8709 | 1.7469 | 1.6055 | 1.4311 |
| 2.5188 | 2.3583 | 2.1868 | 1.9998 | 1.8983 | 1.7891 | 1.6691 | 1.5325 | 1.3637 | 1.0000 |

## Reference

Hayslett, H. T., Jr. (1968). "Statistics Made Simple." Doubleday, Garden City, New York.

# AUTHOR INDEX

Numbers in italics refer to the pages on which the complete references are listed.

G



H



I



J



K

## R



## S



## T

# SUBJECT INDEX

N



O



P

V



W



X



Y